Spectroscopic Instrumentation
Fundamentals and Guidelines for Astronomers

More information about this series at
http://www.springer.com/series/4175

Thomas Eversberg • Klaus Vollmann

Spectroscopic Instrumentation

Fundamentals and Guidelines for Astronomers

 Springer

Thomas Eversberg
Schnörringen Telescope Science Institute
Waldbröl, Germany

Klaus Vollmann
Schnörringen Telescope Science Institute
Waldbröl, Germany

SPRINGER-PRAXIS BOOKS IN SPACE EXPLORATION

ISBN 978-3-662-44534-1   ISBN 978-3-662-44535-8 (eBook)
DOI 10.1007/978-3-662-44535-8
Springer Heidelberg New York Dordrecht London

Library of Congress Control Number: 2014953769

Cover illustration: Larger image: This illustration shows the three spectra produced simultaneously by the new efficient X-shooter instrument on ESO's Very Large Telescope. X-shooter can record the entire spectrum of a celestial object (in this example a distant lensed quasar) in one shot – from the ultraviolet to the near-infrared– with great sensitivity and spectral resolution. This unique new instrument will be particularly useful for the study of distant exploding objects called gamma-ray bursts, among the most energetic events in the Universe, which fade rapidly in brightness in matter of hours after their appearance. The rainbow colours applied to the spectra indicate X-shooter's wide spectral coverage and are meant for illustrative purposes only. The majority of the wavelengths covered are in fact invisible to the human eye. Credit: ESO. Smaller image: The X-shooter NIR spectrograph optical layout (Vernet et al. 2011, reproduced with permission ©ESO).

Printed on acid-free paper

Springer is part of Springer Science+Business Media (www.springer.com)

To
  Anthony F. J. Moffat
  &
  Hans-Siegfried Nimmert
  Teachers, Motivators & Friends

Preface

To design spectroscopic instruments for astronomy, one needs appropriate tools. However, the necessary skills are not sufficiently aggregated in the literature so far. This is confirmed by repeated discussions with scientists and engineers. We wanted to meet this gap with an extensive summary and analysis. For the entire understanding this includes, inter alia, wave and imaging optics, fiber optics, detectors, and considerations about appropriate data reduction and analysis. In particular, we will highlight the parameters for the two most common fundamental spectroscopic systems in astronomy—standard long-slit and echelle spectrographs. We do this in complete mathematical detail and perform all necessary calculations by example. We attempt to avoid the unfortunately common phrase "As you can easily see!" Therefore, the book can be used as an introduction to spectroscopy in appropriate university lectures and seminars. Thus we will not only appeal to design engineers for optical instruments but also professional astronomers and their students as well as advanced amateur astronomers. However, we emphasize that our text is really nothing more than a collection and organized combination of already existing publications. On the other hand, we have also tried to work through mathematical and physical approaches for the reader to make the understanding as easy as possible. We do not claim to be the first who present the appropriate theoretical and practical methods. Therefore, it is important to take this opportunity to recognize the many pioneers of instrument development, on whose shoulders we stand.

We explicitly point out that our book represents only an introduction to the topic. In particular, our considerations on refractive lenses and their aberrations, which may play a central role for spectroscopy at relatively small telescopes, provide only a first insight into the calculation strategies and consequences for the interpretation of lenses. The reader is especially here urged to consult further and much more extensive works. This is also more or less true for all other considerations, which we discuss in various chapters. Apart from the theoretical data reduction methods, the practical aspects (software and hardware) are today particularly subject to rapid changes. Appropriate texts are quickly aging. One can capture the enormous depth of the spectroscopic world by only going beyond our introductory discussions.

For the sake of completeness we also show potential spectroscopic astrophysical applications. Of course, our outline of the spectroscopy of massive stars is only a field of many, and we have chosen this topic only because of our own expertise. Nevertheless, we believe that after reading the book, a transfer to observation targets other than point sources (sun, nebulae, galaxies) is easily possible. With our considerations, the reader should be able to fully calculate, design, and build a spectrograph for his/her own purposes. We will repeatedly indicate that the instrumental choice always depends on the corresponding applications or targets. The ultimate selection must be done by the observer.

Waldbröl, Germany   Thomas Eversberg
July 2014   Klaus Vollmann

Contents

List of Figures

List of Tables

Chapter 1
Prologue

A Short Story

In 1985 Klaus Vollmann was walking with a friend through the city of his home town when the friend suddenly said, "There's Thomas". He had discovered the motorcycle in front of a gambling house and now they went in. Klaus already knew that this Thomas Eversberg was also interested in astronomy and he wanted to meet this guy. But this motorbike was standing in front of a gambling house (!) and he discovered the guy now in a bright green leather suit in front of a slot machine. Gosh! Anyway, the conversation quickly came to astronomy and why bother with this nonsense? Thus began a friendship which started the morning after an unsuccessful observation of Comet Halley in the local mountains, by making a detailed plan over breakfast to construct a joint telescope. Klaus wanted to do photometry of minor planets, so he needed a long focal-length telescope. Thomas, though, believed this to be a waste of time and was much more interested in deep-sky observations of galaxies, for which he needed short focal-length optics. Both decided to circumnavigate this problem by a compromise and building a Newton–Cassegrain telescope, starting with the mounting. Fortunately, after their physics studies, both Thomas and Klaus became spectroscopists, although Thomas was never quite clear what Klaus really did with his rocket experiments in the upper atmosphere, and Klaus repeatedly questioned the various methods of massive-star spectroscopy, which intrigued Thomas.

© Springer-Verlag Berlin Heidelberg 2015
T. Eversberg, K. Vollmann, *Spectroscopic Instrumentation*, Springer Praxis Books,
DOI 10.1007/978-3-662-44535-8_1

1.1 Ulysses

When the Ulysses[1] probe, launched in 1990, flew over the solar poles during the
years 1994 and 1995 (see Fig. 1.1), it discovered a hitherto unknown phenomenon.
Earlier models of the solar wind assumed a radially symmetric outflow of homogeneous material. Now the Ulysses instruments found strong deviations from spherical
symmetry in both the velocity and the density of the wind. A radially outflowing
wind-flow at the equator of about 400 km/s shows an abrupt change to high speeds
of around 700 km/s at higher latitudes (Fig. 1.2). In addition, coronal holes show up
mainly at the equator, which produce alternating fast and slow wind-streams from
the surface. The solar rotation leads to collisions between these currents at different
speeds along spiral co-rotating interaction regions. Surprisingly, even a simple star
of medium-small size and luminosity like the Sun reveals in such an obvious way, a
deviation from spherical symmetry and contradicts the picture of a simple sphere of
ideal gas with a spherically symmetric wind. Of course, the clue is rotation. Now,
our Sun is only a star of medium-small mass and brightness. For the most massive
and brightest stars in the Universe, the current state of stellar research suggests a
much more dramatic situation. We will see in Chap. 14 at the end of this book that

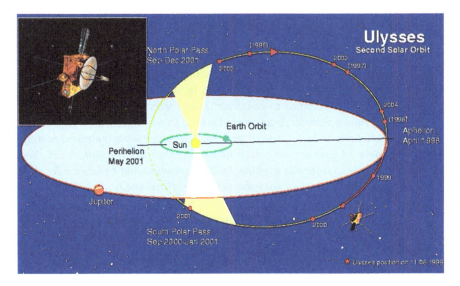

Fig. 1.1 The solar orbit if the Ulysses probe. To leave the ecliptic, a swing-by maneuver was
performed using Jupiter

---

[1]After Ulysses, the hero of the Odyssey, who discovered many worlds and came home with a
treasure of experience.

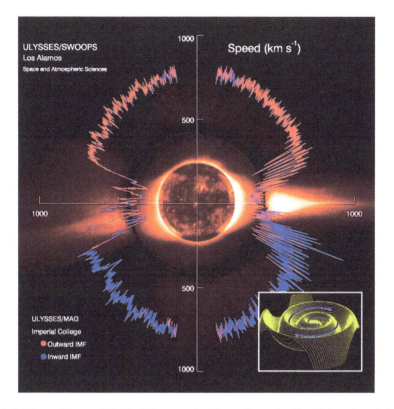

Fig. 1.2 Observations made by the Ulysses probe. The outflow velocities of the solar wind vary from about 400 km/s at the equator to about 700 km/s at the poles (McComas et al. 1998). The wind produces density fluctuations in the form of a spiral over the equator (*inset*)

their large masses and thus extreme luminosities represent machines which not only produce heavy elements in their interior, but also winds of enormous densities and speeds. These winds also have a decisive influence on the immediate environments of the stars, as well as on the structure of entire host galaxies. Massive stars dominate in their control of interstellar and galactic evolution and it is their light that we see first in galaxies.

However, even the closest massive stars are so far away from us that it is challenging to get information about their nature via imaging techniques. Even with the largest telescopes or with multi-telescope arrays, where interferometric techniques can be readily applied, we mostly only see points of light for massive stars, and only in the very closest cases can we get images of limited quality of their surroundings or even their surfaces, using interferometric techniques. In order to analyze the light from cosmic objects, particularly at large distances where the stars remain unresolved, the method of stellar spectroscopy is the work-horse of astronomy. This situation prevails only since the discovery of lines in the spectrum of the Sun by Fraunhofer in 1813, along with the basic works of Bunsen and Kirchhoff on "spectral analysis". By measuring the atomic absorption and emission lines in stars, nebulae, galaxies and other cosmic objects, along with the determination of motion in the direction of observation by the Doppler effect, we now can obtain a quantitative and effective spatial image of these objects in the Universe. In short, spectroscopy is the main key for understanding of the cosmos (Fig. 1.3).

In contrast, the detailed understanding of spectroscopic techniques and their correct applications are not subject of the astronomical undergraduate training but only relatively late during post graduate studies. Only fundamental wave optics including, e.g., wave optics, light dispersion and diffraction are part of the undergraduate studies. However, developing state-of-the-art spectroscopic instrumentation (e.g., sophisticated spectrographs, multi-object spectroscopy, fiber technique) require scientists with a technical understanding to optimize such instruments and to proceed to even better future design. This is not only valid for the telescope dynasoars of some meters aperture and more but also for small telescopes which are necessary to perform long-term investigations of up to a few months. Small telescopes have to be taken into special account because thousands are available all over the world. That are amateur astronomers eager to learn as much as possible in astronomy and are able to perform spontaneous observations at any time.[2] With the recent great technological achievements in engineering and data extraction we

Fig. 1.3 Joseph v. Fraunhofer (1787–1826)/Robert Bunsen (1811–1899)/Gustav Kirchhoff (1824–1887)

---

[2]This is highlighted by Christophe Martayan (Eversberg 2011, discussion): *"Currently we are closing the small professional telescopes and let us imagine that ηCar explodes as a SN. It will be so bright that its luminosity will prevent the observations with the VLT or future ELTs. This kind of case will give a huge opportunity to amateurs for doing the best observations and provide scientific data of the event(s)."*

are currently in the so-called "golden age of astronomy". Professional telescopes are becoming larger and larger and have already reached 10 m sizes. The chemical wizardry of extracting information from photographic plates has been replaced in almost all areas by digital CCD cameras, with linear and efficient detection. But this affects also astronomy with small and even very small telescopes, since this development is associated by a an enormous reduction in cost (the likely dominating factor for a limited budget). Although smaller telescopes have been shut down at many professional observatories, apertures of up to 50 cm are no longer so rare in the amateur domain. And these aficionados can use CCD cameras of high efficiency without significant financial efforts. With few exceptions, the same technologies are in principle available for everybody. Here are some examples:

- Because of extraordinary technological developments, the observational results of many amateur astronomers have reached a level that has been seen only in professional astronomy not many years ago. Those who cannot imagine this to be true have only to look to some classical recordings of well-known galaxies (e.g., from the 2.5 m Mount Wilson telescope) and compare them to today's amateur images of the same objects, which were obtained with 30 cm aperture telescopes but with modern CCD cameras (Fig. 1.4).
- The quality of investigations of planets with small telescopes is so high, even surpassing the above example. Seeing effects caused by the Earth's atmosphere, which so far have prevented a close examination of the planets, are now all but completely eliminated with fast-reacting WebCams. The pictures are now only limited by the diffraction behaviour of the respective optics. It happens more and more today that one finds small telescope recordings of Mars and other

Fig. 1.4 Modern CCD exposure of Messier 51 taken by Daphne and Tony Hallas (astrophoto.com) with a 37 cm telescope (*left*) and a historical photo from a photografic plate taken with the 2.5 m Hooker Telescope at Mount Wilson (*right*)

planets for which one might first believe that they had been taken by the Hubble Space Telescope (Fig. 1.5).

• In former times, synchronous and three-phase motors were used to control professional telescopes. Astronomers worked with circles and star maps from books. Today, even very small telescopes are using stepping motors controlled by electronic units and digitally driven by star catalogues stored in a computer (known as the GoTo technique). And autoguiders can track the telescope at high precision. Pointing software with appropriate professional algorithms can deliver excellent positioning of the telescope even for very small equipment by taking inaccuracies in the mount and the bending behavior of the mechanical system into account. All these techniques can now be bought "off the shelf" for relatively little money.

One might now conclude that small telescopes significantly below 1 m aperture can generally provide data that meet professional standards. However, this is not often the case. Even today an approach to professional levels (or even a collaboration with professionals) in the areas where large telescopes are required, e.g., deep-sky and spectropolarimetry (see Sect. 15) is problematic due to limited financial resources. In addition, 20 cm aperture telescopes already reach a resolution on the order of the typical seeing disk. Observations of relatively low-brightness objects are not taken quickly enough with small telescopes to compensate for the seeing by webcam observations. Telescopes with larger apertures can only usefully exploit this, however, if adaptive optics are applied (in this case, the long exposure times for deep-sky objects plays a minor role).

This is different, however, in spectroscopy, and in particular for the line profile analysis of stars, mostly massive stars. Unlike deep-sky objects, one can detect, analyze and interpret spectral changes in the emission lines of the winds of these objects on time-scales from minutes to years, even with small telescopes.

Fig. 1.5 Combined webcam images of Mars taken by Sebastian Voltmer (astrofilm.com) with a Celestron C14 telescope

Because of their high surface brightness they can be investigated even from urban environments as long as they are bright enough to acquire a reasonable spectral quality. The frequent lack of proper imaging techniques for such stars (no pretty pictures, often considered as a disadvantage) is outweighed by the uniqueness of individual spectroscopic observations, the subsequent possible physical analysis and a potential worldwide collaboration with professional and amateur astronomers, in the sense of citizen science. Fortunately, there is great professional interest in the analysis of the winds from massive stars, whose time-varying emission lines are attributed to dynamic phenomena in the atmospheres. Long-term observations over years often cannot be easily performed by professional astronomers due to limited observation time. Moreover, such campaigns are less interesting for many scientists—they deliver too few publications!

For the serious amateur astronomer who is interested in scientific applications and problems,[3] spectroscopy of stars is thus highly attractive. First, one can learn to apply quite sophisticated scientific techniques and analytical methods and second, one can work with measurement systems that include all facets of optical physics. This includes geometric and wave optics, lenses, mirrors, and their aberrations, as well as dealing with metals and mechanics. In short, beside its importance for our understanding of distant objects spectroscopy offers an extensive range of activities, which enables unique astronomical applications. However, such a fairly complex combination of different fields of work is probably the reason why spectroscopy is only catching on slowly among amateur astronomers. Respect for (and even fear of) this astronomical technique, however, is usually quite high.

With this book want to illuminate spectroscopic fundamentals and various facets of respective instrumentation. However, we also want put astronomical spectroscopy into perspective and show that it is available to very small observatories with relatively modest means. In this vein, we comprehensively describe all the necessary mathematical tools and present their implementation in two spectrograph types. In addition, we present the most important variants and several practical examples for astronomical spectroscopy from student to professional and discuss fiber optics and CCD cameras for various equipment designs. In a comprehensive summary we illuminate the techniques of data reduction. After a brief outline of the significance and determination of measurement errors, we introduce one of the most popular spectroscopic targets, the massive stars and their spectacular stellar winds. Finally we introduce some fundamentals of spectropolarimetry representing the next application step for large telescopes and then present a perspective on potential future applications for the small telescopes.

---

[3]In this book we distinguish between hobby and amateur astronomers. The former operate their passion in terms of aesthetical enjoyment of nature, without any scientific approach. This includes, e.g., casual astro-photographers even when they attain fairly extraordinary results. True amateur astronomers, on the other hand, pursue scientific questions and want to fullfil research needs and approaches similar to those in professional research.

Chapter 2
Fundamentals of Standard Spectroscopy

A Short Story

Thomas had been invited to the meeting of the SPECTROSCOPY section of the German Amateur Astronomy Association to give a talk about the winds of massive stars. Klaus wanted to hear it as well, and during the trip to Heidelberg he asked if Thomas believed whether amateurs are able to understand a mathematical derivative, at least. Thomas did not know and he became concerned. During the meeting some amateur astronomers discussed recently obtained Solar spectra and the respective optical depths. After Thomas had given his talk the first question was about the velocity law of stellar winds and its derivative. Question answered!

2.1 The Law of Diffraction

By formulating the diffraction law, René Descartes and Willebrord van Roijen Snell opened for the first time the door to a quantitative investigation of light. Still today, scholars learn their description of light as a kind of pressure in a solid medium with the consequence that in the case of reflection, the incoming angle is equal to the outgoing angle to the normal. Snell formulated the law of reflection which carries his name—Snell's law, according to which, each wave will be turned away by passing a transition between two media of different phase velocities. This is depicted in Fig. 2.1. Van Roijen Snell was also able to show that a wavefront[1] in a medium of diffractive index n_1 falling under the angle δ_1 into another medium of a diffraction

---

[1]A wavefront is a plane of equal wave phases. This front can be a curved or flat.

© Springer-Verlag Berlin Heidelberg 2015
T. Eversberg, K. Vollmann, *Spectroscopic Instrumentation*, Springer Praxis Books,
DOI 10.1007/978-3-662-44535-8_2

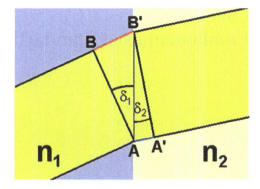

Fig. 2.1 Refraction of a light beam at the border between two media of different refractive indices

index n_2 will leave the border under the angle δ_2 into a medium of diffraction index n_1.[2] The index of diffraction is the ratio of the speed of light in the vacuum and the speed of light in the respective medium $n = c_0/c$. That is $\sin \delta_1 = \overline{BB'}/\overline{AB'} = c_1 \cdot t$ and $\sin \delta_2 = \overline{AA'}/\overline{AB'} = c_2 \cdot t$. Hence, we have the relation

$$\frac{\sin \delta_1}{\sin \delta_2} = \frac{c_1 \cdot t}{c_2 \cdot t} = \frac{n_2}{n_1}$$

or written in a more popular form

$$n_1 \cdot \sin \delta_1 = n_2 \cdot \sin \delta_2. \tag{2.1}$$

Later Fermat showed that the wavefront chooses the path which is the fastest in time. Later, Augustin Jean Fresnel made a quantitative analysis of the transmitted and reflected parts. The Fresnel equations, named after him, calculate the ratio between the reflected and transmitted amplitudes of an incoming beam (see Appendix B.3). The refractive index n is different for different materials and is also wavelength dependent. As a result, white light can be separated into its colors by refraction. In the following we will discuss the optical characteristics of a plane-parallel plate and a prism by a direct application of Snell's law to gain insight for use in a spectrograph.

2.2   On the Geometrical Optics of a Prism

Why can a prism be used in a spectrometer? This can be understood by considering the path of a ray through a plane parallel plate (Fig. 2.2). The beam with angle of incidence α strikes a flat glass surface of refractive index $n > n_{air}$ and will be

---

[2]Here we neglect the reflexion, which actually occurs as well.

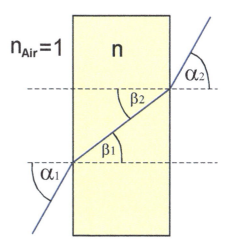

Fig. 2.2 Refraction of light in a plane parallel plate

respectively refracted. At each boundary we apply Snell's law. The beam hits the plane parallel plate and is re-directed according to Eq. 2.1. The law of refraction, generally applied to both boundaries, is then

$$n_1 \cdot \sin \alpha_1 = n_2 \cdot \sin \beta_1 \qquad (2.2)$$

and

$$n_2 \cdot \sin \beta_2 = n_1 \cdot \sin \alpha_2. \qquad (2.3)$$

Because the plate is surrounded by air the indices of refraction are $n_1 = 1$, $n_2 = n$ and for Snell's equations we have

$$\sin \alpha_1 = n \cdot \sin \beta_1 \qquad (2.4)$$

$$\sin \alpha_2 = n \cdot sin\beta_2. \qquad (2.5)$$

We discuss a plane parallel plate with $\beta_2 = \beta_1$ and hence:

$$\sin \alpha_1 = n \cdot \sin \beta_1 \qquad (2.6)$$

$$\sin \alpha_2 = n \cdot \sin \beta_1. \qquad (2.7)$$

These equations can obviously only be valid if $\alpha_2 = \alpha_1$. Therefore, because $\beta_2 = \beta_1$, the beam leaves the plate with angle of incidence α_1. The beam of incidence only experiences a parallel shift and this shift depends on the thickness of the plane parallel plate. The refractive index varies with wavelength and, hence, the angle β_1 also changes with wavelength. However, the product of the refractive index and the sine of the refraction angle remains constant and, hence, the angle α_2 remains independent of the wavelength. Only the position where the beam leaves the glass changes. The thicker the glass plate, the larger the separation of the colours (Fig. 2.3). Obviously, it would be more favourable if the beams of different colours diverged after leaving the dispersive element. The geometric separation of the wavelengths would then become larger with distance from the element. However, this is only possible if $\beta_2 \neq \beta_1$, i.e., if both planes of the plate are not parallel. This result thus requires the geometric form of a prism for the dispersive element.

When illuminating a prism with a laser (monochromatic light), the direction of the beam leaving the prism will be different from that of the incident beam. Of course, the question is if the new direction of the laser beam depends on the direction of the incident beam. The goal of the following calculations is to describe the dependence of the angle δ on the angle of diffraction γ and the measurable angle of incidence α_1 (Fig. 2.4). Again, we start with the Snell's law. Applied to the two surfaces of a prism and analogous to the calculations for a plane parallel plate, the refraction law is now

$$\sin \alpha_1 = n \cdot \sin \beta_1 \tag{2.8}$$

$$\sin \alpha_2 = n \cdot \sin \beta_2. \tag{2.9}$$

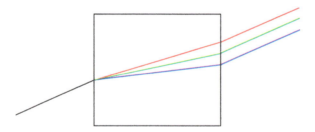

Fig. 2.3 Effect of a plane parallel plate at various wavelengths

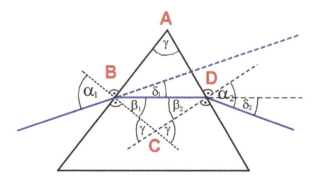

Fig. 2.4 Diffraction of light in a prism

In addition, from Fig. 2.4 we obtain the following angle relations

$$\delta_1 = \alpha_1 - \beta_1 \tag{2.10}$$

$$\delta_2 = \alpha_2 - \beta_2 \tag{2.11}$$

$$\gamma = \beta_1 + \beta_2. \tag{2.12}$$

With these angle relations the overall angle $\delta = \delta_1 + \delta_2$ can now be reformulated to

$$\delta = \delta_1 + \delta_2 = \alpha_1 - \beta_1 + \alpha_2 - \beta_2 = \alpha_1 - \alpha_2 + \gamma. \tag{2.13}$$

With Eq. 2.12 $\beta_2 = \gamma - \beta_1$ and from the 2nd Snell equation (2.3) with $n_1 = 1$ (air) we now get

$$n \cdot \sin(\gamma - \beta_1) = \sin \alpha_2.$$

This equation can now be reformulated with the goniometric relation

$$\sin(\alpha - \beta) = \sin\alpha \cdot \cos\beta - \cos\alpha \cdot \sin\beta$$

to

$$n \cdot (\sin\gamma \cdot \cos\beta_1 - \cos\gamma \cdot \sin\beta_1) = \sin\alpha_2.$$

With the trigonometric identity (here for the angle β_1) $\sin^2\beta_1 + \cos^2\beta_1 = 1$ we reformulate this to $\cos\beta_1 = \sqrt{1 - \sin^2\beta_1}$. With the 1st Snell equation we now get

$$\sin\beta_1 = \frac{\sin\alpha_1}{n}$$

and by substituting this into the last expression for $\cos\beta_1$

$$\cos\beta_1 = \sqrt{1 - \frac{\sin^2\alpha_1}{n^2}}$$

we apply simple algebra and get

$$\sin\alpha_2 = \sin\gamma \cdot \sqrt{n^2 - \sin^2\alpha_1} - \sin\alpha_1 \cdot \cos\gamma.$$

By substituting this into Eq. 2.13 we finally get the beam deflection δ for a prism in air with the angle of diffraction γ and the angle of incidence α_1

$$\delta(n, \gamma, \alpha_1) = \alpha_1 + \arcsin(\sin\gamma \cdot \sqrt{n^2 - \sin^2\alpha_1} - \sin\alpha_1 \cdot \cos\gamma) - \gamma. \tag{2.14}$$

Depicting $\delta = \delta(n, \gamma, \alpha_1)$ graphically (Fig. 2.5) one can see that the function has a minimum at a specific angle of incidence. In our example we used an angle $\gamma = 30°$ and a refraction index of $n = 1.5$. The minimum deflection δ for an angle of incidence of $\alpha_1 = 15.9°$ is about $22.8°$. For the same refractive index and a prism of a $60°$ refraction angle, the minimum deflection is about $50°$. To estimate the exact angle of incidence with minimum deflection we have to calculate the first derivative of the above equation with respect to the angle α_1 and set it to zero. We obtain

$$\delta_{min} = 2\alpha_1 - 2\beta_1. \qquad (2.15)$$

The estimation of the minimum deflection angle δ_{min} and, hence, the derivative of the function $\delta(n, \gamma, \alpha_1)$ with respect to α_1 is somewhat problematic because it contains the derivative of the arcsin function. Generally, the derivative of this function is

$$\frac{d}{dx} \arcsin\left(f(x)\right) = \frac{f'(x)}{\sqrt{(1 - f(x)^2)}}.$$

The argument of this transcendental function arcsin is

$$f(x) = \sin \gamma \cdot \sqrt{n^2 - \sin^2 \alpha_1} - \sin \alpha_1 \cdot \cos \gamma$$

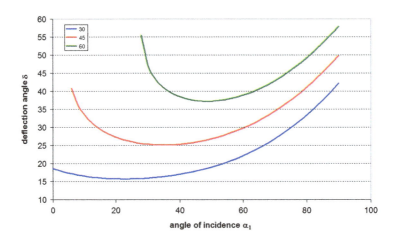

Fig. 2.5 Deflection angle δ over the angle of incidence α_1 for a $30°$, $45°$ and $60°$ prism

and by itself has a rather complicated character. The complete derivative of the function, hence, introduces additional terms. For this reason we follow a simpler way for the mathematical operation, which often can be found in the literature. As further to the above equations, we calculate $\frac{d\delta}{d\alpha_1}$ by using the starting equation $\delta = \alpha_1 - \alpha_2 + \gamma$ (2.13) and find its derivative with respect to α_1 (α_2 is a function of α_1). We obtain

$$\frac{d\delta}{d\alpha_1} = 1 + \frac{d\alpha_2}{d\alpha_1}.$$

We are looking for the minimum of δ, and hence, we set the result to zero and find $\frac{d\alpha_2}{d\alpha_1} = -1$. The derivative of the 3rd angle relation (2.12) with respect to β_2 delivers the relation $\frac{d\beta_1}{d\beta_2} = -1$ for a constant angle γ. Now we take Snell's equations $\sin\alpha_1 = n \cdot \sin\beta_1$ and $\sin\alpha_2 = n \cdot \sin\beta_2$ for our problem again. The derivative of the first one with respect to α_1 delivers

$$\frac{d(\sin\alpha_1)}{d\alpha_1} = n \cdot \frac{d(\sin\beta_1)}{d\beta_1} \cdot \frac{d\beta_1}{d\alpha_1} \Rightarrow \cos\alpha_1 = n \cdot \cos\beta_1 \cdot \frac{d\beta_1}{d\alpha_1}.$$

We perform the derivative of the second function with respect to α_2, as well, and obtain the two relations

$$\cos\alpha_1 = n \cdot \cos\beta_1 \cdot \frac{d\beta_1}{d\alpha_1} \tag{2.16}$$

$$\cos\alpha_2 = n \cdot \cos\beta_2 \cdot \frac{d\beta_2}{d\alpha_2}. \tag{2.17}$$

By dividing these two equations we find $\frac{\cos\alpha_1}{\cos\alpha_2} = \frac{\cos\beta_1}{\cos\beta_2} \cdot \frac{d\beta_1}{d\alpha_1} \cdot \frac{d\alpha_2}{d\beta_2}$ and the product of the two derivatives delivers 1. This is because the above calculations deliver $\frac{d\beta_1}{d\alpha_1} \cdot \frac{d\alpha_2}{d\beta_2} = \frac{d\alpha_2}{d\alpha_1} \cdot \frac{d\beta_1}{d\beta_2} = (-1) \cdot (-1) = 1$, and hence, $\frac{\cos\alpha_1}{\cos\alpha_2} = \frac{\cos\beta_1}{\cos\beta_2}$. With $sin^2\alpha + cos^2\alpha = 1$ we now get by substitution

$$\frac{1 - \sin^2\alpha_1}{1 - \sin^2\alpha_2} = \frac{1 - \sin^2\beta_1}{1 - \sin^2\beta_2}$$

and again by using Snell's law $\sin\beta_{1 or 2} = \frac{\sin\alpha_{1 or 2}}{n}$

$$\frac{1 - \sin^2\alpha_1}{1 - \sin^2\alpha_2} = \frac{n^2 - \sin^2\alpha_1}{n^2 - \sin^2\alpha_2}.$$

Because of $n \neq 1$ this relation has only one solution for α_1 and α_2 if $\alpha_1 = \alpha_2$ (the second trivial solution for $n = 1$ corresponds to the absence of the prism). Snell's equations directly deliver $\beta_1 = \beta_2$. We have shown that the minimum deflection corresponds to the symmetric path through the prism. For this case the angle of incidence and the angle refraction] are the same and we find for the minimum deflection angle $\delta_{min} = 2\alpha_1 - 2\beta_1$ (see Eq. 2.15). From the 3rd angle relation (2.12) we get the minimum deflection angle $(\beta_1 = \beta_2)$ $\gamma = 2\beta_1$ or $\beta_1 = \frac{\gamma}{2}$, and hence, $\delta_{min} = 2\alpha_1 - \gamma$ or $\alpha_1 = \frac{\delta_{min}+\gamma}{2}$. Substituting into the 1st Snell equation we find

$$\sin\left(\frac{\delta_{min} + \gamma}{2}\right) = n \cdot \sin\left(\frac{\gamma}{2}\right).$$

Re-arranging to isolate the refraction index n we obtain the so-called Fraunhofer equation

$$n = \frac{\sin\left(\frac{\delta_{min}+\gamma}{2}\right)}{\sin\left(\frac{\gamma}{2}\right)}. \tag{2.18}$$

This last equation (2.18) illustrates how to estimate the refractive index of an unknown prism material from its known refraction angle γ and the measured deflection δ_{min}. In addition, we will need this equation to estimate the angular dispersion of a prism.

2.3 Principles of Wave Optics

Beside geometrical optics we need some basic knowledge about wave optics to understand prism and grating spectrographs. The following introduction will guide the reader through the subject, although it cannot replace the extensive literature about optics.

One of the most important effects of wave optics we have to get familiar with, is the diffraction of light. Diffraction is a property of light which is not describable via geometrical optics. We will not discuss the complete mathematical description of this phenomenon but only present important results of diffraction optics for spectroscopy. There are a number of more detailed texts (e.g. Hecht and Zajak 2003) and the interested reader should consult these publications for more information.

Beside various mechanical tools the armamentarium necessary for the design of a spectrograph consists of some fundamental physics, often used in astronomical optics, which we consider here in detail. To give an idea on the order of magnitude of some important optical parameters, we will discuss fundamental equations and illustrate them with example calculations. In addition we will derive formulae out of these fundamental equations, which are important for the construction of a spectrometer.

2.3.1 Interference Phenomena

Many optical phenomena require their description via the wave nature of light. Electromagnetic waves show interference effects, analogous to water waves, which can be cancelled or reinforced. If the low level part of a water wave (negative deflection) meets the high level of another wave (positive deflection) of the same wavelength, the respective amplitudes add to each other to a yield a new litude. For instance, if both amplitudes have the same absolute size, i.e., the litude values are the same with opposite sign, their sum vanishes at this position. The overall deflection at this position is zero for this specific example. Of course, all other litude combinations can create a new water wave. In this context this is known as the linear superposition or the superposition principle of waves. This principle is generally valid for all wave phenomena and electromagnetic waves are no exception.

A well known example of such interference of light are the Newtonian rings seen in transparent glass plates (Fig. 2.6) introducing light path differences by an air wedge between the plates. Depending on the path difference both waves can then constructively interfere with each other (light rings) or destructively interfere (dark rings).

Interference, and hence wave phenomena, can basically best be noticed by regions of changing intensity (transitions between light and dark areas). This is completely analogous to the varying amplitudes of water waves. The existence of interference patterns is a clear sign of the wave character of light.

2.3.2 The Huygens–Fresnel Principle

The diffraction of light is based on the interference of numerous *elementary waves* and can be phenomenologically described as a deviation from geometrical optics which is based on a distribution of the wave front. This perturbation can be introduced by, e.g., a pinhole. This pinhole then introduces new elementary waves. If we backside illuminate a pinhole by monochromatic light the respective light spot

Fig. 2.6 Newtonian rings as an example for interference patterns of electromagnetic waves (from Wikipedia-Ulfbastel – Licensed under Creative Commons Attribution-Share Alike 3.0 via Wikimedia Commons)

[from, e.g., a laser] on a screen behind the pinhole will possibly have (depending on the pinhole-screen distance) a complicated and unexpected structure, with respect to the laws of geometrical optics. For instance, one can observe that light waves can illuminate areas even in the geometrical shadow of obstacles (baffles or optical slits). The pinhole characterizes the influence of diffraction on many instruments. This is because the circular pinhole geometry, hence, influences a circular part of the wavefront. We speak about a plane wave if the areas of identical wave phases are planes. An important property of a plane wave is its linear propagation. The original spherical wave of a star, visible in the sky, becomes a plane wave for large distances. Because the distance of a star is much larger than its diameter, we can to very good approximation assume that the light arriving from distant stars is a plane wave (see Fig. 2.7).

Huygens claimed that every point of a wave front can be considered as the origin of a set of elementary waves (secondary spherical waves). The new wave front then comprises these new elementary waves. Fresnel additionally assumed that the resulting field is a phase and litude superposition of all created elementary waves. Hence, diffraction is always coupled with interference, except in the case of a truly elementary wave.

It is not easy to understand why a wave front can be understood as the origin of many elementary waves. However, at the end of the nineteenth century Kirchhoff was able to show that the Huygens Principle is the direct result of the general wave equation. The derivation of this principle would exceed the framework of this book and we simply adopt it as a useful rule for our considerations to describe diffraction phenomena.

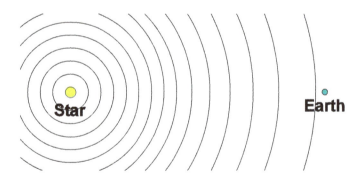

Fig. 2.7 Pictorial of a plane wave front for a star at a large distance

2.3.3   Fraunhofer Diffraction for a Slit and a Pinhole

When light passes through an aperture, we generally can observe diffraction and interference patterns.[3] We consider the diffraction at a simple optical slit with a width b which is close to the light wavelength and, hence, much smaller than its height. Fraunhofer diffraction requires a plane wave front, i.e., the light source as well as the observational plane should either be placed sufficiently far away from the slit or the plane-parallel wave fronts are created by optics. The calculation of the interference pattern, i.e., the intensity distribution in the observation plane, is based on the mathematical picture of all elementary waves created by the slit and their linear superposition at a point P positioned in the observation plane. This calculation will be performed for all geometrical positions, yielding an analytical equation for the intensity distribution of the vector E of the electromagnetic wave across the observation screen. Mathematically this is the integration (summation) of all elementary waves. For the derivation we start with the general definition of a spherical wave[4] starting from point P_0

$$E_0(\vec{r}) = \frac{A_0}{|\vec{r}|}e^{i(\omega t - \vec{k}\vec{r})} \tag{2.19}$$

with E_0 the field strength at point $\vec{r}$ and A_0 the field strength per unit area at point P_0. At the position of the diaphragm, located in the $x - y$-plane of the coordinate origin the amplitude has a strength

$$a(\xi, \eta, 0) = \tau(\xi, \eta) \cdot E_0(\xi, \eta, 0) \tag{2.20}$$

with the transmission function τ of the diaphragm for the coordinates $x = \xi$ and $y = \eta$. We now consider the part of the field strength dE which is generated in each of the infinitesimal-area elements $dA = d\xi d\eta$ of the diffracting aperture. This field is then again spread as a spherical wave by the Huygens principle and is then observed at point P, with

$$dE = \frac{a}{|\vec{s}|}e^{i(\omega t - \vec{k}\cdot\vec{s})}dA. \tag{2.21}$$

---

[3]If the incoming light is in a parallel beam (plane wave front) one has Fraunhofer (far-field approximation) diffraction. That is the case if the light source is located at a large distance or if it is virtually shifted into infinity by a lens. The more general case of a non-parallel incoming beam is called Fresnel (near-field approximation) diffraction.

[4]A comprehensive presentation of the theory of electromagnetic waves can be found in Hecht (2003).

The constant a denotes the initial field strength in the diffracting aperture and $\vec{k}$ the wave vector,[5] whose sum $|\vec{k}| = 2\pi/\lambda$ is defined in the argument of the complex exponential function. According to the Huygens principle the E-vector at point P consists of many elementary waves $e^{-i\vec{k}\vec{r}}$ (see Sect. 2.3.2). According to the principle of superposition the E-field at P is a superposition (integration) of all individual elementary waves. After the integration over all waves we obtain for the electric field strength at point P the double integral:

$$E(x, y) = \int \int \frac{a(\xi, \eta, 0)}{|\vec{s}|} e^{i(\omega t - \vec{k}\cdot\vec{s})} d\xi d\eta. \tag{2.22}$$

We now replace a by the expression for the incoming spherical wave and get

$$E(x, y) = A_0 e^{i\omega t} \int \int \tau(\xi, \eta) \frac{e^{-i\vec{k}\vec{r}}}{|\vec{r}|} \frac{e^{-i\vec{k}\vec{s}}}{|\vec{s}|} d\xi d\eta. \tag{2.23}$$

The integral in Eq. 2.23 can generally only be estimated numerically. However, under certain conditions one can develop $|\vec{r}|$ and $|\vec{s}|$ into Taylor series in order to find simple analytical solutions. If (x_0, y_0, z_0) and (x, y, z) are the coordinates of the points P_0 and P and if $(\xi, \eta, 0)$ are the coordinates of the point Q in the aperture, we can obtain the vectors $\vec{r}$ and $\vec{s}$

$$r^2 := |\vec{r}|^2 = (x_0 - \xi)^2 + (y_0 - \eta)^2 + z_0^2 \tag{2.24}$$
$$s^2 := |\vec{s}|^2 = (x - \xi)^2 + (y - \eta)^2 + z^2 \tag{2.25}$$

and the distances of both points P_0 and P from the coordinate centre, according to the geometric definition in Fig. 2.8

$$r'^2 := |\vec{r}'|^2 = x_0^2 + y_0^2 + z_0^2 \tag{2.26}$$
$$s'^2 := |\vec{s}'|^2 = x^2 + y^2 + z^2. \tag{2.27}$$

If we include these distances into the equations for $|\vec{r}|$ and $|\vec{s}|$ one finds

$$r = r' \sqrt{1 - \frac{2(x_0\xi + y_0\eta)}{r'^2} + \frac{\xi^2 + \eta^2}{r'^2}} \tag{2.28}$$

$$s = s' \sqrt{1 - \frac{2(x\xi + y\eta)}{s'^2} + \frac{\xi^2 + \eta^2}{s'^2}}. \tag{2.29}$$

[5]The wave number $|\vec{k}| = 2\pi/\lambda$ is defined analogously to the angular frequency $\omega = 2\pi f$.

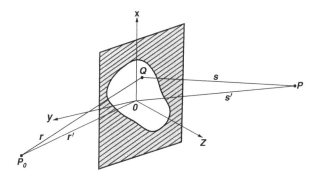

Fig. 2.8 Diffraction at a diaphragm (we use the geometrical definitions from Born and Wolf 1959). The origin of coordinates is in the $x - y$ plane of the diaphragm. The light point source is at the point P_0 with the coordinates $\vec{r}' = (x_0, y_0, z_0)$, the observation point P has the coordinates $\vec{s}' = (x, y, z)$. The arbitrary point Q inside the diaphragm is described by the coordinates $\vec{q} = (\xi, \eta, 0)$. For the vectors we have the correlations $\vec{r}' = \vec{q} + \vec{r}$ and $\vec{s}' = \vec{q} + \vec{s}$

If we now assume that the aperture size is small compared to r' and s' we may develop the functions for r and s into Taylor series of the variables ξ/r', η/r', ξ/s' and η/s' and obtain the linear approximation

$$r \approx r' - \frac{(x_0\xi + y_0\eta)}{r'} + \frac{\xi^2 + \eta^2}{2r'} - \cdots \tag{2.30}$$

$$s \approx s' - \frac{(x\xi + y\eta)}{s'} + \frac{\xi^2 + \eta^2}{2s'} - \cdots \tag{2.31}$$

Depending on the distance one has to take the first or second order of the Taylor series into account. For large distances (the so-called far-field) one finds simplifications up to the linearization (1st order) of the square-root function. Taking only the linear terms of ξ and η of this approximation into consideration one speaks about the so-called Fraunhofer diffraction. On the other hand, if located in the immediate vicinity of the aperture, one also considers higher terms of the square-root (up to the 2nd order). A detailed discussion of the conditions for Fraunhofer diffraction can be found in Born and Wolf (1959).

For the treatment of Fraunhofer diffraction we now assume in a first step that the aperture is illuminated by a plane wave front with a propagation direction perpendicular to the aperture plane. This ensures that the complex exponential of Eq. 2.23 no longer depends on the aperture variables ξ and η and we can separate $\frac{e^{-ikr}}{r} = $ const from the integral. Then we consider the Taylor expansion for s only in the second exponential function. For the absolute value of $\vec{s}$ we use with sufficient accuracy s'. After using the approximation for s the complex exponential function is separated into three components

$$e^{-i|\vec{k}|s} \rightarrow e^{-i|\vec{k}|s'} \cdot e^{\frac{-i|\vec{k}|(x\xi + y\eta)}{s'}} \cdot e^{\frac{-i|\vec{k}|(\xi^2 + \eta^2)}{2s'}} . \tag{2.32}$$

The first exponential function provides only a constant phase factor, while the second linearly depends on the transverse coordinates of the aperture plane and the observation plane. The last exponential function is only dependent on the coordinates of the diffracting aperture and converges in the far field approximation to against the value of 1. However, if this factor increases sufficiently compared to the second exponential function due to a shorter distance to the observation plane it is called Fresnel diffraction. In the case of the far field, the electric field strength at point P is

$$E(x, y) = \int_{-\infty}^{+\infty} \int_{-\infty}^{+\infty} A(\xi, \eta) e^{\frac{-i|\vec{k}|(x\xi + y\eta)}{s'}} d\xi d\eta, \tag{2.33}$$

where all constants have been added to the aperture function $A(\xi, \eta)$. In a last step we now introduce the wave vector $\vec{k}$ with its components k_x and k_y, as follows:

$$k_x := |\vec{k}| \cdot \frac{x}{s'} = |\vec{k}| \sin \Theta \tag{2.34}$$

$$k_y := |\vec{k}| \cdot \frac{y}{s'} = |\vec{k}| \sin \Phi, \tag{2.35}$$

here Θ and Φ are the diffraction angles in the $x - y$-plane. Substituting these variables in the last integral we obtain the so-called Fresnel–Kirchhoff diffraction integral,

$$E(k_x, k_y) = \int_{-\infty}^{\infty} A(\xi, \eta) e^{-i(k_x \xi + k_y \eta)} d\xi d\eta. \tag{2.36}$$

Mathematically, one calls this calculation rule, i.e., the integration of the product of a function A with a complex wave function $e^{-i\vec{k}\vec{r}}$ as a Fourier transform of the function A into $E(\vec{k})$. Thus we obtain the basic finding of Fourier optics:

The amplitude function of the outgoing wave in an observation point corresponds to the Fourier transform of the transmission function of the illuminated object.

Here, the Fourier transform in N-dimensional space is defined as

$$\mathscr{F}(f(\vec{x})) := F(\vec{k}) = \frac{1}{(2\pi)^{\frac{n}{2}}} \int_{-\infty}^{\infty} f(\vec{x}) e^{-i\vec{k}\vec{x}} d^n \vec{x}. \tag{2.37}$$

In summary, therefore, we see that by this calculation rule the intensity of the diffraction pattern $I = E^2$ at a distance R can be determined from a known diaphragm function A. Conversely, it is also possible to determine the aperture function from the diffraction pattern by inverse transformation. The amplitude function of the outgoing wave in an observation point corresponds to the Fourier transform of the transmission function of the illuminated object. With this knowledge, we now turn to the case of a one-dimensional slit of width b. This is defined by

$$f(x) = \begin{cases} 1, & \text{for } |x| \leq b/2 \\ 0, & \text{for } |x| > b/2. \end{cases} \tag{2.38}$$

Thus we obtain the diffraction function using the one-dimensional Fourier transform:

$$\begin{aligned} \mathscr{F}(f(x)) = F(k_x) &= \frac{1}{\sqrt{2\pi}} \int_{-\infty}^{\infty} f(x) e^{-ik_x x} dx \\ &= \frac{1}{\sqrt{2\pi}} \int_{-b/2}^{b/2} e^{-2\pi i k_x x} dx = \frac{1}{\sqrt{2\pi} i k_x} [e^{-ik_x x}]_{-b/2}^{+b/2} \\ &= \frac{1}{\sqrt{2\pi} i k_x} (e^{-ik_x b/2} - e^{ik_x b/2}) = \frac{1}{\sqrt{2\pi} i k_x} (-2i \sin(k_x b/2)) \\ &= \frac{b}{\sqrt{2\pi}} \frac{\sin(k_x b/2)}{k_x b/2}. \end{aligned}$$

The intensity $I(k_x)$ is the square of the absolute value of the amplitude and hence of the Fourier transform $F(k_x)$:

$$I(k_x) = |F(k_x)|^2 = \left(\frac{b}{\sqrt{2\pi}} \frac{\sin(k_x b/2)}{k_x b/2} \right)^2.$$

We now introduce the x-component of the wave vector $k_x = k \sin \Theta$ (Eq. 2.34) and we obtain

$$I(k) = \left(\frac{b}{\sqrt{2\pi}} \frac{\sin(k \sin \Theta \cdot b/2)}{k \sin \Theta \cdot b/2} \right)^2. \tag{2.39}$$

This intensity distribution is a periodic function with maxima and minima (see Fig. 2.9), whose minima can be found at zero level of the amplitude function $F(k_x)$:

$$k \sin \Theta_{\text{Min}} \cdot b/2 = n \cdot \pi,$$

where the order n can acquire the values $n = \pm 1, \pm 2, \ldots$. The angular position of the minima can be transformed into a corresponding geometrical position (depending on the distance) and is given by the following equation:

$$\sin \Theta_{\text{Min}} = n \cdot \frac{\lambda}{b}. \tag{2.40}$$

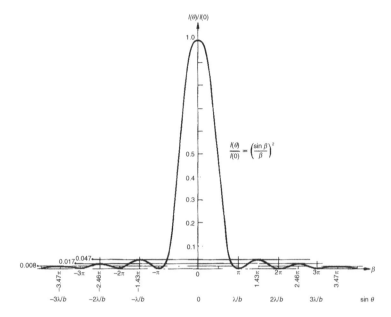

Fig. 2.9 Diffraction at an optical slit (Hecht 2003)

For instance, illuminating a slit of $30\,\mu m$ width using a laser of $630\,nm$ wavelength gives a 1st intensity minimum at the position

$$\sin\Theta_{Min} = \pm\frac{\lambda}{b} = \pm\,\frac{630\cdot10^{-9}m}{30\cdot10^{-6}m} = \pm0.021$$

For small angles we recall the useful approximation[6] $\sin\phi \approx \tan\phi \approx \phi$. Using this approximation and the conversion factor from radians to degrees for the angle of the 1st minimum we find $\pm0.021 \cdot 180/\pi = \pm1.2°$. To calculate the respective geometrical position of the 1st minimum we simply multiply the angle in radians with the distance between the slit and the projection plane. For instance, if the distance is $l = 1,000\,mm$ we find the minimum at Δx:

$$\Delta x = \pm0,021 \cdot 1000\,mm = \pm21\,mm$$

---

[6]Note that the angle should be expressed in radians. As an example we calculate the trigonometric function for an angle of $10°$, corresponding to $10° \cdot \pi/180° = 0.1745329$ radians. The main trigonometric functions are then $\sin(10°) = 0.1736482$ and $\tan(10°) = 0.1763269$. Comparing with the angle in radians, it is obvious that one can replace the trigonometric function by the angle itself. The accuracy of this approximation is better than 1% for angles smaller a $10°$.

In this example both 1st order minima can be found 21 mm to the left and right of the main maximum. The example is a relatively simple method for the estimation of the slit width using a laser, which plays an important role in spectroscopy (see also Sect. 4.1.2).

A two-dimensional slit can be described by the product of two separate functions of two corresponding slit equations 2.38:

$$f(x) = \begin{cases} 1, & \text{for } |x| \le b_x/2 \\ 0, & \text{for } |x| > b_x/2 \end{cases}$$

and

$$g(y) = \begin{cases} 1, & \text{for } |y| \le b_y/2 \\ 0, & \text{for } |y| > b_y/2. \end{cases}$$

Then we obtain for the product $h(x, y) = f(x) \cdot g(y)$ the two-dimensional Fourier transform:

$$H(k_x, k_y) = \frac{1}{2\pi} \int_{-\infty}^{+\infty} \int_{-\infty}^{+\infty} h(x, y) e^{-i(xk_x + yk_y)} dx dy$$

$$= \frac{1}{\sqrt{2\pi}} \int_{-\infty}^{+\infty} f(x) e^{-ixk_x} dx \cdot \frac{1}{\sqrt{2\pi}} \int_{-\infty}^{+\infty} f(x) e^{-iyk_y} dy$$

$$= F(k_x) \cdot G(k_y),$$

or more generally

$$\mathscr{F}_2[(f(x) \cdot g(y)] = \mathscr{F}_1(f)(k_x)) \cdot \mathscr{F}_1(g)(k_y)), \tag{2.41}$$

where the indices of the Fourier transform $\mathscr{F}_1$ and $\mathscr{F}_2$ specify the dimension of the Fourier transform. For a two-dimensional slit we then obtain

$$H(k_x, k_y) = \frac{b_1}{\sqrt{2\pi}} \frac{\sin(k_x b_x/2)}{k_x b_x/2} \cdot \frac{b_1}{\sqrt{2\pi}} \frac{\sin(k_y b_y/2)}{k_y b_y/2}. \tag{2.42}$$

For the case of a rectangular diaphragm one observes the diffraction minima at

$$\sin \Theta_{\text{Min}} = n_1 \cdot \frac{\lambda}{b_x} \tag{2.43}$$

$$\sin \Phi_{\text{Min}} = n_2 \cdot \frac{\lambda}{b_y} \tag{2.44}$$

with $n_1, n_2 = \pm 1, \pm 2, \ldots$ and the slit widths in the x and y direction b_x, b_y. Considering the simple long slit (Eq. 2.40) for comparison, the result for the rectangular diaphragm as a two-dimensional slit is not surprising (see Fig. 2.10). This can be extended to the estimation of the diffraction pattern from a circular diaphragm (Fig. 2.11). The mathematical treatment is equivalent to that of a slit but one has to introduce polar coordinates because of the circular shape of the aperture. The diffraction pattern then follows a Bessel function. For the intensity depending on the diffraction angle θ one obtains[7]

$$I(r) \propto \left(\frac{J_1((kD \sin \Theta)/2)}{(kD \sin \Theta)/2} \right)^2, \qquad (2.45)$$

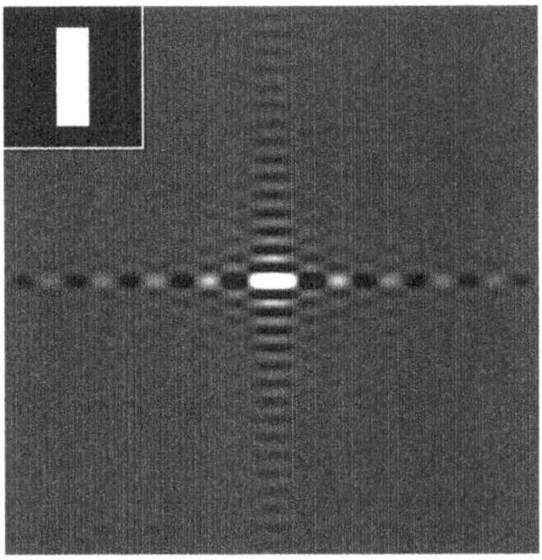

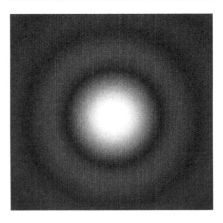

Fig. 2.10 Diffraction pattern of a rectangular diaphragm (from Wikipedia-Anton-Licensed under Creative Commons Attribution-Share Alike 3.0 via Wikimedia Commons)

Fig. 2.11 Diffraction pattern of a point source in a telescope (from Wikipedia-Licensed under Creative Commons Attribution-Share Alike 3.0 via Wikimedia Commons)

---

[7]A detailed derivation can be found in Hecht (2003).

with J_1 the Bessel function of first kind and first order. For the first diffraction minimum in the circular case, we find

$$\sin \Theta_{1,Min} = 1.22 \cdot \frac{\lambda}{D}, \tag{2.46}$$

with D the diaphragm diameter. The angle of the 1st minimum is 22 % larger than that of a slit or a rectangular diaphragm. Let us now assume that an astronomer observes a star with a telescope of 20 cm aperture. The star emits a spherical wave which reaches the earth essentially as a plane wave front due to the large distance. Because of diffraction at the telescope optics the stellar image in the focal plane will not have an arbitrarily small diameter, as perhaps expected, but shows an angular diameter of the so-called Airy disk[8] according to Eq. 2.46. If our Sun were at a distance of, e.g., one light-year, the diameter would not be $0.5°$ ($1{,}800''$) but about $1800'' \cdot 150 \cdot 10^6\, \text{km}/9.5 \cdot 10^{12}\, \text{km} = 0.028''$. We will now try to answer the question if one could still observe spots on the Sun at such a distance. As a first step, we consider how large the stellar image appears in the focal plane of the telescope. According to Fig. 2.11 the vast majority of the intensity can be detected in the center of the diffraction pattern. We use Eq. 2.46 for the estimation of the pattern size which obviously computes roughly the half-intensity angular diameter of the stellar image. The diffraction pattern (called Airy disk, as mentioned above) then has an angular diameter of $2.44 \cdot \frac{\lambda}{D}$, with D the diameter of the telescope mirror or the lens, respectively. As an example, we now calculate the angular diameter ϕ of the star in the telescope at the wavelength of 550 nm.

$$\phi = 2.44 \cdot \frac{\lambda}{D} = 2.44 \cdot \frac{550 \cdot 10^{-9}\text{m}}{0.2\text{m}} = 6.71 \cdot 10^{-6}\text{rad}$$

Expressed in arcseconds this is:

$$\phi \cdot \frac{180°}{\pi} \cdot 3600''/° = 6.71 \cdot 10^{-6}\text{rad} \cdot 206265''/\text{rad} = 1.38''$$

where the angular diameter ϕ must be expressed in radians. Thus, the Sun at a distance of one light-year imaged in the focal plane of a telescope is about 50 times smaller than the diameter of the Airy disk. Hence, it will be utterly impossible to identify details on its surface. Even the solar disk would not be resolved. In fact two stars with angular separation of $0.028''$ would produce inseparable Airy disks in the telescope. The question of when two stars (point sources) can be separated will be answered in the next chapter.

We note that the diameter of the seeing disk produced by turbulent layers in the Earth's atmosphere and depending on conditions very close to the telescope have typical sizes between $1''$ and $2''$ (worse for low altitudes and poor dome seeing) and

---

[8]After the mathematician and astronomer Sir George Biddell Airy—1801–1892.

hence are far larger than the diffraction pattern in the telescope. The geometrical resolution of a telescope would then be limited by atmospheric disturbances and not by diffraction. Depending on the local atmospheric turbulence (seeing), the image quality and resolution can only be improved by a larger aperture if the telescope diameter does not exceed about 10–15 cm. For larger telescopes (without adaptive optics) the greater diameter does not increase the resolution but only the light-collecting efficiency.

To obtain the diameter of the Airy disk d in mm one has to multiply the angular diameter ϕ with the focal length of the telescope. For our example we take a f/10 system with a focal length of $f = 2{,}000$ mm:

$$d = \phi \cdot f = 6.71 \cdot 10^{-6} \cdot 2000 \,\text{mm} = 0.0134 \,\text{mm}$$

or 13.4 μm. Note that the diameter of the Airy disk depends on the wavelength of the incoming light. In the blue (400 nm) the Airy disk diameter is 9.7 μm and in the red (650 nm) it is 15.8 μm. Hence, the difference in the optical wavelength range is already 60 %, which influences the resolution of telescopes as long as they are diffraction limited.[9]

2.3.4 Spectral Resolution and Resolving Power

The above described diffraction patterns of light generally introduce limited resolution of optical instruments. For the estimation of the spectral resolution of spectrometers we now consider a quantitative specification.

The basic idea behind the definition of resolution is to find an expression for detectable structures. For instance, the separation of close binary stars is an equivalent question in imaging astronomy. Both stars emit spherical waves which reach the Earth (almost) as a plane wave front. Both stars produce an Airy disk in the focal plane of the telescope, which both overlap (see Figs. 2.12 and 2.13). The definition of resolution is both arbitrary and reasonable. One avenue is the so-called Rayleigh criterion, suggested by Lord Rayleigh in 1878. It asks when two objects with partially overlapping Airy disks can be considered as separated. He assumed that two objects can just be separated when the central maximum of one diffraction pattern is positioned in the 1st minimum of the other pattern (see Fig. 2.14)

With this definition and the approximation for small angles, the distance of both maxima of the diffraction pattern $\Delta\alpha$ is again given by Eq. 2.46 $\Delta\alpha \approx 1.22 \cdot \frac{\lambda}{D}$. For our example with a 20 cm telescope we obtain a resolution of 0.69″ for 550 nm, which is below the average seeing of most sites on Earth.

---

[9]If the geometrical resolution is limited by the local seeing the diameter of the seeing disk is wavelength independent.

Fig. 2.12 Definition of the Rayleigh criterion

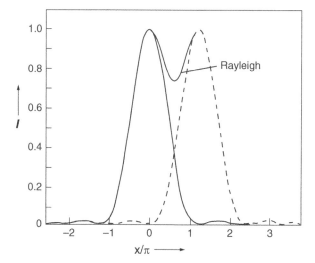

Fig. 2.13 Airy disk of a binary for the Rayleigh criterion (from Wikipedia-Geek3-Licensed under Creative Commons Attribution-Share Alike 3.0 via Wikimedia Commons)

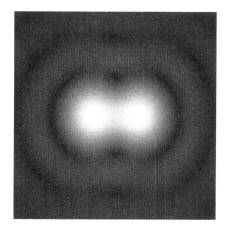

Fig. 2.14 Presentation of the Rayleigh criterion: The diffraction patterns of two point sources can be considered separated when the maximum of one of the two patterns coincides with the 1st minimum of the other

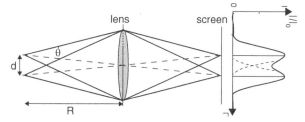

An alternative definition of the resolution is the so-called Sparrow criterion. It takes into account the capability of the human eye to distinguish intensity differences: Two light sources can be separated if their combined intensity function along the line joining their centres shows a minimum. If both sources have the same intensity the Sparrow criterion delivers a minimum separation if

$$\Delta\alpha = 0.95 \cdot \frac{\lambda}{D}.$$

Quantitatively, the Sparrow criterion is obviously similar to the Rayleigh criterion. However, we will use the less optimistic Rayleigh criterion for consideration of the resolution of a spectrograph.

With the above expressions taken from diffraction theory we are now able to understand and derive the most important properties of spectrometers.

2.4 The Prism Spectrograph

The archetype of a spectrograph uses the property of glass to separate white light into various wavelengths (colours) by deflecting (refracting) different wavelengths into different directions. The "splitting" occurs because of the change in light velocity during the transition from one medium to the other. The velocity change is wavelength dependent. An exact explanation of this process can be found in the book "QED - The strange theory of light and matter" by Feynman (1985). The working principle of a prism spectrograph is shown in Fig. 2.15.

The converging light coming from the light source will be made parallel by a lens, called the collimator. It guarantees that the spectral resolution reaches the theoretical maximum. Later we will elaborate on the necessity of the collimator in a grating spectrograph in more detail. After the collimator the light reaches the

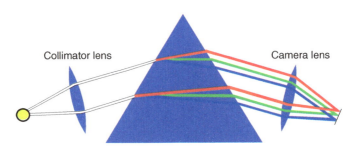

Fig. 2.15 Prism spectrograph principle. The light source (e.g., illuminated slit) is parallelized by a collimator, diffracted by a prism and imaged onto a sensor by camera optics

prism. It separates the light into its colours and the following camera optics makes an image in the focal plane, where it can be detected by a film or a CCD. An ideal linear refraction angle with wavelength cannot be achieved due to the optical properties of a prism. For this reason an accurate wavelength calibration for the complete spectral range must be performed.

2.4.1 Properties of a Prism Spectrograph

With the above theoretical considerations we now can begin to design a spectrograph. Basically there are two design strategies for the construction of a prism spectrograph for astronomical use:

1. One uses direct star light falling onto the prism, which disperses it and images the result with camera optics or a telescope onto the focal plane. The advantage of this method is that one already has a plane wave front coming from the star which makes collimator optics unnecessary.
2. One first images the star with a telescope, collimates the beam and disperses the light with a prism. The spectrum will then be imaged onto the focal plane by camera optics.

For the first case the prism should be larger than the telescope aperture D. The dispersion limiting element is then the aperture of the telescope. If the star only emits monochromatic light, we would obtain a focal Airy disk defined by Eq. 2.46 of diameter $2.44 \cdot \frac{\lambda}{D} \cdot f$. The parameter f is the focal length of the telescope. In the more realistic case of non-monochromatic light, the Airy disk would change its position in the focal plane depending on how the refraction changes with wavelength. The geometrical separation of the two respective wavelengths and, hence, two different refractions, can be calculated again by the Rayleigh criterion. Both Airy disks can be distinguished if the separation is $1.22 \cdot \frac{\lambda}{D} \cdot f$. This means that the geometrical resolution is always connected to a corresponding spectral resolution.

2.4.2 The Angular and Linear Dispersion of a Prism

A prism provides the opportunity to separate light into its various wavelength components. The degree of separation is represented by the so-called dispersion relation and we distinguish between angular and linear dispersion. In the following we will discuss in detail how to estimate the angular dispersion of glasses, i.e., the dispersion of a wavelength range per unit angle.

The refractive index of most glasses is a function of wavelength. The Fraunhofer equation (Eq. 2.18) for the minimum angular difference from the last chapter can be used to estimate the angular dispersion of a prism. From the Fraunhofer equation we get

$$n \cdot \sin\left(\frac{\gamma}{2}\right) = \sin\left(\frac{\delta_{min} + \gamma}{2}\right)$$

or

$$\delta_{min} = 2 \cdot \arcsin\left[n \cdot \sin\left(\frac{\gamma}{2}\right)\right] - \gamma.$$

Now we use this equation to derive the angular dispersion $\frac{d\delta_{min}}{d\lambda}$. There is no explicit wavelength dependence. The only wavelength dependent parameter is the index of refraction $n = n(\lambda)$. With this knowledge we have to apply the chain rule from differential calculus and thereby get

$$\frac{d\delta_{min}}{d\lambda} = \frac{d\delta_{min}}{dn} \cdot \frac{dn}{d\lambda}.$$

We obtain the derivative $\frac{d\delta_{min}}{dn}$ from the equation for δ_{min} by using the above formula for the differential quotient of the arcsin function

$$\frac{d\delta_{min}}{dn} = \frac{2 \cdot \sin\frac{\gamma}{2}}{\sqrt{1 - n^2 \cdot \sin^2\frac{\gamma}{2}}}.$$

In addition we have (Fig. 2.16)

$$\sin\frac{\gamma}{2} = \frac{b}{2a}$$

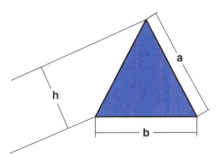

Fig. 2.16 The prism as diffraction limiting element

and according to Eq. 2.18

$$\sqrt{1 - n^2 \cdot \sin^2 \frac{\gamma}{2}} = \sqrt{1 - n^2 \cdot \sin^2 \beta_1} = \sqrt{1 - \sin^2 \alpha_1} = \cos\alpha_1 = \frac{h}{a}.$$

Hence, we get the simplified expression for $\frac{d\delta_{min}}{dn}$

$$\frac{d\delta_{min}}{dn} = \frac{2 \cdot \sin \frac{\gamma}{2}}{\sqrt{1 - n^2 \cdot \sin^2 \frac{\gamma}{2}}} = \frac{2 \cdot \frac{b}{2a}}{\frac{h}{a}} = \frac{b}{h}$$

and the result for the angular dispersion in radians per wavelength unit is

$$\frac{d\delta_{min}}{d\lambda} = \frac{b}{h} \cdot \frac{dn}{d\lambda}. \tag{2.47}$$

By adding the focal length f_2 of the camera $d(\delta_{min} \cdot f_2) = dx$ we get the linear dispersion of the prism in mm per wavelength unit.

$$\frac{d(\delta_{min} \cdot f_2)}{d\lambda} = \frac{dx}{d\lambda} = \frac{b \cdot f_2}{h} \cdot \frac{dn}{d\lambda}.$$

Because the Fraunhofer equation describes the minimum deflection angle δ_{min}, and, hence, the symmetric path through the prism, the derived formula for the angular and linear dispersion does not represent the general case. The general derivative however, would introduce very complex terms, due to the non-linear nature of the arcsin function which deviates just slightly from the symmetric case. Figure 2.17 shows the diffraction index of prism materials versus wavelength. A table of refraction indices of various glasses versus wavelength can be found in the Appendix C.

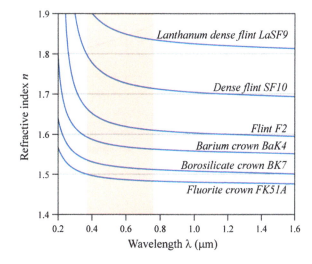

Fig. 2.17 Non-linear refraction index of different prisms versus wavelength (from Wikipedia-DrBob-Licensed under Creative Commons Attribution-Share Alike 3.0 via Wikimedia Commons)

2.4.3 The Wavelength Dependence of the Refraction Index:
The Sellmeier Equation

To calculate the angular dispersion of a prism, one needs the dependence of the refractive index on the wavelength $\frac{dn}{d\lambda}$. The wavelength-dependent refractive index n can be parameterized using the so-called Sellmeier equation. From the findings of Cauchy, W. Sellmeier published in 1871 the following formula. The necessary coefficients are provided by the glass manufacturers for each type of glass. The accuracy of the parametrization in the wavelength range from 0.4 to 1.0 μm is of the order of $5 \cdot 10^{-6}$.

$$n(\lambda) = \sqrt{1 + \frac{B_1\lambda^2}{\lambda^2 - C_1} + \frac{B_2\lambda^2}{\lambda^2 - C_2} + \frac{B_3\lambda^2}{\lambda^2 - C_3}}. \qquad (2.48)$$

As an example we list the coefficients for Borosilicate glass BK7.

$B_1 = 1.03961212$; $B_2 = 0.231792344$; $B_3 = 1.01046945$; $C_1 = 6.00069867 \cdot 10^{-3}\,\mu m^2$; $C_2 = 2.00179144 \cdot 10^{-3}\,\mu m^2$; $C_3 = 103.560653\,\mu m^2$.

One can see that the Sellmeier equation uses the wavelength in μm according to the coefficient units. Figure 2.18 shows the refractive index and dispersion depending on wavelength, for BK7.

The dispersion $\frac{dn}{d\lambda}$ between 400 and 700 nm varies by a factor 5 and becomes smaller with increasing wavelength. At H_α (656.3 nm) the dispersion of BK7 is about $3.16 \cdot 10^{-5}\,nm^{-1}$, at H_δ (410.2 nm) $12.8 \cdot 10^{-5}\,nm^{-1}$.

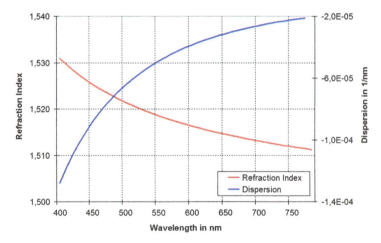

Fig. 2.18 Refractive index and dispersion $\frac{dn}{d\lambda}$ for BK7

2.4.4 The Spectral Resolution of a Prism

We will now consider which spectral resolution we typically can achieve with a prism spectrograph. For that we first discuss the question of how the above diffraction effects limit the spectral resolution. The parallel wavefront falling onto the prism is geometrically limited by the size of the prism surface; this area can be considered as a slit aperture (Fig. 2.16). For that reason the refraction of light at different wavelengths in different directions is diffraction limited. According to Eq. 2.40 the Rayleigh criterion for a rectangular prism surface is:

$$\Delta\delta_{min} = \frac{\lambda}{h}.$$

The spectral resolving power is defined by $R = \frac{\lambda}{\Delta\lambda}$. We put this equation, redistributed to λ, into the equation for the maximum spectral resolution of a prism R and obtain

$$R = \frac{\lambda}{\Delta\lambda} = h \cdot \frac{\Delta\delta_{min}}{\Delta\lambda}.$$

Obviously the maximum resolving power increases with the width of the collimated beam and hence with the size of the prism. We now replace $\frac{\Delta\delta_{min}}{\Delta\lambda}$ by the result for the angular dispersion of Eq. 2.47 and obtain

$$R = \frac{\lambda}{\Delta\lambda} = h \cdot \frac{\Delta\delta_{min}}{\Delta\lambda} = h \cdot \frac{b}{h} \cdot \frac{dn}{d\lambda} = b \cdot \frac{dn}{d\lambda}$$

where $\frac{dn}{d\lambda}$ is the behaviour of the refractive index depending on the wavelength, b, the edge length of the prism and $\Delta\lambda$, the minimum distinguishable wavelength difference.

The spectral resolving power of a prism is proportional to its edge length.

As an example we calculate the spectral resolving power of a completely illuminated prism of 100 mm edge length b and get for red wavelengths around H_α a maximum spectral resolution of

$$R = \frac{\lambda}{\Delta\lambda} = 100\,\text{mm} \cdot 10^6\,\text{nm/mm} \cdot 3.16 \cdot 10^{-5}\,\text{nm}^{-1} = 3160$$

The smallest resolvable wavelength element is $\Delta\lambda = 656.3\,\text{nm}/3160 = 0.208\,\text{nm}$ or about 2.1 Å. In the blue range (H_δ) the dispersion is about a factor 4 higher. Here, the maximum resolving power is 12,800 or 0.5 Å.

We see that the spectral resolution is equivalent to the geometrical resolution for binary stars and limited by the finite size of the prism and the diffraction of light. This is not surprising because the spectral resolution is in fact a result of a geometrical decomposition. Not completely illuminating the prism will compromise the theoretical resolution of a spectrograph. In this case we have to consider the decreased effective edge length for our calculations. In contrast, we can increase the resolution for a given glass material by increasing the edge length b of the prism. This made prisms very large (and heavy) in earlier times, when prisms were preferentially used in astronomy.

At this point we note that the theoretical resolution of a prism can only be achieved under ideal circumstances. We recall that the geometrical resolution of a 20 cm telescope is about $0.69''$. This corresponds to a resolution of $1.38''$ for a 10 cm prism. The seeing conditions in different regions vary between $3–4''$ (urban environments and can reach values below $1''$ at extraordinary places and the above calculated spectral resolution of a slitless prism spectrometer cannot be achieved under urban environmental circumstances. The implementation of a slit would increase the theoretical resolution. However, the price to pay for this would be a significant loss of light due to the larger seeing disk in the sky. In the next chapter we will show how to generally calculate the resolution of a more adaptable grating spectrograph.

2.5   The Grating Spectrograph

The so-called grating spectrograph works completely analogously to the prism spectrograph. Figure 2.19 shows the working principle with mirror optics as collimator and camera. Neglecting the fact that we now no longer need colour-corrected lenses (mirrors are completely achromatic), we have a so-called diffraction grating as dispersive element. The present example shows a grating that is rotatable in the dispersion direction to adjust the instrument to different wavelength regions. This is necessary if the whole spectral range should be observable and if the grating has such a high dispersion that only small wavelength ranges can be covered by the spectrograph optics.

Today almost all spectrographs use optical gratings as dispersion elements. Prisms do not have the regularly necessary dispersion, and hence, spectral resolution

Fig. 2.19 Principle of the grating spectrograph with mirrors as collimator and camera (from Wikipedia-Kkmurray-Licensed under Creative Commons via Wikimedia Commons)

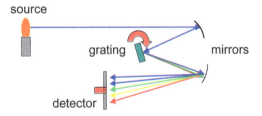

(a special exception is discussed in Chap. 5). An additional advantage is the almost linear dispersion, independent of the wavelength.

A grating uses the wave nature of light and its constructive and destructive interferences (see Sect. 2.3.1). If we consider a glass plate with a series of parallel grooves or lines and allow a parallel beam of light to pass through it, the transmitted wave fronts interfere with each other and are lified (constructive interference) or attenuated (destructive interference). This is depicted in Fig. 2.20. A monochromatic collimated beam which hits a dispersion grating at an angle α (here in transmission) will be refracted to an angle β.[10] The detailed situation is depicted in Fig. 2.21. One can see that integer numbers of the respective infalling wavelength will be lified. The angle β is different for each different wavelength. By considering these simple geometrical relations we obtain the equation for constructive interference.

$$n \cdot \lambda = d \cdot (\sin \alpha + \sin \beta) \tag{2.49}$$

This is the so-called grating equation and it describes which angles of specific incoming beam wavelengths and their integer numbers are necessary to introduce constructive interference. At specific angles β we, hence, find only specific wavelengths which fulfill the grating equation.

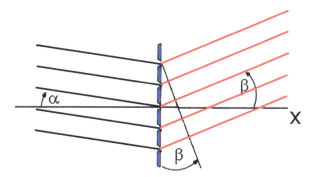

Fig. 2.20 Beam geometry at a diffraction grating. A parallel beam infalling at angle α will be diffracted to the angle β by the transmission grating. This angle is only valid for one single wavelength (e.g. β_{red}), other beams are diffracted to another direction, valid for their wavelengths

---

[10]The expression "'refraction'" should not be understood as a simple direction change of the incident light. Due to Huygens' principle the grating grooves are the sources of elementary waves which will be globally lified and attenuated due to constructive and destructive interference, respectively, for specific wavelengths. The constructive interference is angular dependent and introduces an apparent direction change of the incoming beam.

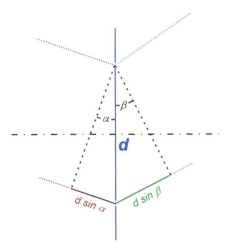

Fig. 2.21 Derivation of the grating equation from geometrical conditions at the grating. The optical path difference between adjacent incoming beams (*left*) passing the grating at two points of distance d is $d \sin \alpha$. The path difference between adjacent refracted beams (*right*) is $d \sin \beta$. The overall path difference is, hence, $d(\sin \alpha + \sin \beta)$. Constructive interference for these two beams occur for integer numbers n of the respective infalling wavelength λ. The condition for constructive interference at a certain wavelength is, hence, $n \cdot \lambda = d \cdot (\sin \alpha + \sin \beta)$

In the above example we considered a transmission grating, but generally this is also valid with certain prefix changes for reflection gratings. We will discuss this in more detail in the following chapter.

2.5.1 Fraunhofer Diffraction for a Grating

In Sect. 2.3.3 we have calculated Fraunhofer diffraction at a slit. Now we will discuss Fraunhofer diffraction at a diffraction grating and consider it as a series of single slits whose diffraction patterns interfere with each other. We start with the general intensity function of a grating and calculate the case of a one-dimensional series consisting of N slits. For the mathematical representation of the grating function we use the property of the Dirac δ-function[11]:

$$g(x) * \delta(x) := \int_{-\infty}^{+\infty} g(x')\delta(x - x')dx' = g(x). \tag{2.50}$$

---

[11]For the definition of the δ-function and their calculation rules we refer to the relevant literature.

Thus[12] we can formulate, for example, a double-slit with the slit function $f(x)$ of Eq. 2.38:

$$f_2(x) = f(x) * \delta(x + d/2) + f(x) * \delta(x - d/2),$$

where the parameter d defines the distance between the two slits. For the functional description of multi-slits (gratings) we need to approximate all slits by the sum of all corresponding δ functions and then fold this sum again with the function $f(x)$. For each slit the corresponding distance $x - nd$ is taken into account:

$$f_N = f(x) * \sum_{n=0}^{N-1} \delta(x - nd),$$

where d is reflecting the periodicity of the grating (grating constant). Considering the folding rule (Appendix B.5.1—convolution in time) and the geometrical series for the sum of all exponential functions we obtain for the Fourier transform of f_N

$$F_N(k_x) = \frac{b}{\sqrt{2\pi}} \frac{\sin(k_x b/2)}{k_x b/2} \cdot \sum_{n=0}^{N-1} e^{-ik_x nd} = \frac{b}{\sqrt{2\pi}} \frac{\sin(k_x b/2)}{k_x b/2} \cdot \frac{1 - e^{-ik_x Nd}}{1 - e^{-ik_x d}}.$$

By squaring the amplitude we again obtain the intensity function

$$I(k_x) = |F_N(k_x)|^2$$

$$= \left(\frac{b}{\sqrt{2\pi}}\right)^2 \frac{\sin^2(k_x b/2)}{(k_x b/2)^2} \left|\frac{1 - e^{-ik_x Nd}}{1 - e^{-ik_x d}}\right|$$

$$= \left(\frac{b}{\sqrt{2\pi}}\right)^2 \frac{\sin^2(k_x b/2)}{(k_x b/2)^2} \left|\frac{\frac{1}{2i}(e^{+ik_x Nd/2} - e^{-ik_x Nd/2}) \cdot e^{-ik_x Nd/2}}{\frac{1}{2i}(e^{+ik_x d/2} - e^{-ik_x d/2}) \cdot e^{-ik_x d/2}}\right|$$

$$= \left(\frac{b}{\sqrt{2\pi}}\right)^2 \frac{\sin^2(k_x b/2)}{(k_x b/2)^2} \cdot \frac{\sin^2(k_x Nd)}{\sin^2(k_x d)^2} \cdot \left|\frac{e^{-ik_x Nd/2}}{e^{-ik_x d/2}}\right|$$

$$= \left(\frac{b}{\sqrt{2\pi}}\right)^2 \frac{\sin^2(k_x b/2)}{(k_x b/2)^2} \cdot \frac{\sin^2(k_x Nd/2)}{\sin^2(k_x d/2)}.$$

The absolute values of the complex exponential functions in the penultimate equation are 1 by definition ($|e^{i\phi}| = 1$).

[12]On the left side of Eq. 2.50 the function $g(x)$ is convoluted with the δ-function. The principle of a mathematical convolution will be discussed in Sect. 2.6.4.

We now replace again the wave number component k_x in the arguments of the sine function by Eq. 2.34 and get for the arguments[13]

$$\delta = \frac{|\vec{k}| \cdot s}{2} \cdot \sin \beta = \frac{\pi \cdot s}{\lambda} \cdot \sin \beta \qquad (2.51)$$

$$\psi = \frac{|\vec{k}| \cdot d}{2} \cdot \sin \beta = \frac{\pi \cdot d}{\lambda} \cdot \sin \beta, \qquad (2.52)$$

with the general definition of the wave vector $|\vec{k}| = 2\pi/\lambda$. The parameters s and d denote the dimension of the single slit and the distance between them (grating constant), respectively, and β the angle of diffraction.

Furthermore, we abbreviate

$$B(\delta) = \left(\frac{\sin \delta}{\delta}\right)^2 \qquad (2.53)$$

$$G(\psi) = \left(\frac{\sin(N\psi)}{\sin \psi}\right)^2 \qquad (2.54)$$

and can then combine to

$$I(\delta, \psi) = I_0 \cdot B(\delta) \cdot G(\psi) \qquad (2.55)$$

with I_0 the incident intensity. The intensity function therefore consists of the product of several components, on the one hand the incident light wave with a given intensity I_0, and on the other the diffraction function $B(\delta)$ of the single slit and the interference or slit function $G(\psi)$ of all N grating grooves.

We now consider the functional behavior of the grating function[14] $I(\psi) = G(\psi)$ and search for the normalization factor so that we have $I(\psi) = 1$ when $\sin \beta = 0$. We calculate by using the L'Hospital rule

$$I(\sin \beta = 0) = \lim_{\sin \beta \to 0} \frac{\sin^2(kNd/2 \sin \beta)}{\sin^2(kd/2 \sin \beta)}$$

$$= \lim_{\sin \beta \to 0} \frac{2\sin(kNd/2 \sin \beta)\cos(kNd/2 \sin \beta)Nkd/2}{2\sin(kNd/2 \sin \beta)\cos(kd/2 \sin \beta)kd/2}$$

$$= \lim_{\sin \beta \to 0} \frac{\left((\cos^2(kNd/2 \sin \beta) - \sin^2(kNd/2 \sin \beta))\right) N^2}{\left(\cos^2(kNd/2 \sin \beta) - \sin^2(kd/2 \sin \beta)\right)}$$

$$= N^2.$$

---

[13]The phases shown are $\alpha = 0$ due to our assumption exclusively for normal incidence on the diffractive element.

[14]We assume the slit width s is infinitely narrow but transparent so that it corresponds to a δ-function so that the diffraction function becomes a constant of value of unity.

This result indicates that only the $\frac{1}{N}$-th fraction of the incident amplitude $A \sim \sqrt{I}$ falls on each single slit. For the grating function we then find the following intensity function (maxima normalized to 1):

$$\frac{I(\sin \beta)}{I(0)} = \frac{\sin^2(kNd/2 \sin \beta)}{N^2 \sin^2(kd/2 \sin \beta)}. \tag{2.56}$$

If we now additionally consider the slit function $B(\delta)$, then the grating function is superimposed by the function $sin(\delta)/\delta$. The order maxima decrease beyond the zeroth order. The grating function is modulated by the sinc function. The narrower the slit, i.e., the larger the grating constant d, the more "broadband" is the modulation (see c in Fig. 2.22). Hecht and Zajac (2003) derives a simple estimate for the intensity of the first secondary maximum beside the maximum of the order. Between each order maximum, significantly smaller secondary maxima can be found. The intensity ratio between the central and the first secondary maximum (normalized to 1) is $I/I(\beta = 0) \approx 1/22$. For the second secondary maximum the value falls to $1/62$.

Hence, a few percent of the incident intensity can be found in the "dark areas" between the order maxima (see d in Fig. 2.22). These secondary maxima are not available for analysis. We now ask at which diffraction angles β the order maxima of the grating intensity function occur. As already shown with L'Hospital, the intensity function $k\frac{d}{2} \sin \beta = 0$ has a central maximum $= N^2$. Hence, this function value can also be found at $k\frac{d}{2} \sin \beta = n \cdot \pi$ (see Fig. 2.22c). Rearranging this equation yields

$$d \sin \beta = n \cdot \frac{2\pi}{k} = n \cdot \lambda \tag{2.57}$$

with $n = 0, \pm 1, \pm 2, \ldots$

This result is the grating equation for the simplest case, the so-called Littrow configuration (incidence angle $\alpha = 0$). Of course, it is possible not only to illuminate the screen vertically, as we have assumed in the derivation of the Fresnel–Kirchhoff diffraction integral, but also at a certain incident angle α. For this general case, the path differences of the individual wave fronts will also change, having an influence on the grating equation. We consider the general form of the Fresnel–Kirchhoff diffraction integral, Eq. 2.23, which can be also be written as follows

$$E(x, y) = A_0 e^{i\omega t} \int \int \frac{\tau(\xi, \eta)}{|\vec{r}||\vec{s}|} e^{-i\vec{k}(\vec{r}+\vec{s})} d\xi d\eta. \tag{2.58}$$

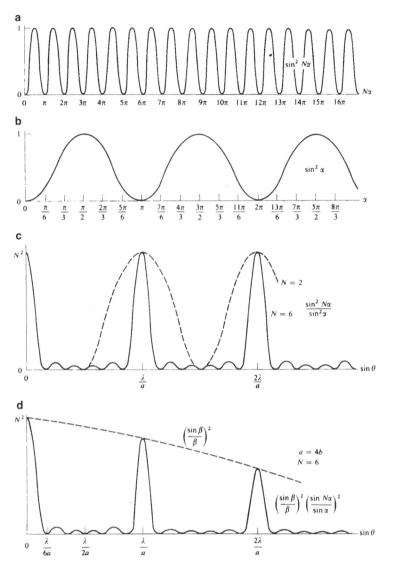

Fig. 2.22 Description of the grating intensity function. (**a**) Intensity function according to the numerator in Eq. 2.56. (**b**) Intensity function according to the denominator in Eq. 2.56. (**c**) Result of Eq. 2.56 for the cases $N = 2$ and $N = 6$. This illustrates that, with an increasing number N of grating grooves the maxima of the diffraction function (i.e., the orders of the grating equation 2.49) become narrower. Note the position of the maxima at $0, \frac{\lambda}{a}, \frac{2\lambda}{a}, \ldots$. With many grating grooves one thus can for example direct the wavelengths of polychromatic light into different directions and separate them spatially. The constant height of each peak represents infinitely narrow grating grooves. (**d**) In reality grating grooves have a finite width and provide a modulating slit function, in addition to the intensity function of Eq. 2.55. This slit function is called the blaze function, which ensures that the intensity of the individual orders decreases with the order number. In our example the slit width has been chosen to be $b = a/4$ in weight and the minimum of the slit function appears exactly at the 4th intensity maximum of the grating function $4\lambda/a$. For higher orders the envelope grows as its sinc function grows again (Hecht 2003)

In the Fraunhofer far-field approximation we now again replace r and s in sufficient approximation by r' and s' and the sum of $r + s$ in the exponential function by the corresponding Taylor expansion (Eq. 2.30):

$$r + s \approx r' + s' - \frac{(x_0 \xi + y_0 \eta)}{r'} - \frac{(x \xi + y \eta)}{s'} + \ldots \qquad (2.59)$$

Rearranging to the aperture coordinates ξ and η delivers

$$r + s \approx r' + s' - \left(\frac{x_0}{r'} + \frac{x}{s'} \right) \xi - \left(\frac{x_0}{r'} + \frac{x}{s'} \right) \eta. \qquad (2.60)$$

We summarize the terms in brackets and reach the general coordinates

$$u = - \left(\frac{x_0}{r'} + \frac{x}{s'} \right)$$

$$v = - \left(\frac{y_0}{r'} + \frac{y}{s'} \right).$$

Using these coordinates and inserting into Eq. 2.58 we obtain

$$E(x, y) = \int_{-\infty}^{+\infty} \int_{-\infty}^{+\infty} A(\xi, \eta) e^{-i |\vec{k}|(u\xi + v\eta)} d\xi d\eta. \qquad (2.61)$$

Structurally, Eq. 2.61 does not differ from Eq. 2.36. The only difference is the additional sum in the generalized coordinates, taking into account an incident beam coming from an arbitrary direction. Therefore, in this case we also have an incident wave vector $\vec{k}^0$ and an outgoing wave vector $\vec{k}$ with the corresponding components in the $x - y$-plane (Fig. 2.8) as follows:

$$k_x^0 = \frac{|\vec{k}| \cdot x_0}{r'} = |\vec{k}| \sin \alpha_x$$

$$k_y^0 = \frac{|\vec{k}| \cdot y_0}{r'} = |\vec{k}| \sin \alpha_y$$

$$k_x = \frac{|\vec{k}| \cdot x}{s'} = |\vec{k}| \sin \beta_x$$

$$k_y = \frac{|\vec{k}| \cdot y}{s'} = |\vec{k}| \sin, \beta_y$$

where α_x, α_y are the angles of incidence and β_x, β_y are the diffraction angles. Note that the absolute value of the wave vector $|\vec{k}|$ occurs in the incoming and the outgoing components, since the wavelength is not changed by the diffraction process. Now let us consider again the one-dimensional case where the y-components y_0, y are each zero. The corresponding components of the two wave vectors are then also equal to zero and the integration is only done for the variable ξ. The wave vector component in the complex exponential function of the diffraction integral in Eq. 2.61 is now $k_x^0 + k_x$ and the grating intensity function becomes, by inserting the diffraction integral and integration,

$$\frac{I(\sin \alpha_x, \sin \beta_x)}{I(0)} = \frac{\sin^2(kNd/2(\sin \alpha_x + \sin \beta_x)}{N^2 \sin^2(kd/2(\sin \alpha_x + \sin \beta_x))}. \tag{2.62}$$

At this point we only consider the one-dimensional case, hence, we can neglect the y-component and delete the index for the x-direction. As above, we obtain for the maxima $kd/2(\sin \alpha + \sin \beta) = n \cdot \pi$, and thus the general grating equation for an incident angle $\alpha \neq 0$ becomes

$$d \cdot (\sin \alpha + \sin \beta) = n \cdot \lambda \tag{2.63}$$

with $n = 0, \pm 1, \pm 2, \ldots$

We point out in particular that the coordinates x and x' can have positive and negative values depending on the direction of the incident wave front. Therefore, the grating equation can be formulated with different signs when using respective angular values:

$$\pm \sin \alpha \pm \sin \beta = \frac{n \cdot \lambda}{d}. \tag{2.64}$$

Depending on the geometric situation the path difference of adjacent waves therefore becomes shorter or longer (Fig. 2.23). This should be realized in practice, especially with respect to the use of reflecting gratings (instead of transmission gratings) because the coordinate signs change after reflection at the grating. Of the four possible theoretical combinations two can be combined because they only differ by their overall sign. They only differ by the incident and exit direction.[15] The two cases

$$+ \sin \alpha + \sin \beta = \frac{n \cdot \lambda}{d} \tag{2.65}$$

$$- \sin \alpha - \sin \beta = \frac{n \cdot \lambda}{d} \tag{2.66}$$

are therefore equivalent, as well as

$$+ \sin \alpha - \sin \beta = \frac{n \cdot \lambda}{d} \tag{2.67}$$

$$- \sin \alpha + \sin \beta = \frac{n \cdot \lambda}{d}. \tag{2.68}$$

---

[15]Usually blazed gratings are used (discussed in the next section) and thus one diffraction direction is preferred. Hence, both possibilities are "restricted" to only one.

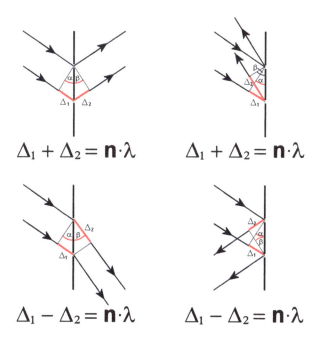

$$\Delta_1 + \Delta_2 = \mathbf{n}\cdot\lambda \qquad \Delta_1 + \Delta_2 = \mathbf{n}\cdot\lambda$$

$$\Delta_1 - \Delta_2 = \mathbf{n}\cdot\lambda \qquad \Delta_1 - \Delta_2 = \mathbf{n}\cdot\lambda$$

Fig. 2.23 Possible optical arrangements for transmission and reflection gratings. Both cases can additionally occur with negative optical path difference. This means, for example, that for the transmission grating (*upper left*) the incident beam paths come from below and leave the grating downwards. Δ_1 and Δ_2 indicate the optical path differences of the incoming beam $d\sin\alpha$ and the diffracted beam $d\sin\beta$, respectively. When adding the path differences ($\Delta_1 + \Delta_2$) they both occur in the same incoming and diffracted beam (*top*, indicated in *red*). When subtracting them ($\Delta_1 - \Delta_2$) they occur in both beams (*bottom*, indicated in *red*)

The positions of the grating order maxima are wavelength dependent and, hence, can be used for the spectral decomposition of light. The grating diffracts light of different wavelengths into different directions. This fact is exploited in spectroscopy. The modulation of the slit function has a disadvantage, though. Higher orders deliver reduced light intensities. Therefore in most cases gratings are only used in the first order $n = \pm 1$. Since the order maxima have a finite width and the spectral resolution is defined by this width, the grating thus provides a finite spectral resolution. We therefore try to estimate this width by using Eq. 2.56.

The grating function is normalized to 1, hence, we are looking for the full width σ at half maximum $I(\sigma/2) = 1/2$ (FWHM), where $\frac{\sigma}{2} = \sin\Theta_{\sigma/2}$ represents the width of the order maximum. We have

$$\frac{I(\frac{\sigma}{2})}{I(0)} = \frac{1}{2} = \frac{\sin^2(kN\frac{d}{2}\frac{\sigma}{2})}{N^2\sin^2(k\frac{d}{2}\frac{\sigma}{2})}$$

or by rearranging and extracting the root

$$\frac{N}{\sqrt{2}}\sin\left(k\frac{d}{2}\frac{\sigma}{2}\right) = \sin(kN\frac{d}{2}\frac{\sigma}{2}).$$

We can approximate the sine function in the denominator by its argument and we obtain

$$\frac{N}{\sqrt{2}}k\frac{d}{2}\frac{\sigma}{2} \approx \sin(kN\frac{d}{2}\frac{\sigma}{2}).$$

This is a transcendental equation of the form $x = \sqrt{2}\sin x$ with $x = Nk\frac{d}{2}\frac{\sigma}{2}$ which can be solved by iteration or by calculating the sine at various nodes and respective interpolation. We, however, choose the approximation of the sine function by the corresponding Taylor series[16] and stop after the second Taylor element, $\sin x \approx x - \frac{x^3}{6}$. Thus we obtain $x/\sqrt{2} \approx x - x^3/6$, i.e., $x \approx 1.33$. With this value for x and using the wave vector $k = 2\pi/\lambda$ we obtain for σ

$$\sigma \approx 2.66\frac{\lambda}{d \cdot \pi \cdot N} = 0.85\frac{\lambda}{d \cdot N}. \tag{2.69}$$

The width of the order maximum is inversely proportional to the number of illuminated grating lines and to the illuminated grating grid width $d \cdot N$.

We will discuss the spectral resolving power associated with the grating equation again in Sect. 2.5.5.

2.5.2 Higher Efficiency with a Blaze Angle

When dispersing light the whole light energy of the respective source is distributed over all orders, i.e. the integer numbers n in Eq. 2.49. Order number $n = 0$ represents normal reflection (without diffraction) and is not spectrally usable. One can now manipulate the grooves so that a larger fraction of the light falls into a specific (mostly the first) order. We have shown that the respective light intensities as well as the geometrical distances between the orders decrease with n. Such blaze gratings are scored by diamonds so that the reflection law is fulfilled with respect to a specific reflection surface, the so-called blaze angle, and for a specific wavelength, the so-called blaze wavelength (Fig. 2.24). For each blaze grating the blaze wavelength or the blaze angle is given. One could now assume that prisms are no longer used because of the advantage of blaze gratings. But this is not the case. Prisms work exclusively in first order and the complete light energy is present there. For that reason prisms show a constant efficiency over all wavelengths. Figure 2.25

---

[16]The error introduced by this procedure is of the order of 5 %. The iterative solution for the argument x delivers a slightly more accurate value of $x = 1.39$.

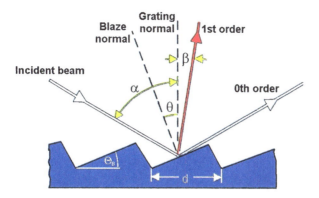

Fig. 2.24 Grating with blaze angle Θ

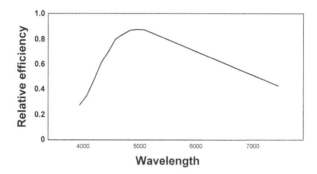

Fig. 2.25 Example efficiency curve of a blazed grating with 5,000 Å blaze wavelength

shows that blazed gratings do not show a constant efficiency over wavelength, in
contrast to prisms.

If we now apply the theory of grating diffraction more generally to the blaze
grating we find from Eq. 2.51 the following specific phase differences with respect to
the illuminated part of the grating facet (the reflecting grating surface perpendicular
to the incident direction):

$$\delta = \frac{k \cdot s}{2} \cdot (\sin(\alpha - \Theta_B) + \sin(\beta - \Theta_B)), \tag{2.70}$$

$$\psi = \frac{k \cdot d}{2} \cdot (\sin \alpha + \sin \beta). \tag{2.71}$$

For ψ, this phase difference is immediately understandable. In the last section
we showed with Eqs. 2.61–2.63 that for any angle of incidence $\alpha \neq 0$, the path
difference with respect to the grating plane $d \cdot \sin \beta$ is replaced by $d \cdot (\sin \alpha + \sin \beta)$.
In the case of the phase difference δ, however, one has to consider that due to the
blazing the grating facette is inclined against the grating plane by the angle Θ_B

and the angles α and β have to be replaced by $\alpha - \Theta_B$ and $\beta - \Theta_B$, respectively. In this context the function $B(\alpha, \beta) = (\sin \delta / \delta)^2$ is described as the blaze function. It is the efficiency distribution which determines the efficiency for the order maxima. This efficiency distribution has the consequence that the signal-to-noise ratio, especially in echelle gratings (see Sect. 5), drops almost symmetrically at the order edges.

The question now arises for which incidence and diffraction angle the blaze function of the grating is at its maximum. Substituting the phase difference δ (Eq. 2.70) in the general blaze function (Eq. 2.53) one obtains

$$B(\alpha, \beta, \Theta_B) = \left[\frac{\sin \left(\frac{1}{2} \cdot k \cdot s \cdot (\sin(\alpha - \Theta_B) + \sin(\beta - \Theta_B)) \right)}{\left(\frac{1}{2} \cdot k \cdot s \cdot (\sin(\alpha - \Theta_B) + \sin(\beta - \Theta_B)) \right)} \right]^2, \qquad (2.72)$$

or with the sinc function

$$B(\alpha, \beta, \Theta_B) = \mathrm{sinc}^2 \left(\frac{1}{2} \cdot k \cdot s \cdot (\sin(\alpha - \Theta_B) + \sin(\beta - \Theta_B)) \right). \qquad (2.73)$$

The blaze function has its maximum when the argument of the sinc function vanishes[17]:

$$\frac{1}{2} \cdot k \cdot s \cdot [\sin(\alpha - \Theta_B) + \sin(\beta - \Theta_B)] = 0$$

or

$$\sin(\alpha - \Theta_B) = \sin(\Theta_B - \beta)$$

and hence

$$\frac{\alpha + \beta}{2} = \Theta_B. \qquad (2.74)$$

The blaze function has its maximum where the average of both incidence angle and diffraction angle are equal to the blaze angle Θ_B.

The wave vector k can be expressed by the wavelength.

$$B(\alpha, \beta, \Theta_B) = \left[\frac{\sin \left(\frac{1}{\lambda} \cdot \pi \cdot s \cdot [(\sin(\alpha - \Theta_B) + \sin(\beta - \Theta_B)]) \right)}{\frac{1}{\lambda} \cdot \pi \cdot s \cdot [(\sin(\alpha - \Theta_B) + \sin(\beta - \Theta_B)]} \right]^2. \qquad (2.75)$$

---

[17]Because of the facet structure of the blaze grating vignetting is introduced when illuminating the grating obliquely. However, this only plays a role for very large blaze and incident angles (see Fig. 2.24). Here we neglect this effect but discuss it for the echelle grating in Chap. 5.

Note that this expression is the general blaze function in wavelength space which is more convenient for the design of a spectrograph.

Furthermore, one can use this general blaze function including the replacement of the wavelength by the grating equation to derive a useful relation for the blaze function in angular space. One then obtains for the argument of the sinc function after inserting the wavelength

$$\frac{n \cdot \pi \cdot s}{d} \cdot \frac{\sin(\alpha - \Theta_B) + \sin(\beta - \Theta_B)}{\sin(\alpha) + \sin(\beta)}.$$

With the trigonometric identity

$$\sin \alpha + \sin \beta_b = 2 \sin\left(\frac{1}{2}(\alpha + \beta_b)\right) \cos\left(\frac{1}{2}(\alpha - \beta_b)\right) \qquad (2.76)$$

we can write

$$\frac{n \cdot \pi \cdot s}{d} \cdot \frac{\sin\left(\frac{1}{2}[(\alpha - \theta_B) + (\beta - \theta_B)]\right) \cos\left(\frac{1}{2}(\alpha - \theta_B) - (\beta - \theta_B)\right)}{\sin\left(\frac{1}{2}(\alpha + \beta)\right) \cos\left(\frac{1}{2}(\alpha - \beta)\right)}$$

or

$$\frac{n \cdot \pi \cdot s}{d} \cdot \frac{\sin\left(\frac{1}{2}(\alpha + \beta - 2\theta_B)\right) \cos\left(\frac{1}{2}(\alpha - \beta)\right)}{\sin\left(\frac{1}{2}(\alpha + \beta)\right) \cos\left(\frac{1}{2}(\alpha - \beta)\right)}.$$

Cancelling down the cos-functions in the numerator and denominator and reshaping the sine function using the addition theorems leads to

$$\frac{n \cdot \pi \cdot s}{d} \cdot \frac{\sin \frac{1}{2}(\alpha + \beta) \cos \theta_B - \cos \frac{1}{2}(\alpha + \beta) \sin \theta_B}{\sin \frac{1}{2}(\alpha + \beta)}.$$

We finally obtain for the blaze function in angular space

$$B(\alpha, \beta, \Theta_B) = \left[\frac{\sin\left(\frac{n \cdot \pi \cdot s}{d} \cdot (\cos \theta_B - \sin \theta_B / \tan \frac{1}{2}(\alpha + \beta))\right)}{\frac{n \cdot \pi \cdot s}{d} \cdot (\cos \theta_B - \sin \theta_B / \tan \frac{1}{2}(\alpha + \beta))} \right]^2. \qquad (2.77)$$

Figure 2.26 shows the comparison between the derived equation and measurements at a real grating. The agreement is sufficiently good that Eq. 2.77 can be used for estimating the overall efficiency of a spectrometer or exposure times of objects at different wavelengths.

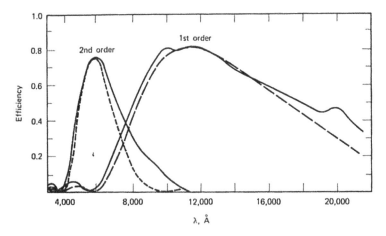

Fig. 2.26 Comparison of computed (*dashed line*) and measured blaze efficiency according to Eq. 2.77 (Gray 1992)

2.5.3 The Wavelength of the "Blaze" at Arbitrary Angle of Incidence

We now determine the wavelength at which the grating has its maximum efficiency. For that we consider the grating equation 2.49 for the case $\alpha = \beta = \Theta_B$ ("Littrow configuration"),

$$\frac{n \cdot \lambda_{\Theta_B}}{d} = 2 \sin \Theta_B, \qquad (2.78)$$

with the so-called blaze wavelength λ_{Θ_B} which is clearly allocated to the blaze angle Θ_B and usually given by the grating manufacturers.[18] Now we illuminate the grating under an angle α which differs from the blaze angle Θ_B. The general grating equation is then

$$\frac{n \cdot \lambda_B}{d} = \sin \alpha + \sin \beta_B$$

with λ_B the present blaze wavelength and β_B the diffraction angle at which the blaze angle appears. We now divide the general grating equation by the Blaze equation

---

[18]For example, for the case of a grating with 1,200 lines/mm with a blaze wavelength $\lambda_{\Theta_B} = 500$ nm a blaze angle of $\Theta_B = 17,5°$ is indicated.

(Eq. 2.78) and obtain the wavelength ratio

$$\frac{\lambda_B}{\lambda_{\Theta_B}} = \frac{\sin\alpha + \sin\beta_b}{2\sin\Theta_B}.$$

According to Eq. 2.74 the blaze function has a maximum when the condition $\Theta_B = \frac{1}{2}(\alpha + \beta_b)$ is fulfilled. That means for the wavelength ratio

$$\frac{\lambda_B}{\lambda_{\Theta_B}} = \frac{\sin\alpha + \sin\beta_B}{2\sin\left(\frac{1}{2}(\alpha + \beta_B)\right)}.$$

With the identity in Eq. 2.76 we then obtain the wavelength of the blaze:

$$\lambda_B = \lambda_{\Theta_B} \cos\left(\frac{1}{2}(\alpha - \beta_B)\right)$$

or with the angle of incidence α

$$\lambda_B = \lambda_{\Theta_B} \cos(\alpha - \Theta_B). \tag{2.79}$$

Note that $\lambda_B < \lambda_{\Theta_B}$ always applies which means that the blaze angle shifts towards smaller wavelengths. For the 1,200 l/mm grating with a blaze wavelength of 500 nm, mentioned in the last footnote, one finds a blaze wavelength $\lambda_B = 492$ nm for an angle difference of $\alpha - \Theta_B = 10°$.

2.5.4 The Angular and Linear Dispersion of a Grating Spectrometer

Now, to estimate the dispersion of a grating, we need to calculate the deviation of the angle β when changing the wavelength λ. To do so, we use the first derivative $d\beta/d\lambda$ of the grating equation. For simplification we can use a constant angle $\alpha = 0$. The grating equation and its first derivative is then

$$n \cdot \lambda = d \cdot \sin\beta$$
$$\frac{d\beta}{d\lambda} = \frac{n}{d \cdot \cos\beta}. \tag{2.80}$$

The angular dispersion $d\beta/d\lambda$ of a grating is indirectly proportional to the order number.

The angular dispersion $d\beta/d\lambda$ of a grating is indirectly proportional to the cosine of the diffraction angle β.

Instead of groove widths one generally uses the reciprocal grating constant g in lines per mm (l/mm).

$$n \cdot \lambda = d \cdot \sin \beta$$

$$\frac{d\beta}{d\lambda} = \frac{g \cdot n}{\cos \beta}. \tag{2.81}$$

For example, a grating with $600\,\text{l/mm}$ has double the angular dispersion of a grating with $300\,\text{l/mm}$. In addition, the angular dispersion remains constant if the grating constant is reduced and the order in which the grating works is increased by the same factor.

The angular dispersion $d\beta/d\lambda$ of a grating is proportional to the grating line density.

As an example we calculate the angular dispersion of an off-the-shelf grating with $g = 1,200$ lines per mm at the wavelength of H_α and for an incoming angle $\alpha = 0$. In 0th order, no dispersion is present. For the first order we use the grating equation 2.49 and form it to the diffraction angle

$$\sin \beta = \frac{1 \cdot \lambda}{d} = \lambda \cdot g = 656.3 \cdot 10^{-9}\text{m} \cdot 1200/10^{-3}\text{m} = 0.78756.$$

We obtain $\beta = 51.96°$ and $\cos \beta = 0.616$. Now we can estimate the angular dispersion around H_α:

$$\frac{d\beta}{d\lambda} = \frac{g \cdot n}{\cos \beta} = \frac{1200/10^{-3}\text{m}}{0.616} = 1.948 \cdot 10^6 \frac{\text{rad}}{\text{m}}$$

or

$$1.948 \cdot 10^6 \frac{\text{rad}}{\text{m}} \cdot \frac{180}{\pi} \cdot 10^{-9}\,m/nm = 0.112°/\text{nm}$$

For calculating the linear dispersion $\frac{dx}{d\lambda}$ we need to multiply the angular dispersion in radians with the focal length f of the optics used and again recalculate it to a standard wavelength unit in nm or Å. For a 135 mm lens, e.g., we obtain

$$\frac{d\beta}{d\lambda} \cdot f = \frac{dx}{d\lambda} = 1.948 \cdot 10^6 \,\text{rad/mm} \cdot 135\,mm \cdot 10^{-9}\,m/nm = 0.263\,\text{mm/nm}$$

The inverse linear dispersion is $1/0.263\,\text{mm/nm} = 3.8\,\text{nm/mm} = 38\,\text{Å/mm}$. A CCD camera, using a chip with 1530 pixels of 9μ size could image a wavelength interval of $38\,\text{Å/mm} \cdot 1530 \cdot 0.009\,\text{mm} = 523\,\text{Å}$.

2.5.5 The Maximum Resolving Power of a Grating

The argument for the estimation of the maximum grating resolving power is equivalent to that of a prism. The diffraction limited element is now the grating dimension. We start with the angular dispersion of a grating calculated by

$$\frac{d\beta}{d\lambda} = \frac{g \cdot n}{\cos\beta}$$

or

$$\Delta\lambda = \Delta\beta \cdot \frac{\cos\beta}{g \cdot n}.$$

For a rectangular grating of width B we have $\Delta\beta = \frac{\lambda}{B}$ (see Eq. 2.40) or, correspondingly, $\Delta\lambda = \frac{\lambda}{B} \cdot \frac{\cos\beta}{g \cdot n}$. For the resolving power we now get

$$R = \frac{\lambda}{\Delta\lambda} = \frac{\lambda \cdot B \cdot g \cdot n}{\lambda \cdot \cos\beta} \approx \frac{B \cdot g \cdot n}{\cos\beta}$$

or

$$R = \frac{N \cdot n}{\cos\beta}$$

with $N = B \cdot g$ the total number of grating lines. For small diffraction angles β we approximate $\cos\beta \approx 1$, or

$$R \approx N \cdot n = B \cdot g \cdot n.$$

The spectral resolving power of a grating is proportional to both, its absolute line number and the grating constant.

It is therefore possible to estimate a minimum grating width for a given spectral resolving power.

An example: If a spectral resolving power of 10,000 is needed and a grating of 1,200 l/mm is used, the grating must be at least 8.3 mm wide according to the above considerations. This number seems to be quite small with respect to the fact that off-the-shelf gratings of 50 mm width are relatively cheap. Such a grating obviously should provide a maximum spectral resolving power R of at least $1,200 \text{l/mm} \times 50\,\text{mm} = 60,000$! Off-the-shelf spectrometers however do not deliver such a high value. To understand the reason for this discrepancy, we note the following in anticipation of the following chapters.

As for a prism spectrograph, the seeing disk on the sky or the optical slit in a spectrograph and the dimension of the CCD pixels (or grain size in a film) limit the spectral resolution. To avoid light loss in the spectrograph and to properly adjust the focal ratio of the collimator lens at a given focal length, oversized gratings are used in practice. The parallel beam after the collimator lens has to be completely covered and dispersed by the (inclined) grating. Hence, a relatively large grating is required. That normally means that the theoretical spectral resolving power of a grating is much larger than that in practice. The general estimation of the spectral resolution will be discussed in Sect. 2.6.4.

In addition, the collimated beam in a grating spectrograph has normally a circular or elliptical cross section. Different parts of the beam illuminate different groove numbers. As a consequence the resolving power is reduced by about 15 % by this effect alone (Erickson and Rabanus 2000).

2.6 Collimator, Camera and Pixel Size

To obtain a spectrum one needs optical elements which on the one hand, prepare the converging telescope light bundle and on the other hand, images the light dispersed by a prism or a grating without image errors. Moreover, the reduceability of the spectral information depends of the proper choice of the imaging CCD chip and its pixels. Therefore, we now examine the corresponding parameters of all respective optical elements of a spectrograph and their adjustment and optimization.

2.6.1 *Reproduction Scale and Anamorphic Magnification Factor*

Normally, collimator and camera optics have different focal lengths. Both sets of optics influence the imaging magnification and reduction in a spectrometer. We will now calculate the spectrometer optics and start with the lens equation.

$$\frac{1}{f} = \frac{1}{b} + \frac{1}{g}. \tag{2.82}$$

f is the focal length of the respective lens, g the object width of the object to be imaged and b the image width. As an introduction we are first interested in the telescope image of a star. The star is at a large distance from the observer and we can consider the object width as infinite to good approximation. From the lens equation we then get ($g \to \infty$)

$$\frac{1}{f} = \frac{1}{b} + \lim_{g \to \infty} \frac{1}{g} = \frac{1}{b}$$

or redistributed:

$$b = f.$$

The image width for an object at infinity is conformal to the focal length of the lens. The real image is then in the focal plane of the lens.

Now we calculate the image size of an extended object at infinity (e.g. the Sun). Here the general definition of the reproduction scale β is the relation of image size B to object size G.

$$\beta = \frac{B}{G}$$

The intercept theorem additionally provides

$$\frac{G}{g} = \frac{B}{b}$$

and hence

$$\beta = \frac{B}{G} = \frac{b}{g}.$$

Now we redistribute the lens equation for the general case to the unknown image width b

$$\frac{1}{f} - \frac{1}{g} = \frac{1}{b}$$

or

$$b = \frac{f \cdot g}{g - f}$$

and get for β

$$\beta = \frac{b}{g} = \frac{f \cdot g}{g \cdot (g - f)} = \frac{f}{g - f}.$$

With this generally valid equation for the reproduction scale we can now return to the original problem of the object at infinity and its size in the focal plane. With the definition equation for β we get

$$B = \beta \cdot G = \beta \cdot \frac{G}{g} \cdot g.$$

The ratio G/g is the tangent of the angular dimension of the object.

$$B = \beta \cdot G = \beta \cdot \tan \phi \cdot g = \frac{f}{g - f} \cdot \tan \phi \cdot g = \frac{f \cdot g}{g - f} \cdot \tan \phi.$$

We factor out g in the denominator and obtain

$$B = \frac{f \cdot g}{g \cdot \left(1 - \frac{f}{g}\right)} \cdot \tan \phi = \frac{f}{\left(1 - \frac{f}{g}\right)} \cdot \tan \phi.$$

By estimating the behaviour of g at infinity we calculate

$$B = \lim_{g \to \infty} \frac{f}{\left(1 - \frac{f}{g}\right)} \cdot \tan \phi = f \cdot \tan \phi.$$

As an example we calculate the image size of the Sun. This can easily be done with the focal length of the telescope by using the above formula. For a telescope of 2,000 mm focal length we get an angular dimension of 30 arcminutes.

$$B_{Sonne} = f \cdot \tan \phi \approx f \cdot \phi = 2000 \, \text{mm} \cdot 0.5° \cdot \pi/180° = 17.45 \, \text{mm}.$$

Of course, one can also consider the inverse image relations: The real object imaged in the focal plane of a lens acts as an object at infinity but a real or virtual image is not present. The light behind the lens consists of a parallel (collimated) beam, which can be imaged again into a new focal plane by another lens. Collimator and camera lenses in a spectrograph act exactly in this sense. The real image of the star in the telescope focal plane is in the focal plane of the collimator, at the same time. The beam leaving the collimator is parallel and thus satisfies the necessary condition for Fraunhofer diffraction at the grating. The beam "diffracted" at the grating is guided into different directions depending on wavelength. The seeing disk or the optical slit seems to come from infinity. The parallel wavefront can be imaged by a lens (camera) into the focal plane, as usual. But now we do not see a star anymore but a spectral strip of various colors.

For the calculation of the image relation in a spectrograph we start with the collimator. A slit (or the seeing disk) of width s is imaged by the collimator (virtually coming from infinity) at the angle $\phi \approx s/f_{col}$. The slit, hence, delivers a real image of size $s' \approx f_{cam} \cdot \phi$ to the focal plane of the camera and we have

$$s' = f_{cam} \cdot \frac{s}{f_{col}}.$$

By varying the slit size one varies the angle ϕ and hence the slit image, as well. The slit image s' in the camera focal plane depends linearly on s. We should note that a change of the angle ϕ by $\Delta\phi$ will change the incoming angle α on the grating by $\Delta\alpha = \Delta\phi$.

Now we additionally place the grating between the collimator and the camera and investigate the consequences for the geometrical optics inside the spectrograph. According to the grating equation 2.49 the wavelength λ (of a single order) appears for a specific combination of incoming angle α and diffraction angle β. For that reason we derive the grating equation for α and β and obtain

$$\frac{d\lambda}{d\alpha} = \frac{d}{n} \cdot \cos \alpha$$

$$\frac{d\lambda}{d\beta} = \frac{d}{n} \cdot \cos \beta.$$

We take the ratio of both derivatives

$$\frac{d\lambda}{d\alpha} \Big/ \frac{d\lambda}{d\beta} = \frac{d\lambda}{d\alpha} \cdot \frac{d\beta}{d\lambda} = \frac{d\beta}{d\alpha}$$

and obtain by division

$$\frac{d\beta}{d\alpha} = \frac{\cos \alpha}{\cos \beta}$$

or in approximation with $\Delta\alpha$ and $\Delta\beta$

$$\Delta\beta \approx \frac{\cos \alpha}{\cos \beta} \cdot \Delta\alpha.$$

The last equation can be interpreted as follows: When changing the angle of incidence α by $\Delta\alpha$, the diffraction angle β changes by $\Delta\beta$. The diffraction angle alteration $\Delta\beta$ is proportional to $\Delta\alpha$. The respective proportional factor $\cos\alpha/\cos\beta$ is called the anamorphic magnification factor A because this factor is only defined in the dispersion direction and not perpendicular to it. The anamorphic magnification factor is the projection ratio of the incident and diffracted beams (see Fig. 2.27). An incoming beam of width x exhibits width $x \cdot \frac{\cos\alpha}{\cos\beta}$ after the diffraction process at the grating.

Perpendicular to the direction of dispersion the grating acts like a mirror and has no influence on the imaging scale. The anamorphic magnification factor operates like a magnification or reduction factor, depending on the resolution of the spectrograph, if it is larger or smaller than 1, respectively, and has to be accounted for.

Because of the specific geometry due to diffraction at the grating the slit does not show the angle $\Delta\alpha = \phi$, but

$$\Delta\beta = \frac{s'}{f_{cam}} = \frac{\cos\alpha}{\cos\beta} \cdot \Delta\alpha = \frac{\cos\alpha}{\cos\beta} \cdot \phi = \frac{\cos\alpha}{\cos\beta} \cdot \frac{s}{f_{col}}.$$

Fig. 2.27 Imaging geometry at the grating with respect to the anamorphic magnification factor. The elementary surface is 1

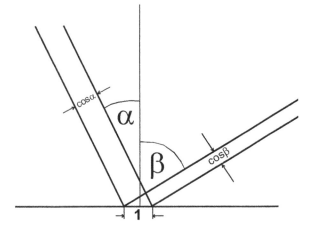

Therefore the size of the slit image (or the seeing disk) s' is calculated by

$$s' = s \cdot \frac{\cos \alpha}{\cos \beta} \cdot \frac{f_{cam}}{f_{col}}$$

2.6.2 The Necessity of a Collimator

Sometimes one can find in astronomical journals suggestions and descriptions for the design of a spectrograph with a transmission grating. The grating is simply fastened into the filter screw of an eyepiece and the resulting spectrum can be observed by eye or with a CCD. This simple method guides the beginner without great effort to the first contact with spectroscopy, including looking at the Balmer series of bright stars or the many absorption lines of our Sun. The attentive reader realizes that for this design a collimator is not used but a spectrum is observable anyway. For this reason we now want to briefly emphasize the function and necessity of a collimator, especially with respect to grating spectrographs, although this is basically also valid for prism spectrographs.

First, a stellar seeing disk is imaged in the focal plane of a telescope. Behind the focus the beams diverge, i.e., the grating is not illuminated by a parallel wavefront. We already described in Sect. 2.3.3 that in this case the requirement for Fraunhofer diffraction (far field approximation) is no longer present. The grating does not operate correctly. In addition, if the beam diverges, the grating size and camera diameter then depend on the respective distance to the telescope focal plane. For such a configuration the image scale is very complicated. By not using a collimator behind the focal plane, the stellar image represents an object in the image width of the camera lense, i.e., according to the lens equation the size of the star imaged in the camera plane strongly depends on the distance between camera and telescope focal plane.

As an example we assume that the camera can be positioned so that the stellar disk in the focal plane cannot be magnified (like a spectrograph with identical collimator and camera focal length), i.e., image and object sizes are identical. For a quantitative consideration we use again the above relation for the image scale

$$\beta = \frac{b}{g} = \frac{B}{G} = \frac{f}{g-f}.$$

β is exactly 1, if $g = 2 \cdot f$

$$\beta = \frac{f}{2 \cdot f - f} = \frac{f}{f} = 1.$$

For an image scale of $\beta = 1$ the image size has to be at least twice the focal length. The overall length of a spectrometer would be increased. Choosing a smaller image size, the diameter of the stellar image in the camera focus would be magnified with larger image size and, hence, the spectral resolution would be reduced. For larger image sizes the focus would be smaller because of $\beta < 1$ but one has to accept again that the dimensions of the spectrograph would become larger.[19]

Now one could imagine placing the grating directly in front of the telescope focus. The divergent beam would not affect the optical path anymore and the instrument dimensions could become slightly smaller. However, there is a specific physical reason why this strategy is not recommended. We recall again the general grating equation (2.49):

$$n \cdot \lambda = d \cdot (\sin \alpha + \sin \beta)$$

and convert this equation to the diffraction angle

$$\beta = \arcsin \left(\frac{n \cdot \lambda}{d} - \sin \alpha \right).$$

The diffraction angle obviously depends on the incoming beam. Until now we did not take the angle of incidence α into account. Now this dependence becomes important because a converging beam reaches the grating in various angles depending on the focal ratio. We now use a grating of 1,200 lines/mm as an example to

---

[19] Additional image aberrations introduced by the diverging beam would cause additional degradation of the spectral resolution.

demonstrate the dimension of this effect. At Hα in first order the diffraction angle
can be calculated by

$$\beta = \arcsin(656.3 \cdot 10^{-9}\text{m} \cdot 1200/10^{-3}\text{m} - \sin\alpha).$$

By plotting the diffraction angle β depending on the angle of incidence α we get
the characteristics as in Fig. 2.28. To estimate the angles of incidence occurring at
the grating we have to consider the dependence of the maximum beam angle α_{max} of
the telescope optics and the f-ratio: $\alpha_{max} = \arctan(D/F)$ (Fig. 2.29). For instance,
for an f/10 telescope $\alpha_{max} = 5.8°$ and for f/20 it is $\alpha_{max} = 2.9°$.

We continue with an f/10 telescope of 2,000 mm focal length. The variation of the
angle of incidence from $\alpha = 0°$ to $\alpha_{max} = 5.8°$ introduces a diffraction angle change
of about 9.5°. For a transmission grating positioned 10 mm in front of the focus we
would obtain a magnification of the Hα focus of 10 mm $\cdot \tan(9.5°) = 1.67$ mm.
Hence, the original telescope focus of about 40 μm at 4″ seeing is greatly increased
at Hα to 1.67 mm by a factor of 43. Of course, the exact factor depends on the optical
elements and their respective distances. But even an advantageous arrangement of
the camera lens will probably not help. An investigation (Federspiel 2002) showed
that for a transmission grating of 207 lines/mm combined with an f/10 telescope, the
theoretical resolution is degraded from about 1 to 25 Å. Hence, the use of a grating
without collimator cannot be recommended.

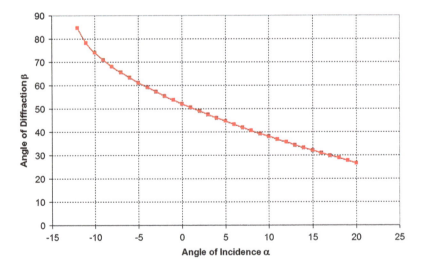

Fig. 2.28 Diffraction angle β versus angle of incidence α for the example in the text

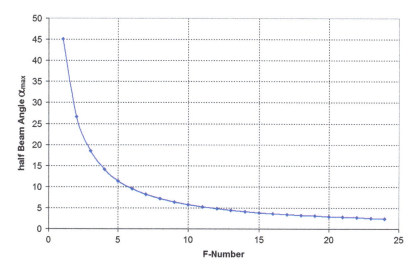

Fig. 2.29 Dependence of the beam angle α_{max} on the f-ratio $\alpha = \arctan(D/F)$

2.6.3   The Spectral Resolution of a Spectrograph

The key parameter for the reduction and interpretation of spectrograph data is not only the efficiency but also its spectral resolution ($\Delta\lambda$). It is often given in the inverse form of resolving power

$$R = \lambda/\Delta\lambda \qquad (2.83)$$

with λ the wavelength where the resolution is estimated and $\Delta\lambda$ the resolvable spectral element. Our goal is to find out how small $\Delta\lambda$ can be at most. For that we recall the principle function of a grating spectrograph. The collimator transforms the divergent bundle behind the telescope focus into a parallel wavefront. For simplification we now assume at first that the star only emits monochromatic light (e.g. from a laser). Then we can neglect the impact of the grating because it only separates different wavelengths geometrically. For this simplification, the grating would only displace the stellar image away from the optical axis. The camera focuses the collimated beam onto the focal plane. The result is an image of the star again but with a magnification factor defined by the combined collimator-camera optics, depending on the ratio of the respective focal lengths. For instance, if the camera has twice the focal length as the collimator, the monochromatic stellar disk in the focal plane has twice the diameter as the image in the primary telescope focus (see Fig. 2.30). Now we can extract the wavelength interval from the smallest diameter of the stellar image. For that we need the angular dispersion of the grating and the angle of the stellar image in the focal plane. We use the equation which we

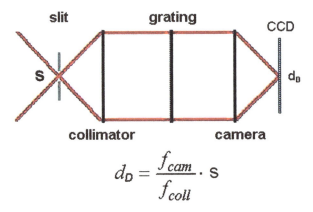

$$d_D = \frac{f_{cam}}{f_{coll}} \cdot s$$

Fig. 2.30 Principle of a grating spectrograph. The optical slit in the telescope focal plane defines a light source (*star*) of constant width in dispersion direction. The focal lengths of collimator and camera as well as the telescope focal length and the slit width s (i.e. the seeing ω with $f_{Tel.} \cdot \tan \omega \equiv s$ without slit) define the spectral resolution. This is also valid for the anamorphic magnification factor $A = \cos \alpha / \cos \beta$ which is 1 if the beam reaches the grating perpendicularly and can, hence, be neglected for our case (see Fig. 2.27)

used above (Eq. 4.4), except for the estimation of the image field of view

$$\Delta \lambda = \frac{p \cdot N}{f_{cam}} \cdot \frac{d\lambda}{d\beta} = \frac{\Delta CCD}{f_{cam}} \cdot \frac{d\lambda}{d\beta}. \tag{2.84}$$

Now however, we do not consider the field of view but conversely the smallest possible diameter of a point object. The fundamental problem however remains the same. We just have to replace $p \cdot N = \Delta CCD$ by the stellar diameter in the focal plane of the spectrograph. We call this diameter d_D and get

$$\Delta \lambda = \frac{d_D}{f_{cam}} \cdot \frac{d\lambda}{d\beta}. \tag{2.85}$$

With the equation for the angular dispersion (Eq. 2.80) we obtain

$$\Delta \lambda = \frac{d_D}{f_{cam}} \cdot \frac{d}{n} \cdot \cos \beta. \tag{2.86}$$

To be complete we note that this deduction is equivalent to the estimation of the maximum spectral resolution of a grating. The angular diameter of the Airy disk has just to be replaced by the angular diameter ω of the star. Now we obtain the spectral resolving power R under consideration of the grating equation for λ.

$$R = \lambda / \Delta \lambda = \frac{\frac{d}{n} \cdot (\sin \alpha + \sin \beta)}{\frac{d_D}{f_{cam}} \cdot \frac{d}{n} \cdot \cos \beta} \tag{2.87}$$

or

$$R = \frac{f_{cam}}{d_D} \cdot \frac{(\sin\alpha + \sin\beta)}{\cos\beta}. \tag{2.88}$$

Figure 2.30 depicts the geometrical context. The diameter of the star in the focal plane d_D is represented by the spectral width at half maximum of an ideal point light source on the sky. For that reason, it is often denoted by "full width at half maximum" (FWHM). According to Eq. 2.6.1 it is calculated by

$$d_D = A \cdot \frac{f_{cam}}{f_{col}} \cdot s = A \cdot \frac{f_{cam}}{f_{col}} \cdot f_{tel} \cdot \tan\omega. \tag{2.89}$$

An example: With a seeing of $4''$ on the sky, a telescope focal length of 3,000 mm, a collimator focal length of 200 mm and a camera focal length of 300 mm d_D is

$$300\,\text{mm}/200\,\text{mm} \cdot 3000\,\text{mm} \cdot \tan(4''/3600'') = 0.087\,\text{mm}$$

Now all parameters for the estimation of the spectral resolution are known.

2.6.4 Spectrometer Function: The Folding Integral

In the last chapter we learned that the spectral resolution directly depends on the diameter of the star in the focal plane of the spectrograph and, hence, also on the imaging quality of the respective optics. The smaller the diameter of the stellar image the better is the spectral (and geometrical) resolution. Broadening of the stellar light inside the Earth atmosphere by seeing and instrumental optics has direct influence on the "Full Width at Half Maximum" and, hence, on the spectral resolution. In addition, limited apertures introduce diffraction effects inside the spectrograph which also have an impact on the width of the stellar disk. In addition, we did not take the pixel size of the CCD into account. If the pixel is larger or smaller than the stellar image, it cannot be neglected for the estimation of FWHM or the spectral resolution.

A stellar point-source will be imaged by the following optical chain and broadened by all single elements:

Point source → *(seeing)* → *telescope diffraction* → *(slit)* → *collimator diffraction* → *grating diffraction* → *camera diffraction* → *pixel size*

We have accounted for the seeing and the broadening of the slit in brackets because they do not affect the broadening chain simultaneously. Either we work without a slit, when seeing plays the role of image broadener or one works with a slit and seeing is eliminated (see Sect. 4.1.2).

We now want to estimate the resolution of a spectrograph <u>with</u> and <u>without</u> a slit. For doing so we have to "add" all broadening effects inside the spectrograph in a proper manner. This addition will be mathematically performed by a folding. It describes an integral operator which delivers a third function out of two others f and i. This folding $(f * i)$ is

$$o(t) = f(t) * i(t) = \int_{-\infty}^{\infty} f(t - \tau) \cdot i(\tau) d\tau. \tag{2.90}$$

This specification is known from general signal theory. If the system function $f(t)$ of an arbitrary linear system is known, we can calculate the output $o(t)$ by folding with the input $i(t)$. The function value f at position t indicates how strong the delayed value t of the weighted function is, thus $f(t - \tau)$ contributes to the value of the integral t.

In our case, the system function of the spectrometer is depicted by the broadening function of an arbitrarily narrow spectral line. In an experiment one could estimate the system function with a broadened laser beam which illuminates a diffusor in front of a telescope with a spectrometer. At the system exit one then measures the spectral width of the line. All optical components contribute to the spectral broadening of the laser source. We now want to know how to calculate the overall broadening of the line.

In the case of a number of processes contributing to spectral line broadening we obtain the overall function $f(t)$ by convolving all the single broadening functions $f_i(t)$.

$$f(t) = f_1(t) * f_2(t) * \ldots * f_n(t). \tag{2.91}$$

As an example we consider a spectrograph with entry and exit slit, for simplification. Both elements have a broadening effect on the narrow laser line and have to be convolved according to the above rule $f(t) = f_1(t) * f_2(t)$. For this case the folding can be easily interpreted geometrically. The slits represent a rectangular function and the convolving integral is

$$f(t) = f_1(t) * f_2(t) = \int_{-\infty}^{\infty} f_1(t - \tau) \cdot f_2(\tau) d\tau. \tag{2.92}$$

The spectrometer function $f(t)$ results from step-by-step shifting of both rectangular functions. For each infinitesimal step the overlapping area will be calculated as an integral. The area increases linearly until both rectangles perfectly match and then decreases to zero. This effect is depicted in Fig. 2.31. The resulting spectrum $o(t)$ at the exit of the system will therefore be calculated by two successive convolutions

$$o(t) = f_1(t) * f_2(t) * i(t). \tag{2.93}$$

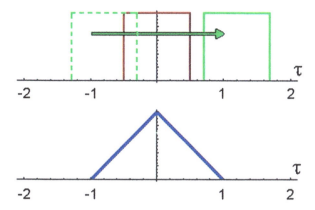

Fig. 2.31 Convolving of two rectangular functions. As a visual analogue we let the function f_1 (*green*) "run" over f_2 (*red*). The integrated area, the integral $f_1 * f_2$, increases and then decreases again

Generally, the calculation of the overall system function is a complex task, especially if the various broadening functions exhibit different forms. The slit is a rectangular function, as already described, similar to a CCD pixel. However, diffraction patterns from the collimator and the camera aperture have the form of a Bessel function of various widths. For mathematical simplification we can use a usual correlation from convolution theory.

It is important to know that the folding of a Gaussian with half width σ_1 with another Gaussian with half width σ_2 is again a Gaussian with half width $\sigma = \sqrt{\sigma_1^2 + \sigma_2^2}$. If we assume for simplification Gaussian features in all optical elements as broadening or diffraction patterns we can then estimate the overall width by applying

$$\sigma_{ges} = \sqrt{\sigma_1^2 + \sigma_2^2 + ... + \sigma_n^2}. \tag{2.94}$$

With our simplification of the problem we now only have to calculate the diameter of all system broadening functions and then add all terms quadratically.

As already remarked further above, the seeing disk on the sky is imaged by the telescope with all its specific parameters. In addition to that, the anamorphic magnification factor and the collimator and camera focal lengths influence the reproduction scale of the optical system, as shown in Fig. 2.27. The diameter of the seeing disk has to be corrected by the magnification ratio of both focal lengths and the anamorphic magnification factor

$$d_D = A \cdot \frac{f_{cam}}{f_{col}} \cdot \tan \omega \cdot f_{tel}. \tag{2.95}$$

The ratio $A = \cos\alpha/\cos\beta$ describes the anamorphic magnification factor of the grating again and the product $s = \tan\omega \cdot f_{tel}$ is the geometrical diameter of the seeing disk ω in the focus of the telescope of focal length f. Without seeing (diffraction limited) we have to apply the width (FWHM) of the diffraction pattern of the corresponding optics for s, hence, $s = 2.44 \cdot \lambda \cdot \frac{f_{tel}}{D_{tel}}$. However, further above we already explained that seeing strongly dominates over diffraction and we can neglect the latter at this point. This is not valid for the collimator, which also produces a diffraction pattern. Its Airy disk is not influenced by the telescope. Normally, the diameter of its Airy disk is of the order of the slit width so that corresponding diffraction patterns become relevant. Then the equation $\sigma_{col} = 2.44 \cdot \lambda \cdot \frac{f_{col}}{D_{col}}$ can be considered as a lower limit approximation[20] (this is similar for the camera). Because the imaging properties of the spectrograph are defined by the focal lengths of its optics we need to consider their ratios and, of course, the anamorphic magnification factor by $A \cdot \frac{f_{cam}}{f_{col}} \cdot d_{col}$.

The diffraction-limited point spread function (PSF) of the grating ($\sigma_{grating}$) will also not be influenced by the telescope; the collimator has already made the beam parallel. By applying the diffraction pattern of a rectangular aperture (see Eq. 2.40) the grating diffraction pattern diameter is

$$\sigma_{grating} = \frac{\lambda}{w} \cdot l \tag{2.96}$$

with l the distance between grating and focal plane and w the aperture (width) of the grating.

As for the grating, the diffraction pattern of the camera (σ_{cam}) does not have to be modified. Both parts can simply be added quadratically. Finally, we have to consider that the pixel size p of the detector has to be quadratically added because of its limited size. We obtain for the **resolution without slit**:

$$d_D^2 = \left(A \cdot \frac{f_{cam}}{\cdot f_{col}} \right)^2 \cdot \left(\tan^2\omega \cdot f_{tel}^2 + \sigma_{col}^2 \right) + \sigma_{grating}^2 + \sigma_{cam}^2 + p^2. \tag{2.97}$$

For the **resolution with a slit** we replace $\tan\omega \cdot f_{tel}$ (the telescope projection of the seeing disk in the sky) by the fixed slit width s and find

$$d_D^2 = \left(A \cdot \frac{f_{cam}}{f_{col}} \right)^2 \cdot \left(s^2 + \sigma_{col}^2 \right) + \sigma_{grating}^2 + \sigma_{cam}^2 + p^2. \tag{2.98}$$

The simplest method to measure the resolution of a real apparatus is performed by a calibration lamp in front of the telescope (without slit) or directly in front of the optical slit. After acquisition of the respective emission line spectrum one measures the line FWHM which already represents the spectral resolution $\Delta\lambda$, from which the resolving power $R = \lambda/\Delta\lambda$ can be calculated.

---

[20]In most cases the lens used here will not function as diffraction limited, i.e., the point spread function will exhibit a larger FWHM than that of the diffraction pattern.

2.6.5 Broadening Processes and the System Function: Multiple Convolution

As already mentioned in Eqs. 2.97 and 2.98 each optical element broadens the spectrometer function. It is now natural to ask for the form of the diffraction function at the output of a spectrometer composed of many optical apertures. Beside the entrance slit and the CCD pixel size all other imaging elements, including the grating, broaden the spectrometer function as well. For that reason we use a simplifying assumption to investigate the impact of a multiple folding, as shown in Eq. 2.91. To find the broadening function we oversimplify that all limiting apertures can be described by a rectangular function of unit width[21] $s = 1$. We therefore consider the M-fold convolution product of the square function:

$$g_M(t) = \text{rect}_1(t) * \text{rect}_2(t) * \ldots * \text{rect}_M(t),$$

where $\text{rect}(t)$ is a rectangular function defined by

$$rect(t) = \begin{cases} 1, & \text{when } |t| \leq 1/2 \\ 0, & \text{when } |t| > 1/2. \end{cases}$$

Since according to the convolution theorem of Fourier analysis the convolution of two functions is the product of their Fourier transforms, we can reduce the convolution of all optical element functions to a product of their Fourier transforms and thus simplify the computing process, as in

$$h(t) * g(t) = \mathscr{F}(h) \cdot \mathscr{F}(g) = H(f) \cdot G(f)$$

With the frequency-dependent Fourier transforms $H(f)$ and $G(f)$ we obtain the Fourier transform of the rectangular function in the frequency domain

$$\mathscr{F}[\text{rect}(t)] = \frac{\sin \pi f}{\pi f}$$

and, hence, for the Fourier transform of the convolution product

$$G_M(f) = \prod_{i=1}^{M} \frac{\sin \pi f}{\pi f} = \left[\frac{\sin \pi f}{\pi f}\right]^M.$$

---

[21]Different widths of the rectangular function only lead to scaling factors in the respective result.

For simplification we use the Taylor expansion of the sine function around $x = 0$,

$$\sin x = x - \frac{x^3}{3!} + \frac{x^5}{5!} - \dots$$

By logarithmizing and implementing the sine expansion we obtain

$$\ln G_M(f) = M \ln \left[\frac{\sin \pi f}{\pi f} \right] = M \ln \left(1 - \frac{(\pi f)^2}{6} + \frac{(\pi f)^4}{120} - \dots \right).$$

For $|f| \ll 1$ one can stop the expansion after the second term. For $x \ll 1$ we can additionally use the Taylor expansion for the logarithm function $\ln(1 + x) \approx x$ and we obtain

$$\ln G_M(f) \approx -M \frac{(\pi f)^2}{6}$$

or

$$G_M(f) \approx \exp \left[-M \frac{(\pi f)^2}{6} \right]. \tag{2.99}$$

To return to the original question of the entire broadening function, we calculate the inverse transform of $G_M(f)$. We get[22]

$$g_M(t) \approx \sqrt{\frac{6}{M \pi}} \exp \left[-\frac{6t^2}{M} \right]. \tag{2.100}$$

The multiple convolution of the rectangular function with itself obviously tends to a Gaussian function.

This result is also known from the stochastics as the so-called "central limit theorem", which states that the sum of M independent and identically distributed random variables $M \to \infty$ converges asymptotically to the normal distribution. The connection to the spectrometer function becomes clearer when we consider that the distribution density of the sum of two stochastically independent random variables X and Y of the probability densities f and g is just the convolution of $f * g$. One can then understand the central limit theorem on the computer by using a tabulator programme generating a sufficient number of columns of uniformly distributed random numbers[23] and by adding line by line the sum of these numbers to a new random number. The histogram of the new variable, already after a few additions, is almost a normal distribution (see Fig. 2.32). This stochastic approach has an intuitive physical meaning for the spectrometer. One can send the photons stochastically

---

[22]Note that this only applies for $|f| \ll 1$ and large M.

[23]The uniform distribution represents our spectrometer slit function.

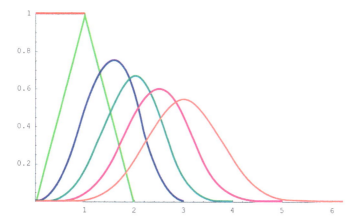

Fig. 2.32 Sequence for the sum of the first six normal distributions (multiple folding of rectangular functions) and their output distributions

through the spectrometer. With a normal distribution across the entrance slit the photon histogram at the output of the spectrometer represents the spectrometer function.

We thus have also confirmed the necessary assumption for Eq. 2.94 that a Gaussian function represents a good approximation for the spectrometer function. Another important result of the above considerations is the conclusion that the spectrometer function becomes smoother with an increasing number of optical elements. Therefore fewer frequencies in the Fourier space are required for the reconstruction of this output function (the Fourier transform of the local spectrometer function is a narrow band in frequency space). We now consider an important property of the two reciprocal functions[24] $g_M(t)$ and $G_M(f)$. For this purpose we recall the general definition of the Gaussian distribution, $f(x) = \frac{1}{\sigma\sqrt{2\pi}} \exp\left[-\frac{x^2}{2\sigma^2}\right]$. This function has the expected value 0 and the standard deviation $\pm\sigma$. If we compare this definition with the argument-factors of the function $g_M(t)$ (Eq. 2.100) and its Fourier transform $G_M(f)$ (Eq. 2.99) we obtain for the standard deviations $\sigma_t = \sqrt{\frac{M}{12}}$ and $\sigma_f = \sqrt{\frac{3}{M\pi^2}}$ and, hence, for the product of these two standard deviations

$$\sigma_t \cdot \sigma_f = \sqrt{\frac{M}{12}} \cdot \sqrt{\frac{3}{M\pi^2}} = \frac{1}{2\pi},$$

or with the angular frequency $\omega = 2\pi f$

$$\sigma_t \cdot \sigma_\omega = 1. \tag{2.101}$$

---

[24]In Fourier analysis, the Fourier transform is commonly referred to as "reciprocal function".

Apparently, the two standard deviations of the Gaussian function and its Fourier transform behave reciprocally, i.e., when the standard deviation of the time function $g_M(t)$ is small, the associated standard deviation of the Fourier transform is large and vice versa. This is the so-called Küpfmüller uncertainty principle of signal processing for time signals, which also applies to spatial signals. The time average $\bar{t}$ and its standard deviation σ_t for non-negative and real functions $f(t)$ are defined according to the statistical definitions

$$\bar{t} = \frac{\int_{-\infty}^{+\infty} tf(t)dt}{\int_{-\infty}^{+\infty} f(t)dt}$$

$$\sigma_t^2 = \frac{\int_{-\infty}^{+\infty}(t-\bar{t})^2 f(t)dt}{\int_{-\infty}^{+\infty} f(x)dx}.$$

(2.102)

For functions with negative or complex values new definitions for the expectance value and the variance for the square-integrable functions of a Hilbert space are introduced in functional analysis[25]

$$\bar{t} = \frac{\int_{-\infty}^{+\infty} t|f(x)|^2 dx}{\int_{-\infty}^{+\infty} |f(x)|^2 dx}$$

$$\sigma_t^2 = \frac{\int_{-\infty}^{+\infty}(t-\bar{t})^2|f(x)|^2 dx}{\int_{-\infty}^{+\infty} |f(x)|^2 dx}.$$

(2.103)

These definitions result in spaces that are mutually reciprocal to the general uncertainty relation. In a more general form this is known as the Heisenberg uncertainty principle:

$$\sigma_t \cdot \sigma_\omega \geq \frac{1}{2}$$

$$\sigma_x \cdot \sigma_k \geq \frac{1}{2},$$

The equality in this inequation is solely valid for the Gaussian function. The reciprocal time variable t and angular frequency ω as well as the position x and the wavenumber k ($k = \frac{2\pi}{\lambda}$) can both occur in combination. σ indicates the respective standard deviations of the variables in the associated Hilbert space.

For the uncertainty principle we consider an example from signal analysis. A periodic sinusoidal signal $f(t) = \sin \omega_0 t$ with $f_0 = 50\,\text{Hz}$ will be sampled with $f_a = 500\text{Hz}$. The signal has a finite length of $\Delta t = 50\,\text{ms}$, i.e. 2.5 periods are sampled. Because of the discrete sampling of $N = 500\text{Hz} \cdot 50\,\text{ms} = 25$ equidistant

---

[25]Physically the observable intensity is the square of the amplitude function.

moments, the Fourier spectrum will have a maximum of N discrete frequencies. The frequency distance in the discrete spectrum of our example will be $\Delta f = \frac{5000\,Hz}{25} = 20\,\text{Hz}$. If we extend the signal by a factor of 4 to a period of $\Delta t = 200\,\text{ms}$, we get $N = 100$ sampling points. In this case, the frequency distance is thus $\Delta f = 5\,\text{Hz}$. The parameters Δt and Δf in the form $\Delta t \sim \frac{1}{\Delta f}$ represent the Heisenberg uncertainty (whose formulation is valid strictly only for the standard deviations[26]). We will discuss the issue of signal sampling and its consequences in Sect. 2.6.6 in detail. A signal with a short time period or a limited spatial extent thus always has a spectrum with a large bandwidth. One can illustrate this fact: a periodic signal with frequency f must be observed at least for one period in order to determine this period with certainty.

We highlight the spectrometer function and its connection to the Heisenberg uncertainty principle again with two examples.

- First, we assume that the spectrometer function is a delta function. (a) In the spatial domain, the detected spectrum is the convolution of the incident spectrum (e.g., a Gaussian emission line) with the spectrometer function. However, the convolution with a delta function reproduces the incident spectrum, i.e., it has not changed through the spectrometer and its optics. (b) Since the δ-function is arbitrarily narrow the Heisenberg relation tells us that in Fourier space the spectrometer function is an unlimited broadband, i.e., a constant with the value 1. After Fourier transformation, the emission line remains a Gaussian function. According to the folding rule (Appendix B.5) the output spectrum is the product of the two transformed functions. Since the Fourier transformed spectrometer function is a constant, changes in the input spectrum in Fourier space do not occur and by inverse transformation one again obtains the measured spectrum.
- In the 2nd example we consider a real spectrometer function of finite width, again in Fourier space. The Fourier transform of the spectrometer function of finite width is a function with a finite bandwidth, according to the uncertainty principle. In this case the incident spectral line is broadened and flattened by folding the spectrometer function. A very narrow spatial spectral line is broad in Fourier space. If a spectral line is much narrower than the spectral function of the system (e.g., a laser line) we obtain almost the spectrometer function at the output. Spectral information located at the edges of the spectral line will then be suppressed by the product by the narrow spectrometer function (the information provided is at the border or outside of the carrier of the system function). There is a further consequence of the uncertainty principle to be treated here (described in Gray 1992 under "Resolution Theorem"): in addition to the spectrometer function being of finite width FWHM, the spectrum is recorded only in a finite range of wavelengths. The measured spectrum is therefore the

---

[26]The dependence of the quantities Δt and σ_t or Δf and σ_f depend on Eq. 2.103. This however in different ways, depending on the signal shape of $f(t)$. Therefore, these variables cannot be used directly in the uncertainty relation. For example, the standard deviation (in our example $\sigma_t = 16\,\text{ms}$) is different from the length of the sine signal ($\Delta_t = 50\,\text{ms}$ in our example).

product of the infinitely extended spectrum and a finite square function. As a result, the Fourier transform of the rectangular function shows a broad sinc spectrum. If the Fourier transform of a spectral line with high-frequency spectral signatures that are particularly close to each other (i.e. similar intensity structures in the spatial domain) they will be smoothed out in the convolution in Fourier space with the broad sinc function. This is also a consequence of the uncertainty principle, which only allows[27] a large Δk for a small Δx

2.6.6 Shannon's Theorem or the Nyquist Criterion

The slit size, the pixel size and the diffraction patterns of all optical parts define the resolution element in the spectrograph focal plane where discrete CCD pixels sample the spectral signal. However, a spectrum is a continuous signal in wavelength whereas the CCD detector consists of discrete elements (pixel). It is therefor necessary to ask for the accuracy a discrete CCD detector can image the real continuous world. Hence, we need to know how many discrete CCD pixels are necessary to accurately identify a continuous resolution element. For instance, a spectrum can only be completely reproduced if it is sampled by a certain number of points over a resolution element as represented by the FWHM of the slit image d_D in the focal plane, i.e., the width of the spectrometer broadening function. If an insufficient number of sampling points is used (the sampling frequency is too small), artifacts are potentially introduced. These artifacts are known as alias effects. Phase-shifted patterns or new structures occur, not existing in the original image (Fig. 2.33). The so-called Nyquist–Shannon theorem (Nyquist 1928; Shannon 1948a; 1948b), often called Whittaker–Kotelnikow–Shannon theorem (Whittaker 1915; Kotelnikov 1933; Whittaker 1935) gives a measure for the accurate approximation of a continuous signal by discrete sampling points. It states that a continuous, band-limited signal needs to be sampled by a sampling frequency of more than $2 f_{max}$ to unambiguously reconstruct the original signal. "Band-limited" means that the spectrum consists only of frequencies in the interval $[f_{min} - f_{max}]$.

The Nyquist–Shannon theorem, mostly called Nyquist criterion, is often directly translated to the spectroscopic domain by assuming that two pixels covering a resolution element is sufficient to reconstruct the original signal. As we will show below, this assumption is not correct. Therefore we briefly illuminate what the theorem is really saying and we will illuminate the consequences for the spectroscopic domain. Finally we present the correct application and its limitations with respect to spectroscopic data.

---

[27]In the time domain, the situation is similar: For a signal that has been recorded only for a certain time Δt it is not possible to detect structures of very low frequencies.

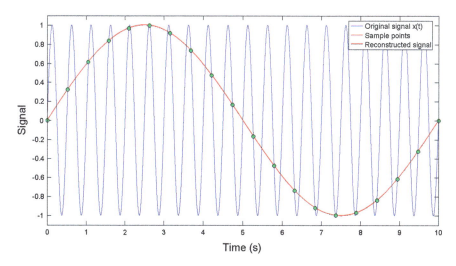

Fig. 2.33 The alias effect for a harmonic signal. A continuous signal (*blue line*) is discretized by an improper sampling frequency which is too low. The output (*circles*) delivers a false signal with an artificially enhanced period, although the amplitude is conserved (*red line*)

The Message of Nyquist–Shannon: Sampling with the Cardinal Series

The Nyquist–Shannon theorem has been introduced in the field of signal theory. It connects continuous, analogue frequency signals with discrete sampling points. When deriving the Nyquist–Shannon theorem one needs to consider the so-called cardinal series of Whittaker (1915),

$$h(t) = \sum_{n=-\infty}^{n=+\infty} h\left(\frac{n}{2f_{\max}}\right) \cdot \frac{\sin \pi(2f_{\max}t - n)}{\pi(2f_{\max}t - n)}. \tag{2.104}$$

This expression formally implies an interpolation or recovery algorithm. It says that the value of a function $h(t)$ can be determined by an infinite series. This series contains the product of the function values $h(\frac{n}{2f_{\max}})$ measured at the sampling points $\frac{n}{2f_{\max}}$ and sinc functions with an argument shifted by the index n at any time t. The minimum sampling frequency in the respective function arguments must be at least $2f_{\max}$. Only with this minimum sampling frequency it is possible to recover the original signal.

To investigate this series in more detail we consider a single summand in this series with the index m. Hence

$$h(t_m) = h\left(\frac{m}{2f_{\max}}\right) \cdot \frac{\sin \pi(2f_{\max}t_m - m)}{\pi(2f_{\max}t_m - m)}.$$

The associated sampling time for this specific case is $t_m = m/2f_{max}$. Thus, the argument of the sinc function vanishes and its function value is 1. The contribution of the series for the indices $-m$ and $+m$ is exactly $h(\frac{m}{2f_{max}})$ (note we sum from $-\infty$ to ∞). Thus, the summand $\pm m$ of the series at the time t_m represents nothing more than the sampling value at the point t_m and thus the value of the desired function. At all other sampling points $n \neq m$ the sinc function is nil (see Fig. 2.34) and thus does not provide any contribution to any other sample points. The sinc function actually acts as a δ-function at the location of the sampling point $n = m$. Only for times t between the sampling points does the sinc function provide a corresponding contribution to the infinite series. However, this applies to all summands of the series. Thus, all unknown intensity values of all sinc functions between the sampling points add to the desired function values. Interestingly the function $h(t)$ is completely determined by the sampling points and their function values. It can be "recovered" by the cardinal series (Eq. 2.104) as shown in Fig. 2.35. This series provides a formal algorithm to interpolate perfectly the unknown function $h(t)$ only from the sampling points at any time t. An appropriate representation of the sampling points for seven sampling points of the cardinal series (randomly selected) function is shown in Fig. 2.35.

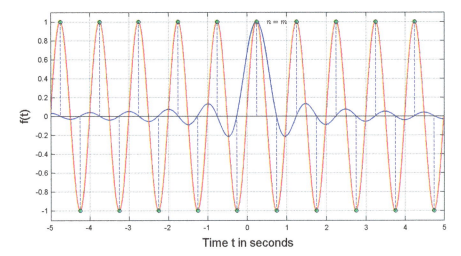

Fig. 2.34 Representation of a single sinc function for a specially selected sampling point. An example sine function (*red*), which is sampled at the Nyquist frequency specifically in its extreme (*green*). In addition, we show a sinc function with the same frequency (*blue*). Only the sampling point at the maximum of the sinc function (index m) has a function value $\neq 0$. All other sampling points are nil at the sinc function. This means that the sinc function with the argument m only contributes at times t between the sampling points and at the point $n = m$ to the cardinal series. Note that only an infinite number of sinc functions can fully represent the unknown function

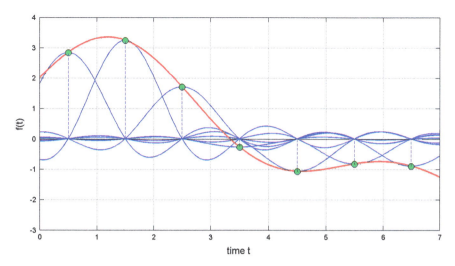

Fig. 2.35 Illustration of the recovery algorithm or the cardinal number, as a weighted superposition of shifted sinc functions. The sinc functions vanish at all sampling points (see nodes) of the function to be reconstructed (*red*), except at the point under consideration and provide the necessary values between the sampling points

Consequences for Spectroscopy

The prerequisite for this recovery algorithm is an adequate sampling frequency ($>$ $2f_{max}$) but also the actual realization of the infinite sum, i.e., an infinite number of sampling points of an infinitely (in time or space) extended function must be considered. Only under these conditions is the Nyquist–Shannon theorem strictly valid and the cardinal series converges towards the unknown function $h(t)$. These conditions, however, can never be fulfilled by any real measurement. No time and position limited signal, i.e. no function with a "finite support"[28] fulfills the condition for a band-limited function. This is also valid for periodic signals, e.g., pure sine oscillations or non-continuous signals. It is a purely ideal consideration for a ideal situation. Already the CCD detector with its limited geometrical extension cannot deliver an infinite series of sampling points. It cannot deliver band-limited signals and Nyquist–Shannon is formally not applicable.

If the sampling frequency is too low, alias effects as already mentioned above occur, in which the signal frequency is incorrectly determined (Fig. 2.33). An illustration of aliasing effects is presented in Fig. 2.36 for different sampling frequencies and interpolation methods. If the sampling theorem is not fulfilled, the signal sampling is called undersampling. One might argue that the finite

---

[28]The mathematician refers to finite support as the non-zero portion of a function, i.e., the function vanishes only outside certain limits. These limits define the interval of the compact support of the considered function.

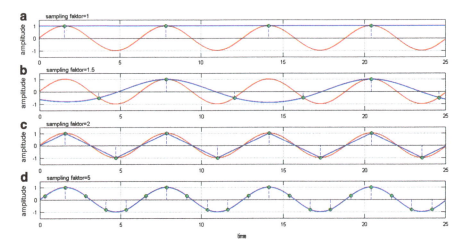

Fig. 2.36 Sampling of a sine wave. (**a**) Sampling a sine wave once per cycle delivers a constant signal. (**b**) Sampling the wave 1.5 times per cycle delivers a lower frequency sine wave (see Fig. 2.33). (**c**) Sampling at twice the sample frequency, i.e. Nyquist rate, delivers even in case of linear interpolation the exact sample frequency but the sine wave form is transformed to a sawtooth wave. However, if the sampling points are right at the zero-crossing of the sample frequency, no signal frequency nor any waveform can be obtained. (**d**) Sampling the wave with many sampling points also delivers the waveform with increasing sampling quality depending on the number of sampling points

spectrometer function always delivers an upper-limit frequency f_{max} because all infinitely steep flanks of a spectral line[29] will be convoluted with the spectrometer function again to finite flank slopes. However, due to the finite dimensions of the CCD chip only a finite spectral interval $\Delta\lambda$ ($\sim \Delta x$) can be considered. Due to the Küpfmüller uncertainty principle $\sigma_x \cdot \sigma_k = 1$ ($k = \frac{2\pi}{\lambda}$)[30] this leads to a finite frequency resolution.

Signal Band Limitations

The application of the cardinal series requires an infinite number of measurement points—the Nyquist–Shannon theorem requires a band-limited signal. Optical spectra, however, as for all measured signals, are not band-limited. For that reason

---

[29]When talking about an infinitely steep line flank we mean a spectral line shape of a box-car signal. However, at the entrance of the spectrometer the line flanks are never infinitely steep, since the physical processes in stars broaden all lines due to pressure and Doppler broadening (but not less than the natural line width) which already leads to a flattening of the flanks.

[30]In a more general form this is known as the Heisenberg uncertainty principle.

interpolation between sampling points plays a crucial role in the reconstruction of the original signal.

A signal with a short time period or a limited spatial extent has a spectrum of large bandwidth. A time- (or spatial-) limited signal $h_{\text{limited}}(t)$ can be expressed mathematically as the product of the initially unlimited function $h(t)$ and a rectangular function $\text{rect}(t)$.

$$h_{\text{limited}}(t) = h(t) \cdot \text{rect}(t).$$

For this case the box-car function is defined by

$$\text{rect}(t) = \begin{cases} 1, & \text{when } |t| \leq T \\ 0, & \text{when } |t| > T. \end{cases}$$

Outside the interval $[-T, T]$ the function h_{limited} is equal to zero. From Fourier analysis it is known that the Fourier transform of a product of two functions is equal to the convolution of their Fourier transform,

$$\mathscr{F}\left(h(t) \cdot g(t)\right) = H(f) * G(f),$$

where f is the frequency $f = \omega/2\pi$. Thus we obtain for the Fourier transform of our spatially limited function $h_{limited}(t)$

$$\mathscr{F}(h_{\text{limited}}(t)) = \mathscr{F}\left(h(t) \cdot \text{rect}(t)\right) = H(f) * \mathscr{F}(\text{rect}(t)).$$

The Fourier transform of the box-car function, however, as further noted above, is the sinc function. Hence, we obtain for the Fourier transform

$$\mathscr{F}(h_{\text{limited}}(t)) = H(f) * 2T \frac{\sin 2\pi Tf}{2\pi Tf}.$$

The convolution indicates the integral

$$\mathscr{F}(h_{\text{limited}}(t)) = H_{\text{limited}}(f) = 2T \int_{-\infty}^{+\infty} H(u) \cdot \frac{\sin[2\pi T(f - u)]}{2\pi T(f - u)} du. \quad (2.105)$$

In a mathematical sense the sinc function has no "finite support", i.e., it does not converge ($\omega \to \pm\infty$) to zero at the edges (Fig. 2.37). Since the function $H(u)$ in the integrand of Eq. 2.105 also does not vanish at the edges, we can conclude that the Fourier transform of a temporally or spatially limited signal does not have a limited support, i.e., it is not band-limited and thus the necessary conditions for the Nyquist–Shannon theorem in general are not fulfilled. In addition, sampling points do not in practice lie exactly on the signal but the measured values are already modified by digitization and the ever-present signal noise. Hence, the retrieval

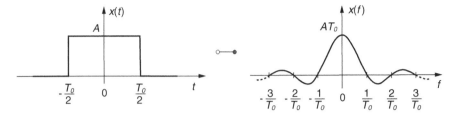

Fig. 2.37 Boxcar function (*left*) and its Fourier transform (*right*), the sinc function. This transform consists of an infinite number of frequencies f and amplitudes $A(f)$, i.e. the transform does not converge to zero and is thus not band-limited. The large side lobes significantly decrease the quality of the frequency spectrum. Thus, in a box-car signal, frequency components are included which are distributed over a large range. Rectangular pulses with steep edges basically have a very broad frequency spectrum

algorithm (cardinal series) of Whittaker (1915) cannot be used in its original sense and Nyquist–Shannon is formally not applicable.[31] A signal sampling by two pixels per resolution element will not deliver the original spectral signal.

2.6.7 Signal Sampling by Appropriate Interpolation

Instead of the cardinal series one could also use high-order polynomials for reconstruction (of degree $(n-1)$ for n sampling points). In general, a very high-degree polynomial fit is not suitable for interpolation. The mathematical problem of a polynomial fit can be attributed to an inversion of the so-called Vandermonde matrix: if the function values $(y_1, y_2, \ldots, y_n)$ at the sampling points $(x_1, x_2, \ldots, x_n)$ should be interpolated by a polynomial fit of $(n-1)$th degree the Ansatz

$$y(x) = a_0 + a_1 x^1 + a_2 x^2 + \cdots + a_{n-1} x^{n-1}$$

for n sampling points delivers the following linear equation system:

$$
\begin{pmatrix}
1 & x_1 & x_1^2 & \cdots & x_1^{n-1} \\
1 & x_2 & x_2^2 & \cdots & x_2^{n-1} \\
1 & x_3 & x_3^2 & \cdots & x_3^{n-1} \\
\vdots & \vdots & \vdots & \ddots & \vdots \\
1 & x_n & x_n^2 & \cdots & x_n^{n-1}
\end{pmatrix}
\cdot
\begin{pmatrix}
a_0 \\
\vdots \\
a_k \\
\vdots \\
a_{n-1}
\end{pmatrix}
=
\begin{pmatrix}
y_1 \\
\vdots \\
y_i \\
\vdots \\
y_n
\end{pmatrix},
$$

---

[31]In practice the substitution of the infinite Cardinal series by a finite sum will introduce a truncation error which depends on the number of the considered summands.

where the polynomial coefficients $(a_1, a_2, \ldots, a_n)$ represent the searched parameters. However, the Vandermonde matrix on the left side of this vector equation becomes less conditioned as the degree of the polynomial is increased. If the entries of the matrix are too far apart, one exceeds the numerical calculation accuracy which leads to difficulties in the inversion of the Vandermonde matrix. This is due to the possibly very different values of the matrix elements, depending on the number of data points and the polynomial order. The problems in finding enough suitable polynomials of high order are all the less surprising when we take into account that the polynomial interpolation and the cardinal series are mathematically completely equivalent for the case of equidistant sampling[32] and, hence, both methods should be seen as equally unsuitable. In practice, for the analysis of spectra one therefore always has to turn to other interpolation methods, e.g., suitable spline functions. The evaluation of spectra and their quality therefore also depends on the interpolation method used.

Pixel Sampling, Interpolation and Signal-to-Noise Ratio

For discrete sampling points (in our case the CCD pixels) the question arises how many pixels are needed and by what method they should be interpolated to reproduce a continuous spectral signal from discrete sampling points. We also need to know the errors that result at a fixed sampling frequency and different interpolation. The strength of this reproduction error[33] initially depends on the form of the incoming signal. The reproduction quality can be estimated by reconstructing the signal by sampling with a suitable algorithm. That is, the sampled signal will be interpolated and is then compared with the original signal. The resulting differences (or their standard deviation σ along the spectrum) from the true function value are a measure of the reproduction quality, similar to the quality of the original spectrum. The latter is defined by the signal-to-noise ratio (SNR or S/N), hence, we can illustrate this reproductive noise also by using an appropriate SNR.

The influence of the interpolation on the sampling results is significant. This is shown in Fig. 2.38. In this example, a noise-free sine function is sampled and reconstructed with linear and cubic spline interpolations. The interpolation of the sampling points with a spline (see Appendix B.4) delivers a much better sampling quality than that of a linear interpolation. A quantitative depiction of a function approximated by a Gaussian emission line is provided in Fig. 2.39. The corresponding consideration for the sampling quality, depending on the pixel sampling per resolution element ($FWHM$), is shown in Fig. 2.40. For example,

---

[32]It should be remembered that the sine function, and thus the sinc function can be developed into a polynomial series.

[33]Compared to the strictly valid cardinal series the interpolation method is always imperfect. For this reason one basically undersamples. However, the term "undersampling" is already used in the context of the Nyquist sampling theorem, and the minimum sampling frequency.

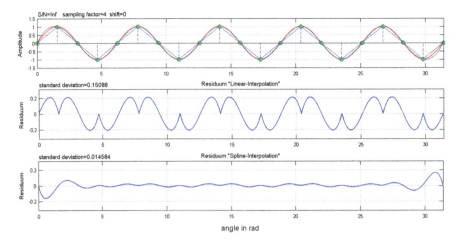

Fig. 2.38 *Top*: Sampling of a noise-free sine function with four sampling points per period, linear and spline interpolation. *Middle*: Standard deviation of the linearly interpolated signal from the original function. *Bottom*: Standard deviation of the spline-interpolated function signal from the original function. The standard deviations from the original signal are about ten times smaller for a spline interpolation than for a linear interpolation

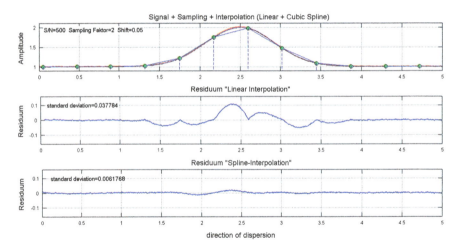

Fig. 2.39 *Top*: Sampling of a slightly noisy Gaussian signal ($S/N_{peak} = 500$) with three sampling points per *FWHM*, linear and spline interpolation. *Middle*: Sampling noise of the linear interpolation. *Bottom*: Sampling noise of the spline interpolation. The standard deviation of the sampling noise in the range $x = 1$ to $x = 4$ is about six times smaller for spline interpolation than for linear interpolation

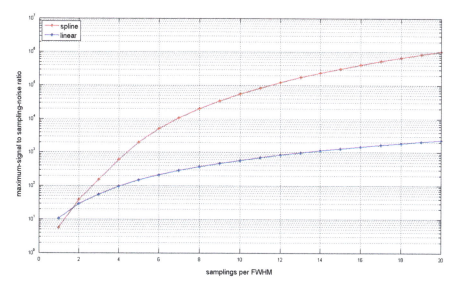

Fig. 2.40 Ratio between the maximum signal of a Gaussian function and the sampling noise depending on the number of sampling points; standard deviation at different interpolated samples without spectral noise. *Blue*: Interpolation with a linear function. *Red*: Interpolation with a spline function. A mean value was calculated from 1,000 different equidistant sampling points used for interpolation. Because of the noise-free Gaussian function the sampling accuracy increases to unlimited signal-to-noise

Fig. 2.40 shows that the interpolation errors are equivalent to a SNR of 30 for a linear interpolation and 40 for a spline interpolation when sampling all resolution elements with only two pixels. Hence, the maximum SNR is limited to these values which is by far not sufficient for the vast majority of astronomical measurements. However, when sampling with more pixels the SNR limits significantly increase. For example, when sampling with five pixels the maximum SNR is limited to about 150 for linear interpolation but to about 2,000 for spline interpolation. In any case, linear interpolation introduces relatively large reproductive noise, limiting the final SNR significantly. For instance, an SNR of about 1,000 would then require sampling with 14 (!) pixels, impractical for most applications. The spline sampling very quickly delivers a much better sampling with increasing number of sampling points (smaller differences from the original signal). For instance, the standard deviation of all sampling differences for a sampling with five pixels is about one order of magnitude better than that for a linear interpolation and increases to two orders when using 10 sampling pixels. For any sampling the cubic spline provides a standard deviation which is two orders of magnitudes better than the respective linear interpolation. If the number of sampling points is doubled, the spline interpolation is improved by a factor of four compared with linear interpolation (the ratio between the SNR of both methods shows exactly a quadratic dependence).

   This analysis, however, does not correspond to reality because all spectral signals are noisy (due to, e.g., photon noise and read-out noise of the camera). Figure 2.41 represents the case of a noisy Gaussian profile with $S/N = 700$ at its peak. At first, the resulting SNR increases for both interpolation methods. However, none of the curves exceed the externally predetermined SNR of 700; that is, the theoretically possible scanning quality is limited by the noise in the signal. Therefore, the reproduction noise cannot be improved by oversampling because of the externally provided threshold. For the present case, the sampling limit is reached already with five pixels for the interpolation with a spline (for a corresponding sampling quality with linear interpolation, 18 pixels are required for the same data quality). If the signal has a higher S/N, correspondingly more sampling points (pixels) are needed, as shown in Fig. 2.40. This can be seen dramatically for a linear interpolation. For a S/N of 2,000 a sampling with 20 pixels would be required whereas only five pixels would be necessary with a spline interpolation. Therefore, for spectra of high S/N one should take many sampling pixels and a correspondingly appropriate interpolation function to ensure sampling with high quality. Since the reproduction error in practical data analysis is always greater than zero, independently from the interpolation function, one always undersamples a theoretically noise-free signal. The original question after sufficient sampling can therefore be answered only with an appropriate boundary condition. As shown above, this boundary condition is the spectral S/N of the spectrum to be analyzed. As long as the standard deviation σ of the noise is smaller than the interpolation error, the latter limits the sampling quality.

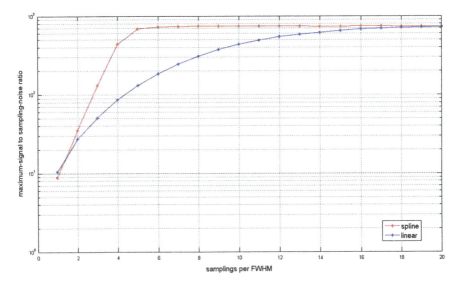

Fig. 2.41 As Fig. 2.40, now for a Gaussian function with spectral noise. *Blue*: Interpolation with a linear function. *Red*: Interpolation with a spline function. The spectral $S/N = 700$ in the line peak limits the sampling quality. It cannot exceed this value

Hence, one can reduce the sampling error by using a finer sampling.[34] On the other hand, if the standard deviation σ of the signal noise is larger, the sampling quality is limited by the associated S/N.

The approximation of a spectral resolution element with a Gaussian is reasonable and, hence, our above considerations are applicable. That means, if a certain S/N is required for a spectral measurement[35] one needs to apply an accurate fitting-function to the data (here a spline) plus a sufficient pixel sampling according to Fig. 2.40. As can be noted above, a linear interpolation is always drastically inferior to a spline interpolation. We therefore recommend to generally use at least the latter.

Our considerations of the optimal sampling initially referred to an ideal sampling, i.e., a delta function as a spectrometer function. For this case, the sampling values exactly mirror the function values at the sampling points. For a real spectrometer function of finite width one needs to average by the applicable convolution within the specified interval defined by the spectrometer function. This of course has implications for the sampling. A simple estimate delivers a solution for this problem: with a spectrometer function of width FWHM, spectral lines of about the same order of magnitude can still be resolved. In other words, the smallest line which can still reasonably be detected has the same width FWHM as the spectrometer function. If we consider Gaussian functions as simplified spectral lines and the spectrometer function we obtain a final spectral line width of $\sigma = \sqrt{\text{FWHM}^2 + \text{FWHM}^2} = \sqrt{2} \cdot \text{FWHM}$ due to the folding rule discussed in Sect. 2.5.1 and in Appendix B.5.1 (convolution in time). If the spectrometer FWHM is sampled according to the Nyquist principle with exactly two pixels the total width σ is already slightly oversampled by $2\sqrt{2} = 2.8$. Now we have an easy way to determine the proper sampling frequency for a real spectrometer, including signal noise.

---

[34] In Nyquist nomenclature this is oversampling.

[35] In fact, this is also valid for astronomical 2D imaging.

Chapter 3
Remarks About Dioptric Imaging Systems

A Short Story

When we had finished our first spectrograph, we enthusiastically took our first stellar spectrum of Regulus. Surprisingly, the intensities at the edges of the spectrum inexplicably fell completely off. We had invested great care into our understanding of the grating and the lens optics—that couldn't be the reason at all. Maybe it was the optics? Eventually we discovered internal vignetting by the aperture stop and then we also discovered that we had not taken the grating dispersion sufficiently into consideration.

3.1 Basic Remarks

As already mentioned, optical imaging elements are necessary to produce a spectrum. On one hand, the divergent beam coming from the slit must be collimated to meet the Fraunhofer condition at the grating. On the other hand, camera optics must image the slit via the dispersed but still parallel beam onto the CCD. In the simplest case there are two spherical lenses to accomplish this task (at this point we do not consider reflective elements like parabolic mirrors). In most cases, spherical refractive elements are used because they can easily be produced and are extremely precise. The two lenses in the spectrometer need to image the slit (in the general case the object) as accurately as possible. In this case, each object point should be transferred into a sharp image point. It is known from geometrical optics that spherical lenses have aberrations that can widen an object point in the imaging process. This widening potentially leads to degraded spectral resolution. The use of simple lenses (simplets) for spectrometer optics may not lead to the desired goal, since the resulting aberrations may be unacceptably large. The requirements for the spectrometer optics can be different and depend in particular on the system f ratio. The greater the f ratio, the smaller is the negative impact of aberrations, although

© Springer-Verlag Berlin Heidelberg 2015
T. Eversberg, K. Vollmann, *Spectroscopic Instrumentation*, Springer Praxis Books,
DOI 10.1007/978-3-662-44535-8_3

the light power is reduced. If a classical spectrograph of medium to high resolution is designed, one needs to image quasi monochromatic light (a narrow spectral band) sharply on the CCD chip. In this case, the influence of chromatic aberrations plays a minor role. For an echelle spectrometer (see Chap. 5) the situation is more complicated: all wavelengths (e.g., in the visible range) should be detected with the same spectral resolution. Besides chromatic aberration, for large object or image field angles the camera lens must also be corrected for off-axis aberrations such as coma, field curvature and astigmatism.[1] The correction requirements for optical spectrograph systems can be quite diverse. In general, the correction of an optical lens system is performed by a suitable combination of spherical lenses. In order to know the prevailing system aberrations the imaging optics must be calculated quantitatively. In the following we will therefore discuss the basic optical design tools to assess the performance of lens systems. However, we cannot illustrate the concepts of advanced optics design. We rather want to show the foundations and the respective access to the literature of this topic.

3.2   Beam Calculation of an Optical System in the Paraxial Area

Before we discuss aberrations, which play an important role in the definition of the spectral resolution, we first deal with the "error-free" optical system and illuminate the ideal imaging process in more detail. This is especially important for understanding the aberrations quantitatively.

The optical system consists of a series of spherical refractive surfaces which determine the specifics of the beam passing through the system, the total focal length, and the imaging scale of the overall optical setup. In order to fully calculate the passage of an incident beam, all construction elements, i.e. any refractive surface of index ν and its curvature radius r_ν, the distances between consecutive surfaces e'_ν and the refractive index between these refracting surfaces n'_ν need to be known.[2] Using the beam angle of incidence and the optical elements one can then determine the beam deflection for each surface, and thus compute the overall system.

For the detailed calculation we start with an arbitrary refracting surface with index ν (Fig. 3.1). In front of the surface the refractive index is n_ν and behind the surface n'_ν. A beam is geometrically determined by its angle relative to the optical axis u_ν and its intersection with it. The distance from the surface apex to this point of intersection is called the intersection length s_ν of the beam on the surface ν. The height of the incident beam on the surface is h_ν. We use the following sign conventions:

---

[1]An optical system having no aberrations is described as corrected.

[2]The primed variables basically denote the corresponding parameters <u>after</u> the refraction procedure.

Fig. 3.1 Diffracting
spherical surface with the
important parameters for ray
tracing calculations (Berek
1970)

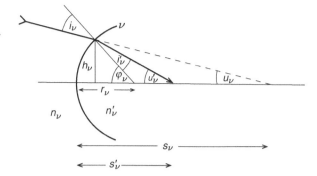

- The intersection length s_ν is positive if the intersection of the incident beam with the optical axis is behind the surface apex (i.e. in the direction of light propagation).
- The incidence height h_ν is positive if h_ν is above the optical axis.
- The angle u_ν is also positive if the incident beam comes from above the optical axis.
- The radius of curvature r_ν of the spherical surface is positive if the center of curvature (the intersection of r_ν with the optical axis) is behind the vertex of the surface. With a positive curvature, the refractive surface has its convex side facing the incident beam.

In addition, we require the angle i_ν between the incident beam on the surface normal as well as the associated angle between the surface normal and the optical axis φ_ν. According to Fig. 3.1 we have

$$\sin \varphi_\nu = \frac{h_\nu}{r_\nu} \qquad \tan u_\nu = \frac{h_\nu}{s_\nu}.$$

We now introduce the concept of the so-called "paraxial" light beam, which is located only an infinitesimal distance from the optical axis. That means h_ν and thus the angle of incidence are small.

$$\varphi_\nu \approx \frac{h_\nu}{r_\nu} \qquad u_\nu \approx \frac{h_\nu}{s_\nu}.$$

For the non-diffracted beam we see from Fig. 3.1 that $\varphi_\nu = i_\nu + u_\nu$ and for the diffracted one $\varphi_\nu = i'_\nu + u'_\nu$. Hence

$$i_\nu = h_\nu \left(\frac{1}{r_\nu} - \frac{1}{s_\nu} \right) \qquad i'_\nu = h_\nu \left(\frac{1}{r_\nu} - \frac{1}{s'_\nu} \right).$$

From Snell's law of refraction $n_\nu \sin i_\nu = n'_\nu \sin i'_\nu$ and assuming small angles i_ν and i'_ν (which as a direct consequence of the small h_ν) we obtain

$$n_\nu \cdot i_\nu = n'_\nu \cdot i'_\nu$$

or with the above calculated angle terms

$$n_\nu \left(\frac{1}{r_\nu} - \frac{1}{s_\nu} \right) = n'_\nu \left(\frac{1}{r_\nu} - \frac{1}{s'_\nu} \right) =: Q_\nu. \tag{3.1}$$

Q_ν in Eq. 3.1 is called the "Abbe invariant of paraxial rays". It can be determined from both the object-side and from the image-side variables. Slightly re-arranged the last equation is

$$\frac{n'_\nu}{s'_\nu} - \frac{n_\nu}{s_\nu} = \frac{n'_\nu - n_\nu}{r_\nu}. \tag{3.2}$$

This is the equation for the intersection length of a paraxial beam path for <u>one</u> refracting surface. This equation can be re-arranged to the desired intersection length (intersection with the optical axis) of the refracted ray s'_ν

$$s'_\nu = \left[\frac{1}{r_\nu} - \frac{n_\nu}{n'_\nu} \left(\frac{1}{r_\nu} - \frac{1}{s_\nu} \right) \right]^{-1} \tag{3.3}$$

and is calculated from the structural elements of the refracting surface and the distance of the object point. In summary the intersection length on the image side in the paraxial region is independent of the beam tilt and therefore independent of the beam height of incidence. All paraxial rays emanating from an object intersect at one single point. This point defines the Gaussian image plane perpendicular to the optical axis. We point out that therefore no image of an extended object can be constructed by the intersection equation 3.3. Only the image plane of an infinitesimally small and hence ideal figure can be estimated.

Considering a second surface of index $(\nu + 1)$ following the first one, one needs the output variables $s_{\nu+1}$, $r_{\nu+1}$, $n_{\nu+1} = n'_\nu$ and $n'_{\nu+1}$ for the calculation. All parameters except $s_{\nu+1}$ are again known for the respective design elements. The transition to the next surface is obtained with the distance to the next surface e'_ν by the simple geometrical relationship (Fig. 3.2)

$$s_{\nu+1} = s'_\nu - e'_\nu. \tag{3.4}$$

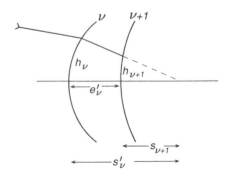

Fig. 3.2 Two diffracting spherical surfaces and their distance e'_ν (Berek 1970)

By successive application of Eqs. 3.3 and 3.4 on all k surfaces of the optical system and for a given object location s_1, one finally obtains the location of the image on the optical axis s'_k after the last refracting surface.

3.3 Paraxial Image Scale and Focal Length of a Lens System

The image location is not the only parameter which determines the imaging process. Another important parameter is the paraxial magnification called reproduction scale. We have already discussed it in Sect. 2.6.1 with respect to the anamorphic magnification factor and the lens equation defined by the ratio of image to object size. We now determine the paraxial magnification for the general case of an optical system with k refracting surfaces. If l_1 is a small object element with the (negative)[3] intersection length $(-s_1)$ and its respective image size l'_1 and image length s'_1, we obtain for the corresponding small angle (as shown in Fig. 3.3)

$$\omega_1 = -\frac{l_1}{s_1} \qquad \omega'_1 = -\frac{l'_1}{s'_1}.$$

Hence, we obtain the magnification ratio (also referred to as a lateral magnification):

$$\beta'_1 = \frac{l_1}{l'_1} = \frac{s'_1 \omega'_1}{s_1 \omega_1}.$$

Due to the law of refraction again $n_1 \omega_1 = n'_1 \omega'_1$ we obtain

$$\beta'_1 = \frac{l_1}{l'_1} = \frac{s'_1 n_1}{s_1 n'_1}.$$

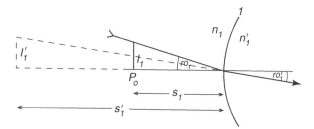

Fig. 3.3 Geometric interpretation of the paraxial image scale (Berek 1970)

---

[3] With respect to the above introduced sign convention.

The last equation is valid for each available surface

$$\beta'_2 = \frac{l_2}{l'_2} = \frac{s'_2 n_2}{s_2 n'_2} \qquad \beta'_3 = \frac{l_3}{l'_3} = \frac{s'_3 n_3}{s_3 n'_3} \qquad \cdots \qquad \beta'_k = \frac{l_k}{l'_k} = \frac{s'_k n_k}{s_k n'_k}.$$

If we multiply all the equations together we get

$$\beta'_1 \cdot \beta'_2 \cdot \ldots \cdot \beta'_k = \frac{l'_1}{l_1} \cdot \frac{l'_2}{l_2} \cdot \ldots \cdot \frac{l'_k}{l_k} = \frac{n_1}{n'_1} \cdot \frac{n_2}{n'_2} \cdot \ldots \cdot \frac{n_k}{n'_k} \cdot \frac{s'_1}{s_1} \cdot \frac{s'_2}{s_2} \cdot \ldots \cdot \frac{s'_k}{s_k}.$$

Due to our convention for the primed parameters we have

$$l'_1 = l_2 \qquad l'_2 = l_3 \qquad \cdots \qquad l'_{k-1} = l_k$$
$$n'_1 = n_2 \qquad n'_2 = n_3 \qquad \cdots \qquad n'_{k-1} = n_k$$

Therefore we obtain for the product of all the individual magnifications

$$\prod_{v=1}^{k} \beta'_v = \frac{l'_k}{l_1} = \frac{n_1}{n'_k} \prod_{v=1}^{k} \frac{s'_v}{s_v}.$$

However, since l'_k/l_1 just determines the paraxial magnification of the overall system, we have:

$$\beta' := \frac{l'_k}{l_1} = \frac{n_1}{n'_k} \prod_{v=1}^{k} \frac{s'_v}{s_v}. \tag{3.5}$$

Hence, the location and size of the image in the paraxial region of the investigated optical system with k diffracting spherical surfaces are determined. This simple result arises from the linearization of the trigonometric functions for small angles.

At this point, we still have to estimate the focal length of the overall system. The system focal length is given by the lens equation in Sect. 2.6.1. This is the image width for the case of infinite object distance. If we insert the object distance $s_1 = \infty$ in Eq. 3.5, we have a magnification $\beta' = 0$. The image of a star would therefore only show its diffraction pattern. If we now further suppose that the object size is $l_1 = G$, we can rewrite Eq. 3.5 with a formal multiplication by s_1 to

$$s_1 \beta' = s_1 \frac{l'_k}{l_1} = \frac{n_1}{n'_k} s_1 \prod_{v=1}^{k} \frac{s'_v}{s_v} = \frac{n_1}{n'_k} s'_1 \prod_{v=2}^{k} \frac{s'_v}{s_v}. \tag{3.6}$$

Using the Eqs. 3.3 and 3.4, the expression on the right-hand side of Eq. 3.6 can be completely attributed to the structural elements and thus represents a constant of the optical system. This constant is called the focal length f:

$$f = s_1 \beta' = \frac{n_1}{n'_k} s'_1 \prod_{v=2}^{k} \frac{s'_v}{s_v}. \tag{3.7}$$

Considering even the angular size of the object at infinite distance $B/s_1 = -\tan \omega_1$, we obtain from Eq. 3.6 for $f = s_1 l'_k / B$

$$l'_k = f \tan \omega_1.$$

The image size can thus be determined from the angular size of the object and the system focal length. This result is already known from Sect. 2.6.1. It is, hence, also valid for the case of an entire lens system.

3.4   The Focal Length of a Single Lens: The Lensmaker Equation

We now derive the often necessary focal length equation of a single lens for two refracting spherical surfaces, surrounded by air. For doing so, we consider in Fig. 3.4 a paraxial incident beam falling onto a lens of central thickness e. For simplification, the incident beam should be diffracted at the first surface so that it intersects the optical axis right at the lens center (favorable choice of the coordinate system). Thus, the intersection lengths within the lens are $s'_1 = +e/2$ and $s_2 = -e/2$.

For this case the refraction indices are

$$n_1 = 1 \qquad n'_1 = n_2 = n \qquad n'_2 = 1$$

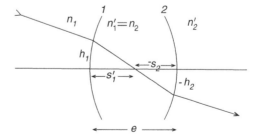

Fig. 3.4 Geometric situation for the derivation of the lensmaker Equation (Berek 1970)

According to Eq. 3.7 we obtain for the focal length

$$f = \frac{n_1}{n'_k} s'_1 \prod_{\nu=2}^{k} \frac{s'_\nu}{s_\nu} = \frac{n_1}{n'_2} s'_1 \prod_{\nu=2}^{2} \frac{s'_\nu}{s_\nu} = \frac{1}{1} s'_1 \frac{s'_2}{s_2}.$$

Because of $s'_1 = s_2$ we obtain $f = -s'_2$ under the condition that $s_1 = \infty$. The task is now to replace the unknown intersection length s'_2 by the structural component of the lens r_1, r_2 and e. We implement the above intersection lengths s'_1 and s_2 in Eq. 3.2 and obtain for the two surfaces the equations

$$\frac{n}{s'_1} - \frac{1}{s_1} = \frac{n-1}{r_1}$$

$$\frac{1}{s'_2} - \frac{n}{s_2} = \frac{1-n}{r_2}. \tag{3.8}$$

These equations can be re-arranged to the already known intersection length s_1 before the first surface and the desired intersection length s'_2:

$$\frac{1}{s_1} = +\frac{2n}{e} - \frac{n-1}{r_1}$$

$$\frac{1}{s'_2} = -\frac{2n}{e} - \frac{n-1}{r_2}.$$

Subtraction of the two equations provides

$$\frac{1}{s'_2} - \frac{1}{s_1} = -\frac{4n}{e} + (n-1)\left(\frac{1}{r_1} - \frac{1}{r_2}\right). \tag{3.9}$$

For the focal length the object distance is infinite, so we have $1/s_1 = 1/\infty = 0$. From the last equation we thus get

$$\frac{1}{s'_2} = -\frac{4n}{e} + (n-1)\left(\frac{1}{r_1} - \frac{1}{r_2}\right) = -\frac{1}{f}. \tag{3.10}$$

We now reformulate the term $4n/e$, which is divergent for the case of a thin lens ($e \to 0$), by multiplying the two invariant equations 3.8

$$\frac{n}{s'_1}\left(\frac{n}{s_2} - \frac{1}{s'_2}\right) = \frac{(n-1)^2}{r_1 r_2}$$

and replace the above defined focal lengths $s'_1 = s_2 = -e/2$ again

$$\frac{4n}{e} = -\frac{2}{s'_2} - \frac{(n-1)^2}{n}\frac{e}{r_1 r_2}.$$

Using again Eq. 3.10 for the reciprocal focal length we finally obtain the focal length of a single lens

$$\frac{1}{f} = (n-1)\left(\frac{1}{r_1} - \frac{1}{r_2}\right) + \frac{(n-1)^2}{n}\frac{e}{r_1 r_2}. \qquad (3.11)$$

Apparently, the focal length of the lens depends on the central separation e between the two refracting surfaces. Decreasing this distance continually, the 2nd summand also gets smaller and we get for the limit $\lim\limits_{e \to 0}\left(\frac{1}{f}\right)$

$$\frac{1}{f} = (n-1)\left(\frac{1}{r_1} - \frac{1}{r_2}\right) \qquad (3.12)$$

the so-called "lensmaker equation". For this limit, only theoretically possible, one refers to "thin lenses" in contrast to the real "thick lenses" for the case $e \neq 0$.

When turning again to the more general equation 3.9 and replacing again $4n/e$, we obtain

$$\frac{1}{s_2'} - \frac{1}{s_1} = \frac{2}{s_2'} + \frac{(n-1)^2}{n}\frac{e}{r_1 r_2} + (n-1)\left(\frac{1}{r_1} - \frac{1}{r_2}\right)$$

or

$$-\left(\frac{1}{s_2'} + \frac{1}{s_1}\right) = (n-1)\left(\frac{1}{r_1} - \frac{1}{r_2}\right) + \frac{(n-1)^2}{n}\frac{e}{r_1 r_2}$$

and hence

$$-\left(\frac{1}{s_2'} + \frac{1}{s_1}\right) = \frac{1}{f}.$$

This is the Gaussian lens equation (here with minus sign according to the above introduced sign conventions[4] as we have previously used with Eq. 2.82.

It now becomes clear at this point that the derivation of the focal length equation for a simple lens is already relatively complex. An analytical treatment of a multi-lens system, which is necessary in the case of a high-order correction level, will therefore be much more difficult. Therefore, today the so-called matrix method is normally used to summarize the refractive process in the form of a linear imaging process and thus to facilitate the calculation. A further possibility to simultaneously evaluate the performance of an optical system are the so called ray-tracing methods. In this case no analytical solution for the respective optical system is derived, but a series of evenly distributed light rays are artificially sent from the object point

---

[4]Given that s_1 and s_2' are negative, this is consistent with the usual lens equation.

through the system. Using the methods outlined above the intersection points in the image plane are then derived. The set of all image points then allows a representation of the light distribution in the image plane. In this context one talks about a "spot diagram". If one performs the calculations from surface to surface, i.e. non-paraxial, exact for larger angles, one also obtains the image errors visible in the resulting spot diagram. The diameter of such a spot can then be used as an estimator for the image aberrations. We will discuss the ray-tracing methods in more detail in Sect. 3.11.

The following terms are important to further systematize the imaging process of a lens. In this context one refers to the so-called basic or cardinal elements of the lens:

- **Foci.** We had already found that the focal length f is a constant of the optical system. For the image-side focal length f there is a conjugated position on the object side that transforms a point-like and divergent light source behind the lens into an axis-parallel beam. This distance is referred to as object-side focal length f'.
- **Principal planes.** If paraxial rays fall onto a finite thick lens, these are united behind the lens image-focus. By extending the two beams within the lens body, we obtain a point of intersection of the two beams. The entirety of all points of incidence for different heights are located on a curved surface. Near the optical axis this surface approaches a plane, which is called the principal plane.[5] Since the lens has a finite thickness and two foci, consequently there are also two principle planes. The two intersections of these planes with the optical axis occur at the main points. The principle planes thus replace the refractive lens surfaces for the structural representation of the beam path and can be located inside or outside of the glass body depending on the lens geometry.
- **Nodal points.** If one wants to also use the central beam for construction to describe the thick lens, one also needs the nodal points N_1 and N_2 (Fig. 3.5). An inclined beam directed to the first node N_1 leaves the system at the same angle starting from the second node N_2. The introduction of the nodal points fully parameterizes the central beam (cardinal or chief ray) through a thick lens.

3.5   Monochromatic Seidel Aberrations

Since the optical system produces wavelength-dependent images (mini spectra), these images must be free from image aberrations. These aberrations, analytically calculated by Philipp Ludwig von Seidel 1855 for the first time, are caused by the nonlinearity of Snell's law at large angles of refraction and by most spherical lenses.

---

[5]At this point it should be clear that the principle plane is only valid for the paraxial zone. Considering the entire lens, the principle plane is a curved surface, which causes the Seidel aberrations.

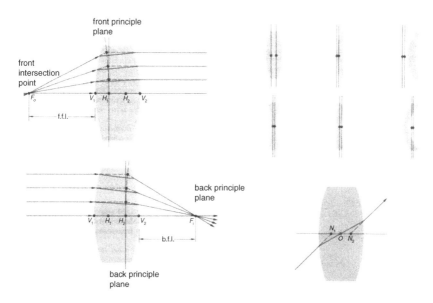

Fig. 3.5 Principle planes and intersection points (Hecht 2003)

First, we introduce a definition of the term "aberration", which arises naturally from the paraxial image. In general aberration is referred to the deviation of a light beam from its ideal image point by an optical system. The reference for the "ideal" image point itself is the image location defined by the Gaussian image. Therefore, the aberration is the discrepancy between the mathematical transformations of the Gaussian imaging and Snell's law of refraction in the image plane. This transformation leads to shifts or to extensions of the image points and thus transforms points in the object plane to figures of confusion in the image plane. This results in scale distortions and loss of contrast, and hence, to loss of resolving power or image quality. Basically, one has to distinguish between "longitudinal" and "transverse" aberration. The longitudinal aberration is the deviation of the image points from the Gaussian plane. This deviation introduces a deviation of the penetration points in the image plane, i.e., perpendicular to the optical axis. This deviation is then visible as blurring in the image, referred to as transverse aberration. Longitudinal and transverse aberrations in connection with spherical aberration are geometrically illustrated in Fig. 12.5. A mathematical definition of the transverse aberration is given in Sect. 3.7 in connection with the calculation of the Seidel aberrations.

They relate to all wavelengths and can degrade the spectral information in the form of significant degradation of the spectral resolving power. Before we begin the detailed discussion of Seidel aberrations, it is useful to first introduce a phenomenological representation of monochromatic aberrations. This allows the reader an easier access to the subsequent mathematical treatment.

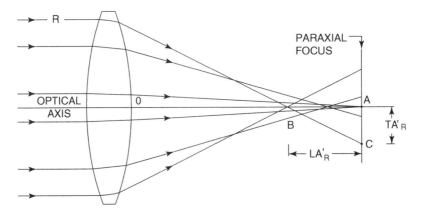

Fig. 3.6 Spherical aberration. Near axis and off-axis rays (R) have different focal lengths. The difference between the paraxial focal length and the focal length of the beam R is LA'_R. One refers in this context to the so-called "longitudinal spherical aberration". In the Gaussian image plane the longitudinal deviation introduces an image blur, the so-called "transversal spherical aberration" TA'_R (*Source*: W.J. Smith 2000, Modern Optical Engineering)

Spherical aberration is generated by the spherical shape of the refractive surfaces and increases with the aperture of the imaging system. This aberration is therefore often referred to as aperture aberration. Near-axis and off-axis rays have different focal lengths and introduce image blurring in the paraxial image plane (Fig. 3.6). If the image center is in focus, image edge points are out of focus. Spherical aberration can be avoided by correction lenses. For mirror optics spherical aberration plays a minor role because of the use of paraboloids.

Coma is also a type of spherical aberration. It is the focal deviation for rays not parallel to the optical axis (Fig. 3.7). Coma is significant especially for point sources and small f-ratios. Seeing disks of stars at the aperture edge are imaged in an oval form with a one-sided fuzzy edge. The blurred edge for (positive) coma is directed away from the image center (optical axis). The rays at the lens edge are further out in the image. For the case of negative coma the edge rays are on the inner side. The cause is the oblique incidence of the object beam on the lens. The principal lens planes occur only in the vicinity of the optical axis. Far outside the paraxial region there are no planes, only curved principle surfaces. As a result the transverse magnifications of axial and non-axial rays differ and, hence, lead to the illustrated aberrations. Telescope coma can in principle be corrected by correction lenses, since it greatly restricts the usable field of view for telescopes with small f-ratios.

Astigmatism of oblique bundles (Fig. 3.8) arises when imaging off-axis points. Even with significant beam-limiting (small aperture), the image error is noticeable, although for the axial rays the condition of paraxial imaging is valid. It is therefore obviously not an aperture error as is coma.

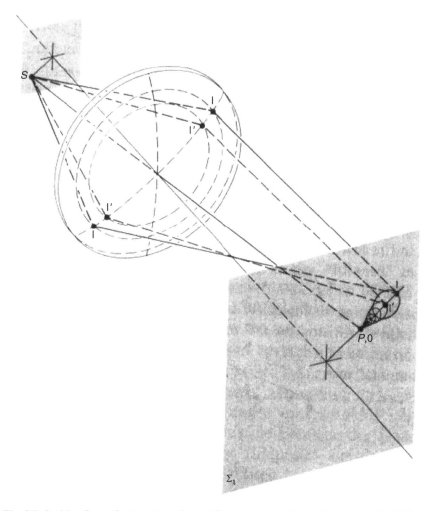

Fig. 3.7 Positive Coma. Seeing disks of stars (S) show an oval form at the image edge (off-axis object points) with a one-sided fuzzy edge. The connection of coma to spherical aberration is obvious (Hecht 2003)

When mapping an off-axis point a beam deformation is performed in the image plane. Instead of a point, two lines are created in the image space, which are called sagittal image line S and tangential image line T.[6] In the Gaussian image plane (which includes, the paraxial focal point F' on the optical axis) off-axis points, therefore, are depicted as blurred elliptical dispersion figures. They are all worse

---

[6]The name sagittal plane is derived from the Latin word "sagitta" = arrow (which is pointing towards the optical axis). Perpendicular to the sagittal plane is the meridional plane. Therefore the tangential focal line is also called meridional focal line.

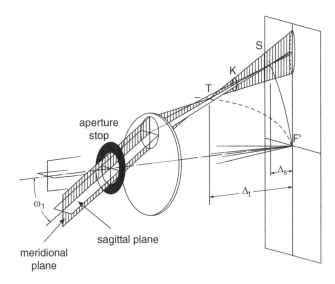

Fig. 3.8 The formation of an astigmatic image and its caustic lines (J. Reiner, Grundlagen der opthalmologischen Optik, 2002, ©Books on Demand GmbH)

the larger the inclination angle ω_1 of the main beams. The distance between the two image lines is a measure of the magnitude of the astigmatism. The larger the distance, the greater the dispersion of the dispersion figures.

The sagittal image line S is always radial to the optical axis. The tangential image line T is perpendicular thereto so that they, runs like a tangent to a cylinder which surrounds the optical axis concentrically. Between the sagittal and tangential image line there is the circle of least confusion (K). For a given lens, the astigmatism depends on the image field angle ω_1. The greater the angle between the principal ray and the optical axis, the larger is the astigmatism. For the angle $\omega_1 = 0$, i.e., in the paraxial image space, no astigmatism is present. Axis parallel incident rays converge at the focal point F'.

As it is apparent from the illustration in Fig. 3.8, the respective set of all image lines provides an image shell which is concentric around the optical axis. These two shells are called sagittal and tangential field curvature. As we will examine in more detail in Sect. 3.5, it is alternatively possible to introduce the so-called "mean field curvature" and additionally astigmatism as the deviation of the two image spheres from this mean field curvature. Astigmatism is thus just another representation form of the meridional and sagittal field curvature.

For a simple biconvex lens, as shown in Fig. 3.8, the tangential image surface is more curved, i.e., the image lines are closer to the lens. The sagittal image surface is less curved and is therefore closer to the paraxial image plane. Between the sagittal and tangential image surfaces, there is another image shell, which is formed by the loci of all circles of smallest confusion. For a graphical depiction of astigmatism of oblique bundles, the distances Δs and Δt of the image lines S and T from the paraxial image plane are plotted as a function of the angle ω_1 in a rectangular coordinate system.

Field curvature without astigmatism is a special case that occurs when the sagittal and tangential image surfaces coincide after correction of astigmatism. Even when astigmatism is fully corrected a mean field curvature remains. However, this curvature does not have different radii in the sagittal and meridional directions. For this case the imaging is no longer astigmatic but stigmatic (point-like). The totality of all the image points is again an image sphere, which is not flat. Since there are often objects that are perpendicular to the optical axis to be displayed in an axis-perpendicular plane (e.g., CCD plane), a blurring figure arises for off-axis points in the image plane. Thus, only a small section of the object will be sharply imaged. The field curvature is regarded as a specific aberration and one refers to it as the so-called Petzval surface and for the associated radius of curvature the Petzval radius.

Distortion arises when the imaging scale of a system is not constant, in particular for off-axis points. Straight lines are sharply imaged, but curved. The more these straight lines are off-axis, the more curved they appear in the image. Another crucial factor is the position of the aperture stop. Distortion occurs in diaphragm systems whose aperture is not positioned at the principal image plane (Fig. 3.9). If the diaphragm is in front of the image principal plane, the image is distorted in barrel form (straight lines at the image edge bent outwardly). If the diaphragm is behind the image principal plane, it is shaped like a pillow (straight lines at the image edge bent inwardly). The remedy for this is a symmetrical lens structure of the lens system wherein the aperture is positioned close to the image-side principal plane. However, this requires a system of fixed focal length, which in turn has an effect on the overall geometrical length. Most lens systems always show a slight distortion: Wide-angle lenses record barrel-shaped and telescopic lenses record pillow-shaped images. The amount of distortion depends on the quality of the lens.

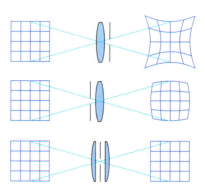

Fig. 3.9 Distortion (from Wikimedia Commons)

3.6 Chromatic Aberrations

In contrast to the Seidel errors, chromatic aberration is caused by the wavelength dependence of the refractive index $n(\lambda)$. This can already be seen from the intersection length equation 3.2 for a refracting surface. We consider again the case of an object point at infinite distance $s_1 = -\infty$ for $n = 1$ and get the simple equation

$$s'(\lambda) = r \frac{n'(\lambda)}{n'(\lambda) - 1}.$$

The intersection length for $s = -\infty$ is equivalent again to the focal length $s'(\lambda) = f(\lambda)$. To get a quantitative idea of the effect, we calculate intersection lengths for different wavelengths and for the standard BK7 glass. The result is shown in Table 3.1. Between the F line and the C line we obtain an intersection length difference $\Delta s' = 1.4989$ mm for a surface curvature radius of $r = 50$ mm. As for a prism, the surface diffracts different wavelengths into different directions. Because of this wavelength dependency for the incident light, a variation of focal length and intersection length for a lens is introduced (Fig. 3.10). Polychromatic images are widened and show a color fringe. In the context of spectroscopy a slit image is thereby potentially broadened and the spectral resolving power is reduced. In standard spectroscopy, generally relatively narrow spectral regions are detected, in particular in the case of high resolution. The requirements for the camera lens can then be relaxed in favor of a compromised focus. For echelle spectroscopy, in which the entire optical spectrum is imaged at high resolution (see Chap. 5), however, this does not apply. In this case, all wavelength ranges have to be imaged without degradation of the resolution element (slit image). If it is necessary to use lenses for spectral imaging instead of mirror optics, one must therefore apply combined multi-lens systems, because otherwise the maximum deviation from the nominal focus between red and blue light (called the secondary spectrum or longitudinal

Table 3.1 Wavelength dependent intersection lengths of a diffracting surface with the curvature radius $r = 50$ mm

Fraunhofer line	Wavelength [nm]	Index of refraction	Intersection length [mm]
F	486.1	1.52282859	145.7164
c	546.1	1.51872202	146.3907
d	587.6	1.51680003	146.7482
C'	643.8	1.51472085	147.1400
C	656.3	1.51432235	147.2153

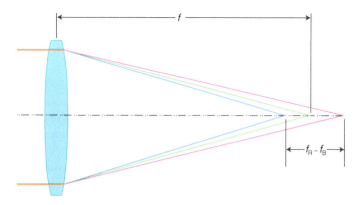

Fig. 3.10 Chromatic aberration of a lens (Witt 2005)

chromatic aberration $\Delta f = f_{C'} - f_{F'}$ with $f_{C'}, f_{F'}$ = focal length for red and blue light) becomes too large.[7]

The deviation from the average focus is calculated on the basis of the considerations for the focal length f or the refractive power of $F = 1/f$ of a thin lens with refractive index n and surface radii r_1 and r_2 as already derived in Sect. 3.4:

$$\frac{1}{f} = (n - 1)\left(\frac{1}{r_1} - \frac{1}{r_2}\right). \qquad (3.13)$$

n is usually given for mid-green wavelengths. To calculate the longitudinal chromatic aberration, we will derive the corresponding difference for the red (C') and the blue (F') wavelength:

$$\frac{1}{f_{F'}} - \frac{1}{f_{C'}} = \frac{f_{C'} - f_{F'}}{f_{C'} f_{F'}} = (n_{F'} - n_{C'})\left(\frac{1}{r_1} - \frac{1}{r_2}\right).$$

Now let f_e be the average focus at green wavelengths between $f_{C'}$ and $f_{F'}$. With the approximation $f_{C'} f_{F'} \approx f_e^2$ we obtain

$$\frac{f_{C'} - f_{F'}}{f_e^2} = (n_{F'} - n_{C'})\left(\frac{1}{r_1} - \frac{1}{r_2}\right). \qquad (3.14)$$

For f_e in turn we use Eq. 3.13

$$\frac{1}{f_e} = (n_e - 1)\left(\frac{1}{r_1} - \frac{1}{r_2}\right). \qquad (3.15)$$

---

[7]Here we use the defined wavelengths of the red cadmium line $C' = 6{,}438$ Å, the blue cadmium line $F' = 4{,}800$ Å and the green mercury line $e = 5{,}461$ Å.

From Eqs. 3.14 and 3.15 we then obtain $\frac{f_{C'}-f_{F'}}{f_e} = \frac{n_{F'}-n_{C'}}{n_e-1}$ or

$$f_{C'} - f_{F'} = \frac{f_e}{(n_e - 1)/(n_{F'} - n_{C'})}.$$

This means that the longitudinal chromatic aberration depends on the intermediate focal length and a constant from the refractive indices for the three red, green and blue wavelengths (which are defined by cadmium and mercury lines). The constant $\nu = \frac{n_e-1}{n_{F'}-n_{C'}}$ is called "Abbe number". It is a measure for the relative dispersion, which is important for making achromatic lens groups. The longitudinal chromatic aberration Δf can thus be estimated by the simple ratio

$$\Delta f = f/\nu. \tag{3.16}$$

The reciprocal of the Abbe number in turn indicates the relative focal length change for the selected wavelengths. Single-lens longitudinal chromatic aberrations Δf are consistently too large for practical applications (Table 3.1). Therefore, multi-lens optics are necessary, which, however, have to fulfill an important condition.

Let us consider the simplest multi-lens optics, an achromat of two lenses. We use Eq. 3.13 for the refractive power $F = 1/f$, consider F for both lenses and then derive the difference of the two refractive powers. We obtain

$$\Delta F = \frac{1}{f} - \frac{1}{f + \Delta f} = \frac{\Delta f}{f(f + \Delta f)}.$$

With $\Delta f \ll f$ we now obtain $\Delta F = \frac{\Delta f}{f^2}$. With Eq. 3.16 we get $\Delta F = \frac{1}{f\nu} = \frac{F}{\nu}$. The total refractive power of two thin touching lenses is the simple sum of the individual refractive powers $F = F_1 + F_2$. From these last two equations, we now impose the achromatism condition: If refractive powers and focal lengths for blue and red wavelengths in an achromat should be identical, the sum of the power differences must be 0, i.e. $\Delta F = \Delta F_1 + \Delta F_2 = 0$ or

$$\frac{F_1}{\nu_1} + \frac{F_2}{\nu_2} = 0. \tag{3.17}$$

Known as the achromatic condition Eq. 3.17 can only be fulfilled if F_1 and F_2 have opposite signs, one lens magnifies and the other de-magnifies, because of $\nu_1 > 0$ and $\nu_2 > 0$. In addition, $\nu_1 \neq \nu_2$ must be fulfilled in order to get a positive refractive power ($F > 0$). The longitudinal chromatic aberration of classical Fraunhofer achromats of flint and crown glass is consistently about f/2,000, i.e. at least 0.5 mm for a focal length of 1,000 mm (Fig. 3.11). To improve this value, one has to resort to assemblies of multiple lenses.

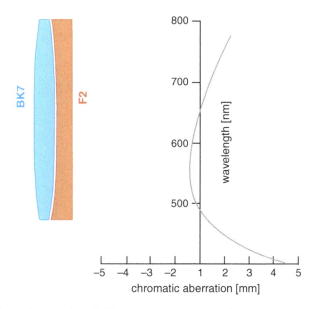

Fig. 3.11 Chromatic aberration of a Fraunhofer achromat (Witt 2005)

The relative dispersion ϑ_e of a glass refers to the dispersion (refractive index difference) between the F' and e line $n_{F'} - n_e$ in relation to the basic dispersion $n_{F'} - n_{C'}$.

$$\vartheta_e = \frac{n_{F'} - n_e}{n_{F'} - n_{C'}} \tag{3.18}$$

Considering now ϑ_e as a function of the Abbe number, i.e. $\vartheta_e = \vartheta_e(\nu)$, we obtain a linear function in whose vicinity we can find the most glass types (Fig. 3.12). The normal line hence represents all glasses whose relative partial dispersion depends linearly on the Abbe number, i.e. they have "normal dispersion". With the relative partial dispersion of Eq. 3.18, we can now determine the longitudinal chromatic aberration for two glasses.

$$\Delta f = \frac{\vartheta_1 - \vartheta_2}{\nu_1 - \nu_2}. \tag{3.19}$$

The fraction $\frac{\vartheta_1 - \vartheta_2}{\nu_1 - \nu_2}$ represents the gradient of the normal line. For glasses with normal dispersion, so-called "normal glasses", this gradient is about 1/2,000 and it cannot be optimized. This can only be achieved with glasses whose partial dispersion and Abbe number are not on the normal line. Equation 3.19 specifies that the minimization of Δf requires a minimum difference of both partial dispersions $\vartheta_1 - \vartheta_2$ and a maximum difference between the two Abbe numbers $\nu_1 - \nu_2$. This requires at least one type of glass with anomalous partial dispersion which is not

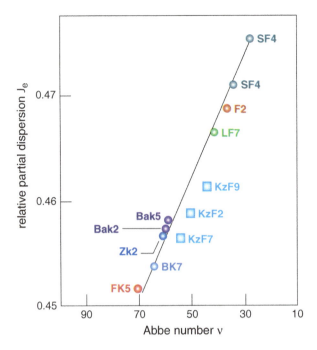

Fig. 3.12 Relative partial dispersion as a function of the Abbe number (Witt 2005)

on the normal line. In other words, the two glasses need to be on a horizontal line in the ϑ - ν diagram. With a clever choice of glass materials the longitudinal chromatic aberration of a two-lens combination can be reduced to 1/20,000 with an aberration curve having two zero-crossings (Fig. 3.13). However, in order to extend the wavelength range of the lowest color error, it usually requires a longitudinal chromatic aberration curve of three or more zero-crossings (Fig. 3.14) or lens combinations with very shallow running curve of longitudinal chromatic aberration. For the determination of the lens glass types we refer at this point to the relevant literature and the corresponding tables and graphs in the Schott company website.

In summary, we want to emphasize once again that the chromatic aberration must be as low as possible to keep the resolving power of a spectrograph almost constant at all wavelengths. The longitudinal chromatic aberration curve $\Delta f(\lambda)$ has to have a reasonably flat course. This can be achieved by the combination of several lenses with refractive powers of different signs (focusing and diverging lens) and of different Abbe numbers. The expression "apochromat".[8] neither relates to the number of inserted lenses nor to the number of zero crossings of the aberration curve, but only to the reduction of longitudinal chromatic aberration.

---

[8]The apochromat is a color corrected lens system for at least three wavelengths.

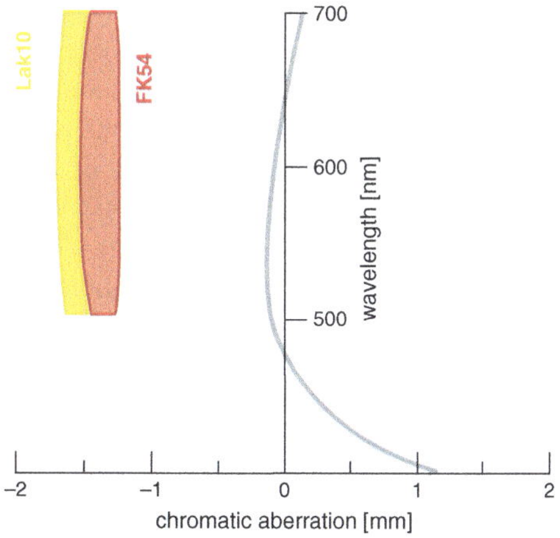

Fig. 3.13 Chromatic aberration of a two-lens apochromat out of FK54 and LaK10 with $f = 1,000$ mm (Witt 2005)

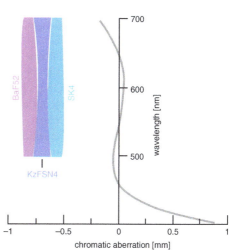

Fig. 3.14 Chromatic aberration of an apochromat out of BaF52, KzFSN4 and SK4 with $f = 1,000$ mm (Witt 2005)

3.7   The Calculation of Seidel Image Aberrations

As already mentioned above, aberrations in optical systems may affect the optical imaging quality. In the computer age, very powerful ray-trace programs are available to the optics design engineer which simultaneously calculate many rays of an optical system, and thus deliver so-called puncture diagrams of optical beams in the image plane. From these diagrams, it is possible to judge the beam distribution in the image plane and, if possible, to improve or optimize the optical system. With the help of the aberration theory developed by Seidel (1856) it is possible to calculate these aberrations analytically. When performing the exact calculations for an object imaged by an optical system, trigonometric functions close to the optical axis occur. In this case, the so-called paraxial region, these functions can be approximated by their angles themselves. In this context, one speaks of the Gaussian image (Gauss

1841). Outside the paraxial range, it is necessary to develop the trigonometric functions in power series up to higher orders and truncate the expansion after reaching the required accuracy. The Seidel error theory was originally based on the consideration of the third order in the power series expansion of the trigonometric functions (for large angles). This geometric area is directly adjacent (as next occurring order of the Taylor series) to the paraxial area and is also referred to as Seidel area.

For each optical beam, the deviation of the intersection point from the Gaussian image point can be expressed by five separate mathematical functions, which are geometrically identified in Sect. 3.5 as the Seidel aberrations. The functions describing the aberration contain all the structural elements of the optical system and the coordinates of the beam in the object plane. Each diffracting surface in the optical system causes a partial aberration. The sum of all resulting deviations indicates the total error of the optical system. The calculation and critical analysis of the Seidel errors allows a detailed assessment of an imaging system and therefore plays a significant role in optics design, despite neglecting the higher terms of the trigonometric functions (higher than 5th order). The advantage of this approximate calculation is that the relative behavior of the aberrations can be determined as a function of physical parameters such as lens diameter, angle of incidence etc. In contrast, ray-trace calculations deliver the individual Seidel parts of the total error of an optical system, but when changing or optimizing the system, the impact of these changes is not immediately clear.

Thus a basic understanding of the underlying Seidel error theory is highly useful, if not necessary, to design the imaging optics of a spectrometer as efficiently as possible. We therefore strongly advise the reader to acquire the fundamentals of the Seidel aberration theory besides the use of ray-trace programs in order to obtain a quantitative approach to their size and dependencies. The large and complex topic of aberrations becomes significant to the optical designer in order to correctly interpret the spectrometer and how the results of ray-trace programs are to be interpreted. We waive the derivation of the expressions for the third-order aberrations, as the necessary calculations are very extensive. In this context we only refer to the study of various recommended publications and books listed at the end of this chapter. We refer in particular, to the work of Schwarzschild in 1905 and Haferkorn in 1986. Berek (1970)[9] mentioned that "the derivation of the Seidel coefficients offers nothing of further interest, but represents only a very skillful overcoming of elementary mathematical complications". In this chapter we therefore present only the final result for the determination of image errors. However, we would like to point out that the treatment of the mirror spectrometer is easier because of only a few reflective surfaces. We therefore show in Chap. 7.5 the derivation of the Seidel aberrations for a single refracting surface. From our point of view, another difficulty are the very different derivations and representations of the Seidel coefficients in the

---

[9]Max Berek was scientific director of the company Ernst Leitz and developed (among other things) the first miniature lens.

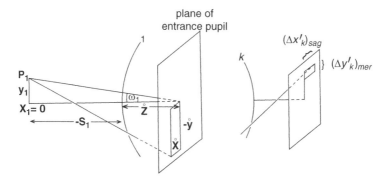

Fig. 3.15 Variables for the position of an arbitrarily inclined ray in the object plane. For the paraxial case the object point P_1 would be imaged in the image point $\bar{P}'_k$. Considering aberrations one obtains the true point P'_k in the image plane with the meridional and sagittal deviation components $(\Delta x'_k)_{sag}$ and $(\Delta y'_k)_{mer}$ (Berek 1970)

literature that can easily lead to misunderstandings. Therefore, below we will give some examples for the concrete application of the Seidel coefficients.

We start with the definition of the imaging error. We consider an arbitrary optical system with an object vertical to the optical axis at a distance s_1 from the 1st of k refracting surfaces (Fig. 3.15). The system possesses the paraxial magnification β'. Starting from an object point with the meridional distance y_1 from the optical axis, the beam spans the angle ω_1 to the optical axis. Ideally imaged, this beam would cross the axis-perpendicular image plane in the meridional distance $\bar{y}'_k = \beta' \cdot y_1$ from the optical axis (this respectively applies, of course, to the sagittal x direction $\bar{x}'_k = \beta' \cdot x_1$). Due to the aberrations of the system, however, geometrical deviations $(\Delta y'_k)_{mer}$ and $(\Delta x'_k)_{sag}$ from the ideal crossing point in the Gaussian image plane occur (see Fig. 3.15 right) in both directions:

$$(\Delta y'_k)_{mer} = y'_k - \bar{y}'_k$$

$$(\Delta x'_k)_{sag} = x'_k - \bar{x}'_k$$

with y'_k and x'_k the meridional and sagittal coordinate of the true intersection point.

Without derivation, we now present the full calculation method for the meridional and sagittal deviation of the intersection points from the Gaussian image location of the considered optical system. For the meridional deviation the overall error is:

$$\frac{(\Delta y'_k)_{mer}}{s_1 \beta'} = \frac{1}{2} \frac{(\mathring{x}^2 + \mathring{y}^2)\mathring{y} s_1^3}{n_1(\mathring{z} - s_1)^3} \sum_{v=1}^{k} I_v - \frac{1}{2} \frac{(\mathring{x}^2 + 3\mathring{y}^2)y_1 s_1}{(\mathring{z} - s_1)^2} \sum_{v=1}^{k} II_v +$$

$$+ \frac{1}{2} \frac{n_1 \mathring{y} y_1^2}{s_1(\mathring{z} - s_1)} \sum_{v=1}^{k} III_v - \frac{1}{2} \frac{n_1^2 y_1^3}{s_1^3} \sum_{v=1}^{k} V_v$$

(3.20)

and for the sagittal deviation:

$$\frac{(\Delta x_k')_{sag}}{s_1 \beta'} = \frac{1}{2} \frac{(\overset{\circ}{x}{}^2 + \overset{\circ}{y}{}^2)\overset{\circ}{x}s_1^3}{n_1(\overset{\circ}{z} - s_1)^3} \sum_{v=1}^{k} I_v - \frac{\overset{\circ}{x}\overset{\circ}{y}y_1 s_1}{(\overset{\circ}{z} - s_1)^2} \sum_{v=1}^{k} II_v +$$

$$+ \frac{1}{2} \frac{n_1 \overset{\circ}{x} y_1^2}{s_1(\overset{\circ}{z} - s_1)} \sum_{v=1}^{k} IV_v$$

$$(3.21)$$

where it is assumed that the z-axis is defined by the optical axis (axis of symmetry) of the system and the x and y coordinates are vertically oriented to it.

The above differences $(\Delta y_k')_{mer}$ and $(\Delta x_k')_{sag}$ from the paraxial case are composed of various proportions. They consist of geometric factors that specify the aperture and object positions and the object size, and of sums whose running index v indicates the respective lens surface and the summands I_v to V_v the corresponding Seidel aberrations (Table 3.2). The prefactors are

- I_v the Seidel coefficient for the spherical aberration,
- II_v the Seidel coefficient for the coma,
- III_v the Seidel coefficient for the meridional image curvature,
- IV_v the Seidel coefficient for the sagittal image curvature and
- V_v the Seidel coefficient for the distortion.

Before the above-mentioned sums there are factors that contain the coordinates of the observed object point x_1, y_1, s_1, the refractive index in the object space n_1 and the location of the aperture stop $\overset{\circ}{x}, \overset{\circ}{y}, \overset{\circ}{z}$ (the circle on the coordinates indicates the aperture). In Table 3.2 all the necessary parameters of the aberration equations are summarized: The so-called "Seidel coefficients" $I_v, II_v, III_v, IV_v, V_v$ are independent of the other parameters $x_1, y_1, s_1, \overset{\circ}{x}, \overset{\circ}{y}, \overset{\circ}{z}$ and are determined only by the characteristic sizes of the optical components of the lens system (see Eq. 3.25).

Table 3.2 Coefficients of the Seidel aberrations

Parameter	Description
n_1	Diffraction index in object space
β_1	paraxial imaging scale
x_1, y_1	Coordinates of the object space P_1
s_1	Object distance to the 1st diffracting surface of the optical system
$\omega_1 = \frac{y_1}{s_1}$	Image field angle for the object side
$\overset{\circ}{x}, \overset{\circ}{y}$	Intersection points of the beam in the plane of the entry pupil
$\overset{\circ}{z}$	Distance of the entry pupil to the diffracting surface
$I_v, II_v, III_v, IV_v, V_v$	Seidel coefficients
$\bar{x}_k', \bar{y}_k'$	Coordinates of the paraxial image point
x_k', y_k'	Intersection point coordinates of the beam with the paraxial image plane
$(\Delta x_k')_{sag}, (\Delta y_k')_{mer}$	Derivation from the paraxial image position

In the expression for the image errors, the object distance occurs before all the Seidel sums. For the camera lens of a spectrometer, the object is at infinity, i.e., for the important special case $s_1 = -\infty$ we obtain the following limits, which must be fulfilled in the above two Eqs. 3.20 and 3.21 and which simplify all calculations. Instead of object size and distance, now the object angle ω_1 occurs:

$$s_1 \beta_1' = f$$

$$\frac{s_1^3}{(\overset{\circ}{z} - s_1)^3} = -1$$

$$\frac{y_1 s_1}{(\overset{\circ}{z} - s_1)^2} = -\tan \omega_1$$

$$\frac{y_1^2}{s_1(\overset{\circ}{z} - s_1)} = -\tan^2 \omega_1 \qquad (3.22)$$

$$\frac{y_1^3}{s_1^3} = -\tan^3 \omega_1.$$

If we set these limit values into the aberration equations of 3rd order and in addition take $n_1 = 1$, we obtain for the meridional and sagittal deviations:

$$\frac{(\Delta y_k')_{mer}}{f} = -\frac{1}{2}(\overset{\circ}{x}^2 + \overset{\circ}{y}^2)\overset{\circ}{y} \sum_{\nu=1}^{k} I_\nu + \frac{1}{2}(\overset{\circ}{x}^2 + 3\overset{\circ}{y}^2) \tan \omega_1 \sum_{\nu=1}^{k} II_\nu -$$

$$(3.23)$$

$$-\frac{1}{2}\overset{\circ}{y} \tan^2 \omega_1 \sum_{\nu=1}^{k} III_\nu + \frac{1}{2} \tan^3 \omega_1 \sum_{\nu=1}^{k} V_\nu.$$

$$\frac{(\Delta x_k')_{sag}}{f} = -\frac{1}{2}(\overset{\circ}{x}^2 + \overset{\circ}{y}^2)\overset{\circ}{x} \sum_{\nu=1}^{k} I_\nu + \overset{\circ}{x}\overset{\circ}{y} \tan \omega_1 \sum_{\nu=1}^{k} II_\nu - \frac{1}{2}\overset{\circ}{x} \tan^2 \omega_1 \sum_{\nu=1}^{k} IV_\nu.$$

$$(3.24)$$

For an infinite object distance, the two expressions 3.20 and 3.21 are simplified and one can recognize in particular the relative dependencies of the image aberrations on the field angle and the aperture of the optical system. The powers of the two parameters for each image aberration are summarized in Fig. 3.16. The equations also indicate that the aberrations linearly increase with increasing focal length f.

Note: We recommend to consequently calculate Eqs. 3.20 and 3.21 for a biconvex lens (Sect. 3.8.2) or Eqs. 3.23 and 3.24 for $s_1 = -\infty$ and $n_1 = 1$) and to compare them with the example results in this chapter. For practical applications, one should at first be familiarized with the various parameters. The calculation is not particularly extensive and can, e.g., be carried out with a spreadsheet program. The

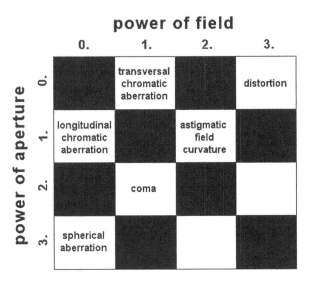

Fig. 3.16 Relative contributions of the Seidel cross aberrations of 3rd order as a function of the field angle and the aperture of the optical system. Beside the Seidel aberrations here the longitudinally and laterally chromatic aberrations are considered

various Seidel sums exclusively include the structural elements of the considered optics (e.g., radii, diffraction indices, distances). The actual calculation of these sums is separately treated in Sect. 3.7.1. This follows a technique of calculation proposed by Seidel, which has been found to be particularly advantageous.

3.7.1   The Calculation of the Seidel Sums

So far, the Seidel sums were only very generally used as contributions of the various image errors in the aberration equations. Thanks to Ludwig Seidel's work, the partial coefficients can be calculated most easily if they are determined indirectly from the elements of the paraxial calculation of a beam (see, e.g., Eq. 3.1), rather than directly from the construction elements of the optics.

We now give a computational scheme that can be used for the calculation of the still missing sums. For a computational calculation of the coefficients, we first calculate the auxiliary variables from the design elements:

$$Q_v = n_v \left(\frac{1}{r_v} - \frac{1}{s_v} \right)$$

$$\epsilon_v = \frac{1}{\left(\frac{h_v}{h_1} \right)^2 \cdot Q_v}$$

$$\delta_v = \sum_{\mu=2}^{v} \frac{e'_{\mu-1}}{\frac{h_{\mu-1}}{h_1} \cdot \frac{h_\mu}{h_1}} \tag{3.25}$$

$$\tau_\nu = \epsilon_\nu + \delta_\nu$$

$$\Delta\left(\frac{1}{ns}\right)_\nu = \frac{1}{n'_\nu s'_\nu} - \frac{1}{n_\nu s_\nu}$$

$$\Delta\left(\frac{1}{n}\right)_\nu = \frac{1}{n'_\nu} - \frac{1}{n_\nu}.$$

For the calculation only the following paraxial parameters are used:

Parameter	Description
n_ν	The diffraction index before the spherical surface ν
r_ν	The curvature radius of the sphere ν
s_ν	The intersection length of sphere ν
h_ν	The incident height onto the sphere ν
n'_ν	The index of diffraction behind the surface ν
e'_ν	The distance between the surfaces ν and $(\nu + 1)$
s'_ν	The intersection length of the diffracted beam

The way to calculate the ratio $\frac{h_\nu}{h_1}$ required for δ_ν arises from the paraxial tracing:

$$\frac{h_\nu}{h_1} = \frac{h_2}{h_1} \cdot \frac{h_3}{h_2} \cdot \frac{h_4}{h_3} \cdots \frac{h_\nu}{h_{\nu-1}} = \frac{s_2}{s'_1} \cdot \frac{s_3}{s'_2} \cdot \frac{s_4}{s'_3} \cdots \frac{s_\nu}{s'_{\nu-1}} = \prod_{\mu=2}^{\nu} \frac{s_\nu}{s'_{\nu-1}}. \tag{3.26}$$

With these variables, which are also necessary for the paraxial calculation of an optical lens system, we can now introduce the parameters A_ν, B_ν, Γ_ν, P_ν and $\square_\nu$, required for the determination of the partial coefficients:

$$A_\nu = \left(\frac{h_\nu}{h_1}\right)^4 \cdot Q_\nu^2 \cdot \Delta\left(\frac{1}{ns}\right)_\nu = \frac{1}{\epsilon_\nu^2} \cdot \Delta\left(\frac{1}{ns}\right)_\nu$$

$$B_\nu = \tau_\nu \cdot A_\nu$$

$$\Gamma_\nu = \tau_\nu \cdot B_\nu \tag{3.27}$$

$$P_\nu = -\frac{1}{r_\nu} \cdot \Delta\left(\frac{1}{n}\right)_\nu$$

$$\square_\nu = \tau_\nu \cdot (\Gamma_\nu + P_\nu).$$

To distinguish the newly defined partial coefficients A_ν to $\square_\nu$ from the Seidel coefficients in the aberration equations, they are called "specific Seidel coefficients". With their help, we now determine the Seidel sums which appear in the formulae 3.20 and 3.21:

$$\sum_{\nu=1}^{k} I_\nu = \sum_{\nu=1}^{k} A_\nu$$

$$\sum_{\nu=1}^{k} II_\nu = \zeta \cdot \sum_{\nu=1}^{k} A_\nu + \sum_{\nu=1}^{k} B_\nu$$

$$\sum_{\nu=1}^{k} III_\nu = 3\zeta^2 \cdot \sum_{\nu=1}^{k} A_\nu + 6\zeta \cdot \sum_{\nu=1}^{k} B_\nu + 3\sum_{\nu=1}^{k} \Gamma_\nu + \sum_{\nu=1}^{k} P_\nu \qquad (3.28)$$

$$\sum_{\nu=1}^{k} IV_\nu = \zeta^2 \cdot \sum_{\nu=1}^{k} A_\nu + 2\zeta \cdot \sum_{\nu=1}^{k} B_\nu + \sum_{\nu=1}^{k} \Gamma_\nu + \sum_{\nu=1}^{k} P_\nu$$

$$\sum_{\nu=1}^{k} V_\nu = \zeta^3 \cdot \sum_{\nu=1}^{k} A_\nu + 3\zeta^2 \cdot \sum_{\nu=1}^{k} B_\nu + \zeta \left(3\sum_{\nu=1}^{k} \Gamma_\nu + \sum_{\nu=1}^{k} P_\nu \right) + \sum_{\nu=1}^{k} \Box_\nu$$

where the new coefficient ζ parameterizes the position of the entrance pupil (aperture) and the object distance:

$$\zeta = \frac{1}{n_1} \frac{s_1 \overset{\circ}{z}}{\overset{\circ}{z} - s_1} = \left(\frac{n_1}{s_1} - \frac{n_1}{\overset{\circ}{z}} \right)^{-1}. \qquad (3.29)$$

All the necessary quantities are now determined and the Seidel aberrations of 3rd order can be calculated.

3.7.2 Discussion of the Different Aberration Contributions

With the above expressions for the Seidel aberrations we will first give a geometric interpretation. Accordingly, we first change the coordinate system to simplistically consider the rotation symmetry. With the introduction of polar coordinates in the aperture image plane

$$\overset{\circ}{x} = \overset{\circ}{\rho} \sin \Psi \qquad (3.30)$$

$$\overset{\circ}{y} = \overset{\circ}{\rho} \cos \Psi \qquad (3.31)$$

we first obtain for the meridional and sagittal deviation:

$$\frac{(\Delta y'_k)_{mer}}{s_1 \beta'} = \frac{1}{2} \frac{s_1^3 \overset{\circ}{\rho}^3 \cos \Psi}{n_1 (\overset{\circ}{z} - s_1)^3} \sum_{\nu=1}^{k} I_\nu - \frac{1}{2} \frac{y_1 s_1 \overset{\circ}{\rho}^2 (2 + \cos 2\Psi)}{(\overset{\circ}{z} - s_1)^2} \sum_{\nu=1}^{k} II_\nu +$$

$$+ \frac{1}{2} \frac{n_1 y_1^2 \overset{\circ}{\rho} \cos \Psi}{s_1 (\overset{\circ}{z} - s_1)} \sum_{\nu=1}^{k} III_\nu - \frac{1}{2} \frac{n_1^2 y_1^3}{s_1^3} \sum_{\nu=1}^{k} V_\nu. \qquad (3.32)$$

$$\frac{(\Delta x'_k)_{sag}}{s_1 \beta'} = \frac{1}{2} \frac{s_1^3 \overset{\circ}{\rho}^3 \sin \Psi}{n_1 (\overset{\circ}{z} - s_1)^3} \sum_{\nu=1}^{k} I_\nu - \frac{1}{2} \frac{y_1 s_1 \overset{\circ}{\rho}^2 \sin 2\Psi}{(\overset{\circ}{z} - s_1)^2} \sum_{\nu=1}^{k} II_\nu +$$

(3.33)

$$+ \frac{1}{2} \frac{n_1 y_1^2 \overset{\circ}{\rho} \sin \Psi}{s_1 (\overset{\circ}{z} - s_1)} \sum_{\nu=1}^{k} IV_\nu.$$

Seidel Coefficient I_ν: Spherical Aberration

In the first step, we consider only spherical aberration, i.e. the term with $\sum_{\nu=1}^{k} I_\nu \neq 0$ and assume that all others are equal to zero. Hence:

$$\frac{(\Delta y'_k)_{mer}}{s_1 \beta'} = \frac{1}{2} \frac{s_1^3 \overset{\circ}{\rho}^3 \cos \Psi}{n_1 (\overset{\circ}{z} - s_1)^3} \sum_{\nu=1}^{k} I_\nu$$

$$\frac{(\Delta x'_k)_{sag}}{s_1 \beta'} = \frac{1}{2} \frac{s_1^3 \overset{\circ}{\rho}^3 \sin \Psi}{n_1 (\overset{\circ}{z} - s_1)^3} \sum_{\nu=1}^{k} I_\nu.$$

We square both parts, add, and obtain the equation:

$$\frac{(\Delta y'_k)^2_{mer}}{(s_1 \beta')^2} + \frac{(\Delta x'_k)^2_{sag}}{(s_1 \beta')^2} = \left[\frac{1}{2} \frac{s_1^3 \overset{\circ}{\rho}^3}{n_1 (\overset{\circ}{z} - s_1)^3} \sum_{\nu=1}^{k} I_\nu \right]^2.$$

This is an equation for a circle of the form $y^2 + x^2 = R^2$, which can also be written as

$$\frac{\sqrt{(\Delta y'_k)^2_{mer} + (\Delta x'_k)^2_{sag}}}{s_1 \beta'} = \frac{1}{2} \frac{s_1^3 \overset{\circ}{\rho}^3}{n_1 (\overset{\circ}{z} - s_1)^3} \sum_{\nu=1}^{k} I_\nu.$$

(3.34)

The parameter y_1 does not occur in this term, i.e., the aberration caused by the first Seidel coefficient is independent of the axial distance of the object point. The aberration is solely dependent on the "incidence height" $\overset{\circ}{\rho}$ of the beams in the aperture plane, and thus from their opening angle. From Eq. 3.34 it also follows that rays falling onto an axis-concentric circle in the aperture plane of radius $\overset{\circ}{\rho}$ also lie on a circle in the corresponding image plane. Since the radius of this circle is independent of the axis distance of the object point y_1, this first image aberration is noticeable in the same way for all object points. Because of the above properties this aberration is referred to as the so-called "aperture aberration" or "spherical aberration", due to the spherical shape of the lenses causing this aberration.

Seidel Coefficient II_ν : Coma

In the second case, we consider only the term with $\sum_{\nu=1}^{k} II_\nu \neq 0$ and assume that all other sums are equal to zero. Thus we obtain for the meridional and sagittal deviation

$$\frac{(\Delta y_k')_{mer}}{s_1 \beta'} = -\frac{1}{2} \frac{y_1 s_1 \overset{\circ}{\rho}^2 (2 + \cos 2\Psi)}{(\overset{\circ}{z} - s_1)^2} \sum_{\nu=1}^{k} II_\nu$$

$$\frac{(\Delta x_k')_{sag}}{s_1 \beta'} = -\frac{1}{2} \frac{y_1 s_1 \overset{\circ}{\rho}^2 \sin 2\Psi}{(\overset{\circ}{z} - s_1)^3} \sum_{\nu=1}^{k} II_\nu .$$

We square again both components, add and obtain in this case:

$$\frac{(\Delta y_k')_{mer}^2}{(s_1 \beta')^2} + \frac{(\Delta x_k')_{sag}^2}{(s_1 \beta')^2} = \left[\frac{1}{2} \frac{y_1 s_1 \overset{\circ}{\rho}^2 (2 + \cos 2\Psi)}{(\overset{\circ}{z} - s_1)^3} \sum_{\nu=1}^{k} II_\nu \right]^2$$

$$+ \left[\frac{1}{2} \frac{y_1 s_1 \overset{\circ}{\rho}^2 \sin 2\Psi}{(\overset{\circ}{z} - s_1)^3} \sum_{\nu=1}^{k} II_\nu \right]^2 .$$

After multiplying this equation can be summarized to

$$\left[\frac{(\Delta y_k')_{mer}^2}{(s_1 \beta')} + \frac{1}{2} \frac{y_1 s_1 \overset{\circ}{\rho}^2}{(\overset{\circ}{z} - s_1)^3} \sum_{\nu=1}^{k} II_\nu \right]^2 + \frac{(\Delta x_k')_{sag}^2}{(s_1 \beta')^2} = \left[\frac{1}{2} \frac{y_1 s_1 \overset{\circ}{\rho}^2}{(\overset{\circ}{z} - s_1)^3} \sum_{\nu=1}^{k} II_\nu \right]^2 .$$

This equation represents the equation of a circle of the form $(y - a)^2 + x^2 = R^2$ with a shift of a in meridional direction. The aberration is proportional to y_1 and disappears accordingly for $y_1 = 0$.

In the image plane we get circles for $y_1 \neq 0$ and a given $\overset{\circ}{\rho}$ as for the aperture aberration whose radii are proportional to $\overset{\circ}{\rho}^2$ and y_1. However, their center points are displaced against the ideal image point. The resulting cometary-shaped figure with different $\overset{\circ}{\rho}$ is shown in Fig. 3.17. This aberration is therefore called "coma aberration" or simply "coma". Due to its similarity to spherical aberration it is also called the "spherical aberration of oblique incidence".

Seidel Coefficient III_ν and IV_ν : Meridional and Sagittal Image Curvature

For the next Seidel errors we now consider only the terms with $\sum_{\nu=1}^{k} III_\nu \neq 0$ and $\sum_{\nu=1}^{k} IV_\nu \neq 0$. All other terms $\sum_{\nu=1}^{k} I_\nu = \sum_{\nu=1}^{k} II_\nu = \sum_{\nu=1}^{k} V_\nu$ are $= 0$. In this case, we obtain for the two aberration components

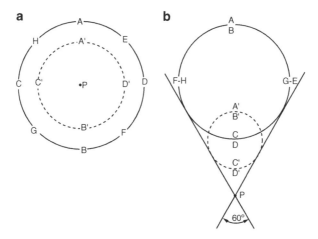

Fig. 3.17 The emergence of coma. Intersection points of various beams in the aperture plane (**a**) and in the image plane (**b**) in each case designated by letters. Note the quadratic dependence of the diameter in the image plane on those in the aperture plane (W.J. Smith, 'Modern Optical Engineering')

$$\frac{(\Delta y'_k)_{mer}}{s_1 \beta'} = \frac{1}{2} \frac{n_1 y_1^2 \overset{\circ}{\rho} \cos \Psi}{s_1 (\overset{\circ}{z} - s_1)} \sum_{\nu=1}^{k} III_\nu$$

$$\frac{(\Delta x'_k)_{sag}}{s_1 \beta'} = \frac{1}{2} \frac{n_1 y_1^2 \overset{\circ}{\rho} \sin \Psi}{s_1 (\overset{\circ}{z} - s_1)} \sum_{\nu=1}^{k} IV_\nu.$$

We multiply the two equations each with $s_1 \beta'$, divide by the respective Seidel sum, square and obtain for the total aberration

$$\frac{(\Delta y'_k)_{mer}^2}{(\sum_{\nu=1}^{k} III_\nu)^2} + \frac{(\Delta x'_k)_{sag}^2}{(\sum_{\nu=1}^{k} IV_\nu)^2} = \left[\frac{1}{2} \frac{n_1 y_1^2 \overset{\circ}{\rho} \beta'}{(\overset{\circ}{z} - s_1)} \right]^2.$$

This equation represents the equation of an ellipse of the form $\frac{x^2}{a^2} + \frac{y^2}{b^2} = R^2$. This aberration also only occurs when the (off-axis) object point $y_1 \neq 0$, and apparently grows disproportionately with y_1^2. The semi-axes of the ellipse are determined by the two Seidel sums $\sum_{\nu=1}^{k} III_\nu$ and $\sum_{\nu=1}^{k} IV_\nu$, the first determining the so-called meridional and the second the sagittal field curvature.[10]

---

[10]Both Seidel sums thus cause optical errors of the same geometric type.

Instead of the separated field curvature one can alternatively introduce the "mean field curvature" $(\Delta s'_k)_{mean}$ and the so-called "astigmatism" $(\Delta s'_k)_{ast}$, where the astigmatism is the deviation from the mean field curvature:

$$(\Delta s'_k)_{mean} = (\Delta x'_k)_{sag} + (\Delta s'_k)_{ast}$$

$$(\Delta s'_k)_{mean} = (\Delta y'_k)_{mer} - (\Delta s'_k)_{ast}.$$

By addition and subtraction of the two equations, we obtain

$$(\Delta s'_k)_{mean} = \frac{1}{2}\left[(\Delta x'_k)_{sag} + (\Delta y'_k)_{mer}\right] \tag{3.35}$$

$$(\Delta s'_k)_{ast} = \frac{1}{2}\left[(\Delta y'_k)_{mer} - (\Delta x'_k)_{sag}\right]. \tag{3.36}$$

Assuming that the absolute values of the trigonometric functions $cos\Psi$ and $sin\Psi$ can be maximum 1, we obtain an estimate for the maximum mean field curvature

$$(\Delta s'_k)_{mean} = \frac{1}{4}\frac{n_1 y_1^2 \mathring{\rho}\beta'}{(\mathring{z} - s_1)}\left[\sum_{v=1}^{k} III_v + \sum_{v=1}^{k} IV_v\right]$$

and respectively for the astigmatism

$$(\Delta s'_k)_{ast} = \frac{1}{4}\frac{n_1 y_1^2 \mathring{\rho}\beta'}{(\mathring{z} - s_1)}\left[\sum_{v=1}^{k} III_v - \sum_{v=1}^{k} IV_v\right].$$

The existing relationship between meridional, sagittal and mean field curvature as well as astigmatism very easily leads in practice to a very important condition: If astigmatism disappears ($\sum_{v=1}^{k} III_v - \sum_{v=1}^{k} IV_v = 0$), all field curvatures are equal. This leads to the important relation for the Seidel coefficient of the field curvature which is now uniform across the image plane and free of astigmatism:

$$\sum_{v=1}^{k} P_v := \sum_{v=1}^{k} III_v = \sum_{v=1}^{k} IV_v = \frac{1}{2}\left[\sum_{v=1}^{k} III_v + \sum_{v=1}^{k} IV_v\right].$$

One can further show that in the case of corrected astigmatism $\sum_{v=1}^{k} P_v$ is exactly the vertex radius inverse value of the curved image field (Fig. 3.18). If this radius is infinitely large, the field would be a flat plane. This is impossible without a correction lens. We will see in the next chapter how this radius of curvature can be estimated in a very simple way and how one can derive a condition for the flat image field.

Fig. 3.18 Graphic depiction of the connection of the variables $(\Delta y_k')_{mer}, (\Delta y_k')_{sag}$ and $(\Delta s_k')_{mean}, (\Delta s_k')_{ast}.$ $\bar{y}_k'$ is the y-coordinate of the ideal (paraxial) intersection point (Picht 1955)

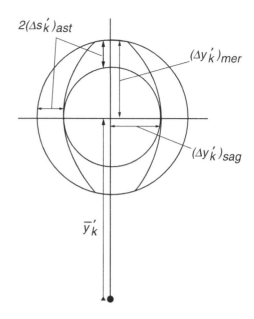

Seidel Coefficient V_ν: Distortion

As a final image aberration, we consider the case that only the term with $\sum_{\nu=1}^{k} V_\nu \neq$ 0 and all others are equal to zero. We get the simple relation

$$\frac{(\Delta y_k')_{mer}}{s_1 \beta'} = -\frac{1}{2} \frac{n_1^2 y_1^3}{s_1^3} \sum_{\nu=1}^{k} V_\nu. \tag{3.37}$$

This error is independent of $\overset{\circ}{\rho}$ and thus not dependent on the opening angle. The error increases with the third power of the axis distance y_1. This 5th Seidel aberration is called "distortion". because this aberration depends on the axis distance y_1 or the angle of incidence ω_1. Thus the image scale changes.

3.8 The Seidel Sums and Their Interpretation

Looking at Eqs. 3.27–3.29 one can directly derive some fundamental properties of the Seidel aberrations and thus procedures for the correction of an optical system:

1. *Since each refracting surface causes specific aberrations, one can compensate for the total error only by corresponding surface contributions of different sign.*
2. *All image aberrations must be corrected separately.*

3. *If the Seidel coefficients for the spherical aberration $\sum A_\nu$, coma $\sum B_\nu$ and astigmatism $\sum \Gamma_\nu$ disappear in a suitable manner, the sagittal and meridional curvature radii correspond to the so-called Petzval radius $r_P = \sum P_\nu$.*

For image aberrations, the Petzval radius r_P provides a specific case, because in the event that all aberrations disappear completely, a curved image field-shell with Petzval radius is maintained. This field sphere is the justification for so-called field curvature. Jozef Maximilian Petzval was the first to give an explanation for this situation and designed the eponymous Petzval lens in 1840.

Furthermore, we are especially interested in the case of infinite object distance $s_1 \to -\infty$ and $n_1 = 1$, important for the case of a spectrometer. For that we obtain for the parameter ζ (Eq. 3.29) by applying the limit

$$\lim_{s_1 \to -\infty} (\zeta) = -\overset{\circ}{z}.$$

The parameter ζ corresponds exactly to the negative distance between aperture and the first refractive surface. Thus we obtain specifically for the Seidel sums:

$$\sum_{\nu=1}^{k} I_\nu = \sum_{\nu=1}^{k} A_\nu$$

$$\sum_{\nu=1}^{k} II_\nu = -\overset{\circ}{z} \cdot \sum_{\nu=1}^{k} A_\nu + \sum_{\nu=1}^{k} B_\nu$$

$$\sum_{\nu=1}^{k} III_\nu = 3\overset{\circ}{z}^2 \cdot \sum_{\nu=1}^{k} A_\nu - 6\overset{\circ}{z} \cdot \sum_{\nu=1}^{k} B_\nu + 3 \cdot \sum_{\nu=1}^{k} \Gamma_\nu + \sum_{\nu=1}^{k} P_\nu \qquad (3.38)$$

$$\sum_{\nu=1}^{k} IV_\nu = \overset{\circ}{z}^2 \cdot \sum_{\nu=1}^{k} A_\nu - 2\overset{\circ}{z} \cdot \sum_{\nu=1}^{k} B_\nu + \sum_{\nu=1}^{k} \Gamma_\nu + \sum_{\nu=1}^{k} P_\nu$$

$$\sum_{\nu=1}^{k} V_\nu = -\overset{\circ}{z}^3 \cdot \sum_{\nu=1}^{k} A_\nu + 3\overset{\circ}{z}^2 \cdot \sum_{\nu=1}^{k} B_\nu - \overset{\circ}{z}\left(3\sum_{\nu=1}^{k} \Gamma_\nu + \sum_{\nu=1}^{k} P_\nu\right) + \sum_{\nu=1}^{k} \square_\nu.$$

Apparently, the aperture setting $\overset{\circ}{z}$ has a massive influence on the image errors. In Sect. 3.12 we will investigate the impact of the stop position in more detail. Without a more detailed analysis, it is reasonable, however, to position the aperture exactly on the first refracting surface, i.e., $\overset{\circ}{z} = 0$. In this case, all contributions introduced by the diaphragm position immediately disappear and we obtain for the Seidel sums the five simplistic equations

$$\sum_{\nu=1}^{k} I_\nu = \sum_{\nu=1}^{k} A_\nu$$

$$\sum_{\nu=1}^{k} II_\nu = \sum_{\nu=1}^{k} B_\nu$$

$$\sum_{\nu=1}^{k} III_\nu = 3 \cdot \sum_{\nu=1}^{k} \Gamma_\nu + \sum_{\nu=1}^{k} P_\nu \qquad (3.39)$$

$$\sum_{\nu=1}^{k} IV_\nu = \sum_{\nu=1}^{k} \Gamma_\nu + \sum_{\nu=1}^{k} P_\nu$$

$$\sum_{\nu=1}^{k} V_\nu = \sum_{\nu=1}^{k} \Box_\nu.$$

Under the above circumstances spherical aberration, coma and distortion are then the specific Seidel coefficients. The image curvatures consist of astigmatism and Petzval curvature.

3.8.1 Average Image Curvature and Astigmatism

If astigmatism (Eq. 3.36) and mean-field curvature (Eq. 3.35) instead of meridional and sagittal field curvature are used, we obtain:

$$(\Delta s_k')_{mean} = 2 \cdot \left(\zeta^2 \cdot \sum_{\nu=1}^{k} A_\nu + 2\zeta \cdot \sum_{\nu=1}^{k} B_\nu + \sum_{\nu=1}^{k} \Gamma_\nu \right) + \sum_{\nu=1}^{k} P_\nu$$

$$(3.40)$$

$$(\Delta s_k')_{ast} = \zeta^2 \cdot \sum_{\nu=1}^{k} A_\nu + 2\zeta \cdot \sum_{\nu=1}^{k} B_\nu + \sum_{\nu=1}^{k} \Gamma_\nu.$$

From Eq. 3.28 one can still derive the following fundamental relations for the coefficients of the meridional and sagittal field curvature

$$\sum_{\nu=1}^{k} III_\nu = 3 \cdot \frac{\left[\sum_{\nu=1}^{k} III_\nu - \sum_{\nu=1}^{k} IV_\nu \right]}{2} + \sum_{\nu=1}^{k} P_\nu$$

$$(3.41)$$

$$\sum_{\nu=1}^{k} IV_\nu = 1 \cdot \frac{\left[\sum_{\nu=1}^{k} III_\nu - \sum_{\nu=1}^{k} IV_\nu \right]}{2} + \sum_{\nu=1}^{k} P_\nu.$$

When astigmatism is eliminated $(\Delta s'_k)_{ast} = 0$ all field curvatures fall together and are $(\Delta s'_k)_{mean} = \sum_{\nu=1}^{k} P_\nu$. Due to the definition of the astigmatism in Eq. 3.36 we obtain the following important conclusions:

1. *Astigmatism results from half the difference of the meridional and sagittal field curvature.*
2. *With an existing astigmatism, the deviation of the meridional field curvature from the Petzval curvature is three times as large as the corresponding deviation of the sagittal field curvature from the Petzval curvature.*
3. *The Petzval sphere defines the field curvature without astigmatism. For this case, meridional and sagittal field curvature are identical.*

These fundamental statements are particularly comprehensive in the set of Eq. 3.39. If astigmatism is corrected in a suitable manner, i.e., $\sum \Gamma_\nu = 0$, only Petzval curvature remains beside spherical aberration and coma. If it is also possible to eliminate spherical aberration, it does not mean that coma simultaneously disappears, because of the difference τ_ν.

Figure 3.19 depicts four representative situations:

1. Positive astigmatism with positive meridional and positive sagittal image curvature.
2. Negative astigmatism with negative meridional and positive sagittal image curvature.
3. Positive astigmatism with positive meridional and negative sagittal image curvature.
4. Negative astigmatism with negative meridional and negative sagittal image curvature.

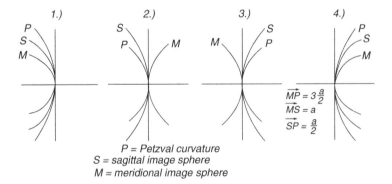

P = Petzval curvature
S = sagittal image sphere
M = meridional image sphere

Fig. 3.19 The different possibilities of inter-dependent positions for the meridional, sagittal and Petzval fields. The quantity a represents the difference between meridional and sagittal imaging errors: $a = \sum_{\nu=1}^{k} III_\nu - \sum_{\nu=1}^{k} IV_\nu$ (Picht 1955)

3.8.2 Example Calculation for a Simplet

For further understanding the procedures, we now present some additional test calculations. We start with a simple biconvex lens with two refracting surfaces. The advantage of this simple example calculation is in particular that in the case of only one lens analytical equations for the aberrations can be specified, and a direct comparison with the Seidel computation scheme is possible. Here, we want to limit our considerations to the comparison of the spherical aberration.

We start with an object point on the y-axis at an infinite distance. This object point causes a beam parallel to the optical axis which penetrates the lens aperture on the radius $\mathring{y}$. The corresponding coordinate on the x-axis for this case is $\mathring{x} = 0$. The longitudinal component of the spherical aberration $\Delta s_{sph.Ab.}$, i.e., the focal length difference between center and edge beams for the case of a simple lens with the specific radii ratio[11] $r_1 : r_2 = 1 : (-6)$ (Menn 2004) is

$$\Delta s_{sph.Ab.} = -0.268 \cdot \frac{D^2}{f} \tag{3.42}$$

where $D = 2\mathring{y}$ specifies the fully illuminated lens aperture.

Dividing the longitudinal error by the f-ratio of the lens, one obtains for the transverse error $\Delta y_{sph.Ab.}$, i.e., the diameter of the entire blur spot:

$$\Delta y_{sph.Ab.} = -0,268 \cdot \frac{D^3}{f^2}. \tag{3.43}$$

It can be seen here that the transverse spherical aberration increases with the third power of the illuminated diameter. We compare this result now with the Seidel aberrations. We consider only the first term in Eq. 3.21, which represents the contribution of spherical aberration in the sagittal direction

$$\frac{(\Delta y'_k)_{mer}}{s_1 \beta'} = \frac{1}{2} \frac{(\mathring{x}^2 + \mathring{y}^2)\mathring{y}s_1^3}{n_1(\mathring{z} - s_1)^3} \sum_{\nu=1}^{k} I_\nu.$$

This equation gives the radius of the corresponding circle of confusion, so for calculating the diameter from this equation we multiply by a factor two. The observed object point is on the y-axis, as already agreed above for simplicity, i.e., $\mathring{x} = 0$ and the aperture should be the lens diameter itself $\mathring{z} = 0$. The meridional component of the aperture aberration disappears under these simplified conditions.

---

[11]This radius ratio determines the minimum spherical aberration for a simple lens. One therefore refers to "best form" lens.

For the case of infinite object distance $s_1 = -\infty$ and the refractive index in object space outside the lens $n_1 = 1$ we obtain from Eq. 3.22 $s_1 \beta' = f$ and thus

$$2 \cdot (\Delta y'_k)_{mer} = -f \cdot \overset{\circ}{y}^3 \sum_{\nu=1}^{2} A_\nu.$$

For an example biconvex lens with radii $r_1 = 58.33$ mm and $r_2 = -350$ mm, i.e., a radius ratio of 1:(-6) we obtain for a refraction index $n = 1.5$ according to Eq. 3.12 a focal length $f = 100$ mm. In Tables 3.3 and 3.4 we present all calculated parameters of the Seidel calculation for these specially chosen lenses.

The coefficient for spherical aberration is thus $\sum A_\nu = 2.14 \cdot 10^{-6}$. If, as in the analytical equation 3.43 $\overset{\circ}{y} = D/2$, we obtain by expanding with f^2

$$2 \cdot (\Delta y'_k)_{mer} = -\frac{D^3}{f^2} \cdot \frac{2.14 \cdot 10^{-6} \cdot f^3}{8} = -\frac{D^3}{f^2} \cdot 0,268.$$

This result is identical to Eq. 3.43. For an f/10 lens we obtain for the spherical aberration a circle of confusion of $2 \cdot (\Delta y'_k)_{mer} = 26.8\,\mu$m. In Sect. 3.11.1 we will further carry out the comparison of the off-axis aberrations associated with the spot diagram.

Note: To compare different optical systems of different focal lengths it makes sense to calculate the specific Seidel coefficients already normalized to the focal length f. Therefore, one determines all the coefficients in the two Tables 3.3 and 3.4 for a focal length of $f = 1$, i.e., all radii and distances are taken with respect to a fraction of the focal length.

Table 3.3 Auxiliary parameters for a thin biconvex lens with $r_1 = 58.33$ mm and $r_2 = -350$ mm. $\nu = 0$ merely reflects the specification of an object point at infinity

ν	r_ν	e'_ν	n'_ν	s'_ν	$\frac{h'_\nu}{h1}$	Q_ν	ϵ_ν	δ_ν	τ_ν	$\frac{s'_\nu}{s_\nu}$
0		0.0	1.0	$-\infty$						
1	58.33	0.0	1.5	-175.0	1.0	$+0.017$	58.33	0.0	58.33	0.0000
2	-350.00	0.0	1.0	-100.0	1.0	-0.013	-77.78	0.0	-77.78	-0.5714

For $\nu = 1$ and $\nu = 2$ the auxiliary parameters are given according to Eq. 3.25

Table 3.4 Specific Seidel coefficient corresponding to Table 3.3 for both surfaces and their sums

ν	A_ν	B_ν	Γ_ν	P_ν	$\square_\nu$
0					
1	1.12e$-$06	6.53e$-$05	3.81e$-$03	5.71e$-$03	5.56e$-$01
2	1.02e$-$06	-7.06e$-$05	5.19e$-$03	9.52e$-$04	-5.56e$-$01
Σ	2.14e$-$06	-1.43e$-$05	1.00e$-$02	6.67e$-$03	0.00e$-$00

3.9   The Impact of Field Curvature: A Simple Example

The star to be investigated will be imaged on the optical axis (paraxial imaging), converted to a parallel beam by the collimator and directed to the grating. Now however, the grating expands the polychromatic beam, i.e., it introduces a field of view which has to be imaged by the camera. The slit image should have the same quality over the whole wavelength range, i.e., it should exhibit the same full width at half maximum (FWHM) in wavelength units. Hence, all image aberrations in the whole optical system should be corrected. The first to be corrected is chromatic aberration (sometimes called "secondary spectrum"). It is normally sufficiently corrected by the use of an achromatic lens (doublet). However, this can be problematic especially for low-resolution spectrographs, where a large or even the whole optical wavelength range is imaged and the use of highly corrected apochromatic lenses is required. When using a doublet, spherical aberration is eliminated as far as possible.[12] However, for tilted beams hitting the camera lens, other aberrations like astigmatism and field curvature cannot be neglected. The aberration which can easily be calculated is field curvature.

We will now consider this aberration in more detail. An off-axis image element P will be astigmatically imaged by a spherical lens into two planes (i.e. will not form a point image). The point on the object side is transferred into two caustic lines B_M and B_S, orientated perpendicularly to each other and having different distances from the lens. The caustic line B_M is positioned in the meridional plane and the caustic line B_S in the sagittal plane (see Fig. 3.8). For eliminating this so-called astigmatism, both planes have to coincide with each other. This can be done with a corresponding correction of the imaging optics and one obtains the so-called Petzval plane. An optical system without spherical aberration (aplanatic system) images object points as image points. The Petzval plane is curved and the imaging or CCD plane should be curved as well, to obtain a sharp image. In contrast to astigmatic planes this image plane is independent of the position and form of the involved lenses. The plane depends only on the corresponding index of refraction and on the focal length. The following condition is valid (Hecht 2003) for a Petzval plane of a number of thin lenses:

$$\frac{1}{r_p} = \sum_{i=1}^{k} \frac{1}{n_i f_i} \tag{3.44}$$

r_p is the curvature radius of the corresponding Petzval plane. Thus, two lenses have a flat Petzval plane if the condition

$$n_1 \cdot f_1 + n_2 \cdot f_2 = 0$$

---

[12]For this reason a doublet is the first choice for a collimator.

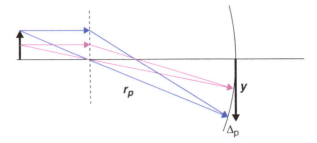

Fig. 3.20 Origin and geometry of image curvature. A flat object is imaged into a curved plane. The Petzval plane is defined if both planes (meridional and sagittal) match because of optical correction. The distance r_p defines the curvature radius of the Petzval plane and y the coordinate of the blurring circle Z in the image plane

is fulfilled. For small values of this so-called Petzval sum we obtain an almost flat image plane. The geometry of the image curvature is presented in Fig. 3.20. With y the distance of the light beam from the optical axis in the focal plane, the Pythagoras theorem provides the simple relation

$$r_p^2 + y^2 = (r_p + \Delta_P)^2$$
$$r_p^2 + y^2 = r_p^2 + 2r_p\Delta_P + \Delta_P^2$$

and respectively

$$y^2 = \Delta_P^2 + 2r_p\Delta_P.$$

We use the approximation that $\Delta_P^2 \ll 2r_p\Delta_P$, i.e.,

$$y^2 \approx 2r_p\Delta_P$$

or

$$\Delta_P = \frac{y^2}{2} \cdot \frac{1}{r_p} \tag{3.45}$$

and find for the distance between Petzval and paraxial focal plane

$$\Delta_P = \frac{y^2}{2} \cdot \sum_{i=1}^{k} \frac{1}{n_i f_i}. \tag{3.46}$$

Note, that Δ_P is the distance between the Petzval and paraxial focal planes in radial direction. For that reason, one actually has to multiply Δ_P with the cosine of the angle between the two r_p. For the case of small distances y (half the width of the film or CCD) with respect to the Petzval radius, this correction is almost 1, though.

As an example we calculate the influence of the image curvature for a simple 200 mm doublet of Edmund Optics (stock no. #45-179). The doublet consists of two glasses BAK4 and SF10. The diffraction index for BAK4 (546.1 nm) is $n_1 = 1.57125$ and for SF10 (546.1 nm) $n_2 = 1.7343$. The radii of the glasses are $R_{11} = 130.48$ mm, $r_{12} = -99.36$ mm, $r_{21} = -99.36$ mm and $r_{22} = -320.20$ mm according to the catalogue. The focal length follows from the conditional equation for thin lenses.

$$\frac{1}{f_1} = (n_1 - 1) \cdot \left(\frac{1}{r_{11}} - \frac{1}{r_{12}} \right) \Rightarrow f_1 = 98.74 \, \text{mm}$$

and for the second lens

$$\frac{1}{f_2} = (n_2 - 1) \cdot \left(\frac{1}{r_{21}} - \frac{1}{r_{22}} \right) \Rightarrow f_2 = -196.19 \, \text{mm}.$$

The overall focal length of the achromat at the above wavelength of 546.1 nm is

$$\frac{1}{f_{total}} = \left(\frac{1}{f_1} + \frac{1}{f_2} \right) = \left(\frac{1}{98.74 \, mm} + \frac{1}{(-196.19 \, mm)} \right)$$

or $f_{total} = 198.79$ mm. Using the indices of refraction and the focal lengths of both lenses we obtain the Petzval radius

$$\frac{1}{r_p} = \frac{1}{n_1 f_1} + \frac{1}{n_2 f_2} = \frac{1}{1.57125 \cdot 98.74 \, \text{mm}} + \frac{1}{1.7343 \cdot (-196.19 \, \text{mm})}$$

i.e. $r_p = 285.18$ mm and with Eq. 3.46 we also obtain the distance of the Petzval plane from the CCD plane depending on the distance from the optical axis y (see Fig. 3.21). According to this equation, the distance Δ_P increases quadratically with y. Imaging a large spectral range by using an achromat combined with a large CCD camera or film will, hence, be somewhat limited. A simple technique to flatten the image plane is the use of a meniscus lens placed directly in front of the focal plane. However, this is almost impossible for normal off-the-shelf CCD cameras, which usually have a large distance between the camera window and the focal plane. For this reason, normal photographic lenses are the first choice. They usually compensate for field curvature and all other aberrations, like astigmatism, within certain limits. On the other hand, a simple doublet is a good choice for a collimator or camera because of the small size. Our calculations have shown that for a distance of only 10 mm from the optical axis the distance to the CCD plane is 175.3 µm. Whether this distance already introduces blurring or not, will be shown in Sect. 3.10.

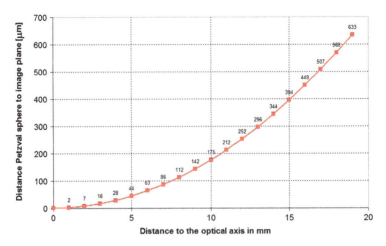

Fig. 3.21 Distance of the Petzval sphere to the CCD plane

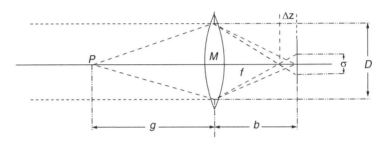

Fig. 3.22 Sketch of the circle of confusion

3.10 Permitted Deviations from Ideal Focus: The Blurring Circle

From photography we know the concept of depth of focus, which is used by photographers as a creative tool. The depth of focus describes a distance interval within which the object is still sharply imaged in the focal plane (Fig. 3.22). Considering lens imaging, the object point P of size g is sharply imaged at distance b behind the lens. With a fixed focal length f of the lens, the lens equation defines the connection of all the various parameters. We now ask what happens to the points positioned at different distances compared to g. For instance, an incident parallel light bundle (virtually at infinity) will be focused at the distance f (the focal length) behind the lens and creates a blurring circle of diameter σ in the image plane at distance g.

Every CCD digital detector has a limited resolution defined by the pixel size (about 5–30 μm). Obviously it is not necessary to make the blurring circle σ smaller than the CCD pixel. This defines an area around the point P at distance g which can be considered as being sharp. In a spectrometer the relations are similar but the bundle from the collimator is always effectively coming from infinity. However, image aberrations introduce focal planes not coinciding with the CCD plane, as e.g., in our example of image curvature. Another example is chromatic aberration. From the corresponding blurring circle and the given pixel size we get a focal zone where the stellar spectrum is sharp. This zone is called "depth of focus". There is an area where the object is sharply imaged, defined by the resolution of the sensor. From Fig. 3.22 and with the theorem of intersecting lines we identify the following correlation

$$\frac{D}{f} = \frac{1}{F\#} = \frac{\sigma}{b - f} = \frac{\sigma}{\Delta z}$$

with $F\#$ the focal ratio. An object with size smaller than g will be sharply imaged behind the defined focal plane. The blurring circle is not only present behind the focal plane but also in front of it and with the same size for identical focal distances. For our spectrograph we obtain an allowed zone for the focal length variation Δz

$$\Delta z = 2 \cdot F\# \cdot \sigma. \tag{3.47}$$

For σ we usually use the pixel size and have to account for effective superpixels in the case of pixel binning. Δz quickly increases with focal ratio and binning.

We now want to consider our example for the image curvature in the last chapter again, assuming an achromat focal ratio (f number) of 4. For simplification, we assume in our example that astigmatism is negligible. Using a CCD camera of 9 μm pixels and twofold binning we obtain a focal interval of $\Delta z = 2 \cdot 4 \cdot (2 \cdot 9\,\mu m) = 144\,\mu m$. Because of image curvature the use of a simple achromat will create a limited and sharply imaged field of view. Using the image center as reference, $2 \times 9\,mm = 18\,mm$ of the CCD camera can be used. Outside this area the sharpness of the spectrum is reduced. For this reason one should always consider the necessary wavelength interval to be used. Meanwhile large CCD chips of small-size picture format (24×36 mm) are available. CCD cameras of such size, however, need lenses with correspondingly corrected fields, i.e., the use of high quality lenses with sufficiently large f numbers to avoid potential vignetting.

3.11   Estimation of the Imaging Performance by Ray-Tracing Methods

There are several possibilities to evaluate the imaging performance of an optical system with respect to image aberrations. In addition to calculation of chromatic and Seidel aberration, it is possible to calculate the exact intersection points by means of raytrace programs. There are many software packages for optic design on the market. The most common contributors and the respective software application fields are give in Table 3.5. Each of these programs have their own technical

Table 3.5 Contributors of raytrace and optical design software.

Software	Contributor	Application
ZEMAX	Focus Software	Analysis + optimization
OSLO	Lambda Research Corp.	Analysis + optimization
Lens View V	Lambda Research Corp.	System database
WINLENS	LINOS Photonics	Analysis + optimization
CODE V	Optical Research Ass.	Analysis + optimization
OpTaliX	Optenso	Analysis + optimization
SOLSTIS	Optis	Analysis + optimization
SYNOPSIS	Optical System Design	Analysis + optimization

advantages and disadvantages. The user should seek information in advance, which of the software packages is useful for him or her. For instance, ZEMAX is a kind of professional optical design software, commonly used in industry and science. The free OSLO version is limited to the calculation of a maximum of ten surfaces (refractive or reflective). However, beginners quickly notice that OSLO and ZEMAX contain extensive functionalities including complex operations. Many of the software packages have a feature that allows optimization of an existing optical system by freely selectable parameters. The user no longer needs to search for the most optimal configuration by manual iteration. A multi-dimensional optimization algorithm performs this task. The following features may be important for the user in addition to the standard functional activities (spot diagrams, etc.):

• Wave front propagation analysis
• Polarization analysis
• Vignetting analysis
• Diffraction analysis
• Tolerance analysis
• Straylight Analysis
• Macro Programming

In general, it should be noted that raytrace programs can hardly reasonably be used without basic knowledge of geometrical and wave optics. Because of the complex relationships and non-linear dependencies, raytracing results are difficult to understand especially for beginners. Some basics of the Seidel aberration theory are not only of great benefit, but almost essential in order to understand the optical system in its entirety.

All in the following presented raytrace results were obtained with the freely accessible version of WinLens 3D. The advantage of this software is already an existing database of all available optical components of the companies LINOS or Qioptics. WinLens3D is also available as a free version "WinLens3DBasic" in which, however, some functions (e.g., optimization) are not accessible.

3.11.1   The Spot Diagram

The frequently used graphical representation of the puncture coordinates of rays can be obtained with trigonometric raytrace calculations. In this context the optician refers to the so-called "spot diagram", where the intersection points of evenly emitted rays from an object point are shown in a chosen image plane. The spot diagram is therefore the quantity of all calculated transverse aberrations in the image plane for a given object height and angle of incidence. The spot diagram shows for a given situation very clearly the specific effects of aberrations. With a sufficient number of projected beams one can estimate the brightness distribution via the distribution of the points of penetration in the image plane. In the paraxial approximation and for ideal imaging, the light rays of the emitting object points would be bundled in a single point of the Gaussian plane. However, in the context of the above-described image aberrations, it comes down to the above-mentioned deviations resulting in characteristic dispersion figures. When calculating spot diagrams, diffraction effects are completely dispensed. The above calculated geometric figures of the image aberrations are superimposed and can, with some experience, be recognized in the spot diagram. In a spot diagram the diameter of the Airy disk is often indicated, i.e., representing the smallest possible spot. The Airy disk defines the spot diameter of a perfect lens system, which is free from aberrations. In this context one speaks of diffraction-limited imaging. If the majority of the beam intersection points for a spot is within the Airy diameter[13] the system is diffraction-limited. The diffraction limit can be simply calculated from the wavelength used and the image-side numerical aperture (NA or inverse f-number) of the optical system. Knowledge of the wave-optical performance is not required (Fig. 3.23).

In Sect. 3.8.2 we showed sample investigations of a special (best form) biconvex lens with 100 mm focal length and in particular its spherical aberration. The circle of confusion was 28.6 μm at infinite object distance. In Fig. 3.24 we calculated the corresponding spot diagram using the raytracing 3D software WinLens. The scaling in x-direction was adapted to the calculated circle of confusion of 28.6 μm. The second plot from the right shows the spot diagram in the paraxial image plane. The diameter of the outermost ring is consistent with the selected image.

For further comparison with the calculated Seidel aberrations, we introduce an off-axis angle $\omega_1 = 2°$ of an infinitely distant object point to examine in detail its impact on the best-form lens. For this case, coma, astigmatism and field curvature have to be taken into account. The spot diagrams in Fig. 3.25 clearly show an elliptical deflection (bottom row) of the concentric circles in the case of on-axis illumination (top row). A slight asymmetry of the ellipses are also noted, which is introduced by a slight comatic portion of the total image error.

---

[13]Which is only possible when neglecting the diffraction phenomenon.

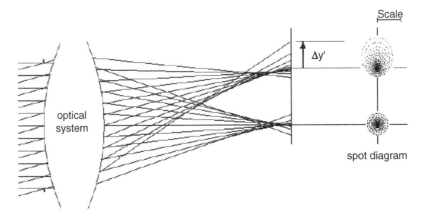

Fig. 3.23 Cross aberration presented by a spot diagram for two object angles (T. Thöniß, "Abbildungsfehler und Abbildungsleistung optischer Systeme", LINOS Photonics GmbH)

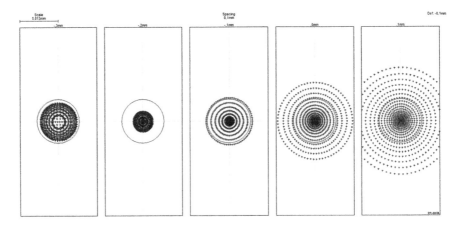

Fig. 3.24 Spot diagrams of the best-form biconvex lens versus "defocus" (position of the current image plane on the optical axis in mm), as calculated in Sect. 3.8.2. The object-side field of view angle is $\omega_1 = 0°$, i.e., only spherical aberration is present. The defocus is indicated above the corresponding spot diagrams. Defocus $= 0$ mm defines the Gaussian image plane. The respective image area is scaled to the calculated Seidel diameter of $26.8\,\mu m$, i.e., in the paraxial image plane the spot diameter corresponds almost exactly ($\approx 25\,\mu m$) to the calculated circle of confusion of spherical aberration. The small deviation of about 7 % is produced by the exact calculation (without approximations) of the intersection points in the image plane, that is, due to influences of higher orders in the trigonometric functions. The smallest circle of confusion is found at the defocus of -0.2 mm. One can clearly see the disproportionate increase of the ring radii due to the h^3 dependence of spherical aberration (nonlinear gradient). The red circle shows the diameter of the diffraction limited Airy disk

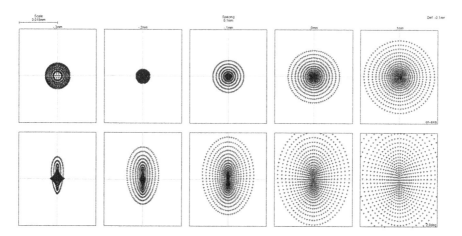

Fig. 3.25 Spot diagrams for the best form biconvex lens calculated in Sect. 3.8.2 versus defocus (on-axis, *top row*) and for an object angle $\omega_1 = 2.0°$ (*bottom*). The concentric circles of spherical aberration have expanded to astigmatic ellipses. The coma, also present in oblique bundles, is only very weak for the best-form lens

For comparison of the Seidel diameter with the spot, all aberration contributions must be added. The proportion of the meridional coma is in our case of infinite object distance, i.e., taking into account the boundary conditions (Eq. 3.22)

$$(\Delta y_k')_{mer} = \frac{3}{2}\mathring{y}^2 f \tan \omega_1 \sum_{\nu=1}^{k} II_\nu .$$

Using all parameters for the $f/10$ lens delivers

$$(\Delta y_k')_{mer} = \frac{3}{2} \cdot (5\,\text{mm})^2 \cdot 100\,\text{mm} \cdot \tan(2°) \cdot (-1.43 \cdot 10^{-5}) = 0.00187\,\text{mm}.$$

Since $(\Delta y_k')_{mer}$ indicates only half the image aberrations, we obtain for the meridional coma an almost negligible contribution of 3.8 μm. In the sagittal direction we get in a similar way, a contribution of 1.2 μm.

The circles of confusion by meridional and sagittal field curvature can be determined completely analogously from the equations given. We obtain for the meridional deviation $(\Delta y_k')_{mer} = 0.0112\,\text{mm}$ and for the sagittal deviation $(\Delta x_k')_{sag} = 0.0051\,\text{mm}$, corresponding to 22.4 and 11.2 μm for the respective circles of confusion. The calculated total Seidel diameter (sph. aberration + coma + astigmatism) in the sagittal direction is 26.8 μm+10.2 μm+1.2 μm= 38.1 μm for the paraxial image plane. Figure 3.25 shows the corresponding spot diagram. The respective image area was therefore scaled in the horizontal axis to 38 μm. In the meridional direction the lens delivers in the same plane an ellipse diameter of 52.9 μm.

Some raytrace programs such as WinLens 3D deliver the image curvature radii or the focal distances instead of the circles of confusion. If, however, one needs the circles of confusion one has to calculated them from the given parameters. For that reason we also provide the necessary equations for completeness. We use Eq. 3.45, as already mentioned in connection to the Petzval radius:

$$\Delta_P = \frac{y^2}{2} \cdot \frac{1}{r_p}.$$

We replace the axis distance y in the image plane by the focal length and the image angle $y = f \cdot \tan \omega_1$ and obtain for the deviation of the Petzval sphere from the Gauss plane

$$\Delta_P = \frac{f^2 \cdot \tan^2 \omega_1}{2} \cdot \frac{1}{n_1' \cdot f} = \frac{(100\,\text{mm} \cdot \tan 2°)^2}{2} \cdot \frac{1}{150\,\text{mm}} = 0.0406\,\text{mm}.$$

The longitudinal parts Δ_S and Δ_T can be calculated from the previously determined circles of confusion using the lens f-number:

$$\Delta_S = 2 \cdot (\Delta x_k')_{sag} \cdot F\# = 2 \cdot 0.0051\,\text{mm} \cdot 10 = 0.1016\,\text{mm}$$

$$\Delta_T = 2 \cdot (\Delta y_k')_{mer} \cdot F\# = 2 \cdot 0.0112\,\text{mm} \cdot 10 = 0.2236\,\text{mm}.$$

For the radii of curvature of the astigmatic image spheres we find thus analogously

$$r_P = n_1' \cdot f = 1.5 \cdot 100\,\text{mm} = 150\,\text{mm}$$

$$r_S = \frac{f^2 \cdot \tan^2 \omega_1}{2} \cdot \frac{1}{\Delta_S} = 60.000\,\text{mm}$$

$$r_T = \frac{f^2 \cdot \tan^2 \omega_1}{2} \cdot \frac{1}{\Delta_T} = 27.273\,\text{mm}.$$

This is the arrangement (1) of the image field sphere in Fig. 3.19 for the biconvex lens.

3.11.2 Longitudinal and Transversal Aberrations

The most common representation of image error used for the evaluation of optical systems is the so-called transverse ray aberration as a function of incidence height, i.e., the radius of the entrance pupil. The transverse aberration shows the effect in the corresponding image and can be easily be calculated from the longitudinal aberration using the f-ratio of the system. To illustrate the transverse aberration, the lateral deviation $\Delta x'$ or $\Delta y'$ of a bundle of rays from the ideal intersection point

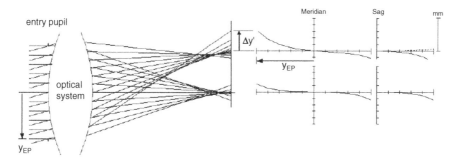

Fig. 3.26 Transverse aberration for two object angles in meridional $\Delta x'$ and sagittal direction $\Delta y'$ versus height of the entrance pupil (T. Thöniß, "Abbildungsfehler und Abbildungsleistung optischer Systeme", LINOS Photonics GmbH)

as a function of the entry pupil coordinates $\overset{\circ}{\rho}^2 = \overset{\circ}{x}^2 + \overset{\circ}{y}^2$ for the meridional and sagittal section plane is given. As a reference the Gaussian plane is usually chosen with the wavelength as an additional parameterization. The functional dependence of the aberrations is then shown for different object angles, as in the sample plot (Fig. 3.26). An ideal point image would therefore provide a horizontal line in both intersecting planes. From the characteristic shape of the functions one can determine the aberrations qualitatively and quantitatively in the existing lens system. From the above functions one can also estimate the quantitative aberrations of the lens system. In addition, from the form of these functions one can also see the composition of the entire aberration.

3.11.3   Field Aberrations

We had already found that image aberrations can be distinguished in on- and off-axis components. For the presentation of the off-axis aberrations one can plot the aberration against the image height or field angle. For example, in WinLens3D one finds the image curvature or the astigmatism, the distortion and the lateral chromatic aberration in the field aberration plot. For the field curvature, the longitudinal spacing of the meridional, sagittal and Petzval surface from the Gaussian plane are shown. The distortion plot includes the relative deviation from the nominal coordinate (vertical central line). To represent the transverse chromatic aberration, one proceeds as follows: One chooses a reference wavelength (e.g., 587 nm, green) and measures for this wavelength the spot position as a function of image height. This procedure is repeated for other wavelengths. Finally, one forms the difference of these image heights for these wavelengths. Figure 3.27 shows an example of the field aberrations for an achromatic lens. The field angle and image height are indicated on the ordinate of the plot.

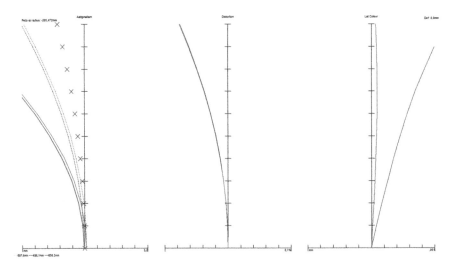

Fig. 3.27 Depiction of the field aberrations in WinLens3D. *Left*: The intersection length differences of the meridional and sagittal field curvature from the ideal image plane for three wavelengths. In addition, the Petzval sphere is presented. *Right*: Transverse chromatic aberration for 486 nm (*blue*) and 656 nm (*red*) against the central wavelength (ordinate)

3.12 Possibilities for the Correction of Aberrations

We note that in Sect. 3.5 the Seidel aberrations arise on the one hand from the trigonometric functions in Snell's Law and on the other hand from the curvature of the spherical lenses.

The chromatic aberration arises, however, from the wavelength dependence of the material refractive index and can be corrected by a combination of two lenses of different material and refractive powers with opposite sign, as already discussed in Sect. 3.6. Lens combinations can also be used for the correction of the Seidel aberrations. It is not possible to eliminate all Seidel aberrations simultaneously with only a few lenses. The number of the refracting surfaces or lenses increases with the desired number of image errors to be corrected. The possibilities of meaningful lens combinations are extremely wide and at first difficult to understand. In the past each successful photographic lens system was thus assigned a special name during its development. In the following we discuss the main tracks to correct optical lens systems. We point out that highly corrected optical systems are generally special solutions, i.e., they are tailored to the problem. New developments cause enormous costs due to individual design and manufacturing. However, in many cases spectrograph developers can also rely on simple systems (e.g., achromatic lenses for collimators) or standard lenses from photographic applications. The challenge lies in the analysis of the limits and possibilities that arise with simple optical systems.

3.12.1   Spherical Aberration

From the Seidel considerations of spherical aberration it became apparent that an object point on the optical axis, which is imaged by a spherical lens, does not provide a point-like image but a circle of confusion in the paraxial image plane. With increasing incidence height the rays are deflected more and more and thus the focal length (intersection width) becomes shorter. The lens refraction power also increases with increasing incidence height. Thus, the lens does not have one, but rather infinitely many focal points. Mathematically, spherical aberration can be expressed by the longitudinal intersection difference $\Delta s'_i = s'_i - s'_0$, with i the respective index of the height of incidence $h_i \ (= \overset{\circ}{\rho}_i)$. For a positive lens this results in a negative aperture error and one refers in this context to an "undercorrected" aperture aberration. Figure 3.28 shows the conditions for a converging and a diverging lens. For the diverging lens the image-side focal length becomes smaller with increasing h. The value is positive because of the just-mentioned definition of the longitudinal aberration. The lens is "overcorrected" for spherical aberration. It can also be seen in Fig. 3.28 that the plotted intersection differences run in opposite directions. To influence or correct the spherical aberration, we have the following opportunities:

- **Setting a system diaphragm (baffling).** Spherical aberration can initially be very easily affected by an aperture stop. Because of the strong dependence of the height of incidence ($\propto h^3$ or $\overset{\circ}{\rho}^3$, see Sect. 3.7.2), one can drastically reduce the aberration by stopping down the aperture of the system, i.e., the system f-ratio becomes correspondingly large. Unfortunately, this also drastically reduces the system light power and also leads to reduced spectral resolving power. In Chap. 6 we will examine this possibility in more detail for the case of a doublet in an echelle spectrometer.
- **Aspheric surfaces.** Another way to correct for spherical aberration is by surfaces of aspheric form, as for parabolic mirrors. The correction, however, can only be carried out for one single object distance. If the lens is used for another distance, some residual aberration remains. Most aspheric lenses are designed

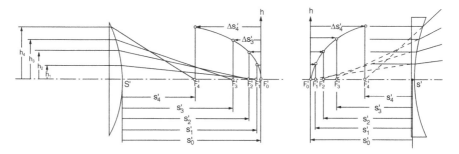

Fig. 3.28 Spherical aberration in a converging and a diverging lens. The object is on the left for both lenses (Reiner 2002, © Books on Demand GmbH)

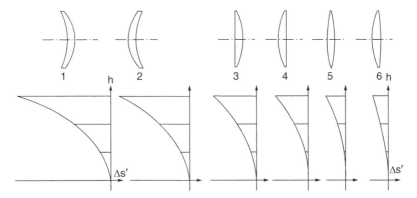

Fig. 3.29 Spherical aberration of collective lenses of equal power but different shape (lens curvature). Incidence height versus intersection difference. On the right is the converging lens of "best form" (smallest spherical aberration) (Reiner 2002, © Books on Demand GmbH)

for an infinite object distance, like almost all achromats commercially available. For the development of a spectrograph this has a certain advantage.

- **Lens bending.** Spherical aberration depends not only on the system aperture but also on the curvature of the lenses. Figure 3.29 shows the spherical aberration depending on the lens shape. The largest aberration can be found for the meniscus lens (1 and 2). For plane and biconvex lenses (3–5) the aberration is significantly reduced. Interestingly, one can choose a bi-convex lens with different radii (6), with minimum spherical aberration ("best form").

  The ratio of the radii depends on the refractive index of the glass. According to Menn (2004) this occurs for a refraction index $n = 1.5$

$$r_1 : r_2 = \frac{7}{12} f : -3.5 f. \tag{3.48}$$

For a focal length of $f = +100$ mm we obtain the radii $r_1 = 58.33$ mm and $r_2 = -350,00$ mm. The ratio of the radii is therefore $1 : (-6)$, that is, the radius of the second surface is considerably greater than that of the first surface. A plano-convex lens with its convex surface facing the object already has a relatively favorable spherical aberration.

Note to coma: Since the coma is directly dependent on the spherical aberration, one can expect that this off-axis error behaves similarly on lens deflection. The two meniscus lenses (1 and 2) in Fig. 3.29 (object at infinity) therefore also show large coma. An single almost plano-convex lens can eliminate coma with respect to a certain object distance. A biconvex lens of the "best form" therefore shows the lowest spherical aberration and very small coma.

- **Power Combinations.** As already indicated in the introduction of this chapter, the signs (+ or −) of spherical aberration in collective and dispersive lenses are different. The usual procedure to eliminate this aberration is, therefore, a

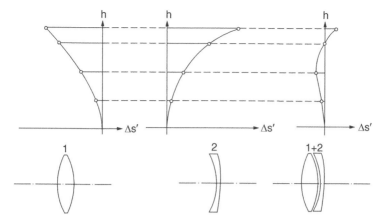

Fig. 3.30 Power combination: Correction of spherical aberration by two lenses of opposite power. The spherical aberration disappears only for a certain height of incidence, i.e., a zone error remains (see total intersection difference on the right) (Reiner 2002, ©Books on Demand GmbH)

combination of two lenses of opposite refractive power. The undercorrection of the collective lens can be corrected by the overcorrection of a diverging lens (see Fig. 3.30). The sums of the radii of the two lenses should not be chosen to be equal, as this would lead to the effect of a plane parallel plate without any aberration for axis parallel incident rays, but also without any imaging effect. One therefore complements a biconvex lens with radii of lowest spherical aberration with a diverging meniscus lens of smaller refractive power. The spherical aberrations compensate each other without having lost the dioptric power of the overall system as far as possible. The spherical aberration varies depending on the incidence height h. One refers in this context to the so-called zone error. The lens combination for a given h is then free of aberrations. The advantage of this method lies also in the fact that the chromatic aberration can be eliminated in the same way. A typical application is the achromatic doublet, in which chromatic and spherical aberrations for the object distance $s_1 = -\infty$ are minimized.

- **Lens splitting.** Instead of one single lens, one can replace it by several lenses of the same total power as the single lens, in order to reduce the angle of incidence on the corresponding refractive surfaces, in particular for large incidence heights (Fig. 3.31).

- **Lens distances.** When allowing variable distances between different lenses of an optical system, one has an additional free parameter for the correction. Simple achromatic lenses typically consist of two cemented lenses. With an air gap between the two components, the correction degree can be significantly increased.

- **Symmetry.** If the optical system is symmetrically constructed out of two parts with the aperture diaphragm in the center of symmetry, asymmetrical aberrations

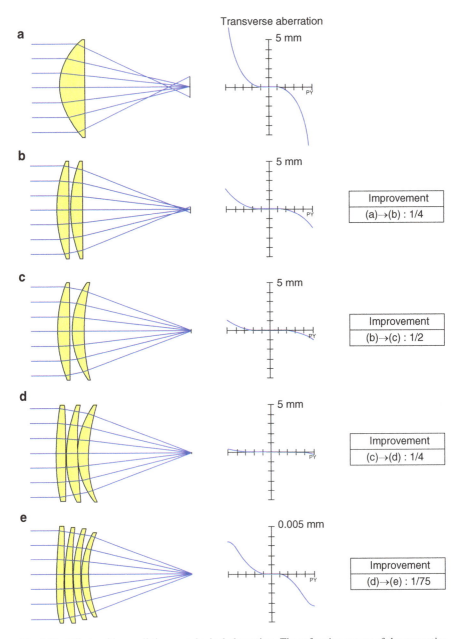

Fig. 3.31 Effects of lens splitting on spherical aberration. The refractive power of the respective overall system remains constant. In the aberration diagrams the transverse spherical aberration is plotted against the height of incidence (Gross et al. 2007)

cancel out. Thus one can particularly influence coma and distortion. Symmetrical combinations are therefore particularly used in systems of high light power. An example for a symmetric system is the well-known Cooke triplet.

3.12.2 The Effects of Aperture Position

In Sect. 3.7.1 we have already noted that for infinite object distance, the axial aperture position (distance between the stop and the first refractive surface $\overset{\circ}{z}$) can have a strong impact on the image aberrations. Thus, the aperture can be positioned in front of or behind the lens. If the lens is the aperture itself (lens mount) all aberration contributions caused by the aperture position vanish according to Eq. 3.39 because $\overset{\circ}{z} = 0$. Depending on the design of the optical system, it may be particularly necessary to eliminate or minimize perturbing aberration components; for example, if coma provides a greater contribution due to the system aperture. Considering Eq. 3.28 reveals the dependency of the Seidel coefficient from the parameter ζ and thus from the location of the entrance pupil of $\overset{\circ}{z}$. The position of the aperture as well as the object distance, therefore, play an essential role in the correction of an optical system. To investigate the influence of the aperture location on the image aberrations, we consider the equations mentioned previously in more detail and examine in particular spherical aberration, coma and astigmatism.

The Seidel sum of spherical aberration introduced by k refracting surfaces

$$\sum_{\nu=1}^{k} I_\nu = \sum_{\nu=1}^{k} A_\nu$$

is independent of ζ. Thus, spherical aberration can apparently not be influenced by the aperture position.

Next, we turn to the elimination of the off-axis errors. For the Seidel sums of coma we obtain from Eq. 3.28

$$\sum_{\nu=1}^{k} II_\nu = 0 = \zeta \cdot \sum_{\nu=1}^{k} A_\nu + \sum_{\nu=1}^{k} B_\nu.$$

Substituting the parameter ζ we obtain

$$\frac{1}{\overset{\circ}{z}} = \frac{1}{s_1} + \frac{1}{n_1} \cdot \frac{\sum_{\nu=1}^{k} A_\nu}{\sum_{\nu=1}^{k} B_\nu}$$

and thus a unique equation for the aperture setting $\overset{\circ}{z}$ at which coma disappears. In this context one refers to the "natural aperture". Apparently the ideal aperture position also depends on the object distance s_1, i.e., the aperture position in a

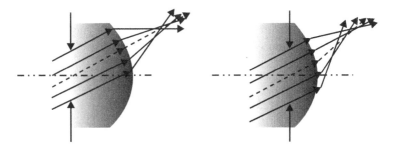

Fig. 3.32 Dependence of coma on the position of the stop. *Left*: the natural stop position. *Right*: a stop position that deviates from the position at left, that leads to coma, due to the asymmetry of the diffracting surface relative to the main beam (here with a *dashed line*). *Source*: T. Thöniß, LINOS Photonics GmbH

photographic lens is a compromise because of the different object distances in reality. In contrast, the predominant object distance in a spectrometer is generally $s_1 = -\infty$ and $n_1 = 1$. Therefore, we obtain specially for spectroscopy

$$\overset{\circ}{z} = \frac{\sum_{\nu=1}^{k} B_\nu}{\sum_{\nu=1}^{k} A_\nu}.$$

To illustrate the natural stop position we consider the following: Rays of off-axis object points reach the optical surfaces under a certain angle. The position of the aperture stop is crucial for the asymmetry of the beams above and below the chief ray[14]) with respect to the curved lens surface. For each optical system one can find the natural aperture position at which the rays are symmetrically refracted at the lens surface with respect to the main beam. For all other aperture positions the beams above and below the different main beam are differently refracted (Fig. 3.32). For this reason coma is often referred to as an asymmetry aberration.

To illustrate the influence of the diaphragm position we consider a symmetrical biconvex lens with coma, that is larger than in a "best form" lens. Figure 3.33 shows the spot diagram of a bi-convex lens with an aperture of $\overset{\circ}{z} = 0$ mm. Calculation of the Seidel coefficients delivers $\sum A_\nu = 3.18 \cdot 10^{-6}$ and $\sum B_\nu = 1.26 \cdot 10^{-4}$. From this one obtains a natural aperture of $\overset{\circ}{z} = +39,6$ mm to compensate for coma. This is illustrated with the spot diagrams in Fig. 3.34. In contrast to Fig. 3.33 here a diaphragm was positioned exactly at the above-calculated distance $\overset{\circ}{z} = +39,6$ mm. Coma has been completely eliminated in the lower off-axis series. In the practical application of the natural stop position, it is advantageous if the Seidel coefficient for spherical aberration $\sum A_\nu$ is not too small, so that the diaphragm does not have to be placed too far away from the lens.

---

[14]A chief ray is defined as the beam starting from an off-axis point through the center of the aperture stop.

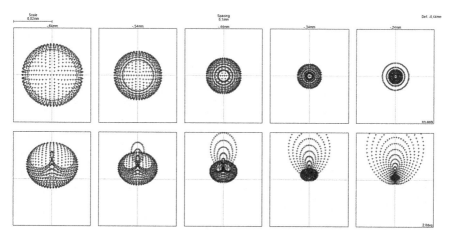

Fig. 3.33 Spot diagram of a bi-convex lens made of BK7 with $f = 100$ mm on the optical axis (*top row*) and below for $\omega_1 = 2°$ (*bottom row*). The spot diagram shows a superposition of spherical aberration, coma and astigmatism

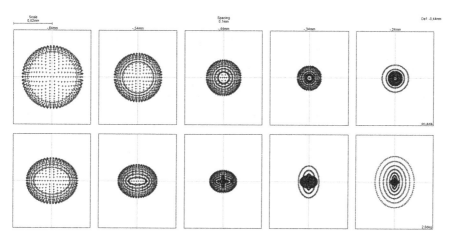

Fig. 3.34 As in Fig. 3.33 but now with elimination of the aberration by the stop position

For the compensation of field curvature and astigmatism there are four different solutions due to the defining equations 3.28, depending on which of the image errors should be eliminated. Without proof we give the solutions again for our important special case $s_1 = -\infty$ and $n_1 = 1$.

• Elimination of meridional field curvature:

$$\overset{\circ}{z} = \frac{1}{\sum_{\nu=1}^{k} A_\nu}\left[\sum_{\nu=1}^{k} B_\nu \pm \sqrt{(\sum_{\nu=1}^{k} B_\nu)^2 - \sum_{\nu=1}^{k} A_\nu \left(\sum_{\nu=1}^{k} \Gamma_\nu + \frac{1}{3}\sum_{\nu=1}^{k} P_\nu\right)}\right]$$

- Elimination of sagittal field curvature:

$$\overset{\circ}{z} = \frac{1}{\sum_{\nu=1}^{k} A_\nu} \left[\sum_{\nu=1}^{k} B_\nu \pm \sqrt{(\sum_{\nu=1}^{k} B_\nu)^2 - \sum_{\nu=1}^{k} A_\nu \left(\sum_{\nu=1}^{k} \Gamma_\nu + \sum_{\nu=1}^{k} P_\nu \right)} \right]$$

- Elimination of mean field curvature:

$$\overset{\circ}{z} = \frac{1}{\sum_{\nu=1}^{k} A_\nu} \left[\sum_{\nu=1}^{k} B_\nu \pm \sqrt{(\sum_{\nu=1}^{k} B_\nu)^2 - \sum_{\nu=1}^{k} A_\nu \left(\sum_{\nu=1}^{k} \Gamma_\nu + \frac{1}{2} \sum_{\nu=1}^{k} P_\nu \right)} \right]$$

- Elimination of astigmatism:

$$\overset{\circ}{z} = \frac{1}{\sum_{\nu=1}^{k} A_\nu} \left[\sum_{\nu=1}^{k} B_\nu \pm \sqrt{(\sum_{\nu=1}^{k} B_\nu)^2 - \sum_{\nu=1}^{k} A_\nu \cdot \sum_{\nu=1}^{k} \Gamma_\nu} \right]$$

From these characteristic solutions for the iris position $\overset{\circ}{z}$, one can see that there are three different possibilities for the elimination of the corresponding image aberration (depending on the radicand R):

1. $R > 0$ yields two real solutions and two possible aperture positions
2. $R = 0$ yields a real solution or diaphragm position
3. $R < 0$ yields no solution or diaphragm position

If the case $R = 0$ can be realized, not only does the corresponding field curvature or astigmatism disappear, but so does also the coma. In Fig. 3.35 the image curvatures T (tangential or meridional) and S (sagittal) are illustrated for a plano-convex lens and three field stop positions. In Fig. 3.36 we show the spot diagrams valid for Fig. 3.35 and $\overset{\circ}{z} = 0$ mm and in Fig. 3.37 the corresponding spot diagrams for $\overset{\circ}{z} = +32$ mm.

In the bottom row of this last figure (off-axis), only spherical aberration is visible. Coma and astigmatism were eliminated by the stop position.

3.12.3 Removal of Petzval Field Curvature Through a Field Flattener

We now examine how to correct the Petzval field curvature by an additional lens. According to Eq. 3.44 the radius of the Petzval image surface is

$$\frac{1}{r_p} = \sum_{i=1}^{k} \frac{1}{n_i f_i}.$$

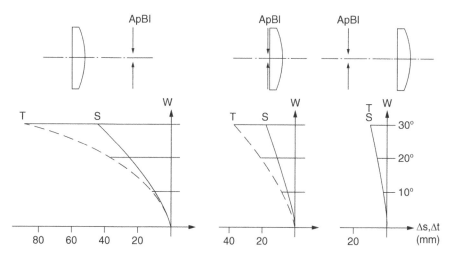

Fig. 3.35 Effects of different aperture positions on the astigmatism of a plano-convex lens with $f = 100$ mm. The flat side faces toward the object. In the example on the right side the meridional T and sagittal image sphere S are identical, the astigmatism is eliminated and only the Petzval radius is valid ($w = \omega_1$) (Reiner 2002)

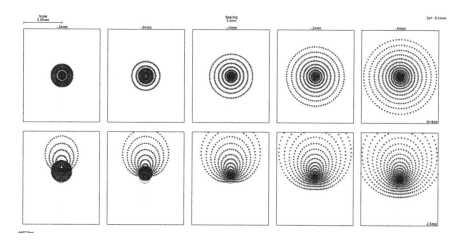

Fig. 3.36 Effects of the aperture position on the spot diagram of the BK7 plano-convex lens of Fig. 3.35 ($f = 100$ mm) whose flat side faces towards the object. The diaphragm position is $\overset{\circ}{z} = 0$ (Fig. 3.35, *center*). The spot diagram shows spherical aberration, coma and astigmatism superimposed. The radii of curvature are $r_T = 34$ mm, $r_S = 73$ mm and $r_P = 161.7$ mm

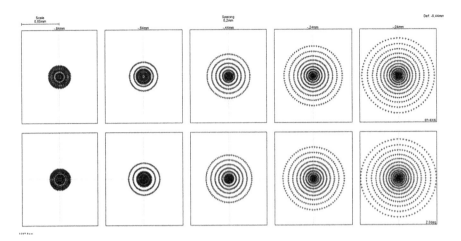

Fig. 3.37 Spot diagram of the plano-convex lens shown in Fig. 3.35 with an aperture position of $\overset{\circ}{z} = +32$ (Fig. 3.35, *right*). The curvature radii here are identical for all three cases $r_P = 161.7\,\text{mm}$. Coma and astigmatism are not present in the spot diagram

If the focal length of the lens to be corrected is f_1 and that of the correction lens is f_2, the total focal length for a composite system of two lenses (see e.g. Hecht 2003) is

$$\frac{1}{f} = \frac{1}{f_1} + \frac{1}{f_2} - \frac{d}{f_1 f_2} \tag{3.49}$$

with d the distance between these two lenses. If we now choose specifically $f_2 = -f_1$ and $n_1 = n_2 = n$, we obtain for the total focal length

$$\frac{1}{f} = \frac{1}{f_1} - \frac{1}{f_1} + \frac{d}{f_1^2}$$

or

$$f = \frac{f_1^2}{d}.$$

The total focal length of such a system (see Fig. 3.38) is positive, independent of the sign of f_1. The Petzval radius is then

$$\frac{1}{r_p} = \frac{1}{nf_1} - \frac{1}{nf_1} = 0 \tag{3.50}$$

or $r_p = \infty$. A correction lens with this function is called Piazzi Smyth lens after its inventor (Piazzi Smyth 1874). If the primary lens is already a multi-lens system consisting of positive and negative refractive powers, one can significantly reduce

Fig. 3.38 Correction of the Petzval field curvature by lenses of equal power and refractive index

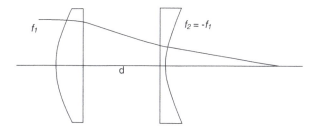

Fig. 3.39 Curved image plane of the Kepler space telescope to correct for field curvature. *Source*: http://www.nasa.gov/mission_pages/kepler/multimedia/images/kepler-focal-plane-assembly.html

or even eliminate the Petzval sum by a convenient choice of the refractive powers. Alternatively, one can as well eliminate the Petzval curvature by an additional Smyth correction lens. The aberrations thereby introduced are low due to the small distance to the image plane. Another way to eliminate the field curvature is shown in Fig. 3.39. In the focus of the so-called Kepler telescope (1.4 m Schmidt system) is an array of 42 CCDs, covering a field of view of $105° \times 105°$. Each CCD sensor has a size of $50 \, mm \times 25 \, mm$ with a resolution of $2{,}200 \times 1{,}024$ pixels (a total of 95 Mega pixels). The CCD array is curved to compensate for the field curvature.

3.12.4   An Achromat Is Necessary

All previously discussed aberrations depend on how the refractive index depends on wavelength. The calculations were carried out so far for a wavelength of 587.6 nm. To demonstrate the effect of chromatic aberration on the Seidel aberrations we consider in Fig. 3.40 the spot diagrams of an $f/10$ plano-convex lens made of BK7 with 100 mm focal length for three different wavelengths and the same for an achromat (Fig. 3.41 of the same focal length. Because of the smaller figures of confusion for the achromat a scaling of 40 μm was chosen. On-axis, the achromatic lens already provides diffraction-limited figures of confusion. For an object angle of $\omega_1 = 2°$ the Airy disk with an average size of $2.44 \cdot 10 \cdot 0.55 \, \mu m = 13.4 \, \mu m$ is already exceeded, though, for the whole wavelength range. The situation can be

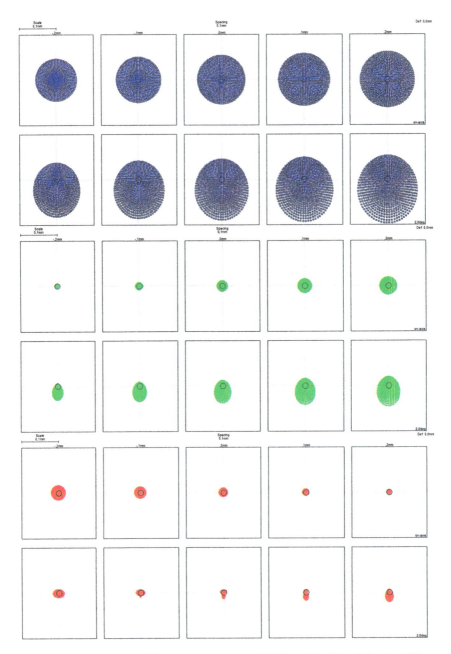

Fig. 3.40 Spot diagrams of a $f/10$ plano-convex lens with 100 mm focal length for three different wavelengths (from *top* to *bottom*: 486.1, 587.6 and 656.3 nm). The object angles were $\omega_1 = 0°$ (*top row*) and $\omega_1 = 2°$ (*bottom row*). The corresponding center spot is located in the Gaussian image plane; the defocus steps are 0.1 mm. While the paraxial red spot shows a nearly diffraction-limited diameter of only 11 μm, the diameter of the blue spots is about 130 μm (the corresponding image section is 200 μm). Achromatization of the lens is hence essential in particular for broadband spectrometer applications

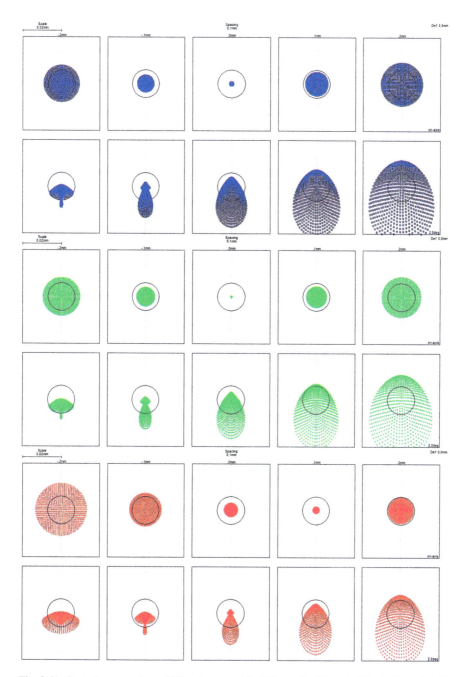

Fig. 3.41 Spot diagrams of an $f/10$ achromat with 100 mm focal length. The horizontal field scaling is 40 μm for the full image. Note that the geometrical scale for the achromatic lens is five times smaller than in Fig. 3.40

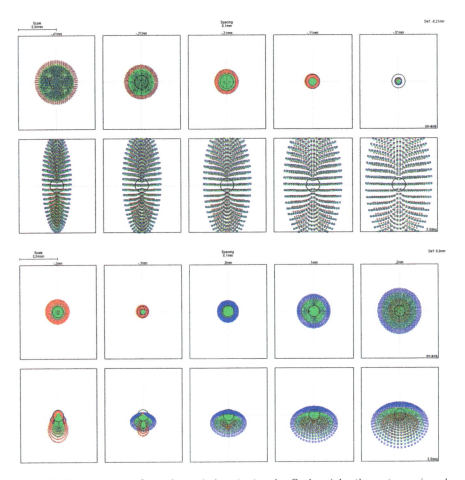

Fig. 3.42 Spot diagrams of an achromatic lens (*top*) and a Cooke triplet (*bottom*) on-axis and for an object angle $\omega_1 = 5°$. All wavelengths are displayed simultaneously in the spots. The defocus steps are again 0.1 mm. None of the lenses is diffraction-limited. If one accepts, however, dispersion figures of about 20–25 μm diameter one can use the triplet for the task. As one can see from the figures of confusion of the triplet, the wavelength-dependent spots still significantly differ. The chromatic aberration of the Cooke triplet is corrected at two wavelengths and thus comparable to the achromat

somewhat improved by a slight defocusing of -0.1 mm, as is visible in the column left of the Gaussian image plane. If we take a half-angle of 2° again, a CCD size of about 7 mm arises for the chosen focal length of 100 mm, in which a spectrum can be mapped almost without aberration. For a focal length twice as long, $f = 200$ mm used in the commercial Lhires spectrograph of the Shelyak company, one hence obtains 14 mm CCD size, i.e., a KAF 1602 sensor can be used.

Finally, we increase the requirements and increase the off-axis angle to $\omega_1 = 5°$, while the f-ratio remains at $f/10$. We compare the spot diagrams of the achromat and a simple Cooke triplet, each with 100 mm focal length (Fig. 3.42). The image sections of the spots were re-adjusted to 40 μm.

3.12.5   Example Considerations for a Commercial Spectrograph

To implement the above considerations, we examine the optical performance of the commercial long-slit spectrograph LHIRES III (Shelyak Instruments, Sect. 8.2.3). This system works with an achromatic lens as camera/collimator in Littrow configuration. To investigate the usable spectral range we used a Celestron C14 telescope and mounted a Sigma-CCD camera (Astroelektronik Fischer) with a KAF6303 chip into the spectrograph. The chip width is 27.65 mm (3,072 pixels of $9\,\mu m$). To decrease the size of the seeing disk the f-ratio of the telescope was reduced by a 0.63 focal reducer to $f/7$. To adjust the spectrograph achromat to this f-number, the standard achromat of 30 mm aperture was replaced by one of 40 mm. The slit width of $23\,\mu m$ corresponds to a seeing of 2 arcsec in the sky and the grating used had a ruling of 1,200 lines/mm. With this configuration we obtained a neon calibration spectrum (Fig. 3.43). The corresponding spectral resolving power was determined for each line by evaluating the FWHM. It is about 973 at 5,850 Å (line A), about 4,459 at the central wavelength of 6,275 Å (line B) and about 1687 at 6,675 Å (line C) (in Fig. 3.43 this is visible in the line widths). The spectral resolving power is thus reduced beyond the central wavelength. To find the cause of this spectral behavior, we consider the aberrations in the form of Seidel parameters (Table 3.6) and the spot diagrams (Fig. 3.44) of the 40 mm achromat (Edmund Optics #47-741-INK). The focal length of the achromat is also 200 mm. The corresponding

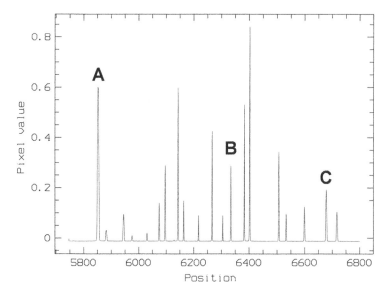

Fig. 3.43 Neon calibration spectrum taken with Lhires. For three lines we determinded the following spectral resolving power: $R_A = 973$, $R_B = 4,459$ (maximum) and $R_C = 1,687$

Table 3.6 WinLens Seidel table for the Edmunds achromat #47-741-INK at $f/5.5$ and 3.2° object angle corresponding to 5,850 Å

	5,850 Å	4,400 Å	6,280 Å	5,300 Å	6,800 Å
Trans Spherical Abn	0.0073	0.0090	0.0071	0.0077	0.0068
Best Axial Defocus	0.0686	0.0841	0.0664	0.0726	0.0642
Sagittal Coma	0.0025	0.0039	0.0023	0.0029	0.0020
Tangential Coma	0.0074	0.0117	0.0068	0.0086	0.0061
T focus Distance	−1.1152	−1.1354	−1.1120	−1.1206	−1.1089
S focus Distance	−0.5231	−0.5302	−0.5219	−0.5250	−0.5207
Petz surf Distance	−0.2270	−0.2277	−0.2268	−0.2273	−0.2266
T surf Radius	−56.0745	−55.0779	−56.2349	−55.8028	−56.3927
S surf Radius	−119.5532	−117.9346	−119.8246	−119.1017	−120.0961
Petz surf Radius	−275.4819	−274.6597	−275.7085	−275.1667	−275.9685
Fractional Distortion	−0.0089 %	−0.0049 %	−0.0095 %	−0.0079 %	−0.0101 %
Chromatic focal shift	−0.2773				
Lateral Chromatic	0.0001				
Secondary Spectrum	−0.3453				
H (Lagrange Invariant)	0.895				

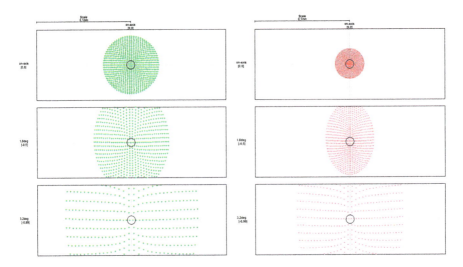

Fig. 3.44 Spot diagrams of the Edmunds achromat #47-471-INK at 585 nm (*left*) and 627.5 nm (*right*)

spots were calculated using the raytrace program WinLens 3D. For the half chip length of 13.825 mm we obtain the (half) field angle arctan(13.825 mm/200 mm) = ±4° for the edges of the spectral range by calculating the inverse trigonometric function.[15] The field angle can be used as an object angle in the raytrace programme because object and image angles are nearly identical for a simple achromat. When calculating the necessary minimum aperture of the achromat (Sect. 4.1.4) we obtain $f/5.5$ as f-ratio. The highest spectral resolving power can be found at 627.5 nm. At this wavelength position the Airy disk of the achromat has size

$$2.44 \cdot 0.6275\,\mu m \cdot 5.5 = 8.4\,\mu m.$$

If one uses this circle of confusion to estimate the spectral resolving power according to Eq. 2.98, we obtain $R = 6,740$. Since the measured resolution is only 4,459, we must assume a defocusing of the optics, which additionally increases the circle of confusion along the spectrum. To obtain the above resolving power of $R = 4,459$, the circle of confusion must be 29 μm. If one calculates the spot diagrams with WinLens 3D one obtains a necessary defocus of +0.21 mm for an on-axis spot of 29 μm diameter. The sagittal diameter of the circle of confusion (dispersion direction) is the reason for the spectral resolving power. This diameter consists of individual image aberrations, as already shown in Sect. 3.7. The diameters of spherical aberration and coma are shown in Table 3.6, but not the astigmatic diameter. Therefore, we continue to consider the sagittal field curvature, i.e., the distance of the field plane to the Gaussian plane, as specified in the table. According to this table, we obtain a sagittal focal line distance of $\Delta_S = 0.5231$ mm for the spectral range at 5,850 Å. For the spot diameter of the astigmatic circle of confusion we divide by the f-ration and obtain 0.5231 mm/5.5 = 0.095 mm. Since all aberrations must be added, we also take the calculated contribution of spherical aberration with 0.0073 mm into account. In addition, we add the coma contribution of 0.0039 mm from the table. Thus we obtain for the sagittal ellipse diameter of the present astigmatism 0.095 mm+0.0073 mm+0.0039 mm = 106.2 μm. Defocusing provides a further enlargement of 0.21 mm/5.5 = 38.2 μm, so that the overall result is a diameter of approximately 145 μm. If we use this diameter we obtain $R = 1,007$ at 5,850 Å for the spectral resolving power.

The resulting spot diagrams are shown in Fig. 3.44 for 6,275 Å (on-axis) and 5,850 Å with the associated object/image angle (Littrow) of 3.2°. For the measurement, line C provides a resolving power of 1687 at 6,675 Å. The off-axis angle is about 3.1° and introduces a sagittal circle of confusion of about 99 μm. Using this value one obtains $R = 1,693$. Overall, a satisfactory consistency of measurement and calculation is present. The achromat provides a sufficiently small spot for all wavelengths up to an image field angle of 2° so that a KAF1602 chip is possible for the Lhires. The spot diagram in Fig. 3.45 shows the corresponding circles of confusion on-axis and for a field angle of 2° as a function of different focus positions.

---

[15]At this point we do not consider the chip diagonal.

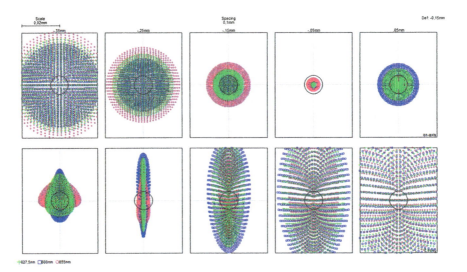

Fig. 3.45 Spot diagrams of the achromat at three wavelengths (6,000, 6,275, 6,550 Å) of a Lhires spectral range of 55 Å when using a KAF1602 CCD. *Top row*: On-axis spots for different defocus positions. *Bottom row*: Spots for the field angle $\omega_1 = 2.0°$. *Center column*: Spots for the best focus position. For both cases the horizontal (sagittal) diameter is about 22 μm and is therefore still sufficiently small not to degrade the spectral resolving power. The achievable resolving power is 5,100 for the entire wavelength range. The off-axis figure of confusion has a relatively large meridional diameter. However, it affects only the direction perpendicular to the dispersion

3.13   Resume

In summary, it can be seen that corrections of an optical system are complex and extremely varied and therefore not easy to understand at first. A correction strategy will always have to be oriented to the specific task. The main distinguishing features are the system aperture or f-ratio and the object angle ω_1. For very small off-axis angles, it will be sufficient to correct the field curvature and spherical aberration. If the off-axis aberrations dominate, the diaphragm position can be used to eliminate coma and astigmatism at the same time. If the requirements for simultaneous correction are not met by the particular optical system one needs to accept a compromise. Other options include three-lens systems, such as the triplet in Fig. 3.46 or especially symmetric systems (double Gauss types) for the correction of coma. In the history of the development of photographic lenses a wide range of solutions have emerged that have received separate names in most cases. Examples are Tessar, Sonnar, etc. Figure 3.47 provides information on the ways to correct for a given f-ratio and object angle.

Fig. 3.46 Optical path in a simple $f/10$ Cooke triplet with 100 mm focal length. Front and rear members are made of SK4 (crown glass) and the center glass of F8 (flint glass)

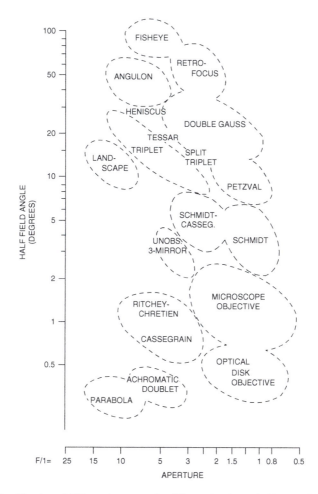

Fig. 3.47 Classification of different lens types for different f-ratios and object angles (W.J. Smith, "Modern Lens Design")

Suggested Reading

- *K. Schwarzschild, Göttingen 1905, "Untersuchungen zur geometrischen Optik. I.", Neunter Theil, Einleitung in die Fehlertheorie optischer Instrumente aufgrund des Eikonalbegriffs, Astronomische Mittheilungen der Königlichen Sternwarte zu Göttingen*
- *K. Schwarzschild, Göttingen 1905, "Untersuchungen zur geometrischen Optik. III.", Elfter Theil, Über die astrophotographischen Objektive, Astronomische Mittheilungen der Königlichen Sternwarte zu Göttingen*
- *H. Haferkorn, Berlin 1986, "Bewertung optischer Systeme", VEB Deutscher Verlag der Wissenschaften*
- *J. Flügge, 1955, Die wissenschaftliche und angewandte Photographie, Band I, Das photographische Objektiv, Springer-Verlag*
- *M. Berek, Berlin 1970, "Grundlagen der praktischen Optik", Walter de Gruyter & Co.*
- *J. Picht, Berlin 1955, "Grundlagen der geometrisch-optischen Abbildung", VEB Deutscher Verlag der Wissenschaften*
- *A.E. Conrady, 1992, "Applied Optics and Optical Design", Part One, Dover Publication Inc.*
- *W.J. Smith, 2000, "Modern Optical Engineering", McGraw-Hill*
- *M. Born, E. Wolf, 1999, "Principles of Optics: Electromagnetic Theory of Propagation, Interference and Diffraction of Light", Cambridge University Press*
- *W.T. Welford, 1986, "Aberrations of Optical Systems", Inst of Physics Pub*
- *M.J. Kidger, 2002, "Fundamental Optical Design", SPIE Press*
- *W.J. Smith, 2005, "Modern Lens Design", McGraw-Hill*
- *R. Kingslake, 2010, "Lens Design Fundamentals", McGraw-Hill*
- *J. Reiner, 2002, "Grundlagen der Ophthalmologischen Optik", Books on Demand*

Chapter 4
Considerations About the Standard Spectrograph Layout

A Short Story

We received our first optical grating as a gift and we started to construct our first spectrograph full of joy and happiness. After completing the task we implemented our booty onto the rotary disk und started with first measurements. We knew fairly well what signal strength to expect but were quite astonished to see a signal that was far too small. The reason was a complete mystery to us so we started to find out why. First, we found out that neither the optical elements nor the CCD camera were potential sources. Eventually Klaus suggested the grating to be the problem. He then discovered that the grating was optimized for ultraviolet light and that it was simply inserted up-side-down.

4.1 Basic Remarks and Requirements

The design of a spectrograph is no task for a single afternoon. Different optical possibilities and other parameters should be taken into account depending on the telescope or fiber optics (Chap. 11) and CCD camera to be used. One can, of course, use optical and mechanical elements which are already available off-the-shelf to keep the spectrograph costs low. But in any case, the design plan should include a proper optical path as well as a mechanical support in form of a reasonable containment which will be mounted to the telescope if fiber optics are not taken into account. And depending on the telescope one needs to take the loading capacity of the mounting for the spectrograph into account (for instance, the High Dispersion Spectrograph (HDS) at Subaru telescope weights 6 tons—see Sect. 8.4.5). It is, hence, important to first discuss the maximum mechanical dimensions and weight of the spectrograph. For big-size telescopes this is done during respective design phases in a planning and management process. For small telescopes below ~1 m

© Springer-Verlag Berlin Heidelberg 2015
T. Eversberg, K. Vollmann, *Spectroscopic Instrumentation*, Springer Praxis Books,
DOI 10.1007/978-3-662-44535-8_4

aperture we recommend to build first a mock-up, e.g., out of wood, even if more work is required. Design changes and technical modifications are then far easier to perform than for versions out of metal. We also recommend 1-to-1 drawings for the optical configuration and to test the position of all components on this drawing. The design of a small spectrograph is a kind of iteration process which should converge in reasonable steps to the final realization. In principle, one can calculate and design any spectrograph by pocket calculator and pencil. Today, though, various design tools (ray-tracing software) are available to quickly develop an optical layout. On the other hand, all optical paths follow geometrical optics (neglecting grating diffraction) and one can even use tools working with geometrical optics only. In any case, for the optical draft one should use a simulation model in which one can quickly adjust any of the large numbers of free parameters and see their impact on the instrumental output. The simulation model should consist of all necessary input parameters, e.g., telescope size and F-number, all the optical elements inside the spectrograph and the corresponding parameter dependencies. Such a simple but very helpful simulation program has been realised by an Excel sheet, containing all free parameters. This is "Simspec", originally designed by Buil (2003) and further updated by Vollmann (2008). In addition, the realization of a spectrometer necessarily depends on the physical phenomena in the investigated targets and, hence, primarily on the required data quality S/N, exposure times t and spectral resolving power R. If these parameters are known one can choose the respective optical components to match them. Different astrophysical problems often require different parameters. S/N and t are directly connected to each other. R however, is defined by the instrumental setup and one can adjust it with respective optical choices. This can be the size and focal length of collimator and camera as well as an interchangeable grating. For spectral broadband measurements of stars low resolution gratings ($\sim$200 lines/mm) are often sufficient whereas high resolution gratings ($\geq$1,200 lines/mm) are required to perform detailed line profile analysis.

4.1.1   Slit Width and Resolving Power

The design of a spectrograph should always be guided by the observing targets. In Chap. 2 we have shown that the basic parameters of the telescope and spectrograph system are coupled and interdependent. They are

1. The telescope aperture,
2. the brightness of the target objects,
3. the available exposure times,
4. the spectral resolving power and
5. the necessary data quality (S/N).

Table 4.1 Typical target and
system parameters at different
telescopes

Telescope	V	Exp. time	R	S/N
IAC80 0.8 m	~8 mag	~1,800 s	~12,000	~200
DAO 1.8 m	~9 mag	~300 s	~4,300	~100

Simply expressed, the first three bullets reflect the question, how many photons are available for the measurements and the fourth bullet on how many resolution elements are these photons distributed. The S/N as the fifth point in turn depends on the first four factors. Hence, we have to adapt the instrument to the requirements of the respective measurements. First, a large telescope delivers more light for analysis. However, the sheer number of photons is not the quality characteristic for data analysis but the spectral contrast (S/N). This has insofar consequences for the instruments since the signal-to-noise ratio depends linearly on the telescope aperture.[1] To achieve the same S/N for two stars of brightness difference of one magnitude, one needs a telescope diameter difference of factor 2.5. The spectral resolving power depends in turn on the spectrograph itself, as we have seen in Sect. 2.6.3. Here, the system must be harmoniously tuned from the slit to the detector pixels. For the exposure time, however, it is important whether one wants to detect short-term effects with sufficient S/N (e.g., stellar wind phenomena as described in Chap. 14) or large-scale time and location information (e.g., galaxies and redshift). However, the exposure time in turn depends on the brightness of the object or of its photon number. Therefore, the development of a spectroscopic instrument requires the consideration of the objective observation program. Although compromises are possible in the direction of poor data quality, but not in the direction of higher quality data. In our initial question of the number of photons per resolution element there is, hence, no clear answer. It depends individually on the technical and scientific constraints. Table 4.1 exemplary shows the appropriate parameters for two different systems. To what extent the necessary spectral resolving power depends on the scientific question is illustrated in Figs. 4.1, 4.2, and 4.3. All figures show the comparison of real high-resolution spectra of different prototype stars taken by the Ultraviolet and Visual Echelle Spectrograph (UVES) at ESO Paranal with accordingly scaled low resolution spectra. It is apparent that the determination of certain parameters depends on what is to be studied. Only then one can clarify which compromises regarding the above boundary conditions (essentially S/N against R) can be accepted. For the examples ζPup and WR185 in Figs. 4.2 and 4.3 one could casually choose very low spectral resolving power (R ~ 1,000) in favor of a correspondingly shorter exposure time.

---

[1]The S/N does not depend on the telescope area but only on its diameter because photon noise follows Poisson statistics as $\sqrt{N_P}$. Then we have $S/N \sim N_P/\sqrt{N_P} \sim \sqrt{N_P} \sim \sqrt{A_T} \sim D_T$, with the number of photons N_P, the area of the telescope optics A_t and its diameter D_T.

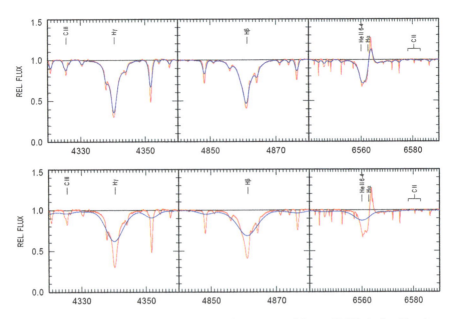

Fig. 4.1 Comparison of a high-resolution UVES spectrum of R = 40,000 (*red*) with a low-resolution spectrum (*blue*) for the A supergiant Deneb = αCyg (A2I). *Top*: R = 5,000. *Bottom*: R = 1,000 (*blue*). In order to obtain useful information about the star, a resolving power of at least 5,000 is necessary because of the narrow absorption lines. A lower resolving power delivers significantly broadened lines (Hamann 2011)

4.1.2   Remarks About the Optical Slit

In stellar astronomy we generally observe point sources. The light wavefront is not only diffracted by a prism or a grating but also by each optical element (see Sect. 2.6.4), hence, also by the telescope. This fundamental behaviour provided by nature implies that we cannot image a stellar point source but rather a diffraction pattern, analogous to diffraction in a prism or a grating. Figures 2.9 and 2.11 depict this behaviour for circular optics. The diffraction intensity can be described by a Bessel function of 1st order (see Appendix B) and one can obtain the diameter of the central diffraction disk. The diffraction effects in a telescope of, e.g., 25 cm aperture provide an Airy disk of about 0.5 arc-seconds (0.5″). The geometrical resolution is defined by the image of the observed object. The resolution defines how separated the two points should be to be detected as two points. One defines that separation is possible if the maximum of the first Airy disk is positioned at the first minimum of the second Airy disk.

The second effect defining the stellar image is atmospheric scintillation (seeing). The origin of this effect is turbulence in the atmosphere on scales of the order of some meters. This distorts the wavefront coming from the star and smears out the stellar image, and on average increases the image diameter. The unit for the half

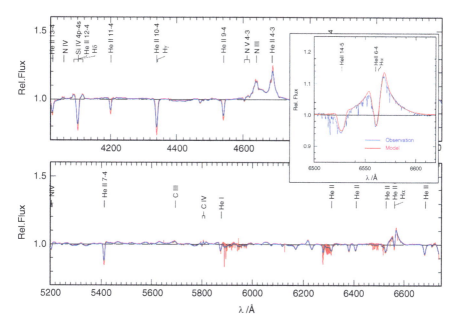

Fig. 4.2 As Fig. 4.1, now for the O-supergiant ζPup (O5Ia) at R = 1,000. All absorption lines of ζPup are strongly broadened because of its rapid rotation ($v \sin i$ = 220 km/s, Schaerer et al. 1997). Broadening is also evident for the He (4,686 Å) and Hα (6,563 Å) emission lines produced in a fast stellar wind of significantly high speeds than 2,000 km/s (Puls et al. 1996, see Sect. 14.7). *Inset*: Comparison of observation (*blue*) with theoretical stellar wind model for the Hα line. Even with a spectral resolution of R = 1, 000, corresponding stellar parameters (e.g., wind speed) can be determined (Hamann 2011)

width of the stellar image (seeing) is arcseconds and is on average about 4″ at urban observing sites. At the best conditions less than 1″ can be achieved. Such places benefit from laminar atmospheric flows and are used for modern observatories (e.g., Chile, Hawaii, Canary Islands). Small telescopes already exhibit diffraction patterns of about 0.5″ and now it is clear that seeing is the main contributor of image distortions and not optical diffraction, assuming optical aberrations are minimal.

As already noted, the image of the object not only defines the geometrical but also the spectral resolution. For the observation of time-dependent line effects it is necessary, however, to work with a fixed and constant spectral resolution. Time dependent fluctuations (seeing) introduce time-dependent variable resolution if the size of the object to be measured is not kept constant in the dispersion direction (the direction where the light is diffracted).

If the image diameter d_2 of the fluctuating seeing disk in the focus of the spectrograph is larger than the stable diffraction effects of all imaging elements, we definitely need an optical slit. That is almost always the case because the seeing disk in the telescope focal plane d_1 becomes larger with the telescope focal length and diffraction effects dominate for focal lengths below 1 m. Observers with telescopes

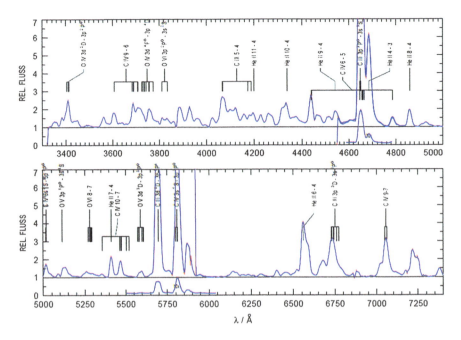

Fig. 4.3 As Fig. 4.1 but now for the Wolf–Rayet star WR185 (WC8) at R = 1,000. In particular, Wolf–Rayet stars with very broad emission lines (Sect. 14.8) very low resolution spectra are sufficient to determine various stellar parameters (Hamann 2011)

of short focal lengths can do without slits.[2] The slit keeps the seeing disk size in the dispersion direction at a constant width by "cutting off" the edges of the stellar image (Fig. 4.4). The slit is, hence, a purely resolution-defining optical element and has to be oriented perpendicular to dispersion direction.

The optical slit defines the spectral resolution.

Figure 4.4 obviously shows that some light in the dispersion direction is lost for data analysis by using a slit. To consider the behavior of the seeing disk on the slit, we have to describe the intensity profile of a stellar image. This profile follows the so-called MOFFAT FUNCTION to good approximation. In contrast to a pure Gauss function it allows for the seeing conditions as well as diffraction effects[3]:

---

[2]Here we do not consider adaptive telescope optics.

[3]The Moffat function (Moffat 1969) describes the behaviour of a stellar image in a photographic emulsion. The behaviour for CCD chips differs somewhat from that but the Moffat function can still be used as a good approximation for seeing and diffraction patterns. It is similar to a Gauss distribution in its core but has broader wings.

Fig. 4.4 The optical slit defines the spectral resolution by cutting off the wings of the fluctuating seeing disk

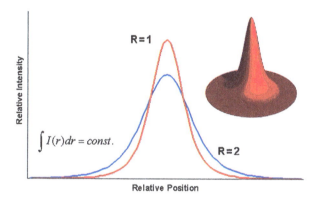

Fig. 4.5 Moffat function with $\beta = 2.5$ (*Red*: $R = 1$, *Blue*: $R = 2$). The integral of both functions is constant

$$I(r) = \frac{I_0}{\{1 + (r/R)^2\}^\beta}.$$

β is a pure form factor which usually lies between 2.5 and 3. With the scale factor R depending on the atmospheric turbulences one can fit the observed profile. The infalling stellar flux remains constant for all conditions and integral over the function remains constant, as well. So, if the function width increases due to seeing, the function intensity decreases. Figure 4.5 shows two Moffat functions with $R = 1$ and $R = 2$ (in microns). The FWHM of this seeing function is an indicator of the seeing and should represent the slit width. In practice the seeing FWHM defines the slit width.

Sometimes the slit is considered as a faceplate (baffle) for the reduction of the sky background and hence for the improvement of the signal contrast (signal-to-noise ratio S/N). However, that is only correct for a faint star and bright moonlight, compared to the slitless case, e.g. objective prism. In contrast to point light sources the sky background is spectrally imaged on the chip, i.e., on each single pixel, as a whole and we have to relate this to the geometrically restricted area of the stellar light. Considering a bright night with moon light (sky background ~19–17 mag per square arcsecond), a grating of 50 mm length and stars of 4 mag, or brighter, the signal-to-noise of our signal is then reduced by less than 10 % if using about 1,000 pixel. This is also valid for stars of about 5.5 mag and a telescope focal length of about 4 m. Hence, the mechanical effort for a baffle is very high in contrast to the expected improvement of the data quality. The construction of a baffle makes only sense if directly designed as an optical slit.

In addition, the slit enables an independent wavelength calibration with a specific lamp (e.g., neon or helium-argon). A direct calibration with the measured spectral lines is basically possible but not reliable if the lines vary in wavelength and, hence, radial velocity. This is especially the case for stars with variable emission lines. This applies especially for Be stars with equatorial material disks and the resulting prominent Hα emissions in their spectra (see Sect. 14.6). The emission is created by material in Keplerian orbits which exhibits cyclical distortions in radial velocity and hence wavelength. For Wolf–Rayet (WR) stars a direct calibration is completely impossible. Their radial-symmetric winds are optically thick, so that their spectra do not show absorption lines but only variable and shifted, broadened emission lines (see Sect. 14.8).

The optical slit introduces a defined and constant spectral resolution and enables one to carry out an independent wavelength calibration on the optical axis. This is essential for high-quality spectroscopic investigations. On the other hand, the slit requires accurate guiding (the target star has to remain on the slit), and leads to potential loss of light (see the Moffat function).

4.1.3   Wavelength Calibration with an Artificial Lamp

At this point we need to illuminate some aspects of the spectral wavelength calibration in more detail. For this task we need a spectrum of an artificial lamp with emission line features at certain laboratory wavelengths. The CCD pixels can then be allocated to these wavelengths available in respective catalogues. We describe the respective data reduction procedure in Appendix A.3.2. Making available such calibration spectra require two preparations. First, an appropriate lamp has to be chosen and second, it has to be positioned at the correct place to avoid systematic errors in the form of wavelength shifts.

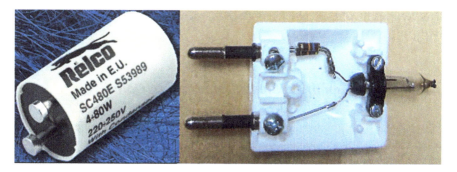

Fig. 4.6 The low-cost glow starter RELCO SC480 (*left*) and its electronical modification for permanent use (*right*). Two 47 kΩ connected in parallel represent the necessary 24 kΩ series resistor (*Courtesy*: RELCO and R. Walker)

In Sect. 4.1.2 we already mentioned the wavelength calibration with neon or helium-argon. In reality, however, there are a number of different lamps with different gas fillings. The fillings need to deliver an appropriate number of calibration lines for an adequate pixel allocation in all wavelength ranges chosen. This is especially valid for echelle spectrographs covering a wide wavelength range. Common gas fillings are mixtures out of, e.g., helium, neon, iron and argon. Usually these are hollow-cathode lamps for at least 400 Euros. A very cheap alternative is a simple glow starter (e.g., RELCO SC480), found in almost every fluorescent lamp (Fig. 4.6, left). The gas mix of He, Ne and Ar, as well as H (probably due to dissociation of remaining water vapour within the bulb) delivers hundreds of emission lines over the entire visible spectrum. Without modification the starter would open and close periodically because of the internal bimetals. For constant permanent use at 230 V a series resistor of about 24–30 kΩ is necessary (Fig. 4.6, right). A respective catalogue and spectra of the RELCO SC480 lamp taken with a SQUES off-the-shelf spectrograph can be found in Appendix E.2. The bulb delivers about 240 evaluable emission lines over the entire visible spectrum. Counting also those in all overlapping echelle orders, this number increases even to approximately 370. That means, calibration lines are sufficiently available over the entire visible spectrum for both, standard spectroscopy in different wavelength ranges and echelle spectroscopy (see Chap. 6).[4]

The second preparation is the positioning of the calibration source on the optical axis of the input beam and collimator and in front of the optical slit. This is necessary to avoid spectral wavelengths shifts due to different light paths between target and calibration lamp. This may be done by positioning the lamp in front of the slit (direct

---

[4]The intensity of the hydrogen Balmer series can vary strongly from between individual starters. Numerous additional lines not listed in the appendix, are generated by alloying elements, coatings and dopants, e.g., tungsten, lanthanum, cerium, hafnium, thorium and others. Their identification is difficult and highly speculative, since for a given line, almost always several plausible determination options exist in the immediate vicinity.

light feed) or by a respective fiber which is externally fed by lamp (indirect light feed).

4.1.4   The Design of Collimator and Camera Optics

Most off-the-shelf gratings are optimized for first order and a specific wavelength blaze and it is sufficient to consider the case $n = 1$ (we will introduce a prominent alternative in Chap. 5) and to estimate the diffraction angle for the blaze wavelength. At this point we should remember Fig. 2.24 again. The incident beam falling on the grating with an incident angle α will be diffracted to the camera by an angle β (both angles with respect to the grating normal).

$$\beta = \arcsin\left(\frac{n \cdot \lambda}{d} - \sin\alpha\right) \tag{4.1}$$

(a) Collimator

Equation 4.1 defines the crucial angle ratios and with it the other parameters can be estimated. First, it is clear that the optical elements should not cause vignetting of the light coming from the telescope. The collimator has to have a corresponding diameter. This is defined by the requirement of identic focal ratios to

$$\frac{F}{D} = \frac{f_{col}}{d_{col}}$$

and hence

$$d_{col} = \frac{D}{F} \cdot f_{col}. \tag{4.2}$$

The grating of geometrical length W must not cause vignetting of the beam. The beam diameter coming from the collimator is smaller than W because of the tilt of the grating. For the collimator diameter we get

$$d_{col} = W \cos\alpha.$$

Now we can estimate the ideal collimator focal length:

$$f_{col} = \frac{F \cdot W \cdot \cos\alpha}{D}.$$

This focal length is the lower limit to avoid vignetting.

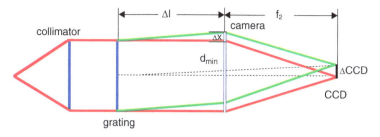

Fig. 4.7 Geometrical situation inside the spectrograph

(b) Camera

The minimum *beam width* is the diameter of the collimated beam at central wavelength and geometrical position at the camera entry lens. Because of the dispersing grating, the camera has to have a larger diameter Δx (spectral field of view). The diameter depends on the angular dispersion $d\beta/d\lambda$, on the length of the vignetting-free wavelength interval $\Delta\lambda$, and on the distance between grating and camera Δl (Fig. 4.7):

$$\Delta x = \Delta l \cdot \tan\left(\frac{d\beta}{d\lambda} \cdot \Delta\lambda\right).$$

The maximum wavelength interval $\Delta\lambda$ can be estimated by the product of the reciprocal linear dispersion (called "*plate factor*") and the size of the CCD chip $\Delta CCD = p \cdot N$

$$\Delta\lambda \approx \frac{d\lambda}{dx} \cdot \Delta CCD = \frac{d\lambda}{d\beta} \cdot \frac{1}{f_{cam}} \cdot \Delta CCD = \frac{d\lambda}{d\beta} \cdot \frac{1}{f_{cam}} \cdot p \cdot N$$

with p the pixel size and N the pixel number of the operating CCD camera in the dispersion direction. The term

$$\frac{dx}{d\lambda} = \frac{d\lambda}{d\beta} \cdot \frac{1}{f_{cam}} \cdot p \tag{4.3}$$

describes the linear inverse dispersion of the spectrograph in the most common units Å/Pixel. The resulting wavelength interval imaged on the CCD $\Delta\lambda$ is then

$$\Delta\lambda = \frac{\Delta CCD}{f_{cam}} \cdot \frac{d\lambda}{d\beta}. \tag{4.4}$$

For the minimum camera dimension d_2 we then get by substitution into the original equation for Δx

$$d_2 = d_{min} + \Delta x = d_{min} + \Delta l \cdot \tan\left(\frac{\Delta CCD}{f_{cam}}\right) \approx d_{min} + \frac{\Delta l \cdot \Delta CCD}{f_{cam}} \qquad (4.5)$$

with d_{min} the diameter of the monochromatic beam given by Eq. 4.2:

$$d_{min} = d_{col} = \frac{D}{F} \cdot f_{col}.$$

4.1.5 Some words on the grating choice

The core element of any spectrograph is its dispersion element, in our present considerations an optical grating (mostly a reflective one). The minimum geometrical grating size for a certain spectrograph is triggered by the F-number of the collimator which should match the F-number of the telescope, the respective telescope and collimator focal length and the incident angle on the grating (later we will present the mathematical relations).

However, the focal length of the telescope and the size of the seeing disk in its focus (slit width), respectively, defines the spectral resolution element. In order to obtain a resolving power adequate for the telescope, the collimator size necessarily increase with the telescope size and thus also the length of the grating. Very large gratings are all custom made by specialized providers (Fig. 4.8). For small telescopes reflection gratings of different sizes and design are available from various manufacturers at relatively low cost. Normally the largest dimensions are 100×100 mm (Zeiss) or 50×50 mm (Edmund Optics and Thorlabs). Other dimensions are, e.g., 12.5×12.5 mm and 30×30 mm. Typical peak efficiencies of blazed gratings in visual light are of the order of 60–70 %, with some manufacturers reaching up to 90 %. The blaze wavelength defines the maximum reflectivity. As a result, this parameter has to be taken into account, depending on the wavelength range to be measured. Gratings with (first-order) blaze wavelengths of 300, 400, 500 and 750 nm can be found on the market so that specific maximum spectrograph efficiencies can be adjusted to various wavelength regions of standard spectrographs (Figs. 4.9 and 4.10). Gratings with larger line densities ($>1,800$ lines/mm) are holographic gratings with efficiencies reduced by about 10–20 % at blaze wavelength but with significantly reduced stray light. The achievable spectral resolution depends on the combination of collimator/camera focal length and the slit width, as we have already described in Chap. 2. For the mechanical construction it might be important to use an interchangeable grating so that various line densities (grating constants) can be applied. However, various line densities introduce various angles of incidence which should be mechanically accounted for. The blaze direction is normally marked on the grating. If this is not the case, one can easily find the correct

Fig. 4.8 The echelle grating for the Potsdam Echelle Polarimetric and Spectroscopic Instrument (PEPSI) at the 8.4 m Large Binocular Telescope (*Courtesy*: AIP)

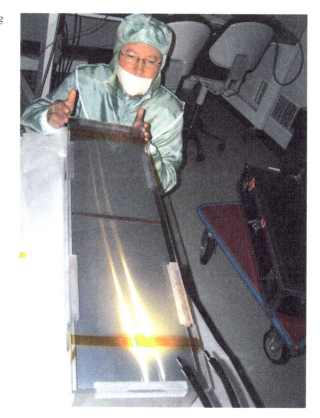

orientation with a laser beam. By directing the beam onto the grating surface one obtains a simple reflection or transmission of the laser source. This is the so-called 0th order. Left and right from this reflection one finds two other reflections, the 1st and the −1st order. The brighter one is the 1st order in the correct blaze direction (Fig. 4.11).

4.1.6 Fixing the Total Angle

For the construction of a spectrograph we opt for optical ray tracing. For the consideration of the grating angular position it is advantageous not to use the diffraction angle β as the design driver but the total angle $\gamma = \alpha - \beta$. γ is the angle between the grating incident beam and the diffracted beam towards the spectrograph camera and the angles α and β are defined with respect to the grating normal. Because of the chosen coordinate system (Fig. 2.23) β is negative and γ is not the sum of both angles but the difference (Fig. 4.12).

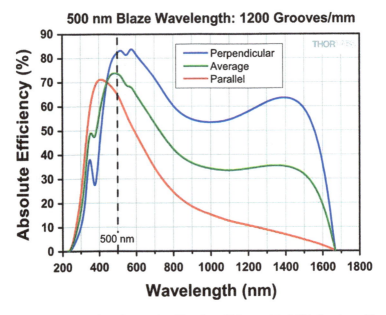

Fig. 4.9 Typical efficiencies of a grating blazed at 500 nm with 1,200 lines/mm (*Courtesy*: Thorlabs). The gratings utilizes an aluminum reflective coating and is measured in the Littrow mounting configuration. The parallel polarization is parallel to the grating grooves

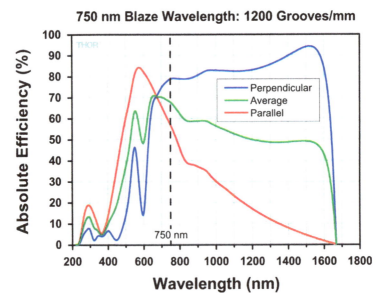

Fig. 4.10 As Fig. 4.9 but for a grating blazed at 750 nm (*Courtesy*: Thorlabs)

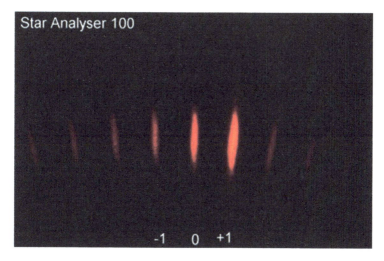

Fig. 4.11 Monochromatic laser diffraction patterns of a blazed Star Analyzer 100 (100 lines/mm) transmission grating. Orders −1, 0 and +1 are indicated (*Courtesy*: Robin Leadbeater)

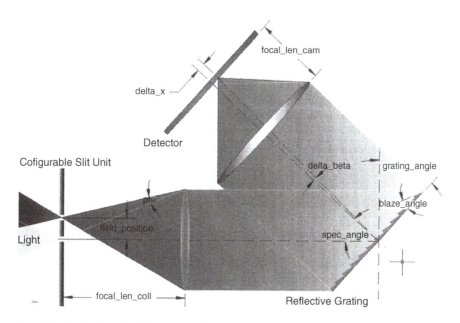

Fig. 4.12 Angles in a classical spectrograph

In the classical spectrograph design concept one uses two different imaging elements (lenses or mirrors) for collimator and camera, each having a different influence on the angle γ depending on the geometrical size. The larger the lenses the larger is the total angle for a given distance collimator to grating and camera to grating distances. That, in addition influences the overall spectrograph size and possibly its weight. This however has to be taken into account for a given telescope mount carrying the spectrograph.[5] In addition to optical considerations one also should keep practical aspects in mind. For instance, for an optical slit to acquire a defined spectral resolution one needs to keep the target star on the slit jaws while tracking the telescope. This is best done by an additional CCD camera and possibly a guide acquire module which additionally requires respective mechanical structure and imaging lenses.

As an alternative to the classical concept, one can use a Littrow spectrograph (see e.g. Sect. 8.2.3). This design has the advantage of a single lens acting both as collimator and camera. The collimated beam falling onto the grating is reflected by the grating at a very small angle γ of only some degrees between the incident and reflected beams. γ has to be chosen to allow sufficient space so that small mirrors or prisms can guide the beam from the telescope focus or to the CCD camera. The Littrow design has various advantages, e.g., small grating width and compact design. However, it is applicable only for lens collimators/cameras resulting in additional necessary care for respective aberration.

However, for a littrow design one often does not have the freedom to decide freely for a spectrograph concept. This is valid if the CCD pixels do not match the size of the order of the seeing disk in the telescope focal plane and, hence, potentially reducing the spectral resolution. The Littrow design is then not applicable and one has to use the standard design by choosing a relatively large camera focal length being different from that of the collimator. To keep the anamorphic magnification factor more or less independent of the grating angle one has to choose as small an angle as possible for γ. But that means increased construction complexity, and the spectrometer becomes larger and larger and one introduces potential problems with the ability of the mounting to support the spectrometer.

4.2 Project "MESSY" Maximum Efficiency Slit SpectroscopY for f/4 Telescopes

Our first choice for a spectrograph design is the analysis of massive stars, whose variable emission lines are a manifestation of dynamical phenomena in their extended atmospheres. Prototypes are Be, O and Wolf–Rayet stars (in Chap. 14 we describe these objects in more detail). O and Be stars have the advantage of being relatively frequent and bright in the sky. This is not the case for the relatively rare

---

[5]At this point we neglect fiber-fed spectrographs.

and challenging[6] but very strong-lined Wolf–Rayet stars. Their apparently brightest representatives in the northern sky can all be found in the Cygnus constellation and have visual brightnesses of only 7–8 magnitudes.[7] To obtain usable spectra from these stars including of their complex winds with our first 32 cm telescope, we had to focus on maximum efficiency (maximum S/N at moderately high spectral resolving power, i.e. small $\Delta\lambda$) for our system. It was our ambition to obtain scientific data quality for line profile analysis even for these objects. In this chapter we will show that such a task is realistically possible if one accounts for correct optical considerations.

To obtain maximum spectral resolution for a given grating dimension the grating has to be completely illuminated by a collimator of large focal length. In addition, the given CCD pixel size has to be matched best, a problem which will be more comprehensible after further considerations. Depending on these requirements we had to first configure all optical parameters of the system path: telescope → spectrograph → CCD pixel size. Then we had to optimize the system in a mathematical iteration process depending on the technical secondary conditions (mainly size and weight). The necessary equations have already been introduced in Chap. 2 and they will become more understandable by applying them to the spectrometer.

A spectrograph cannot be regarded as a system independent from the telescope or from the CCD. The telescope focal length and focal ratio as well as the pixel size of the CCD determine the optics of the instrument. If this is not taken into account the spectroscopic efficiency is reduced. We now had to basically choose a spectrograph design. Our telescope was a combined Newton–Cassegrain system. Our f/16 Cassegrain focus has a long focal length and, hence, the focal lengths of the collimator and camera in the spectrograph would be large, as well, resulting in an immense size. This would require to build a separate spectrograph not mounted at the telescope, but connected via fiber optic link to the telescope focus. However, the transmissivity of fiber optics (focal-ratio degradation (FRD), reduced surface sensitivity, challenging mechanical adjustment) is strongly reduced and even professional users are currently reaching a maximum efficiency of about 70 % (see Chap. 11). In addition, fiber optics require some know-how. Our requirements for high efficiency, hence, excluded fiber optics. Alternatively, a Littrow spectrograph in the Newton-focus was discussed. Because of their compact design in which a single lens acts as collimator and camera, Littrow spectrographs are widely used by amateurs (a professional design is discussed in Sect. 8.2.4). Unfortunately, the mechanical effort of making a Littrow spectrograph increases significantly via the installation of an optical slit, however, and makes the design relatively complex (see Sect. 8.2.3). For our f/4-Newton we opted for a slit spectrograph of classical

---

[6]In the context of spectrograph design the expression "challenging object" is ambiguous. This might be an 8 mag star for a 30 cm telescope or a 20 mag star for a 10 m one, depending on the spectral resolving power R, the spectral contrast S/N and exposure time t.

[7]The apparently brightest WR star γ Velorum is a southern object in the constellation Vela.

design: telescope focus → optical slit → collimator → grating → tilted mirror → camera → CCD. The tilted mirror between grating and camera is used to reduce the geometrical size. In addition, we added a second CCD to capture and guide the star on the slit inside the spectrograph. After discussing with colleagues, experienced in the field of massive stars, realistic specifications for our 32 cm telescope were as follows:

1. Limiting magnitude about 8 mag.
2. Inverse linear dispersion ≈ 0.4 Å/pixel.
3. $S/N \geq 100$.
4. Maximum exposure 30 min.

The limiting magnitude is defined by the brightest available WR target stars in the northern sky. The relatively high spectral dispersion and S/N are required for proper line-profile analysis, which is necessary to detect discrete structures in the wind of WR and Be stars. Short exposure times are additionally required to track turbulent structures (clumps, discrete absorption components) in WR winds. Of course this does not exclude observing fainter stars during longer exposures.

4.2.1   Calculating the Parameters of the Spectrograph

As an example and for better understanding, we calculate the parameters of our spectrograph so that the reader can use our approach as a guideline for his own design. The computation can be modified according to different parameters or concepts. We start with our f/4 telescope of 317.5 mm aperture and a focal length of 1,270 mm. We used the CCD camera OES Megatek with a Tektronix 1024 chip with 1,024 × 1,024 pixels of size 24 × 24 μm. The spectrometer should have a resolution element of no larger than $\Delta \lambda = 0.8$ Angstrom. The target objects have a brightness of up to 7–8 mag. At first we calculate the size of the seeing disk in the telescope focus, assuming a typical seeing of 4 arc seconds in continental urban environments.[8]

$$d = \tan \omega \cdot f_{tel} = \tan(4''/3600'') \cdot 1270\,\text{mm} = 0.0246\,\text{mm}$$

or 24.6 μm.[9] For a slit spectrometer one would then select a slit width in the range of 12–30 μm, depending on the seeing conditions or what spectral resolution should be achieved. However, one should note that for slit widths below 20 μm diffraction

---

[8]Professional observatory sites like Hawaii, Chile and the Canarian Islands deliver seeing conditions significantly below 1 arcsec.

[9]For small angles we have $\tan \omega \approx \omega$ so that $d = 4''/3600 \cdot 180°/\pi \cdot 1270\,\text{mm} = 4''/206265 \cdot 1270\,\text{mm} = 0.0246\,\text{mm}$

effects from the slit itself can degrade the spectral resolution. In addition, for such small slit widths, dust and dirt inside the spectrometer can pose problems.

The next spectrometer component is the collimator. Because the focal ratio of the telescope is f/4 the collimator necessarily has to have the same or a smaller focal ratio. A smaller collimator f-ratio, e.g., f/3 is allowed because no vignetting effects would occur. The entire beam would still be collimated by the lens. A larger collimator f-ratio, e.g., f/5 would introduce light loss due to the larger beam diameter coming from the telescope. No high imaging quality is needed since the stellar light will remain on the optical axis of the lens, i.e., aberrations play a minor role. To reduce chromatic aberration, an achromatic lens, however, should be mounted in any case. Moreover, in this case spherical aberration and coma are also largely eliminated. The focal length of the collimator defines inter alia the spectral resolution of the spectrograph, but also the beam diameter and thus the dimension of the grating. Standard gratings of up to 50×50 mm are available off-the-shelf for relatively low prices. Thereafter the price rises proportional to the surface area, at least. For larger beam diameters one needs appropriate optics for collimation and imaging the spectrum (camera).

For our spectrometer we have chosen a standard 50×50 mm grating. With this grating size we obtain the highest spectral resolution defined by the collimator focal length. Smaller resolutions can be obtained by changing the grating (fewer lines per millimeter) or diminishing the collimator focal length. For a focal ratio of f/4 and grating size of 50 mm the collimator focal length should be 200 mm with the grating positioned perpendicular to the optical axis. However, since the grating is always illuminated under a certain angle, the projected size will be correspondingly smaller than 50 mm. The angle at which the grating has to be illuminated, however, depends on the total angle $\gamma = \alpha - \beta$, with α the angle of incidence and β the diffraction angle. This angle is defined by the design of the spectrograph and depends particularly on the dimensions of the optical components. The larger this angle is chosen to be, the more space is required for the spectrograph, but with the disadvantage that the grating size must increase with fixed collimator aperture. In addition, the anamorphic magnification factor then decreases with increasing angle γ. As a consequence the full width at half maximum FWHM decreases according to Eq. 2.98 and the corresponding spectrometer resolution increases according to Eq. 2.88. If we choose a small total angle γ (Littrow configuration), the variation of the minimum grating dimension during rotation will be correspondingly small. For our example we have chosen 15° as the smallest possible total angle γ due to the geometric positions of collimator and camera and not to shadow the respective light beams at the grating (see Fig. 4.13). The beam coming from the grating is then folded by a mirror behind the grating, both to realize a small angle and to make the construction of the spectrometer somewhat symmetrical.

In general, the efficiency of gratings above 1,200 lines/mm drops quite fast with increasing line numbers. Since we want good signal-to-noise ratio (S/N) combined with high spectral resolving power R, we have opted for a 1,200 l/mm grating and

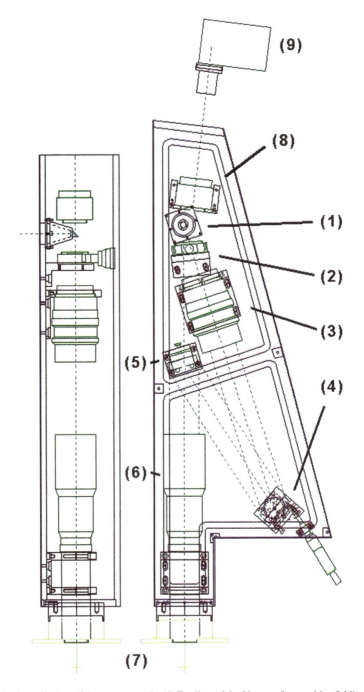

Fig. 4.13 Overall view of the spectrograph. (*1*) Feeding of the Newton focus with a folding prism, (*2*) slit, (*3*) collimator, (*4*) reflection grating, (*5*) folding mirror, (*6*) camera, (*7*) CCD flange, (*8*) guiding optics, (*9*) guiding CCD

tried to keep the spectral resolution at a relatively high level with appropriate focal lengths of collimator and camera. First, we calculate the incidence and diffraction angle by using the grating equation (2.49):

$$\frac{n \cdot \lambda}{d} = \sin \alpha + \sin \beta.$$

Because of $\gamma = \alpha - \beta$ we find

$$\frac{n \cdot \lambda}{d} = \sin \alpha + \sin(\alpha - \gamma).$$

Using the addition theorem[10] $\sin(\alpha - \gamma) = \sin \alpha \cos \gamma - \cos \alpha \sin \gamma$ we obtain

$$\frac{n \cdot \lambda}{d} = \sin \alpha (1 + \cos \gamma) - \cos \alpha \sin \gamma.$$

In addition, we use the addition theorems (a) $1 + \cos \gamma = 2 \cos^2 \left(\frac{\gamma}{2}\right)$ and (b) $\sin \gamma = 2 \sin \frac{\gamma}{2} \cos \frac{\gamma}{2}$ and modify the grating equation to

$$\frac{n \cdot \lambda}{d} = 2 \sin \alpha \cos^2 \left(\frac{\gamma}{2}\right) - 2 \cos \alpha \sin \frac{\gamma}{2} \cos \frac{\gamma}{2}.$$

By factoring out $2 \cos \frac{\gamma}{2}$ and combining with the addition theorem (a) we get

$$\frac{n \cdot \lambda}{d} = 2 \cos \frac{\gamma}{2} \sin(\alpha - \frac{\gamma}{2}).$$

To obtain the angle of incidence α from the previously established parameters γ, λ and d, we rewrite the above to

$$\sin \left(\alpha - \frac{\gamma}{2}\right) = \frac{n \cdot \lambda}{2d \cos \frac{\gamma}{2}} = \frac{n \cdot g \cdot \lambda}{2 \cos \frac{\gamma}{2}}.$$

The grating with 1,200 l/mm, as almost all other gratings, is optimized (blazed) for the first grating order ($n = 1$). As already established we use $\gamma = 15°$ and, together with $\lambda = 550\,\text{nm} = 0.55 \cdot 10^{-3}\,\text{mm}$, we obtain

$$\sin(\alpha - \frac{\gamma}{2}) = \frac{1200/\text{mm} \cdot 0,55 \cdot 10^{-3}\,\text{mm}}{2 \cos(\frac{15°}{2})} = 0.33285.$$

---

[10]Various collections of goniometric relations can be found in the relevant pages of the internet.

We find for the angle $\alpha = 26.94°$ and for the diffraction angle $\beta = \alpha - \gamma = 11.94°$. With the help of the now familiar angle of incidence, we can calculate the maximum beam diameter d_1 behind the collimator:

$$d_1 = w \cdot \cos \alpha = 50 \, \text{mm} \cdot \cos(26.94°) = 44.57 \, \text{mm}.$$

Because of our f/4 system the collimator can therefore have a maximum focal length of $4 \, \text{mm} \times 44.57 = 178.3 \, \text{mm}$. To avoid vignetting and for security reasons, we should select a sufficiently fast collimator lens (i.e. f/2.8 or f/3.5); we therefore decided for an M42 lens with 180 mm focal length and a focal ratio of f/3.5. The grating size is thus slightly undershot by the collimated beam in the red ($\sim$7,000 Å) and slightly overshot in the blue light ($\sim$4,000 Å). The anamorphic magnification factor is $A = \cos \alpha / \cos \beta = \cos(26.94°) / \cos(11.94°) = 0.911$, and thus the beam diameter in front of the camera is increased to $44.57 \, \text{mm}/0.911 = 48.91 \, \text{mm}$. This diameter d_{min} is necessary to image the beam onto the CCD camera without any light loss at the respective wavelength. The necessary diameter to image the spectrum without vignetting has already been calculated in Sect. 4.1.4:

$$d_{cam} = d_{min} + \Delta l \cdot \tan \left(\frac{\Delta CCD}{f_{cam}} \right).$$

At this point we need to highlight the great importance of this equation. The 2nd term enlarges the necessary camera diameter and thus the F-number of the camera depending on the grating—camera distance Δl. This distance must therefore be kept as small as possible. If not, we potentially introduce vignetting into the system because of camera apertures being too small. In the next section we will also see that the camera focal length f_{cam} is not a freely selectable parameter. Hence, compensation by larger camera focal lengths is not a meaningful solution for the problem. In addition, we once again want to point out that the wavelength interval, imaged on the CCD camera, is indirectly proportional to the camera focal length f_{cam}:

$$\Delta \lambda = \frac{d\lambda}{d\beta} \cdot \frac{1}{f_{cam}} \cdot p \cdot N.$$

If the camera focal length is too large we will introduce an unnecessary limitation of the wavelength interval $\Delta \lambda$. Which interval is still tolerated here, must be decided for each individual case.

To determine the size of the camera diameter d_{cam} we still need its focal length, which results from the required spectral resolution. Because of this, one might suppose that the focal length of the camera and thus the resolution can be freely selected. However, this is not the case. The Nyquist–Shannon theorem (Sect. 2.6.6) requires that each spectral element is covered by a specific number of pixels depending on the achieved signal-to-noise ratio S/N. If fewer pixels cover a spectral element necessary for a certain S/N we have *undersampling* and for the case of more than necessary pixels we have *oversampling*, which however only results in higher achievable S/N. Many more sampling pixels cause a limited wavelength interval $\Delta\lambda$ and therefore loss of information. Loss in S/N can be strong for the oversampling case and can subsequently be corrected by pixel integration (pixel binning), although the loss in wavelength coverage remains. On the other hand, if undersampling is present, spectral information is lost and cannot be recovered by mathematical methods. However, we have seen in Sect. 2.6.6 that a reliable sampling needs at least some pixels because of the instrument function, which broadens each spectral element.

According to our specification $S/N \geq 100$ and our considerations about the Nyquist–Shannon criterion we could accept a sampling of 2–3 pixels (see Fig. 2.40).[11] To determine the camera focal length, we use Eq. 2.97 and divide the width of the spectral element d_D by the pixel size. The size of $(d_D/p)^2$ must be at least $2^2 = 4$ according to the Nyquist theorem[12]:

$$\left(\frac{d_D}{p}\right)^2 = \left[\left(A \cdot \frac{f_{cam}}{f_{col}}\right)^2 \cdot \left(\tan^2\omega \cdot f_{tel}^2 + \sigma_{col}^2\right) + \sigma_{grating}^2 + \sigma_{cam}^2 + p^2\right] / p^2 \geq 2^2 = 4.$$

Theoretically one could convert the equation to isolate f_{cam}, which is not easy, however, due to the quadratic terms. In addition, having the focal length f_{cam} in d_{cam} does not make the task easier. We do not focus on this problem but calculate the ratio (d_D/p) iteratively. It is important, though, that about 2–3 pixels per FWHM are available. Also, the camera focal length cannot be chosen arbitrarily, because we have to rely on focal lengths which are commercially available. We use all the necessary parameters and obtain $d_2 = 118.7$ mm for a selected focal length of, e.g., $f_{cam} = 180$ mm and a separation $\Delta l = 500$ mm. This would correspond to a camera focal ratio of 1:1.52. For calculating the spectral resolution (FWHM of the slit image on the detector) we still need the Airy disks of the lenses and the grating:

$$\sigma_{col} = 2.44 \cdot 0.55\,\mu\text{m} \cdot 4 = 5.368\,\mu\text{m}$$

$$\sigma_{grating} = 0.55\,\mu\text{m}\frac{180\,\text{mm}}{50\,\text{mm}} = 1.98\,\mu\text{m}$$

$$\sigma_{cam} = 2. \cdot 0.55\,\mu\text{m} \cdot 1.52 = 2.04\,\mu\text{m}.$$

---

[11]For brighter stars delivering significantly high S/N more pixels per resolution element are mandatory.

[12]At this point and for simplicity of the following calculation we refer to a sampling with only 2 pixels. For an accurate consideration we refer to Sect. 2.6.6.

Hence, we obtain for d_d/p

$$\sqrt{\left[\left(0,91 \cdot \frac{180}{180}\right)^2 \cdot (24,6^2 + 5,368^2) + 1,98^2 + 2,04^2 + 24^2\right]/24^2}$$

$$= 33,3/24 = 1,39.$$

i.e., fewer than 2 pixels per spectral element, which means unwanted undersampling. The focal length is therefore too small. In addition, a camera lens with f = 180 mm and an F-number of f/1.52 is essentially unavailable on the market.

We calculate all the parameters again but now for a camera focal length of f_{cam} = 500 mm and obtain 2.86 pixels/FWHM or a sampling factor[13] of 1.43. The minimum focal ratio for this focal length is f/7.03 and is in the range of available lenses. The FWHM (d_d) is 68.7 μm. Using Eq. 2.88 we calculate the spectral resolving power to be R = 5957. From a detailed examination of the FWHM calculation we now understand that the seeing disk of 24.6 μm and the relatively large pixels of 24 μm limit the spectral resolution, compared to its theoretical value.[14]

Based on our requirements, we can now calculate the complete system. Our parameters are:

- Telescope aperture: D = 317.5 mm
- Telescope focal length: f = 1,270 mm
- Collimator focal length: f_{coll} = 180 mm
- Camera focal length: f_{cam} = 500 mm
- Grating constant: g = 1,200 l/mm

With these values, we should obtain a spectral resolution of about 0.8 Å for two 24 μm pixels and a seeing of 4″. The typical achievable S/N should be about 150 within 1,800 s exposure time for a star of 8 mag.

---

[13]The number pixels/FWHM divided by 2 is sometimes called the sampling factor. The optimum sampling factor is 1.0 = 2 pixels/FWHM.

[14]There is another reason to choose a sampling factor somewhat larger than 1. All previous calculations are based on an average seeing of 4″, to optimally detect faint objects. Some nights the air is very calm and the seeing can be, for example, 2″, or the observer wants to detect a very bright star with higher resolution, without changing the grating. In this case the slit width has to be reduced to, e.g., 12 μm and the sampling factor is automatically reduced. The choice of about three pixels per spectral element allows one to very easily run the spectrometer in two modes without undersampling. Because of stochastic variations in the seeing disk, the authors also assume that slight undersampling does not directly cause significant alias effects.

4.2.2 Optical and Mechanical Design

The easiest choices for a collimator and a camera are standard camera lenses, if allowed by the grating size. With an active area of 50×50 mm this was possible in our case and we could chose a 180 mm camera lense as collimator as well as a 500 mm camera lense as a camera. Our spectrograph design is shown in Fig. 4.13.

First, the focus of the Newtonian telescope is directed into the optical axis of the spectrograph by a folding prism of 10 mm edge length (1) and imaged onto the slit (2) and then towards the collimator (3). The Zeiss slit (Fig. 4.14) has an accuracy of $10\,\mu m$ per scale line on the vernier. This is sufficient for a stellar image of $24\,\mu m$ in the telescope focus for a seeing disk of 4 arcsec (Fig. 4.15).[15] To measure different wavelength intervals of the entire spectrum the grating must be rotatable (4). In addition we wanted to be able to use a grating with 600 and another with 1,200 lines per mm. To achieve this, the following angular ranges result from the incident beam angle (α) with respect to the direction of the grating normal. The grating should be adjustable within the following ranges:

- 600 l/mm in 1st order: 14.5–19.7°
- 1,200 l/mm in 1st order: 21.5–32.6°
- 1,200 l/mm in 2nd order: 36.5–65.4°

Fig. 4.14 Our polished optical Zeiss slit

---

[15]We note that the market may deliver cheaper adjustable slits. Their quality, though, is unknown to us. A very simple and especially for low-cost spectrographs applicable adjustable slit, however, can easily be self constructed. For this task the slit jaws must be guaranteed to remain parallel. This can be provided by two razor blades or even two metal plates with respective sharp edge phases and fixed on an appropriate carrier. The setting of the slit width can then be measured by a thickness gauge or with a laser as described by Eq. 2.40.

Fig. 4.15 Reflexion grating
with 638 lines per mm

We obtain a spectral resolution of 0.28 Å at Hα in the 2nd order. With a total angle of 15° (angle between the incident and the diffracted beam), we achieve a perfect geometric beam coverage even at extreme grating positions at 4,000 and 7,000 Å. The ratio between the diameters of incident and diffracted beam (anamorphic magnification factor) is therefore 0.9, so that the illuminated grating area remains almost constant over the entire spectral range. The following mirror (5) then reflects the beam parallel to the optical axis of the telescope in the direction of the camera (6). The detector is a Megatek CCD camera with a Tektronix TK 1024 chip and a pixel size of 24 μm and the guider is an AlphaMaxi with Kodak KAF400 chip and 9 μm pixels. To use the MegaTEK spectroscopically we designed the final flange (7) to attach the CCD directly to the spectrograph casing. The camera lens is crucial for the spectral image and one can adapt it to other pixel sizes by changing the focal length.

4.2.3 Telescope Guiding

In order to guide the telescope in spectroscopic mode with the zoom lens (8) and the guiding CCD camera (9) we use the stellar reflexions on either side of the slit jaws. Their reflective faces are inclined by 12°, so that the guiding camera can see them without obscuring the light path. For that case it is essential that the jaws reflect a sufficient amount of light from the outer stellar intensity function. The half-width of the seeing disk should exactly correspond to the slit width. In the case of 4″ seeing this corresponds to 24 μm for our Newton telescope. This should now be covered by a sufficient number of pixels of the guiding camera. To ensure this, we use a simple 50 mm lens with M42 connector (8) as microscope lense in front of the guider to increase the slit image about 10 times and to focus it on our guiding camera (9). Thus we obtain a sufficient 26 pixel ($9\,\mu$m) coverage for the $24\,\mu$m slit. This lens, however, must be positioned sufficiently close to the slit. For the calculation of the optical conditions, we again try the formula for the magnification β, derived in Chap. 2:

$$\beta = \frac{b}{g} = \frac{f}{g - f}.$$

For a desired reproduction scale of $\beta = 10$ we get

$$f = 10 \cdot (g - f)$$

or

$$g = \frac{11}{10} \cdot f.$$

With the chosen focal length of $f = 50$ mm, we obtain $g = 55$ mm for the object distance (distance from the slit to the 50 mm lens). The guider has a chip size of 6.8×4.8 mm (756×530 pixels $9\,\mu$m each). The relative image field in the telescope focus is 206,265 arcsec per 1,270 mm = 162 arcsec per millimeter or 2.7 arcmin per mm, i.e., with a magnification factor of $\beta = 10$ we obtain 0.27 arcmin per millimeter in the microscope focus. Hence, this yields an absolute field of view of 1.84×1.31 arc minutes, re-calculated to the dimensions of the guider CCD. This is demanding but just enough to find the star on the slit with the microscope. One realizes the difficulty of the task: On one hand we want to sufficiently resolve the seeing disk for the guiding algorithm, and on the other hand the image field is reduced. For this reason the GoTo telescope mount has to be able to ensure an appropriate positioning accuracy of about 1–2 arc minutes. To ensure a short distance between telescope and spectrograph and to avoid vignetting of the beam, we have designed the prism support (1) in the form of a hollow cone (the "volcano", Fig. 4.19). The cone adjustment is done by appropriate screws. For maximum reflection towards the guiding camera the highly reflective slit has been twisted by 12° into the guider direction.

Figure 4.16 shows details of the opened slit jaws under the microscope. The upper part of Fig. 4.16 shows the lamp reflection from about 80 cm distance on the slit opened by $10\,\mu$m. The lower part of the figure shows the same situation depicted with an additional Barlow lens (Fig. 4.17). With the polished slit we lose about two magnitudes of stellar light and, hence, we can track stars of about 10th mag. This means we can easily guide our faintest target stars of about 8 mag on the slit.

Fig. 4.16 10 μm slit imaged
by the guider. With (*bottom*)
and without (*top*) Barlow
lense, respectively

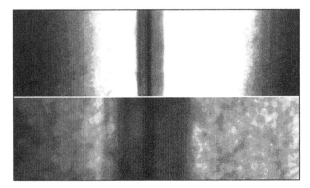

Fig. 4.17 950 μm slit
imaged by the guider. The
structure at the upper right is
a dust grain

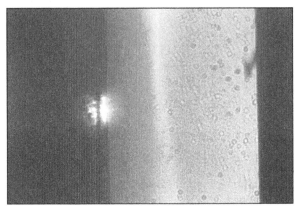

By using the wings of the seeing disk for guiding on the slit, about 40 % of the
light, depending on the seeing conditions, is lost for analysis.[16]

4.2.4   Construction and First Results

In order to ensure that the spectrograph can be attached to the telescope with
sufficient stability and to guarantee sufficient optical stability, we have chosen an
aluminum base plate of 10 mm thickness. We have learned that although cast plates
have high flatness, they are difficult to process because of enclosed cavities. Better

---

[16]One might now think that fiber optics of 70 % transmissivity then becomes attractive again.
However, to avoid mechanical flexure we wanted to guide inside the spectrograph. Flexure effects
could potentially be avoided by fibers. But then one might guide using the outer regions of the Airy
disk outside the fiber aperture and one would potentially loose light in both the dispersion and the
spatial directions. As a result, the efficiency of using fibers is lower than using a slit.

are stretched plates which are also very flat. All the spectrograph elements (lenses, grating, folding mirror, slit etc. plus all respective holders) are mounted on such a plate. At each corner of the base plate, columns support the cover plate, which is identical to the base plate, but milled out for better accessibility of the optical elements. All spectrograph sides and the cover plate have thin cover sheets made of aluminum. Figure 4.18 shows the overall view, and Fig. 4.19 shows the feeding optics (the "volcano"), which feeds light from the Newtonian focus to the slit. The "volcano" with the attached prism forms the optical access to the spectrograph and is adjustable over a number of screws so that the telescope focus is properly redirected to the slit. The optical grating mount allows vertical and azimuthal adjustment (Figs. 4.20 and 4.21).

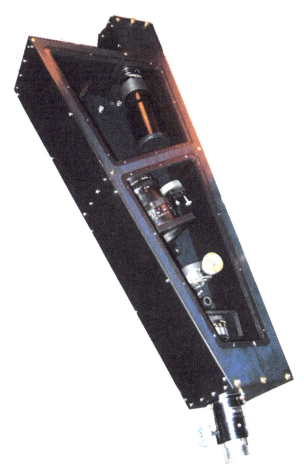

Fig. 4.18 Overall view of the open spectrograph

Fig. 4.19 The feeding optics (the "vulcano")

Fig. 4.20 The grating mount with micrometer screw

The latter is done via a rotary shaft whose angular position is adjustable by means of a calibrated micrometer screw. To ensure a perpendicular pressure point of the micrometer screw on the rotary plate at all angular positions, we designed a corresponding geometric form. This form should correspond to a cycloid[17] for our range of angles; however, it is sufficiently approximated by a circular cutout. The slit can be shifted by about 5 mm towards the collimator direction along the optical axis. This provides an additional opportunity to focus precisely on the slit. Figure 4.22 shows the complete spectrograph at the telescope.

---

[17]A cycloid of radius r follows the Cartesian coordinates $x = r(t - \sin t)$ and $y = r(1 - \cos t)$, with t the parameter in radians.

Fig. 4.21   The grating with 638 lines/mm on the mounting

Our first spectrum has been obtained with the 638 lines/mm grating using a Neon lamp in the lab (Fig. 4.23). The measured inverse spectral dispersion of 0.77 Å/pixel agrees well with the value of 0.75 Å/pixel from the calculation for our 638 lines/mm grating. For this first spectrum, we derived a FWHM of about 2.4 pixels for the emission lines. The correct focusing was performed on a star. For the guiding on the slit we have first examined the reflectivity and the quality of the slit jaws by using the bright star Vega. Figure 4.24 shows the star wandering over the slit corresponding to the telescope movement. The slit was illuminated by a few LEDs for better orientation. The slit width was 15 μm. The top panel shows the star positioned next to the slit and provides a maximum signal of about 24,000 counts for an exposure time of 1 s. The bottom image shows the star positioned exactly on the slit opening and provides a maximum signal of 1,200 counts. The reflected light can be measured by the guider, and hence, sufficient guiding accuracy on the slit is provided. The visible point-source reflections are presumably from polishing holes from the non-perfect polishing procedure—a phenomenon that is very well known to mirror grinders. To avoid the guiding algorithm being confused by these holes, the slit jaws should be polished as carefully as possible.

The first recorded spectrum is shown in Fig. 4.25. The star Regulus in the Hα wavelength region was exposed for 60 s at 50 μm slit width in low resolution with the 638 lines/mm grating. Note the significantly decreasing sensitivity function towards small pixel numbers (blue wavelengths). This is vigetting in the camera lens. According to Eq. 4.5 it is too small for the dispersed beam.

Fig. 4.22 The finished
spectrograph with attached
MegaTEK-CCD at the
Newtonian focus. The focuser
for the guider can be seen at
the upper system end

4.2.5   Vignetting in a Lens System

When looking at the spectrum in Fig. 4.25, it is noticeable that the intensity at the
left edge drops sharply. Because the respective spectral region is in the vicinity of
the Hα line and the Megatek CCD camera used (see Fig. 10.2) has a flat maximum in
this wavelength range, the intensity decrease cannot be due to the spectral response
function of the CCD chip, but only due to vignetting in the optical system. The
general effect of vignetting is a more or less non-homogeneous intensity distribution
in the image plane of an optical system with a homogeneous brightness distribution
in the object plane. In a retrospective analysis of the recorded spectra and the

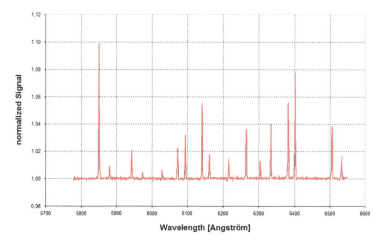

Fig. 4.23 The first Neon spectrum

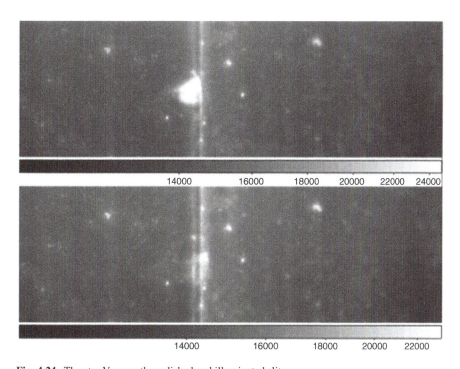

Fig. 4.24 The star Vega on the polished and illuminated slit

188 4 The Standard Spectrograph Layout

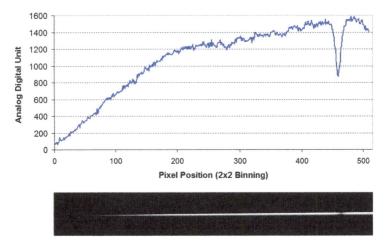

Fig. 4.25 Raw spectrum of regulus with 60 s exposure time. The significantly decreasing sensitivity function towards small pixel numbers is introduced by vigetting in the camera lens

optical components of the spectrometer, it was found that camera lens vignetting, associated with the relatively large 24 × 24 mm chip Megatek CCD camera and can therefore be held responsible for the intensity drop at the edge of the spectra. When using standard camera lenses with standard connectors (M42 thread or bayonet) one can eliminate this problem with a lens of smaller F-number. When planning a spectrometer it is therefore particularly important to make sure that all the optical components are designed with sufficient apertures. Basically, one has to distinguish between two types of vignetting:

• Artificial vignetting by socket parts and
• natural vignetting.

Vignetting by socket parts is generally due to the length of multi lens systems and can largely be influenced by an appropriate design of lens sizes. Natural vignetting, however, occurs in each optical system. It is a continuous decrease in brightness from the image center to the image edge and is caused by the photometric properties of light, i.e., the radiation exchange between mutually inclined surfaces.[18] If one images an object with the object angle ω_1 using any optical system from, the intensity in the image plane is generally (O'Shea 1985):

$$I(\omega_1) = I(0) \cdot \cos^4 \omega_1 \qquad (4.6)$$

---

[18]Object and CCD plane exchange radiation via the imaging optics. Since this is obliquely incident radiation, the emitting and the receiving surfaces are inclined against each other. Taking into account the r^2-law for the intensity one obtains a cos^4 dependence overall.

Fig. 4.26 Contrast-enhanced image of the twilight sky with a digital SLR and a 150 mm $f/5.6$ telephoto lens. The intensity decrease from the image center to its corners is about 20 %. The human eye is a logarithmic receiver. The intensity gradient is registered as only a small difference because of the enormous dynamic range of the retina

with $I(0)$ the intensity on the optical axis at the image center. Natural vignetting plays a significant role especially for wide angle lenses. Figure 4.26 exemplifies vignetting of a 150 mm zoom lens taken with a digital SLR camera. The object was the (almost homogeneous) twilight sky. The lens used has a field angle of approximately 10° (half diagonal.[19]) With Eq. 4.6 we thus can calculate a natural vignetting of about 6 % in the image corner (94 % of the center value). In our example spectrum, however, the difference in brightness between the image center and image corner is approximately 20 %. It must therefore be assumed that in addition to the ever-present natural vignetting we have also artificial vignetting by the internal structure of the lens.

In principle, a certain amount of system vignetting is no particular problem in astronomical spectroscopy, since in the subsequent data reduction all the effects of this kind are calibrated out by the continuum rectification (see Sect. 12.3.5). However, if the vignetting becomes very strong, the corresponding S/N can be significantly reduced, particularly for faint objects, i.e., parts of the spectrum cannot be evaluated. The exposure time of the corresponding object may be increased in this case. By examining vignetting we also noted that the position of the deflection mirror between grating and camera lens was not optimal. The beam behind the grating on the optical path to the camera diverges so that the necessary camera aperture increases and thus the camera F-number (and also the image field to be corrected) needs to be increased. The order of grating and mirror therefore plays an important role and should generally not be realized as in the present example but

---

[19]The ratio of half CCD diagonal (from the center to the corners of the image) and the focal length is the field angle.

vice versa. In general, the distances behind the grating should be kept as short as possible because the geometrical camera size can be minimized. This is particularly important for echelle spectrometers, as in this case, the entire detector field is used. That means the diagonal field angle of the chip increases.

Conclusion

Although our material costs were kept relatively low, the device described here fulfills all the usual professional requirements in spectroscopic studies of stars with regard to an optimal adaption to the system: telescope–spectrograph–CCD including a defined spectral resolution delivered by a slit. It can be argued that the effort of placing a slit inside a spectrograph is rather big for a small spectrograph and that spectroscopy is also possible without a slit. This is entirely correct, of course, and was already presented in the introduction. However, if one intends to meet the requirements outlined previously and one wants to use them for advanced work, one has to realize that a defined spectral resolution is essential to achieve valid and repeatable results.

As often for small telescopes, a basic problem in the development of a spectrograph arises from limited financial resources. For our case this meant that we had to choose a Newtonian configuration with a short focal length in conjunction with cheap 50 mm optics (collimator, grating, mirror, camera) and maximum efficiency of the overall system. In retrospect, it should be noted that the resulting very small seeing disk on the optical slit, necessary for a defined spectral resolution, introduces problems to be solved. In addition, for best adjustment of all optics the optical axes of the telescope and spectrograph should be orientated plane-parallel. Achieving this in a Newtonian system is not trivial. In this respect, easily manageable solutions for axisymmetric telescope foci with larger focal lengths (e.g., Cassegrain) but with larger spectrograph apertures should not be necessarily discarded.

Due to our access to a professional workshop with CNC tools, we were able to achieve a high quality mechanical design of our device. We want to stress, though, that our goal of maximum efficiency can also be achieved by a simpler manufacturing process (see also Chap. 6). The decisive factors for a well-defined and well-working system are rather a careful analysis and adaptation to all the important parameters such as target objects, telescope and CCD camera.

Suggested Readings

- *C. Palmer, Diffraction Grating Handbook, latest edition freely available in the web (e.g., http://www.gratinglab.com/Information/Handbook/Handbook.aspx)*
- *M.C. Hutley, Diffraction Gratings, Techniques of Physics Vol. 6, Academic Press, 1982*
- *E.G. Loewen, Diffraction Gratings and Applications, Marcel Dekker Inc., New York 1997*
- *J. James, Spectrograph Design Fundamentals, Cambridge University Press, New York 2007*

Chapter 5
Fundamentals of Echelle Spectroscopy

A Short Story

We are always looking for diverse optical elements on the internet. One day we purchased a large concave mirror. During the contact process, the seller said that he still had other stuff in his basement and he could certainly send it in order for us to check if useful. He would sell it for 100 Euro. We agreed. A few days later Klaus called Thomas via telephone and said he should come over immediately. When Thomas arrived, there was a collection of optical and electronic devices lying on Klaus' kitchen table, whose purpose was hard to guess. After a while Klaus asked: "What is the most valuable item on the table?" Thomas looked confused again over the parts but discovered a metal frame on which a large echelle grating and a large off-axis mirror collimator were attached. Both were worth about 5000 Euro and we suddenly had a new working field!

5.1 High Orders

As in Chap. 2 we start with the grating equation 2.49:

$$n \cdot \lambda = d \cdot (\sin \alpha + \sin \beta).$$

The angle of incidence α and the diffraction angle β are two parameters that satisfy the interference condition. Both angles are located in a plane which is perpendicular to the grating surface and the direction of the grooves. We now generalize the grating

© Springer-Verlag Berlin Heidelberg 2015
T. Eversberg, K. Vollmann, *Spectroscopic Instrumentation*, Springer Praxis Books,
DOI 10.1007/978-3-662-44535-8_5

Fig. 5.1 Definition of the
angles in the general grating
equation. N = Grating
normal, N_B = Normal of the
blaze surface, γ = Off-axis
angle

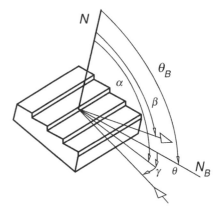

equation by adding the so-called off-axis angle γ, describing the tilt of this plane
against the grating normal (Fig. 5.1). Thus, due to the shortening effect of the off-
axis angle,[1] the grating equation in a general form is now:

$$n \cdot \lambda = d \cdot \cos \gamma \cdot (\sin \alpha + \sin \beta). \tag{5.1}$$

The optical path difference on the right side of the equation is a linear function of the
order number n. We now consider the case of a constant diffraction angle $\beta = \beta_c$.
Then the right side of the grating equation represents a constant and the associated
wavelength $\lambda = \lambda_c$ is only inversely proportional to the order number n:

$$\lambda_c(n) = \frac{1}{n} \cdot d \cdot \cos \gamma \cdot (\sin \alpha + \sin \beta_c). \tag{5.2}$$

The change of the central wavelength as a function of the order number is given
by the derivative with respect to n, i.e. $d\lambda_c/dn \sim 1/n^2$. With the approximation
$\Delta\lambda_c \sim 1/n^2 \cdot \Delta n$ and the particular choice of $\Delta n = 1$ (adjacent orders) this
proportionality means that the distances between the central wavelengths of adjacent
orders decrease with $1/n^2$. Hence, the wavelength intervals will overlap in high
orders. We test this with a concrete example using the grating equation for the
simple case $\alpha = 0°$ and $\gamma = 0°$, hence, $n\lambda = d \sin \beta$. To observe as many
orders as possible, we consider a grating with a relatively small grating constant
of 150 lines/mm ($d = 1/150$ mm $= 0.00666$ mm). For larger grating constants the
diffraction angle quickly increases at higher orders over 90° and would be diffracted
into the grating body. This can be easily tested using the grating equation. For this
simple case, it is

---

[1]The off-axis angle γ is the angle between the incident beam and the plane spanned by the grating
normal N and the blaze normal N_B . The path difference is shortened because the incident beam
is projected onto the diffracted beam (see also James 2007).

$$\sin \beta = \frac{n \cdot \lambda}{0.006666 \, \text{mm}}.$$

The wavelength range to be investigated in our example covers visible light from 400 to 700 nm, and all orders causing diffraction angles smaller than 90° should be considered. For the 5th order and a wavelength of 400 nm = 0.0004 mm the grating equation is

$$\sin \beta = \frac{5 \cdot 0.0004 \, \text{mm}}{0.006666 \, \text{mm}} = 0.30.$$

This corresponds to a diffraction angle $\beta = 17.5°$. All other angles can be calculated accordingly and are shown in Table 5.1. The first order covers the angular range 3.4–6.0°, the second order 6.9–12.1°. No overlap occurs in this case. For the second and third orders however, one can already observe an overlap of 1.7°, which further increases with increasing order number. The overlap of the 8th and 9th orders already shows an overlap of 85 %. This situation is shown in Fig. 5.2

Table 5.1 Overlap of the first 9 orders for a grating of 150 lines per mm and a wavelength range of 400–700 nm

Order number	β	β
n	$\lambda = 400 \, \text{nm}$	$\lambda = 700 \, \text{nm}$
1	3.4°	6.0°
2	6.9°	12.1°
3	10.4°	18.4°
4	13.9°	24.8°
5	17.5°	31.7°
6	21.1°	39.1°
7	24.8°	47.3°
8	28.7°	57.1°
9	32.7°	70.9°

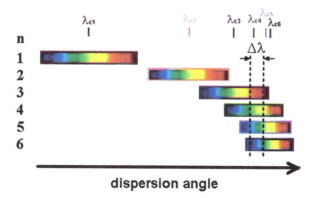

dispersion angle

Fig. 5.2 Qualitative description of wavelength intervals in different orders. The *colours* indicate various central wavelengths. The detection window $\Delta \lambda$ is defined by the optics and the CCD chip is also indicated

again. In each order the whole spectrum is imaged from blue to red. However, the central wavelengths are geometrically displaced and each order provides another wavelength interval $\Delta\lambda$ for a fixed geometric area. Through an appropriate choice of different orders within a certain geometric window one can therefore in principle reach all wavelengths in the optical spectrum.

In spectroscopy the change to a higher order is an effective means to increase the dispersion without changing the grating. Based on measurements in the first grating order and according to Eq. 5.1, the second order provides twice the dispersion, the third order triples it, etc. Then the observable wavelength interval is shortened accordingly, due to the fixed field of view (Table 5.1).

The resulting disadvantage that certain wavelengths cannot be reached can be compensated for by rotation of the grating. However, this technique is not always applicable or optimal. Standard gratings are usually blazed for a specific wavelength in first order which corresponds to proportionally shorter wavelengths in higher orders. One the one hand, the grating efficiency (i.e. the emerging light flux relative to the incident flux) is reduced with increasing order number, which means a corresponding decrease in the signal-to-noise ratio $(S/N)^2$ for a fixed exposure. On the other hand, the corresponding wavelength intervals approach each other in higher orders and finally overlap, as already described. By appropriate filtering, adjacent orders can be suppressed, but in turn the data quality may be decreased to lower S/N, depending on the details of the filters. In summary, the following should be highlighted:

- The light flux decreases with increasing order number.
- The distance between the central wavelengths is inversely proportional to the square of the order number.
- In high orders the wavelength intervals overlap, but are spectrally shifted against each other.
- The different orders each cover the entire spectrum.

The last two bullets have an interesting consequence:

If it would be possible to somehow separate different wavelength intervals of the entire spectrum from each other, which completely overlap in high orders, one could detect the whole spectrum in a single shot without rotating the grating.

5.2 The Echelle Spectrograph

The solution for the above problem (the wavelengths mix in superimposed orders), necessarily results from the fact that each order covers its own specific wavelength interval (Fig. 5.2) for a given direction. This disadvantage of the grating is now

---

[2]The signal-to-noise ratio is the quotient of useful signal to the background noise. In photography this is the contrast. A detailed description of the S/N can be found in Sect. 10.7.

Fig. 5.3 Example design of an echelle spectrograph. In this case, a prism is used as a cross-disperser

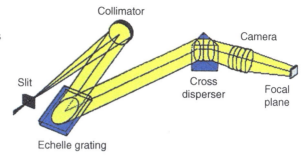

used to advantage in the echelle system. If a second grating or a prism can be positioned after the first grating, its dispersion direction, however, perpendicularly to the dispersion direction of the first dispersing element, the various orders can then be separated perpendicularly to the first dispersion direction (Fig. 5.2). When the first is an echelle grating working at very high orders for high-resolution work, this technique is called echelle spectroscopy (French: échelle = ladder) and the second (normal) grating or prism is referred to as the so-called cross-disperser. An example of the basic design is shown in Fig. 5.3. With this arrangement, the overlapping orders of the first grating are separated perpendicularly to the grating dispersion direction and imaged by a camera lens. An example is shown in Fig. 6.33 for an echelle spectrum of the sun taken with the design of Fig. 5.3.

Because each order images another portion of the entire spectrum, it is thus possible to image the entire visual spectrum with a single shot if the design and adjustment of all optics are correctly chosen. In addition, in contrast to the conventional grating spectrograph, no moving or rotating parts are needed. The echelle grating operates in high orders and is thus usually strongly tilted with respect to the optical axis, due to the necessarily large diffraction angle, as seen in Table 5.1. Each order has its central wavelength at the geometric center (Eq. 5.2), which will play an important role for our considerations. Since all central wavelengths are located exactly one above the other, it is now reasonable to align the blaze angle to this central wavelength. Thus, <u>all</u> orders would have maximum efficiency in contrast to the standard grating.

5.3 The Echelle Grating and Its Dispersion

An echelle is a diffraction grating with a relatively small grating constant of about 30–300 l/mm. According to the grating equation it is used at a large angle of incidence in high orders (typically 10–150). Figure 5.4 shows the geometric situation. The blaze angle Θ_B is between the flank normal (z-axis) and the grating normal. As for all gratings the grating equation 5.1 is also valid for the echelle:

$$n \cdot \lambda = d \cdot \cos \gamma \cdot (\sin \alpha + \sin \beta).$$

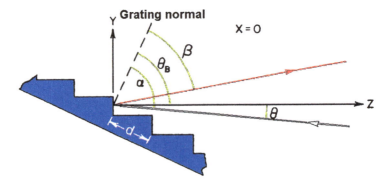

Fig. 5.4 Blazed grating in high order. The x-axis is parallel to the grating lines (Schroeder 1970)

Thus we derive the general angular dispersion

$$\frac{d\beta}{d\lambda} = \frac{n}{d \cdot \cos\gamma \cdot \cos\beta} \tag{5.3}$$

in radians/mm. Using the grating equation 5.1 we obtain

$$\frac{d\beta}{d\lambda} = \frac{\sin\alpha + \sin\beta}{\lambda \cdot \cos\beta}. \tag{5.4}$$

It should be noted, that the angular dispersion does not depend on the off-axis angle γ and, interestingly, not on the grating constant d and the order number n, but only on the angles and the wavelength.[3] If we now choose the blaze angle Θ_B so that $\alpha = \beta = \Theta_B$.[4] Using these angles we obtain the relation

$$\frac{d\beta}{d\lambda} = \frac{2}{\lambda} \cdot \tan\Theta_B. \tag{5.5}$$

Large angular dispersions require large blaze angles.

Note the non-linear influence of the blaze angle in Eq. 5.5 because of the tangent function. The blaze angle of an echelle grating is typically 63.4° or 75.9°. The terms "R2 grating" and "R4 grating" often used in the literature are explained by the fact that the tangent of the blaze angles 63.4° and 75.9° are 2 and 4, respectively. The angular dispersion is a factor of 2 larger for an R4 grating than for an R2 grating.

---

[3]This is of course due to the grating equation which allocates exactly one diffraction angle to the chosen ratio n/d at constant incidence angle α.

[4]This so-called Littrow configuration, which cannot be realized for echelle in practise, delivers relatively simple results. However, it only represents specific cases, which however provide important hints on all common configurations.

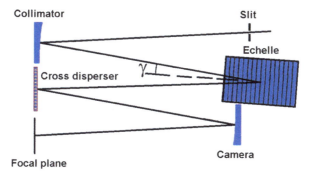

Fig. 5.5 Echelle with a central beam. The off-axis angle γ represents the half angle between the incoming and outgoing beam at the echelle grating (Schroeder 1967)

For a general description of an echelle spectrometer we now distinguish two different configurations. First, there is the so-called Θ configuration corresponding to a standard spectrograph arrangement. The collimated beam falls onto the echelle grating at a small angle $\Theta \neq 0°$. This is necessary to provide geometrical access of the outgoing beam to the cross-disperser and the subsequent camera detector unit (see Fig. 5.4). Incoming and outgoing beams lie in one plane which is defined by the grating normal and the dispersion direction. The angle γ is zero in this arrangement. Another option is the γ configuration where $\Theta = 0°$ and the off-axis angle gamma is $\gamma \neq 0°$. The angle γ lies in a plane perpendicular to Θ. Both configurations have certain advantages and disadvantages, which we will analyze later. The γ configuration has already been discussed above and is shown again in Fig. 5.5, now with mirrors as optical imaging elements.

5.4   The Geometrical Extent of the Echelle Orders

The orders delivered by an echelle spectrometer create a certain image field that needs to fit on the chip of a CCD camera. The geometrical extension of the individual orders must therefore be known. The angular extent of an echelle order $\Delta\beta$ can be calculated using the grating equation 5.1 for two consecutive orders. One calculates the respective diffraction angles for these two orders at a fixed wavelength and determines their difference. For the two orders we have

$$n \cdot \lambda = d \cdot \cos\gamma \cdot (\sin\alpha + \sin\beta_n)$$

and

$$(n + 1) \cdot \lambda = d \cdot \cos\gamma \cdot (\sin\alpha + \sin\beta_{n+1}).$$

The difference of both equations is

$$\lambda = d \cdot \cos \gamma \cdot (\sin \beta_{n+1} - \sin \beta_n).$$

We generally have $d \sin \beta / d\beta = \cos \beta$. The derivative of $\sin \beta$ with respect to β, however, can also be approximated with the difference quotient:

$$\frac{d}{d\beta} \sin \beta \approx \frac{\sin \beta_{n+1} - \sin \beta_n}{\beta_{n+1} - \beta_n} = \frac{\sin \beta_{n+1} - \sin \beta_n}{\Delta \beta}.$$

The difference between the two sine functions in the above equation can then be replaced by $(\sin \beta_{n+1} - \sin \beta_n) \approx \cos \beta \cdot \Delta\beta$ and Eq. 5.6 becomes $\lambda = d \cdot \cos \gamma \cdot \cos \beta \cdot \Delta\beta$. By re-arranging we obtain for the angular difference

$$\Delta\beta = \frac{\lambda}{d \cdot \cos \gamma \cdot \cos \beta}. \tag{5.6}$$

With a camera focal length f_{cam} we therefore have a geometrical order length in the focal plane

$$l = f_{cam} \cdot \Delta\beta.$$

To somewhat simplify this general result for $\Delta\beta$, we re-arrange the grating equation 5.1 to $\frac{n \cdot \lambda}{d \cdot \cos \gamma} = \sin \alpha + \sin \beta$ and replace α and β by a combination of Θ_B and Θ. According to Fig. 5.1 we have $\alpha = \Theta_B + \Theta$ and $\beta = \Theta_B - \Theta$. One can see that these equations describe a simple reflection at the grating. Every grating facet is the origin of Fraunhofer diffraction and additionally acts like a mirror. The path difference of the reflected and diffracted beams of successive facets[5] is described by the following grating equation with the angles Θ_B and Θ.[6] Substituting, we obtain

$$\frac{n \cdot \lambda}{d \cdot \cos \gamma} = \sin(\Theta_B + \Theta) + \sin(\Theta_B - \Theta).$$

With appropriate angle addition theorems[7] we obtain

$$\frac{n \cdot \lambda}{d \cdot \cos \gamma} = 2 \cdot \sin \Theta_B \cdot \cos \Theta. \tag{5.7}$$

---

[5] The grating facet is the reflecting surface with its normal z (see Fig. 5.4).

[6] The Blaze direction is thus defined as the preferred direction.

[7] $\sin(\alpha + \beta) = (\sin \alpha \cos \beta + \cos \alpha \sin \beta)$ and $\sin(\alpha - \beta) = (\sin \alpha \cos \beta - \cos \alpha \sin \beta)$.

We can implement this result into the above Eq. 5.6 for the diffraction angle difference and with $\beta = \Theta_B - \Theta$ we obtain

$$\Delta\beta = \frac{2 \cdot \sin\Theta_B \cdot \cos\Theta}{n \cdot \cos(\Theta_B - \Theta)}. \tag{5.8}$$

The angular extent thus depends only on the blaze angle, the selected Θ and the order number n. In addition, the order extension decreases with increasing order number. As an example, we calculate the geometric lengths of the orders $n_1 = 34$ and $n_2 = 53$ for a R2 grating with 79 lines/mm and for $\Theta = 5°$:

$$\Delta\beta_{34} = \frac{2 \cdot \sin(63,4°) \cdot \cos(5°)}{34 \cdot \cos(63,4° - 5°)} = 0,1001$$

and

$$\Delta\beta_{53} = \frac{2 \cdot \sin(63,4°) \cdot \cos(5°)}{53 \cdot \cos(63,4 - 5)} = 0,0642.$$

Note that these results are in radians. If the 34th order is the lowest imaged order, the spectral field of view is therefore $0.1001 \cdot 180/\pi = 5.7°$. To determine the CCD dimensions, we calculate as an example the order length of the 34th order in millimeters, assuming a camera focal length of $f_{cam} = 200$ mm: $l_{34} = 200$ mm $\cdot$ $0.1001 = 20.02$ mm.

If one uses a γ configuration ($\Theta = 0$ and $\gamma \neq 0$) instead of Θ configuration, the order width is then reduced by almost 20 percent, because of the missing influence of the angle Θ in the above Eq. 5.8. This fact may have important implications for the necessary apertures of the subsequent optical elements .

As with the order width, the angular dispersion can now be written with the echelle parameters Θ_B and Θ. We obtain

$$\frac{d\beta}{d\lambda} = \frac{\sin\alpha + \sin\beta}{\lambda \cdot \cos\beta} = \frac{2 \cdot \sin\Theta_B \cdot \cos\Theta}{\lambda \cdot \cos(\Theta_B - \Theta)}. \tag{5.9}$$

The angular dispersion varies explicitly with the angle of incidence Θ.

5.5 Central Wavelength and Order Number

We now turn to a fundamental property of the echelle grating—the constancy of the product of central wavelength and order number. If Eq. 5.7 is re-arranged to $\lambda \cdot n$ we obtain

$$\lambda \cdot n = 2 \cdot d \cdot \cos\gamma \cdot \cos\Theta \cdot \sin\Theta_B = \text{const.} \equiv K. \tag{5.10}$$

The right part of this equation represents a constant for a given spectrometer. The product of wavelength and order number is thus a constant K in wavelength units. However, this applies only to the central wavelengths, since we have chosen the blaze angle as a preferred direction. Hence, we obtain the following relation mentioned at the beginning of the Sect. 5.1:

$$\lambda_c(n) = \frac{K}{n}. \tag{5.11}$$

The central wavelengths are inversely proportional to the order number n.

Thus, in high orders we observe shorter wavelengths and smaller angular extents. The order length increases with the wavelength. Considering the spectral distance between two successive orders $\Delta\lambda_c$, we obtain

$$\Delta\lambda_c(n, n+1) \equiv \lambda_c(n) - \lambda_c(n+1) \Rightarrow \Delta\lambda_c(n, n+1) = \frac{K}{n} - \frac{K}{n+1} \approx \frac{K}{n^2}. \tag{5.12}$$

The order distances in wavelength units decrease with $1/n^2$.

5.6 The Spectral Extent of the Echelle Orders

After considering the geometric extension of the echelle orders, we now turn to the spectral range accessible with an echelle spektrometer. The so-called "free spectral range" $\Delta\lambda$ of a grating defines the wavelength range for a given order, which does not overlap with an adjacent order. The angular ranges of a certain order and its adjacent one are then identical. Overlap takes place if, for example, the long-wavelength end $\lambda + \Delta\lambda$ of order n has the same diffraction angle β as the wavelength λ in order $n + 1$. For this case λ is unconstrained here. For the mathematical description we take again the general grating equation 5.1 for the two orders mentioned:

$$n \cdot (\lambda + \Delta\lambda) = d \cdot \cos\gamma \cdot (\sin\alpha + \sin\beta)$$

and

$$(n + 1) \cdot \lambda = d \cdot \cos\gamma \cdot (\sin\alpha + \sin\beta).$$

Hence, we have $n \cdot (\lambda + \Delta\lambda) = (n + 1) \cdot \lambda$. Of course, a similar relationship for the short-wavelength end $\lambda - \Delta\lambda$ of order n can be estimated, as well: $n \cdot (\lambda - \Delta\lambda) = (n - 1) \cdot \lambda$. Both conditions deliver the free spectral range

$$\Delta\lambda = \frac{\lambda}{n}.$$

Apparently, the free spectral range decreases at higher orders. According to the grating equation, a grating with half the groove density works at twice the order numbers and thus also possesses half the free spectral range. If the same wavelength interval is observed (e.g., visible light), the order number must be doubled because of the reduction of the free spectral range. The important question how many orders are visible for a given wavelength interval and grating, can be answered with Eq. 5.10. As an example, we again use the above-mentioned configuration with $\Theta_B = 63.4°$, $\Theta = 5°$, $\gamma = 0°$ and $d = 1/(79\,\text{lines/mm})$. The constant K is calculated to be

$$\text{K} = 2 \cdot 1/79\,\text{mm} \cdot 10^6\,\text{nm/mm} \cdot \cos(5) \cdot \sin(63,4) = 2,255 \cdot 10^4\,\text{nm}.$$

If we now determine the orders in which we observe the wavelength $\lambda = 400\,\text{mm}$, we obtain by substitution in Eq. 5.11

$$n = \frac{\text{K}}{\lambda_c} = \frac{2.255 \cdot 10^4\,\text{nm}}{400\,\text{nm}} = 56.4.$$

For the selected grating, light of 400 nm is thus imaged near the center of the 56th order and analogously 700 nm is imaged near the center of the 32th order. The visible spectral range from 400 to 700 nm is thus divided into 24 orders, all of which must be imaged by the CCD chip.[8]

Using the above equations for the order width $\Delta\beta$ (Eqs. 5.6 and 5.8) we can now express the free spectral range $\Delta\lambda$ again with the echelle parameters:

$$\Delta\beta = \frac{\lambda}{d \cdot \cos\gamma \cdot \cos\beta} = \frac{2 \cdot \sin\Theta_B \cdot \cos\Theta}{n \cdot \cos\beta}$$

and hence

$$\lambda = \frac{2 \cdot d \cdot \cos\gamma \cdot \sin\Theta_B \cdot \cos\Theta}{n}.$$

We multiply with λ on both sides $\lambda^2 = \frac{\lambda}{n} \cdot 2 \cdot d \cdot \cos\gamma \cdot \sin\Theta_B \cdot \cos\Theta$ and thus obtain the free spectral range:

$$\Delta\lambda \equiv \frac{\lambda}{n} = \frac{\lambda^2}{2 \cdot d \cdot \cos\gamma \cdot \cos\Theta \cdot \sin\Theta_B}. \tag{5.13}$$

---

[8]More precisely, in the range of orders 32–56 we observe the wavelengths 395.61–704.69 nm.

The free spectral range covers the wavelength interval between two adjacent orders at identical viewing angle β without overlap of the wavelength intervals. One can now choose β so that it lies at the order ends. However, since β at the beginning of an order is the same as β at the end of the previous order, the free spectral range of this order is expressed accordingly.

The result for the free spectral range can also be derived from the relation 5.12. Replacing K again by $\lambda_c \cdot n$ one obtains

$$\Delta\lambda_c(n, n+1) \approx \frac{\lambda_c \cdot n}{n^2}$$

or

$$\lambda_c \approx n \cdot \Delta\lambda_c(n, n+1) \tag{5.14}$$

as an immediate consequence of 5.11.[9] The wavelength difference between two successive and non-overlapping orders multiplied by the order number therefore gives the central wavelength λ_c. On the other hand, if one knows the central wavelengths of the orders, one can very easily determine their spectral lengths.

5.7   Tilted Lines

Considering an enlarged portion of Fig. 5.11, we notice that all absorption lines are tilted with respect to the order normal (Fig. 5.6). To understand this behavior it requires some geometric abstraction because it is due to the fact that the two dispersing elements (echelle and cross-disperser) work in two different dimensions. If a straight slit is oriented parallel to the grating lines of the echelle, a beam, e.g., coming from one end of the slit falling on the grating center has a very small but finite angle γ relative to the optical axis (Fig. 5.5). We will now show that the tilting of the monochromatic slit image (e.g. narrow spectral lines) is caused by just this so-called off-axis angle. Since we have two dispersing elements in the spectrometer which need to be considered separately, we introduce the indices "e" for echelle and "c" for the cross-disperser. First we consider the echelle grating and the diffraction angle change β_e as a function of the off-axis angle γ_e, i.e. the derivative of the diffraction angle function $\beta_e(\gamma_e)$ to the variable γ_e. For the derivation we begin again with the general equation 5.1 for the echelle grating

$$\frac{n_e \cdot \lambda}{d_e} = \cos\gamma_e \cdot (\sin\alpha_e + \sin\beta_e)$$

---

[9]One can also obtained this result from the comparison of the dispersion equation and the angular extent of one order (Eq. 5.6). It is also possible to derive the angular extent $\Delta\beta$ from the Eqs. 5.14 and 5.4.

Fig. 5.6 Tilt of the slit image or spectral lines. Detail of an echelle spectrum of the sun taken with a Perkin-Elmer echelle spectrograph of the authors

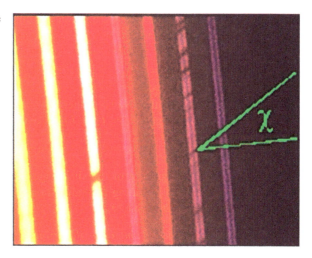

and form this expression to $\sin \beta_e$

$$\sin \beta_e = \frac{n_e \cdot \lambda}{d_e \cdot \cos \gamma_e} - \sin \alpha_e.$$

Now we calculate the derivative with respect to γ_e and separate the desired expression $\frac{d\beta_e}{d\gamma_e}$ with the chain rule:

$$\frac{d(\sin \beta_e)}{d\gamma_e} = \frac{d(\sin \beta_e)}{d\beta_e} \cdot \frac{d\beta_e}{d\gamma_e} = \frac{n_e \cdot \lambda}{d_e} \cdot \frac{d}{d\gamma_e}(\cos \gamma_e)^{-1}.$$

The single inner and outer derivatives yield

$$\cos \beta_e \cdot \frac{d\beta_e}{d\gamma_e} = \frac{n_e \cdot \lambda}{d_e} \cdot (-1)(\cos \gamma_e)^{-2} \cdot (-\sin \gamma_e) = \frac{n_e \cdot \lambda}{d_e} \cdot \frac{\sin \gamma_e}{\cos^2 \gamma_e}.$$

Hence

$$\frac{d\beta_e}{d\gamma_e} = \frac{n_e \cdot \lambda}{d_e} \cdot \frac{\tan \gamma_e}{\cos \beta_e \cdot \cos \gamma_e}.$$

Again inserting the grating equation 5.1 we finally obtain

$$\frac{d\beta_e}{d\gamma_e} = \tan \gamma_e \cdot \frac{\sin \alpha_e + \sin \beta_e}{\cos \beta_e}. \tag{5.15}$$

We now consider the geometric interpretation of the computed derivative. The angle γ_e spans a plane that is perpendicular to the β_e plane. Then the slit length has the direction of the γ_e coordinate (see Fig. 5.1). Due to this orthogonality the derivative $d\beta_e/d\gamma_e$ is tangent to the gradient $\tan \chi$ at the diffraction angle β_e:

$$\frac{d\beta_e}{d\gamma_e} = \tan \chi = \tan \gamma_e \cdot \frac{\sin \alpha_e + \sin \beta_e}{\cos \beta_e}.$$

If the derivative is not zero, the result is a tilted slit image with respect to the original orientation. If we choose again $\alpha_e = \beta_e = \Theta_B$, we obtain

$$\tan \chi = 2 \cdot \tan \gamma_e \cdot \tan \Theta_B. \tag{5.16}$$

The cause for the misalignment of the slit image, and thus of all the resolution elements in the spectrum including the calibration lines, is therefore the off-axis angle γ_e and the large blaze angle Θ_B.

The inclination χ of the slit image, i.e. the resolution elements, increases with increasing blaze and off-axis angles.

An R2 grating with $\tan \Theta_B = 2$ supplies $\tan \chi = 4 \cdot \tan \gamma_e$. Even a small off-axis angle, for example $\gamma_e = 5°$, already causes a tilt angle χ of almost 20°. Figure 5.7 shows this tilt as a function of γ_e and Θ_B, respectively. Figure 5.6 shows the tilt χ of the line elements in practice as enlarged portion of Fig. 5.11 at Hα.

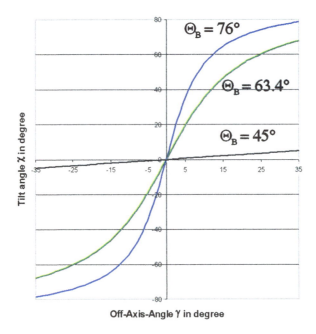

Fig. 5.7 Tilt of the slit image, depending on the blaze angle Θ_B. The *blue, green* and *black lines* represent $\Theta_B = 76, 63.4$ and $45°$

At first glance, this tilting seems a disadvantage, especially since it is sensitive to the angle γ_e especially at large blaze angles, as they are provided by echelle gratings. One could avoid the tilt through a rotation of the slit by the angle χ, but then one would reduced the spectral resolution because the slit width projected in the dispersion direction would increase. To avoid this, one would have to reduce the slit width by a factor of $\cos \chi$. This, however, would reduce the light flux and thus consequently also the overall efficiency of the spectrometer (a quantitative approach is provided by Eq. 5.28 in Sect. 5.11). One solves this problem by keeping in mind that χ is wavelength independent according to Eq. 5.16. The tilt is the same for all wavelengths and does not follow the curvature of the orders. Figure 5.8 illustrates this situation qualitatively. As a consequence, this effect can be compensated for by a simple rotation of the CCD camera by the angle χ. However, this method is limited, of course, by the rotation of the pixel rows, since the orders are no longer orthogonal and thus perhaps a larger CCD chip is required. However, one should realize the important fact that Eq. 5.16 is only an approximation, which is strictly valid only for a certain direction, e.g., the central wavelengths. The general relationship is still illustrated by Eq. 5.16, i.e. the wavelength position within the orders and the corresponding diffraction angle $\beta_e = \beta_e(\lambda)$ will have a certain influence on the tilt. The tilt angle will therefore somewhat vary along the orders. This inclination angle change can be very small, depending on the grating, but can also be up to $10°$. We will discuss this point separately in Chap. 6.

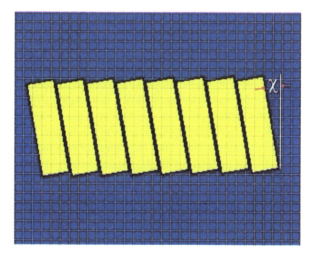

Fig. 5.8 Principle depiction of the slit image tilt with respect to the orthogonal orientation of the CCD chip and its pixels

We now determine the effect of γ_e on the shape of the entire slit image $\beta_e(\gamma)$.[10]
We use again Eq. 5.15

$$\frac{d\beta_e}{d\gamma_e} = \tan\gamma_e \cdot \frac{\sin\alpha_e + \sin\beta_e}{\cos\beta_e}$$

and approximate $\tan\gamma_e$ by γ_e. This is possible because of the small angular extent
of the slit in most cases:

$$\frac{d\beta_e}{d\gamma_e} \approx \gamma_e \cdot \frac{\sin\alpha_e + \sin\beta_e}{\cos\beta_e}.$$

To obtain β_e we integrate this derivative

$$\int \frac{d\beta_e}{d\gamma_e} d\gamma_e \equiv \beta_e(\gamma_e) = \frac{\sin\alpha_e + \sin\beta_e}{\cos\beta_e} \cdot \int \gamma_e d\gamma_e$$

and obtain

$$\beta_e(\gamma_e) = \frac{\sin\alpha_e + \sin_e\beta_e}{\cos\beta_e} \cdot \frac{1}{2}\gamma_e^2 + C.$$

For the diffraction factor before the integral, we use again $\alpha_e = \beta_e = \Theta_B$ and get

$$\beta_e(\gamma_e) = \frac{1}{2} \cdot \gamma_e^2 \cdot 2 \cdot \tan\Theta_B + C = \gamma_e^2 \cdot \tan\Theta_B + C.$$

For the R2 grating, for example, we therefor obtain $\beta_e(\gamma_e) = 2 \cdot \gamma_e^2 + C$. The
image of the complete slit is thus generally a parabola. The still unknown integration
constant C is determined by considering the diffraction angle β_e on the optical axis.
We index γ_e specifically on the optical axis with an additional zero, i.e. $\gamma_{e,0}$. For a Θ
configuration we have $\gamma_{e,0} = 0$, i.e. $\beta_e(\gamma_{e,0}) = 2 \cdot \gamma_{e,0}^2 + C = C = \beta_{e,0}$ if $\beta_{e,0} \neq 0$
due to the chosen coordinate system. Thus, the integration constant is determined.
According to Eq. 5.16 the inclination angle is then $\tan\chi = 0$. At the center the
line is, hence, orientated perpendicular to the order. The slit has a (small) γ_e angle
expansion and provides thus a parabolic spectral line, even in the Θ configuration.
However, a star has only an extremely small angular size (resolution limit) and in
this case the parabola of stellar spectra is not visible. The curvature of the slit image
is only visible at full illumination of the slit, e.g., for calibration measurements with
line emission lamps.

---

[10]We consider the *entire* slit image and its position in the spectrum because a beam falling onto the
grating at angle γ_e represents a certain geometric position in the slit.

Fig. 5.9 Distortion of the spectral image of a fiber aperture. Sector from a Neon spectrum obtained with a Perkin-Elmer echelle spectrograph of the authors

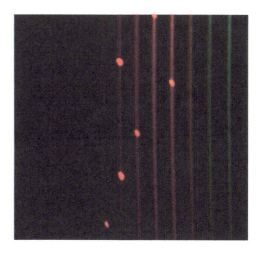

For a γ configuration γ_e has the finite value $\gamma_{e,0}$. Hence, also χ on the optical axis has a finite value, which is given by $\tan \chi = 2 \cdot \tan \gamma_{e,0} \cdot \tan \Theta_B$. The γ configuration delivers lines tilted by the angle χ and, in addition, the lines have a parabolic form. For the integration constant we determine in this case the defining equation $\beta_e(\gamma_{e,0}) = \beta_{e,0} = 2 \cdot \gamma_{e,0}^2 + C$. Thus $C = 0$.

As already mentioned above, the slit is widened by a factor $1/\cos \chi$ because of the inclination. Of course, this is not only valid for a slit but for any extended aperture, including a pinhole, a fiber output or the seeing disk. Such round apertures are deformed into ellipses (Fig. 5.9).

5.8 Curved Orders

Another effect in echelle spectra is the tilt and curvature of the orders against the dispersion direction of the echelle in the form of a curve, as shown in Fig. 5.11. The cause of this tilting is the cross-disperser, whose dispersion direction is generally perpendicular to that of the echelle. Because of this orthogonality, the tangent to the respective order is calculated to be

$$\frac{d\beta_c}{d\beta_e} = \tan \Psi$$

with Ψ the gradient angle of the tangent to the order (Fig. 5.10). One obtains the desired derivation from the conversion of Eq. 5.3 with $d\beta = \frac{n}{d \cdot \cos \gamma \cdot \cos \beta} \cdot d\lambda$ for echelle and cross-disperser to be

$$\frac{d\beta_c}{d\beta_e} \equiv \tan \Psi = \frac{d_e n_c}{d_c n_e} \frac{\cos \gamma_e}{\cos \gamma_c} \frac{\cos \beta_e}{\cos \beta_c}. \tag{5.17}$$

Fig. 5.10 Orientation of the echelle dispersion direction resulting from the echelle grating and cross-disperser for a fixed wavelength. The *black curve* represents part of an order

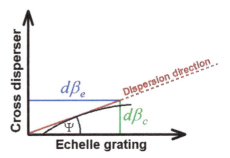

The indices e and c again indicate the echelle grating and the cross-disperser.[11] Equation 5.17 arises from the fact that both dispersion elements (grating and prism) are perpendicular to each other. Hence, the situation for a fixed wavelength, i.e. for a fixed point in the two-dimensional spectrum, can be represented graphically by Fig. 5.10.

For further considerations we now assume a grating cross-disperser. The grating equation applies again, now indexed by c:

$$\frac{n_c \cdot \lambda}{d_c} = \cos \gamma_c \cdot (\sin \alpha_c + \sin \beta_c).$$

α_c and β_c are the angle of incidence and the diffraction angle again, and γ_c is the off-axis angle of the cross-disperser. In this context we need to make clear the following important fact: The diffraction angle β_e of the echelle grating is parallel to the diffraction direction of the cross-disperser and thus automatically acts as its variable off-axis angle γ_c. If the optical axis of the spectrometer is perpendicular to the cross-disperser grating and has the corresponding diffraction $\beta_{e,0}$, then the off-axis angle has the value $\gamma_c = \beta_e - \beta_{e,0}$. The calculation of the curvature of the orders can now be performed as fully equivalent to the curvature of the spectral lines. We obtain

$$\frac{d\beta_c}{d\gamma_c} = \tan \gamma_c \cdot \frac{\sin \alpha_c + \sin \beta_c}{\cos \beta_c}$$

and by an additional approximation of the tan function and subsequent integration

$$\beta_c(\gamma_c) = \frac{\sin \alpha_c + \sin \beta_c}{\cos \beta_c} \cdot \frac{1}{2}\gamma_c^2 + C.$$

----

[11]Note that $d\lambda$ cannot be indexed and thus formally cancels out in the formula.

The orders are thus also parabolically curved. Depending on the order width (or grating constant) the curvature is clearly visible, since the change of the diffraction angle β_e is significantly larger than the γ_e extent of the slit.

The tilt and curvature of the echelle orders is determined by the behavior of the cross-disperser.

The tilt and curvature of the orders is no problem for subsequent data reduction because corresponding reduction software is able to collapse all the orders depicted in two dimensions to a one-dimensional spectrum regardless of its trace form (see Chap. 12).

5.9 Remarks About the Cross-Disperser

To separate the individual echelle orders from each other, it is sufficient to place a cross-disperser in the beam path between the echelle grating and the camera optics. If this cross-disperser has an angular dispersion $d\beta_c/d\lambda$, then the distance between two adjacent orders i and j after their separation is calculated for a camera lens of focal length f_{cam} to be

$$\Delta x_{ij} = f_{cam} \cdot \Delta \lambda_{ij} \cdot \frac{d\beta_c}{d\lambda} \tag{5.18}$$

with $\Delta\lambda_{ij}$ the wavelength difference between the central wavelengths of the two adjacent orders. Note the two indices i and j, which relate to non-constant differences between the central wavelengths of different orders. Therefore the distances Δx_{ij} can be greatly different. These considerations apply to both a grating and a prism as cross-disperser. Before one decides on a design, one should take both the efficiency and the dispersion behaviour into consideration. On the one hand, dispersion gratings possess a sufficiently high linear dispersion. This ensures that even in extreme wavelength ranges, the orders are sufficiently well separated. Because of the partly lower and also non-linear dispersion of prisms, the camera optics should then be appropriately selected to adequately separate the orders. The order distances are reduced because of the prism dispersion $dn/d\lambda$ decreasing with λ in the red wavelength range (see Figs. 2.17 and 5.11). The orders can even overlap in the worst case. This so-called "cross-talk" between adjacent orders is, of course, undesirable and would make reliable data reduction impossible. On the other hand, a prism has a constant efficiency over all wavelengths. The use of a grating cross-disperser can cause significant intensity degradation providing a lower signal-to-noise ratio S/N at extreme orders. When choosing a grating or prism, these aspects must be necessarily considered. If a prism cross-disperser sufficiently separates the wavelengths important to the observer, it is preferable in place of a grating (Fig. 5.12).

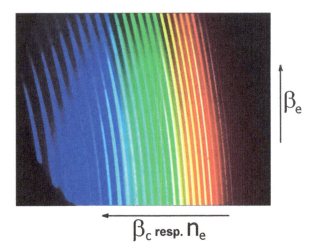

Fig. 5.11 Order orientation with a prism cross-disperser

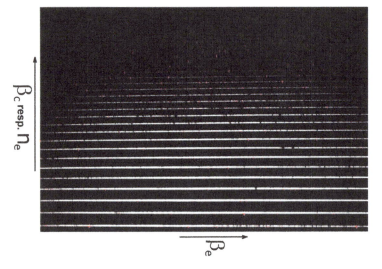

Fig. 5.12 Orientation of echelle orders with a grating cross-disperser. Spectrum of Aldebaran plus Thorium lines in *red* (Avila et al. 2007)

5.10 The Spectral Resolving Power of an Echelle Spectrometer

Using again $\alpha = \beta = \Theta_B$, the theoretical resolving power of the echelle grating $R = \lambda/\Delta\lambda = n \cdot N$ (N = total number of grooves) can be expressed again with the blaze angle. By simply re-arranging Eq. 5.3 we obtain

$$n = \frac{d\beta}{d\lambda} \cdot d \cdot \cos\gamma \cdot \cos\Theta_B. \qquad (5.19)$$

The total number of grating lines (or grooves) is the product of the grating size W and the grating constant g. In addition, we have $d = 1/g$ and we obtain from Eq. 5.9 (with $\gamma = 0$) and Eq. 5.19

$$R = \frac{\lambda}{\Delta\lambda} = n \cdot N = \frac{2W}{\lambda} \cdot \sin\Theta_B. \qquad (5.20)$$

With a large blaze angle we achieve a high spectral resolution.

Since we are working at very large angles β, Fig. 5.4 now illustrates the relationships better than Fig. 2.24. Thus we need gratings with a large blaze angle to obtain large angular dispersion and high resolution. Such gratings are available from various manufacturers. In echelle mode all orders are geometrically stacked above each other. Positioning the blaze direction onto the geometric center of the respective orders and identifying this as the central wavelength, all other central wavelengths are necessarily blazed as well. Therefore, the echelle grating has a significant advantage compared to standard gratings. Because of the tangent function in Eq. 5.5 one can dramatically increase the angular dispersion with increased angle of incidence.

In Sect. 2.6.3 we have already found a relation for the spectral resolving power. In the following we will derive an appropriate equation for the echelle spectrometer. We use again the general equation 2.88:

$$R = \frac{f_{cam}}{d_D} \cdot \frac{(\sin\alpha + \sin\beta)}{\cos\beta}.$$

We replace the trigonometric functions in the right side of the equation again by the diffraction factor lambda $\lambda \cdot d\beta/d\lambda$ (see Eq. 5.4) and obtain

$$R = \frac{f_{cam}}{d_D} \cdot \lambda \cdot \frac{d\beta}{d\lambda}$$

or specifically for the echelle with Eqs. 5.3 and 5.5

$$R = \frac{f_{cam}}{d_D} \cdot \frac{2 \cdot \sin\Theta_B \cdot \cos\Theta}{\cos(\Theta_B - \Theta)}. \qquad (5.21)$$

Of course, the spectral resolving power R explicitly dependents on the angle of incidence Θ. If we compare this last expression with the angular expansion $\Delta\beta$ (Eq. 5.8) we can also write

$$R = \frac{f_{cam}}{d_D} \cdot n \cdot \Delta\beta = \frac{l_n}{d_D} \qquad (5.22)$$

with l_n the geometric length of order n. The spectral resolving power of the echelle is therefore defined by the ratio of order length and full width at half maximum (FWHM) of the spectral element d_D. As for the standard spectrograph, d_D is determined again using the convolution of all broadening elements (Eq. 2.98):

$$d_D = \sqrt{\left(A \cdot \frac{f_{cam}}{f_{col}}\right)^2 \cdot \left(s^2 + \sigma_{col}^2\right) + \sigma_{grating}^2 + \sigma_{cam}^2 + p^2}. \tag{5.23}$$

Because of complexity, the conditions are not obvious and must be calculated individually. Therefore, we derive a simple approximation for the resolving power with known echelle parameters. In this approximation we neglect all elements in Eq. 5.23, except the slit width s. For ideal optics and not too large f ratios, including adequate sampling (see Nyquist criterion in Sect. 2.6.6), optical and pixel diffraction effects can be neglected compared to the slit:

$$d_D \approx s \cdot A \cdot \frac{f_{cam}}{f_{col}} = s \cdot \frac{\cos \alpha}{\cos \beta} \cdot \frac{f_{cam}}{f_{col}}. \tag{5.24}$$

With $\alpha = \Theta_B + \Theta$ and $\beta = \Theta_B - \Theta$ we obtain

$$d_D \approx s \cdot \frac{\cos(\Theta_B + \Theta)}{\cos(\Theta_B - \Theta)} \cdot \frac{f_{cam}}{f_{col}}$$

and hence for R

$$R \approx \frac{f_{col}}{s} \cdot \frac{\cos(\Theta_B - \Theta)}{\cos(\Theta_B + \Theta)} \cdot \frac{2 \cdot \sin \Theta_B \cdot \cos \Theta}{\cos(\Theta_B - \Theta)}.$$

Thus we obtain as the approximation for the spectral resolving power of an echelle system:

$$R \approx \frac{f_{col}}{s} \cdot \frac{2 \cdot \sin \Theta_B \cdot \cos \Theta}{\cos(\Theta_B + \Theta)}. \tag{5.25}$$

We use Eq. 5.25 to calculate the spectral resolving power for different angles of incidence Θ on an R2 grating with $\theta_B = 63.4°$, a collimator focal length of $f_{col} = 100$ mm and a slit width of $50\,\mu$m (Table 5.2).

When the angle Θ increases, the diffraction factor $\frac{2 \cdot \sin \Theta_B \cdot \cos \Theta}{\cos(\Theta_B - \Theta)}$ decreases (see Eq. 5.21). At the same time, the reciprocal anamorphic magnification factor rises strongly, so that the spectral resolving power for $\Theta = 0°$ to $\Theta = 10°$ increases from $R = 8{,}000$ to $R = 12{,}358$. The angle Θ obviously plays an important role and must be considered in the analysis. As noted below (see Sect. 6.3.3), the grating efficiency in particular strongly decreases and the width increases with increasing Θ. Depending on the camera optics, a compromise will be perhaps necessary here. In

Table 5.2 Resolving power of an echelle spectrometer with $f_{col} = 100$ mm and a slit width of 50 μm at different angles Θ

Angle Θ	Diffraction factor $\lambda \cdot d\beta/d\lambda$	Anamorphic magnification factor A	Resolving power R
$-10°$	6.18	2.09	5,914
$-9°$	5.85	1.93	6,075
$-8°$	5.56	1.78	6,245
$-7°$	5.30	1.65	6,423
$-6°$	5.06	1.53	6,610
$-5°$	4.85	1.42	6,809
$-4°$	4.65	1.33	7,018
$-3°$	4.47	1.23	7,241
$-2°$	4.30	1.15	7,478
$-1°$	4.14	1.07	7,730
$0°$	4.00	1.00	8,000
$+1°$	3.87	0.93	8,289
$+2°$	3.74	0.87	8,601
$+3°$	3.62	0.81	8,937
$+4°$	3.51	0.75	9,301
$+5°$	3.40	0.70	9,697
$+6°$	3.31	0.65	10,129
$+7°$	3.21	0.61	10,604
$+8°$	3.12	0.56	11,128
$+9°$	3.04	0.52	11,709
$+10°$	2.96	0.48	12,358

this context it is still important that R is independent of the off-axis angle γ which has therefore no effect on the resolving power.

Considering Table 5.2 further, one notices that the grating can obviously be operated in two modes—with positive or negative angle of incidence Θ. Depending on the application one should decide which configuration is more favourable. We will return to this issue again at the end of the chapter. To get a shorter equation for the resolving power, we apply the technically non realizable case $\Theta = 0$ in Eq. 5.25 and get

$$R \approx 2 \cdot \frac{f_{col}}{s} \cdot \tan \Theta_B$$

i.e. in the case of an R2 grating we obtain an estimate for the spectral resolving power with the easy-to-remember formula

$$R \approx 4 \cdot \frac{f_{col}}{s}.$$

For an R2 echelle system with $f_{col} = 100$ mm in Littrow configuration and a slit width of $50\,\mu$m at $\Theta = 0$ the resolving power is R = 8,000 (see also Table 5.2).

So far, we have assumed a variable spectral resolving power via an adjustable slit width. If one chooses the slit width to be smaller than the focal diameter of the seeing disk of the star one loses light, of course. To make use of the entire stellar light flux, we need to replace the slit width s by the seeing disk. Then we use the relation

$$s = \frac{S['']}{3600} \cdot \frac{\pi}{180} \cdot f = \frac{S[''] \cdot f}{206.265}$$

with $S['']$ the seeing in arcseconds and f the focal length of the telescope in mm. If we insert the seeing disk into Eq. 5.25 we obtain

$$R \approx \frac{206265}{S['']} \cdot \frac{f_{col}}{f} \cdot \frac{2 \cdot \sin \Theta_B \cdot \cos \Theta}{\cos(\Theta_B + \Theta)}$$

and for the special case of an R2 grating in Littrow configuration

$$R \approx \frac{825060}{S['']} \cdot \frac{f_{col}}{f}.$$

5.11   The Total Efficiency of the Echelle Spectrograph

To characterize a spectrometer, inter alia, the so-called total efficiency is used. It is defined by the product of the resolving power and the light throughput. Both parameters should, of course, be as large as possible to ensure a maximum number of photons within a spectral element $\Delta\lambda$ during the exposure time. Schroeder (1970) and Hutley (1982) derived a general relationship for the overall efficiency from a two-dimensional consideration. For this purpose we first consider how the intensity I is generally calculated.

The intensity is the product of the radiance B in [W/cm^2 sr] and the etendue[12] E of the optical system in [cm^2 sr]. For the etendue we shall use the well-known approximation equation which applies for the case of small solid angles Ω:

$$E \approx \frac{A_1 \cdot A_2}{r^2} = A_1 \cdot \Omega.$$

A_1 is the radiating area and A_2 the receiving area at distance r. For a spectrometer system one can identify the following parameters for the calculation of the etendue on the telescope side:

---

[12]The etendue is defined by the product of the receiving surface A and the solid angle Ω.

- The telescope primary mirror is the emitting surface $A_1 = D^2 \cdot \pi/4$
- The slit is the receiving surface $A_2 = s \cdot h$
- The telescope focal length f is the distance $r = f$

With s the slit width and h the illuminated part of the slit length (diameter or FWHM of the seeing disk in the case of a point-source) again. The parameters on the spectrometer side can be determined as follows:

- The slit is the emitting surface $A_1 = s \cdot h$
- The collimator is the receiving surface $A_2 = d_{col}^2 \cdot \pi/4$
- The collimator focal length f_{col} is the distance $r = f_{col}$

Because of etendue conservation we can write

$$D^2 \cdot \frac{\pi}{4} \cdot \frac{s \cdot h}{f^2} = d_{col}^2 \cdot \frac{\pi}{4} \cdot \frac{s \cdot h}{f_{col}^2} \tag{5.26}$$

or

$$\left(\frac{D}{f}\right)^2 \cdot \frac{\pi}{4} \cdot s \cdot h = \left(\frac{d_{col}}{f_{col}}\right)^2 \cdot \frac{\pi}{4} \cdot s \cdot h.$$

Between parentheses in the last equation one finds the reciprocal of f-number of the telescope and of the collimator, respectively. The etendue conservation is therefore completely equivalent to the f-number conservation. As mentioned at the beginning of this section, the spectrometer efficiency is the product of the light throughput and the spectral resolving power R. The throughput is the intensity transported through the spectrometer[13] in W or $photons/second$ and is thus proportional to the etendue of the system. We first use the general equation determining the spectral resolution

$$R = \frac{f_{cam}}{d_D} \cdot \lambda \cdot \frac{d\beta}{d\lambda}$$

and the approximation for d_D

$$d_D \approx s \cdot A \cdot \frac{f_{cam}}{f_{col}} = s \cdot \frac{\cos \alpha}{\cos \beta} \cdot \frac{f_{cam}}{f_{col}}$$

(see Eq. 5.24) and obtain by substituting and using the f-number conservation $D/f = d_{col}/f_{col}$

$$R = \frac{d_{col}}{\frac{s}{f} \cdot \frac{\cos \alpha}{\cos \beta} \cdot D} \cdot \lambda \cdot \frac{d\beta}{d\lambda}. \tag{5.27}$$

---

[13]For simplicity, the transmission of the spectrometer is neglected here.

We now multiply the etendue on the telescope side and get the general result for the overall efficiency:

$$R \cdot E = \frac{d_{col}}{\frac{s}{f} \cdot \frac{\cos \alpha}{\cos \beta} \cdot D} \cdot \lambda \cdot \frac{d\beta}{d\lambda} \cdot D^2 \cdot \frac{\pi}{4} \cdot \frac{s \cdot h}{f^2}.$$

The unit of the overall efficiency $R \cdot E$ is [cm^2 sr], and thus identical to the unit of the etendue. Considering again for simplicity the Littrow case $\alpha = \beta$ (see also Schroeder 1970), we obtain

$$R \cdot E = \frac{d_{col} \cdot \pi}{4} \cdot D \cdot \frac{h}{f} \cdot \lambda \cdot \frac{d\beta}{d\lambda}. \qquad (5.28)$$

For a given telescope-spectrometer system, the product of resolving power R and etendue E is constant. The spectrometer efficiency can only be improved if either the collimator diameter or the blaze angle is enlarged.

This general equation 5.28 with the illuminated slit length h is important for the case of extended objects. For the spectroscopy of stellar objects (point sources) we have $h \cdot s = s^2 \cdot \frac{\pi}{4}$ and we obtain

$$R \cdot E = \frac{d_{col} \cdot \pi^2}{4^2} \cdot D \cdot \frac{s}{f} \cdot \lambda \cdot \frac{d\beta}{d\lambda}.$$

We replace the diffraction factor $\lambda \cdot \frac{d\beta}{d\lambda}$ for the above case $\alpha = \beta$ again by $2 \cdot \tan \Theta_B$ and the seeing disk s by the seeing S in arcseconds:

$$s = \frac{S['']}{3600} \cdot \frac{\pi}{180} \cdot f = \frac{S[''] \cdot f}{206.265}.$$

Hence, we get

$$R \cdot E = 5,98 \cdot 10^{-6} \cdot d_{col} \cdot \tan \Theta_B \cdot D \cdot S[''].$$

The proportionality to D, and in particular to $S['']$ is surprising at first glance, but can be explained by the fact that the resolution is inversely and the etendue is directly proportional to the square of these two parameters. Metaphorically speaking: Since the product $R \cdot E$ is constant for a given optical system, a large R can only be achieved with a small seeing disk or a narrow slit. Again, in order to achieve maximum system efficiency, the diameter of the collimated beam (or of the echelle grating) and the blaze angle should be a maximum. Since the blaze angle for echelle gratings is very large, such gratings provide high system efficiencies.

With the best possible echelle gratings and their high blaze angles we achieve high system efficiencies.

In some cases one can find different results for the efficiency in the literature, e.g., in Pyo (2003).

5.12   Comparison Between Echelle and Standard Spectrographs

In order to compare the angular dispersion of an echelle and a standard grating, we consider again Eq. 5.9 for the echelle. Since for the echelle the term $\lambda \cdot d\beta/d\lambda = 2 \cdot \tan \Theta_B$ obviously only depends on the blaze angle Θ_B, we compare the diffraction factor $\lambda \cdot d\beta/d\lambda$ for the two grating types (Fig. 5.13). For an echelle R2 grating this is simply a constant of value 4. Comparing this with blazed gratings of 600 and 1,200 l/mm, which are typically used in conventional spectroscopy, shows that an echelle grating has a significantly higher dispersion than that of a regular grating, due to high order numbers used. Conversely, this means that an echelle grating needs shorter camera focal lengths than a standard grating for the same dispersion and thus correspondingly has more light power. For the case that the echelle grating has a similar efficiency as a standard grating, the echelle would have a clear advantage. The difference between the angles $\alpha - \beta = 2\Theta$ should be kept small, since any deviation from the blaze angle has a negative effect on the grating efficiency and thus in a loss of light in the spectrometer (see Sect. 5.13.1). On the other hand, a positive Θ delivers a greater spectral resolving power. However, Θ must be positive

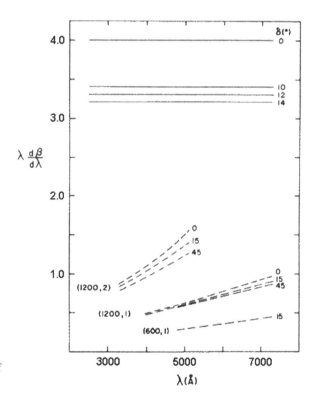

Fig. 5.13 Diffraction factor for echelle and standard gratings at different angles of incidence (Schroeder 1970)

in any case, because $\Theta = 0$ cannot be realized in this configuration due to the corresponding impossible back-reflection.

At all wavelengths an Echelle grating has a significantly higher angular dispersion than a normal grating.

According to Eq. 5.25 we have

$$R = \frac{f_{col}}{s} \cdot \frac{2 \cdot \sin \Theta_B \cdot \cos \Theta}{\cos(\Theta_B + \Theta)}.$$

Moreover, we know from Eq. 5.10 that $\lambda \cdot n$ is a constant:

$$\lambda \cdot n = 2 \cdot d \cdot \cos \gamma \cdot \cos \Theta \cdot \sin \Theta_B.$$

The term $2 \cdot \cos \Theta \cdot \sin \Theta_B$ appears in both equations. Therefore, we can write

$$R = \frac{f_{col}}{s} \cdot \frac{\lambda \cdot n}{d \cdot \cos \gamma \cdot \cos(\Theta_B + \Theta)}.$$

All parameters in this equation except λ and n are constant for a given spectrometer. We hence obtain the important proportionality

$$R \sim \lambda \cdot n$$

i.e. the resolving power of an echelle spectrograph, e.g., at the central wavelengths (i.e. from order to order) is constant. Within an order this is not the case, of course, since n = const. and λ varies. Accordingly, this is the same for a standard spectrograph, which is operated in only one single order in its simplest form. We consider again the definition of the spectral Resolving power $R = \lambda/\Delta\lambda$. Then we get, with the proportionality for R

$$\frac{\lambda}{\Delta\lambda} \sim \lambda \cdot n$$

or

$$\Delta\lambda \sim \frac{1}{n}.$$

The spectral resolution of an echelle system increases with the order number. However, we already know this from Chap. 2. Each order of an echelle spectrograph therefore corresponds to a low-resolution standard spectrograph.

5.13   The Blaze Efficiency of an Echelle Grating

In the last section of this chapter we deal with the efficiency distribution of the echelle orders. We start again with the general intensity equation 2.55 of a grating, which is composed of the product of the diffraction function B of the single slit, the interference function G of all grating grooves N and the incident intensity I_0:

$$I(\delta, \psi) = I_0 \cdot B(\delta) \cdot G(\psi). \tag{5.29}$$

B and G change their intensity depending on their phases δ and ψ:

$$B(\delta) = \left(\frac{\sin \delta}{\delta}\right)^2$$

$$G(\psi) = \left(\frac{\sin(N\psi)}{\sin \psi}\right)^2.$$

For the general case they are

$$\delta = \frac{k \cdot s}{2} \cdot (\sin(\alpha - \Theta_B) + \sin(\beta - \Theta_B))$$

$$\psi = \frac{k \cdot d}{2} \cdot (\sin \alpha + \sin \beta)$$

with $k = 2\pi/\lambda$ again denoting the wave vector. The parameters s and d denote the dimension of the single slit and the distance between single slits (grating constant), respectively, the angular size α the angle of incidence on the grating and β the diffraction angle, with the illuminated part of the facet (perpendicular to the direction of incidence) s^*. In this context, the function $B(\alpha, \beta) = (\sin \delta/\delta)^2$ is called the blaze function. It is the (envelope) efficiency distribution, which determines the echelle efficiency for the order maxima. This efficiency distribution is the reason why the signal-to-noise ratio decreases nearly symmetrically at the order edges. A theoretical calculation of the blaze function is possible with Eq. 2.77 from Chap. 2.

The angular extent of an echelle order $\Delta\beta$ (Eq. 5.8) is typically only a few degrees and each wavelength within a corresponding order will occur in the vicinity of the blaze efficiency maximum. This applies to each order. An echelle is therefore blazed for relatively broad wavelength ranges[14] and therefore shows superior efficiency compared to standard gratings (see Fig. 5.14). [t]

> *An echelle grating works near the blaze angle at all wavelengths, and has a superior efficiency, in comparison to other grating types.*

Because of this diffraction effect the signal-to-noise ratio is reduced at the order edges, of course. The whole situation is illustrated in Fig. 5.15. In addition,

---

[14] According to the diffraction function of the rectangular slit, the angle range is $\frac{\lambda}{s}$.

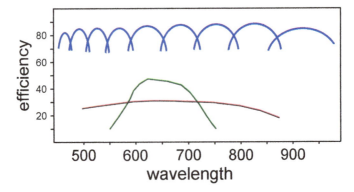

Fig. 5.14 Basic efficiency comparison between a holographic grating (*red*), a blazed grating (*green*) and an echelle grating (*blue*) over the entire visible spectrum. The echelle grating works close to the Blaze angle at all wavelengths and has a superior efficiency compared to the other two grating types. The green curve may well show a different form with respect to the efficiency. The width of all single echelle order efficiencies reflects the corresponding order length

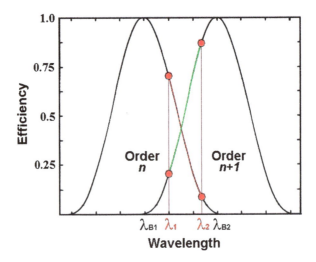

Fig. 5.15 Sketch of the efficiencies in two adjacent orders. Identical wavelength intervals in the orders are indicated in *red* and *green*. Both wavelengths λ_1 and λ_2 are imaged in the two orders n and $n + 1$. The two blaze peaks of the orders are positioned at λ_{B_1} and λ_{B_2}

the variable angular resolution in each order is problematic for subsequent data reduction. According to Eqs. 5.3 and 5.4 the angular dispersion changes by a few percent within one order. Therefore, calibration lamps with only a few lines (e.g., Neon), are useless. A sufficient number of calibration lines as well as a solid algorithm for wavelength calibration are necessary for each order (see Sects. 12.9.1 and 12.9.2).

Because of the nonlinear dispersion within the orders, many lines are needed for wavelength calibration.

These drawbacks, the variable efficiency performance and the non-linear dispersion, are the price to pay for the advantageous two-dimensional format of the echelle spectrograph and its highly effective capture of the entire spectral range.

5.13.1   The Shadowing

As we have already mentioned in Sect. 5.3, the echelle must be operated either in Θ or γ configuration to make the diffracted light available for the camera. However, for the case of a Θ configuration, shadowing effects can occur due to the stepped arrangement of the grating facets. The available blaze surface is then limited for a given blaze angle Θ_B (Fig. 5.16). The restriction of the blaze surface has a negative effect on the diffracted light flux and thus directly on the efficiency of the grating and the wavelength dependence of the blaze function. Starting with the blaze surface $s = d \cdot \cos \Theta_B$ we calculate the fraction s' of the illuminated area for the angle Θ (Fig. 5.16). The facet length, i.e. the length of a groove step in the grating, is $x = s \cdot \tan \Theta_B$. For the non-illuminated part of the blaze surface Δs we have

$$\Delta s \equiv s - s' = x \cdot \tan \Theta$$

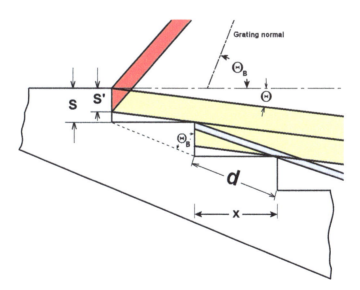

Fig. 5.16 Effective blaze surface as a function of the incident beam angle Θ (*yellow*) and the resulting beam diameter (*blue* and *red*) for a single order. The *blue beam* represents a large angle of incidence Θ and therefore illuminates a smaller area of the facet

and we obtain

$$\Delta s = s \cdot \tan \Theta_B \cdot \tan \Theta.$$

Hence, the effective blaze area s' is

$$s' = s \cdot (1 - \tan \Theta_B \cdot \tan \Theta). \tag{5.30}$$

Obviously, s' becomes smaller with increasing Θ. If we now substitute $s = d \cdot \cos \Theta_B$ in Eq. 5.30, we have

$$s' = d \cdot \left(\frac{\cos \Theta \cos \Theta_B - \sin \Theta_B \sin \Theta}{\cos \Theta} \right).$$

Using the appropriate addition theorem for angles,[15] we obtain as an alternative result for the illuminated part of the facet s'

$$s' = \frac{d \cdot \cos(\Theta_B + \Theta)}{\cos \Theta} = \frac{d \cdot \cos \alpha}{\cos \Theta}$$

or

$$s' \cdot \cos \Theta = d \cdot \cos \alpha.$$

Due to the facet shadowing, one would now suggest that the overall efficiency $R \cdot E$ of the spectrometer changes with incident angle Θ. However, this is not the case. Because of the angle of incidence Θ in an echelle spectrograph, only the facets s' influence the dispersion. Therefore, we can assume in a simplified way, that the spectrometer side etendue is limited by the grating. We obtain this "effective etendue", i.e. the shadowing-reduced effective grating area from Eq. 5.30 by summation over all facets:

$$W_l' = W_L \cdot (1 - \tan \Theta \tan \Theta_B)$$

with W_L the grating length in the dispersion direction. With the grating width W_B perpendicular to the dispersion direction, as well as the considerations for the etendue in 5.11 (Eq. 5.26), the spectrometer side etendue is calculated to be

$$E = W_B \cdot W_L' \cdot \frac{s \cdot h}{f_{col}^2}.$$

---

[15] $\cos(\alpha + \beta) = (\cos \alpha \cos \beta - \sin \alpha \sin \beta)$.

Using Eq. 5.25 for the resolving power, the overall efficiency of the spectrometer is

$$R \cdot E = \frac{f_{col}}{s} \cdot \frac{2 \cdot \sin \Theta_B \cdot \cos \Theta}{\cos(\Theta_B + \Theta)} \cdot W_B \cdot W_L \cdot \frac{s \cdot h}{f_{col}^2} \cdot (1 - \tan \Theta \tan \Theta_B).$$

Expanding and using again the addition theorem for $\cos(\alpha + \beta)$ we get the relation

$$R \cdot E = 2 \cdot W_B \cdot W_L \cdot \frac{h}{f_{col}} \cdot \tan \Theta_B$$

which does not depend on Θ. This result is completely equivalent to Eq. 5.28, i.e. the system overall efficiency is also constant for a "shadowed" grating by oblique light incidence.

We still now have the question of how the shadowing affects the blaze function. Substituting the phase difference δ (Eq. 2.70) in the general blaze function (Eq. 2.53), we obtain the sinc function 2.73

$$B(\alpha, \beta, \Theta_B) = \mathrm{sinc}^2 \left(\frac{1}{2} \cdot k \cdot s \cdot (\sin(\alpha - \Theta_B) + \sin(\beta - \Theta_B)) \right).$$

For oblique light incidence and corresponding shadowing, the facet segment s must be substituted. For the illuminated part of the facet we have $s^* = s' \cdot \cos \Theta = d \cdot \cos \alpha$ and thus for the blaze function

$$B(\alpha, \beta, \Theta_B) = \mathrm{sinc}^2 \left(\frac{1}{2} \cdot k \cdot d \cdot \cos \alpha \cdot (\sin(\alpha - \Theta_B) + \sin(\beta - \Theta_B)) \right). \quad (5.31)$$

By analogy to the intensity distribution of an Airy disk, the effective Blaze surface (which depends on the angle of incidence) generates a non-constant efficiency function. The larger Θ the smaller the effective facet surface and correspondingly wider is the associated blaze function.

The efficiency within the individual orders becomes broader with increasing incidence angle Θ, by analogy to a diffraction function.

We finally obtain for the angular blaze function (Eq. 2.77)

$$B(\alpha, \beta, \Theta_B) = \left[\frac{\sin \left(\frac{n \cdot \pi \cdot s^*}{d} \cdot (\cos \theta_B - \sin \theta_B / \tan \frac{1}{2}(\alpha + \beta)) \right)}{\frac{n \cdot \pi \cdot s^*}{d} \cdot (\cos \theta_B - \sin \theta_B / \tan \frac{1}{2}(\alpha + \beta))} \right]^2 \quad (5.32)$$

which we now can use for a comparison between a shadowed and unshadowed blaze surface. As shown in Fig. 5.17 the blaze function significantly decreases if the grating is illuminated under a certain angle Θ. We will further discuss the consequences of this important effect in Chap. 6.

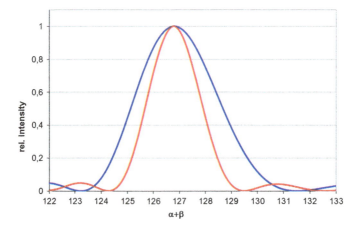

Fig. 5.17 Comparison of two normalized blaze functions for the Littrow case $\Theta = 0°$ (*red*) and $\Theta = 10°$ (*blue*) depending on the total angle $\alpha + \beta$. The grating constant of the R2 grating is 31.6 1/mm. We consider the order of $n = 90$. For the Littrow case without vignetting we have $s = d \cdot \cos \Theta_B$ and for the case with vignetting we have $s^* = d \cdot \cos \alpha$

Mathematical résumé

1. With a large blaze angle we obtain a large angular dispersion.
2. The central wavelengths are inversely proportional to the order number n.
3. The distance between the orders measured in units of wavelength is proportional to $1/n^2$.
4. The inclination χ of the slit image (in resolution elements) grows with increasing blaze and off-axis angle.
5. The tilt and curvature of the echelle orders is determined by the behaviour of the cross-disperser.
6. With a large blaze angle we achieve high spectral resolution.
7. With possibly geometrically large echelle gratings and their large blaze angles we achieve high system efficiencies.
8. For all wavelengths an echelle grating has a significantly higher angular dispersion than a normal grating.
9. For all orders and wavelengths an echelle grating works near the blaze angle. In comparison to other grating types, it has a superior efficiency.
10. Because of the nonlinear dispersion within the orders, many calibration lines are needed for all orders and wavelengths.
11. The efficiencies within the individual orders are analogous to a diffraction function and become larger with increasing angle of incidence θ .
12. The blaze function is at a maximum when the average of incidence and diffraction angles is just equal to the blaze angle Θ_B.

Recommended Readings for Echelle Spectroscopy

- D.J. Schroeder, *Design Considerations for Astronomical Echelle Spectrographs*, Astronomical Society of the Pacific, 1970, Vol. 82, 1253–1275
- T.S. Pyo, *IRCS Echelle Spectrograph and Data Handling*, Subaru Telescope National Astronomical Observatory, May 2003
- R.G. Bingham, *Grating Spectrometer and Spectrographs Re-examined*, Quarterly Journal of the Royal Astronomical Society, 1979, Vol. 20, 395–421
- P. Jacquinot, *The Luminosity of Spectrometers with Prisms, Gratings, or Fabry-Perot Etalons*, Journal of the Optical Society of America, 1954, Vol. 44, 761–765
- M.C. Hutley, *Diffraction Gratings, Techniques of Physics Vol. 6*, Academic Press, 1982
- E.G. Loewen, *Diffraction Gratings and Applications*, Marcel Dekker Inc., New York 1997
- J.F. James, *Spectrograph Design Fundamentals*, Cambridge University Press, New York 2007

Chapter 6
Considerations for Designing an Echelle Spectrometer

A Short Story

After we had "captured" the echelle grating and the corresponding collimator mirror from the internet, these parts were still not sufficient for the construction of a complete echelle spectrograph. The cross-disperser, the camera optics and the slit assembly at the grating rack were missing. The seller was called. After a few days a large cross-disperser prism and camera optics landed on Klaus' kitchen table. Unfortunately, the seller had thrown away the slit unit and we had to bite the bullet. Thomas just said: "A slit does not make the whole spectrograph and problem solvers are never discouraged".

6.1 General Comments on the Design

The attentive reader will have noticed that the critical components of an echelle spectrograph differ only slightly from those of a standard spectrograph. Collimator, camera and CCD must be adapted to the telescope. In both systems, the grating is located in the parallel beam behind the collimator. Up to that point, the only difference is the strong tilt of the echelle grating and the use of high orders. The only additional element is the cross-disperser. If one uses a transmission grating as cross-disperser, rotates it by 90° in the optical spectrograph axis and positions it directly in front of the camera optics, the grating orders are then geometrically separated perpendicular to the first grating direction and the detector can image many of them together. The cross-disperser only needs a small number of lines per millimeter. As a bonus, the mechanical complexity is even reduced because the constructive device for a rotatable grating is eliminated. The perfect spectrograph is only limited by the seeing and the resolving power at the detector. The aberrations of the optical elements should be below this limit. This is achieved, for example, with

© Springer-Verlag Berlin Heidelberg 2015
T. Eversberg, K. Vollmann, *Spectroscopic Instrumentation*, Springer Praxis Books,
DOI 10.1007/978-3-662-44535-8_6

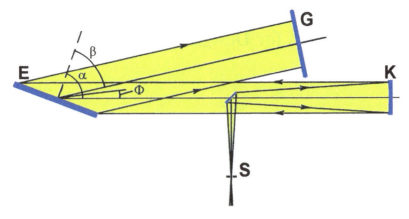

Fig. 6.1 Layout of an echelle spectrograph. S = Optical slit, K = Parabolic collimator, E = Echelle grating, G = Grating cross-disperser (Schroeder 1970)

very large focal ratios of the order of f/20 to f/30. This however makes the seeing disk correspondingly large for large telescope apertures. For faster instruments (f/10 or smaller), possible aberrations must be taken into account and must be corrected by appropriate optics. An exemplary design proposed by Schroeder (1970) suggests as simple a structure as possible and is particularly suitable for obstructing mirror telescopes[1] (Fig. 6.1).

In the present example the stellar seeing disk passes the optical slit S, and is imaged onto a parabolic mirror collimator K via deflection optics. Because of the obstructing secondary mirror in the telescope the deflection optics (mirror or prism) do not vignet and offer the advantage of an axisymmetric orientation of the collimator. If the beam from the telescope, however, strikes the collimator obliquely, potential aberrations occur which have to be corrected depending on the angle of incidence. Special and expensive off-axis collimating optics are therefore unnecessary for the above proposal. The light reflected from the collimator is then diffracted by the echelle E towards the cross-disperser G. The cross-disperser can be a transmission or reflection grating, depending on the position at which the camera is attached. Note that a reflection grating might create geometrical problems when designing the spectrograph because of a given distance to the camera lens which influences the required camera f-ratio. In this respect a transmission grating is the better alternative. Of course, a coated cross-disperser prism is also possible and would provide the greatest efficiency for optimal illumination.

---

[1]In most cases, spectrographs are used with mirror telescopes and their obstructing secondary mirror(s). As a consequence one can advantageously use a spectrograph with central tilted mirror without additional vignetting and diffraction effects.

In Fig. 6.1 we have waived the presentation of the camera lens, but we want to explicitly note that this lens crucially affects the data quality. In contrast to the classical spectrograph we now want to record a spectrum over the entire optical window and simultaneously in high resolving power. In addition to the correction of the usual aberrations (spherical aberration, off-axis aberration and field curvature) the camera lens must therefore also be as free of chromatic aberrations as possible. If this requirement is not sufficiently satisfied, the resolving power is reduced at wavelengths affected by chromatic aberration due to blurred slit imaging. For uniform spectral analysis one therefore has to adjust the resolving power at all other wavelengths. This can be done after data reduction by interpolation of the respective wavelength-to-pixel allocation (so-called re-binning). Lenses for small standard spectrographs are usually easy to find and can be adapted to any CCD pixel size p with respect to their focal length. However, if imaging optics need to have high image quality, which is normally required for echelle spectrographs, the requirements for the correction of these lenses increase. If one wants to work in the infrared wavelength range with standard camera lenses, one should use old lenses from the time before digital cameras. They do not have coatings blocking infrared light.

We now consider the various optical parameters of an echelle spectrograph in more detail[2] and start with an example echelle grating with 79 l/mm, having a blaze angle $\Theta_B = 63.4°$ (R2 grating) and an angle of incidence $\Theta = 6°$, which normally can easily be realized mechanically. The most important parameters are

- the collimated beam diameter and the resulting echelle size,
- the necessary aperture of the camera
- the spectral coverage,
- the order separation and
- the image field size.

Because of the large angle of incidence α, echelle gratings are usually made rectangular but not quadratic. The long side of the echelle is denoted by the echelle length W. It is fully illuminated, if the collimated beam has diameter $d_{col} = W \cdot \cos \alpha = W \cdot \cos(\Theta_B + \Theta)$. Then we obtain $d_{col} = 0.35W$. We thus can determine the necessary echelle length W from the telescope f-ratio. The width d_{min}, i.e., the minimum diameter of the camera for the monochromatic diffracted beam is

$$d_{min} = W \cdot \cos \beta = d_{col} \cdot \cos(\Theta_B - \Theta) / \cos(\Theta_B + \Theta). \qquad (6.1)$$

To get a concrete idea of the dimensions of the optical components, we consider the often used mirror systems of small telescopes at f/10. We now consider specifically a collimator of 100 mm focal length. According to Eq. 6.1 we then need a grating

---

[2]A complete description of all parameters can be found in Chap. 5.

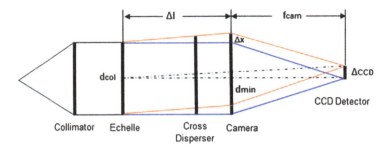

Fig. 6.2 Geometric conditions in the spectrometer. The necessary size of the cross-disperser is calculated analogously to the camera diameter d_{cam}. In this simplified representation the effect of the anamorphic magnification is neglected. Therefor $d_{min} = d_{col}$

length $W = 29$ mm for the collimated beam of $d_{col} = 10$ mm and thus obtain a minimum beam width of

$$d_{min} = \cos(63.4 - 6)/\cos(63.4 + 6) \cdot d_{col} = 1.53 \cdot 10\,\text{mm},$$

which is 15.3 mm. The minimum beam width is the diameter of the collimated central-wavelength beam at the camera entry lens. Because of the echelle (and the cross-disperser) dispersion, the camera diameter must be larger, though (spectral field of view). The respective factor Δx depends on the angular dispersion $d\beta/d\lambda$, the size of the imaged wavelength interval $\Delta\lambda$ without vignetting and the distance echelle-camera Δl (Fig. 6.2). In connection to Eq. 4.5 we have already derived the necessary minimum camera diameter d_{cam}:

$$d_{cam} = d_{min} + \Delta x = d_{min} + \Delta l \cdot \frac{\Delta CCD}{f_{cam}}$$

or with the echelle parameters already given by Eq. 6.1

$$d_{cam} = d_{col} \cdot \frac{\cos(\Theta_B - \Theta)}{\cos(\Theta_B + \Theta)} + \Delta l \cdot \frac{\Delta CCD}{f_{cam}}. \tag{6.2}$$

It is necessary that the cross-disperser be placed between the echelle grating and the camera. This might cause a great distance Δl under certain circumstances. For example, if one chooses a CCD camera with a KAF 1602 chip of $1{,}530 \times 1{,}024$ pixels of $9 \times 9\,\mu\text{m}$ size ($\Delta CCD = 13.77$ mm) at a distance $\Delta l = 100$ mm between echelle and camera of focal length of $f_{cam} = 90$ mm, one obtains a minimum aperture of 30.6 mm for the camera optics, i.e., f/3 or less ("faster"). Correspondingly, for $\Delta l = 150$ and 200 mm one obtains a minimum focal ratio of f/2.3 and f/1.9, respectively. The required camera focal ratio becomes smaller with increasing distance and the imaging optics must be designed accordingly. These are no simple requirements for objective lenses, as they also must have apochromatic

Fig. 6.3 Central wavelength for orders 27–58 for an echelle grating with 79 l/mm, a blaze angle of 63.4° and $\Theta = 6°$

properties. In any case, one builds the echelle spectrograph as short as possible to keep the dimensions of the camera as small as possible. If a reflection grating is used as a cross-disperser, the camera diameter must be adjusted because of the additional angles and thus corresponding geometrical conditions (anamorphic magnification factor). Diffraction occurs in the dispersion direction and the diffracted beam has an elliptical cross-section. Grating and camera must therefore be adapted accordingly to image the echelle spectrum without vignetting. The entire optical spectrum is now displayed in the individual orders. Figure 6.3 shows the central wavelength for each order of the above R2 grating according to Eq. 5.11.

The free spectral range also decreases with the order number (see Eq. 5.13) as well as the geometric length of red (low order number) to blue (high order number) wavelengths. Figure 6.4 shows the free spectral range $\Delta\lambda$ in different orders for the above echelle grating (left) and the corresponding geometric lengths (Eq. 5.8) for a camera of focal length $f_{cam} = 90$ mm (right). The highest order (n = 27) has a geometric length of 11.5 mm on the CCD and fits well with the KAF1602 dimensions of 9.7 mm × 13.8 mm.

Of course, the order separation must be large enough so that an overlap perpendicular to the dispersion direction (cross-talk) is avoided. For a grating cross-disperser this separation decreases from red (low orders) to blue wavelengths (high orders). This also applies for the free spectral range and for the geometric lengths. For a prism cross-disperser it goes in the opposite sense. The respective distances depend on the grating constant of the cross-disperser. Figure 6.5 shows the distances between adjacent orders perpendicular to the dispersion direction. The corresponding peak intensities in the center of the spectrum lines are used as a reference. The cross-disperser has 208 l/mm, the focal length of the camera lens is 90 mm and again the pixel size is $p = 9\,\mu$m. We thus achieve a minimum order separation of 14 pixels and can thus separate orders by up to 6 pixels with no

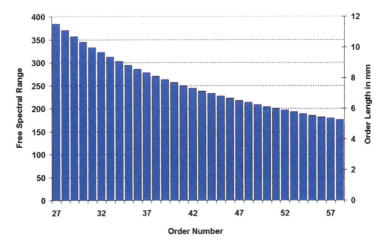

Fig. 6.4 Left axis: Free spectral range $\Delta\lambda$ in Å for all echelle orders (g = 79 l/mm) and the first order of a grating cross-disperser (g = 208 l/mm). *Right axis*: Geometric length of the corresponding orders for a camera of 90 mm focal length

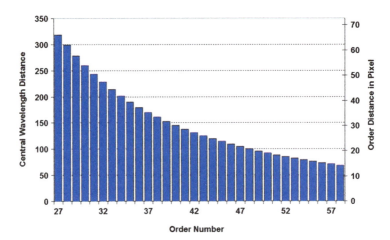

Fig. 6.5 Left axis: Central wavelength spacing of the orders in Å. *Right axis*: Separation of the orders measured perpendicular to the echelle dispersion direction in pixel differences for adjacent orders and a grating cross-disperser with 208 l/mm

overlap. Adding up all the pixel spacings of the orders, the configuration shown here requires a geometrical extension of approximately 9 mm and the CCD is therefore also sufficiently large.

Note: With regard to the design of the optical elements, there is a fundamental difference between standard and echelle spectrographs. A standard spectrograph is not limited by the size of the CCD sensor. For small CCDs and high resolving power one has only a small spectral region, but the grating can be rotated to desired wavelength ranges. In principle, the optimal adjustment of the spectrograph (with respect to the sampling factor) is always possible by the choice of appropriate focal lengths.[3] For an echelle the situation is fundamentally different. The spectrum must be completely imaged on the CCD chip, i.e., there exists a maximum camera focal length which cannot be exceeded to map even the highest order. On the other hand, one also wants to simultaneously achieve the maximum resolving power. Camera focal length and sensor size must therefore be matched. The size of the CCD is thus an additional and also limiting factor in the design of an echelle spectrograph.

6.2   Requirements for the Optical Elements

Due to the large wavelength range to be imaged in echelle spectroscopy, the resulting larger camera field of view and the chromatic aberrations, the imaging optics have to fulfill additional requirements. These requirements will now be examined in more detail. For doing so, we will again illuminate the necessary "depth of focus" (see Sect. 3.10). Considering the conditions at the telescope, spectrometer and CCD camera, this simple concept delivers a first estimate of the allowed point spread functions of collimator and camera. However, the reader should be aware that all the considerations presented here provide only the theoretical focal range of the optics in the image plane. That does not mean that the corresponding optics can satisfy the above requirements, i.e., accurate imaging inside the blurring area. Therefore, the imaging properties need to be supplemented by ray-tracing calculations of the optical elements involved. In Sects. 6.2.1 and 6.2.2 we will derive analytical relations for the sole purpose of giving the reader an idea of what requirements of the imaging elements have to be accounted for by the designer of an echelle spectrometer. The derived relationships are quite general and can therefore also be used for classical long-slit spectrographs.

---

[3]For example, for large CCD pixels, the focal length of the camera needs to be correspondingly larger not to undersample the spectrum

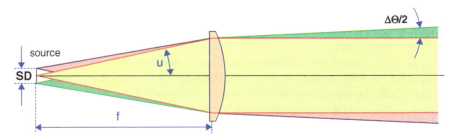

Fig. 6.6 Divergence of the collimated beam at a finite seeing disk

6.2.1 Collimator

The collimator has to transform the stellar light focused by the telescope into a parallel beam. The light beam has to be parallel to exclusively produce Fraunhofer diffraction at the grating (2.6.2). We now require the maximum residual divergence of the collimated beam created by a defocus not to degrade the spectral resolving power. In addition, the finite seeing disk causes an increase in the beam divergence. A seeing disk in the telescope focus with geometric diameter SD will introduce a geometric divergence of the collimated beam. This divergence will have the angle $\Delta\Theta$ (Fig. 6.6).

The divergence angle is given by

$$\tan\frac{\Delta\Theta}{2} = \frac{SD}{2 \cdot f_{col}}$$

or because $\Delta\Theta \ll 1$

$$2 \cdot \tan\frac{\Delta\Theta}{2} \approx \Delta\Theta = \frac{SD}{f_{col}} = \frac{SD}{d_{col} \cdot F\#}. \tag{6.3}$$

Even for the case of ideal focusing we have to accept a certain divergence of the collimated beam because of the size of the seeing disk. For example, we consider a focal 50 μm seeing disk and a collimator focal length of $f_{col} = 100$ mm. This introduces a beam divergence of $\Delta\Theta = 0.05\,\text{mm}/100\,\text{mm} \cdot 180/\pi = 0.029°$. This divergence has the consequence that different portions of the collimator beam have different angles of incidence at the grating. This in turn will introduce a superposition of different wavelengths at the same pixel position of the CCD. The spectral resolving power is thus necessarily reduced.[4]

We now estimate the influence of the angle of incidence Θ on the wavelength again via the echelle constant $K = \lambda \cdot n = 2 \cdot d \cdot \cos\gamma \cdot \cos\Theta \cdot \sin\Theta_B$ (Eq. 5.11).

---

[4]We have discussed the same issue in a slightly different context in Sect. 2.6.2.

This is done by calculating the corresponding central wavelengths behind the grating using K for Θ and $\Theta + \Delta\Theta$ and from that their difference. Alternatively, one may calculate the derivative with respect to Θ:

$$\frac{d\lambda}{d\Theta} = -\frac{2d\cos\gamma\sin\Theta\sin\Theta_B}{n} = -\lambda\tan\Theta.$$

In the linear approximation we obtain for the wavelength shift

$$\Delta\lambda \approx -\lambda\tan\Theta \cdot \Delta\Theta. \tag{6.4}$$

Using the spectral resolving power $R = \frac{\lambda}{\Delta\lambda}$ this equation can be written in the more compact form of the maximum spectral resolving power R_{max}:

$$R_{max} \leq \frac{1}{\tan\Theta \cdot \Delta\Theta}. \tag{6.5}$$

For different $\Delta\Theta$ we obtain Fig. 6.7 on a double logarithmic scale. The divergence $\Delta\Theta$ is introduced by the finite seeing disk. This divergence is different for different wavelengths. This means that, after dispersion by the grating, the largest shift is found at longer wavelengths according to Eqs. 6.3 and 6.4. For example, if we assume a required resolving power $R = 20,000$ at 700 nm, we get for an incident angle $\Theta = 5°$ the maximum allowed wavelength shift (Eq. 6.4) $\Delta\lambda = 0.35$Å. From

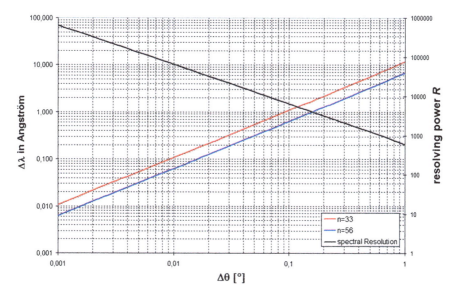

Fig. 6.7 Impact of the angle of incidence $\Theta = 5°$ on the spectral resolving power. The wavelength difference $\Delta\lambda$ has been calculated for the orders n = 33 and n = 56. The *black curve* indicates the corresponding maximum spectral resolving power

the plot or Eq. 6.4 we can derive a maximum $\Delta\Theta = 0.033°$ and using Eq. 6.6 we derive the maximum allowed seeing disk:

$$SD_{max} = \Theta_G^{max} \cdot d_{col} \cdot F\#. \tag{6.6}$$

For f/10 and $d_{col} = 25$ mm we obtain the maximum seeing disk or slit width:

$$SD_{max} = 0,033° \cdot \pi/180° \cdot 25\,\text{mm} \cdot 10 = 144\,\mu\text{m}.$$

If the full seeing disk should be used spectroscopically, we can go a step further and specify the maximum focal length of the telescope at a given seeing in arcseconds instead of the maximum seeing disk. In Eq. 6.4 we can replace the geometric focal seeing disk SD by the telescope focal length, the seeing in the sky in arc seconds and $\Delta\lambda$ and λ by the resolving power. We then obtain

$$F_{max} = \frac{206265 \cdot d_{col} \cdot F\#}{R \cdot S['] \cdot \tan\Theta}. \tag{6.7}$$

Equation 6.7 shows that bad seeing and required high resolving power may introduce problems for large telescopes. An echelle spectrograph with a small collimator diameter $d_{col} = 25$ mm and $\Theta = 5°$ used at $2''$ seeing in the sky and a desired $R = 40{,}000$ just limits the useful telescope mirror diameter to a maximum of $\frac{F_{max}}{F\#} = D_{max} = 736$ mm.

  Note: If telescope and collimator focus do not match perfectly, an additional enlargement of the blurring circle is introduced. Because the seeing disk limits the spectral resolving power, one therefore has to pay attention to the exact focusing of the collimator. The available geometric freedom becomes thereby even smaller with decreasing f-ratio of the telescope system.

  In Chap. 2 we have introduced the concept of the depth of focus. Due to the wide range of wavelengths of echelle spectrographs and due to the necessary colour correction of the optics, we need to examine this concept in more detail. First we deal with the necessary depth of focus (DoF) of the collimator. For the collimator we have:

$$F\#_{Tele} = \frac{F}{D} = F\#_{col} = \frac{f_{col}}{d_{col}}. \tag{6.8}$$

In addition, the following important relationship applies, which we have also already used in Chap. 2

$$F\#_{Tele} = \frac{DoF_{col}}{2 \cdot SD} \tag{6.9}$$

with DoF_{col} the Depth of focus (Eq. 3.47) of the collimator and SD the blurring circle allowed in this case, i.e., the seeing disk[5] at the telescope focus. As long as

---

[5]The variable SD is also the slit width, if it is smaller than the seeing disk.

the seeing disk of the telescope is greater than the blurring circle of the collimator, no deterioration of the slit image or the seeing disk will occur (in this case the beam is sufficiently collimated). Overall, we obtain for the allowed depth of focus

$$DoF_{col} = 2 \cdot F\#_{Tele} \cdot SD. \tag{6.10}$$

The seeing disk in μm is calculated to

$$SD = \frac{F\,[\text{mm}]}{206.265} \cdot S[''] \tag{6.11}$$

and hence

$$DoF_{col} = 2 \cdot F\#_{Tele} \cdot \frac{F\,[\text{mm}]}{206.265} \cdot S[''].$$

If refractive optics are used for the collimator in the above proposal (Fig. 6.1) instead of a concave mirror, apochromatic lenses must be used depending on the f-ratio of the telescope (and thus of the collimator). As the stellar disk is located on the optical axis of the spectrometer input, a field correction of the collimator is not required. Therefore, besides chromatic aberration, only spherical aberration must be sufficiently corrected. In addition, the seeing disk of the star is usually much larger than the Airy disk of the telescope, and thus the effect of the imaging depth at the collimator has a positive effect. For an f/10 system and a 50 μm seeing disk, the allowed depth of focus before and after the collimator focus is 500 μm each, in total 1 mm. Between 400 and 700 nm an achromat with $f_{col} = 250$ mm has a typical (paraxial)[6] longitudinal chromatic aberration of 600 μm to 800 μm (see Fig. 6.10). The foci of the different wavelengths differ by this amount. This has the consequence that the effective focal range of ±500 μm is also limited. As an example, we consider the spot diagrams of a 40 mm/250 mm achromat (here Qioptics 322294000) for different focus positions and three wavelengths illuminated with an f/10 telescope beam (Fig. 6.8). The spot diagrams show the intersection points for the three wavelengths 427 nm (blue), 557 nm (green) and 687 nm (red) at five different axial focus positions (−0.25, 0, 0.25, 0.5, 0.75 mm). The coloured arrows indicate the focal positions of the respective wavelengths. The green focus is at about 0 mm, the red at 0.24 mm and the blue one at 0.8 mm. According to our assumption (f/10 system, 50 μm seeing disk) we obtain a $DoF = ±500$ μm. The lower central spot diagram shows the optimum focus for all wavelengths. It is at about 0.4 mm. All other lower spot diagrams show the respective foci of all wavelengths at a symmetrical distance of 0.1 mm. Note that at the two positions 0.3 and 0.5 mm (second and fourth diagram) the allowed 50 μm for the seeing disk in the telescope and collimator focus are exceeded. That is, the effective usable focal range for this example is only ±0.1 mm instead of ±500 μm because of chromatic

---

[6]The effect of chromatic aberration changes with the illuminated lens diameter.

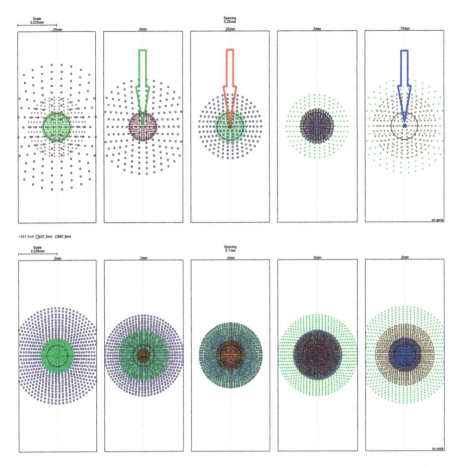

Fig. 6.8 Spot diagrams of a 250 mm/40 mm achromat (Qioptics 322294000) for different focus positions at f/10. The *arrows in the upper row* indicate the almost ideal foci of the three wavelengths. For the green focus the position is $x_{557\,nm} = 0$ mm, for the red focus this is $x_{687\,nm} = +0.24$ mm and for the blue $x_{427\,nm} = +0.8$ mm (the blue focus is represented at $+0.75$ mm and is therefore 0.05 mm in front of its optimal position). This results in a longitudinal chromatic aberration of $+0.8$ mm. Therefore, the best defocus is at $x = +0.4$ mm (*bottom row, center*). As a result, an allowable focal range of $+0.3$ to $+0.5$ mm remains, in which the spots do not exceed the seeing disk of 50 µm

aberration. For smaller focal lengths of the collimator this usable focus range is expanding accordingly. When the lens is fully illuminated, however, the aperture ratio is now only 250 mm/40 mm = 6.25 instead of f/10. Thus, the theoretically possible DoF range decreases to ± 312.5 µm and the spherical aberration increases. Due to the large longitudinal chromatic aberration there is no overlapping region anymore and the smallest spot for all wavelengths is already larger than 50 µm (see Fig. 6.9). Then we introduce an additional divergence of the collimated beam, as already discussed, which results in loss of resolving power. For the reasons

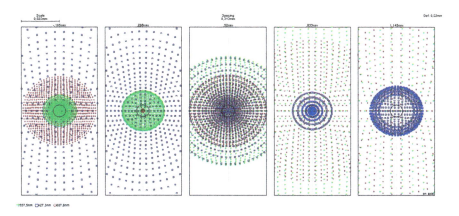

Fig. 6.9 Spot diagrams of the 250 mm/40 mm achromat (Qioptics 322294000) now for f/6.25 with full illumination. This allowed seeing disk of 50 µm is already exceeded for the best focal position of +0.52 mm. For this aperture ratio the use of an achromatic lens as a collimator is obviously not possible

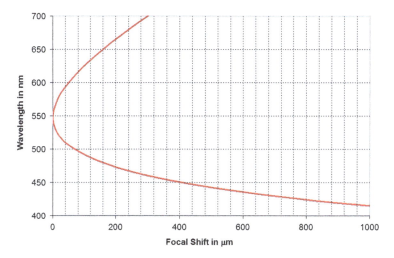

Fig. 6.10 Paraxial longitudinal chromatic aberration of a 250 mm/40 mm achromat (Qioptics 322294000). The focal shift is 0 µm at 550 nm and 719.5 µm at 427.3 nm

mentioned, we can use a simple achromat doublet as a collimator for telescopes of relatively large f-ratio. For telescopes with small f-ratios such as f/6 to f/4 the situation must be closely examined.

We note that the problem of chromatic aberration and the associated restriction of imaging depth does not exist for parabolic mirrors used as in Fig. 6.1. However, if a plane mirror at the spectrograph entrance is used and its vignetting effect should be avoided, the mirror collimator must be illuminated off-axis, which in turn introduces image errors (Fig. 6.10).

6.2.2 Camera

The requirements for the camera optics are much more stringent than for the
collimator. This becomes immediately clear by a simple example: We consider a
pixel size $p = 9\,\mu m$ and a camera focal ratio of f/4. According to Eq. 6.10 the
diffraction-limited depth of focus is then only 72 μm. This is because the camera
blurring circle is no longer determined by the very large seeing disk but by the pixel
size of the CCD camera or the slit width. If the longitudinal chromatic aberration is
larger than 72 μm, it will limit the spectral resolving power.[7] We now consider the
focal range of the camera in detail and begin, analogous to Eqs. 6.8 and 6.9 again,
with the camera f-ratio and the imaging depth of focus DoF_{cam}:

$$F\#_{cam} = \frac{f_{cam}}{d_{cam}}$$

$$F\#_{cam} = \frac{DoF_{cam}}{2 \cdot p \cdot bf}.$$

Of course, in the calculation the binning factor bf must be taken into account since
it significantly affects the size of the blurring circle. Thus we obtain

$$DoF_{cam} = \frac{2 \cdot f_{cam} \cdot \text{pixelsize} \cdot bf}{d_{cam}}.$$

For d_{cam} we obtained a good approximation with Eq. 6.2

$$d_{cam} = d_{col} \cdot \frac{\cos(\Theta_B - \Theta)}{\cos(\Theta_B + \Theta)} + \frac{\Delta l \cdot \Delta CCD}{f_{cam}}$$

or

$$d_{cam} = \frac{d_{col} \cdot f_{cam} \cdot \cos(\Theta_B - \Theta) + \Delta l \cdot \Delta CCD \cdot \cos(\Theta_B + \Theta)}{\cos(\Theta_B + \Theta) \cdot f_{cam}}.$$

Thus we obtain for DoF_{cam}

$$DoF_{cam} = \frac{2 \cdot f_{cam}^2 \cdot p \cdot bf \cdot \cos(\Theta_B + \Theta)}{d_{col} \cdot f_{cam} \cdot \cos(\Theta_B - \Theta) + \Delta l \cdot \Delta CCD \cdot \cos(\Theta_B + \Theta)}. \qquad (6.12)$$

We use again $\theta = 0$ (Littrow configuration) in order to get a simple expression:

$$DoF_{cam} = \frac{2 \cdot f_{cam}^2 \cdot p \cdot bf}{d_{col} \cdot f_{cam} + \Delta l \cdot \Delta CCD}. \qquad (6.13)$$

---

[7]This is no problem in principle, as long as the realized resolving power is sufficient.

This equation represents an approximation for the general depth of focus DoF_{cam} of the camera and reflects the relative dependencies in a simplified way. Accordingly, the depth of focus decreases with the distance Δl between echelle grating and camera. For manipulating the imaging depth during the spectrograph design, one uses Eq. 6.12 for an accurate calculation and changes the parameters available. According to this equation the angle of incidence Θ at the grating also has a non-negligible influence: With increasing angle the imaging depth increases, as well, because the anamorphic magnification factor becomes effective in this case. In Fig. 6.11 we present the numerical results of Eq. 6.12 for two angles of incidence Θ, assuming an f/10 telescope system and a collimator focal length of F_{col}=250 mm.[8] The functions in the two plots are shown for 1×1, 2×2 and 3×3 binning. Thus, with a camera focal length of 250 mm one can reach a maximum DoF of 187 μm with 1×1 binning. These are requirements that can only be achieved with optical correction. For short camera focal lengths between 50 and 100 mm corrected apochromatic lenses are required, since the f-ratios are very small. For longer focal lengths with multiple CCD pixel binning, it may be possible to use a simple achromat as camera lens (with some loss in spectral resolving power). In Sect. 6.6 we will investigate a corresponding example spectrograph in more detail.

Within one echelle order the wavelength changes only slightly. The order is almost "monochromatic", i.e., there are no large wavelength intervals, which would disturbingly widen the 2D spectrum. Thus, within one echelle order the spectral resolution remains almost constant. The other Seidel aberrations must also be

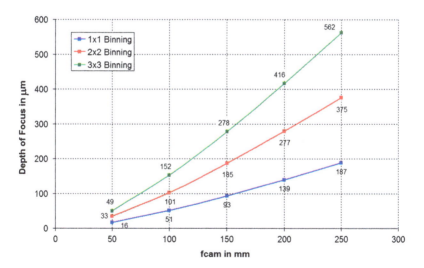

Fig. 6.11 The binning-dependent depth of focus vs. f_{cam} for $\Theta = 0°$ and $\Theta = 6°$

---

[8]We have chosen this focal length, as the largest gratings available at the most popular manufacturers have dimensions of 25×50 mm, matching an f/10 system.

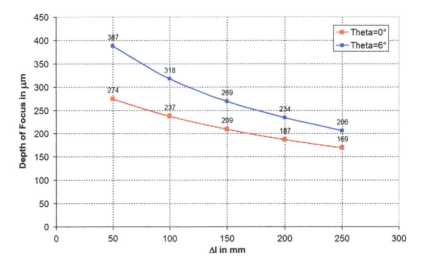

Fig. 6.12 The depth of focus as a function of the distance from the echelle camera

sufficiently corrected for the entire CCD image field. The requirements for the camera optics increase with the beam diameter, which is determined by the beam diameter after the collimator, and thus by the collimator focal length. Therefore, it is important to limit the beam diameter by a collimator focal length as short as possible (if this is allowed by the required resolving power). Hence, the distances between the echelle grating and the camera should be kept as small as possible. Nevertheless, because of the required aperture of cross-disperser and camera a structurally minimum distance for Δl will be necessary. Figure 6.12 shows the dependence of the depth of focus on the distance for two angles of incidence. The DoF significantly decreases because of the decreasing camera f-ratio with increasing Δl. The problem thus represents an optimization task. This optimization task boils down to achieving spectral resolving power R as high as possible, while keeping the resulting aberrations as small as possible.

6.3 The Choice of the Echelle Grating

The performance of a spectrometer is determined among other things by the spectral resolving power. In Sect. 5.10 we had given a simple approximate formula for the resolving power of an echelle with an R2 grating:

$$R = 4 \cdot \frac{f_{col}}{s}.$$

This equation is valid for arbitrary slit widths s. However, to use the "whole" starlight, we should use the size of the seeing disk from Eq. 6.11 instead of the slit width. Hence

$$R = 4 \cdot \frac{f_{col} \cdot 206265}{F \cdot S['']}$$

with F again the focal length of the telescope and S the seeing disk in arc seconds. The focal length of the collimator follows from the f-ratio of the telescope and the required diameter of the collimated beam d_{col}. If we additionally replace the telescope focal length F by the f-ratio and lens diameter, we finally obtain

$$R = 4 \cdot \frac{d_{col} \cdot 206265}{D \cdot S['']} = 825060 \cdot \frac{d_{col}}{D \cdot S['']}. \qquad (6.14)$$

Standard echelle sizes of 12.5×25 and 25×50 mm are available on the market. For example, if we consider a 25 cm telescope, we obtain the resolving power for the 12.5 mm grating at a seeing of 4 arc seconds

$$R = 825060 \cdot \frac{12{,}5\,\text{mm}}{250\,\text{mm} \cdot 4''} = 10313.$$

Accordingly the 25 mm grating achieves twice the resolving power. If one can, one often opts for the greater spectral resolving power and thus also for the larger grating. The simple relationship illustrates the "'advantage" of a small lens diameter, which allows one to achieve relatively large resolving power with small gratings and camera diameters. If the lens diameter increases, then all other apertures of the optical components have to increase, as well.

6.3.1   Effects of Line Density on the Spectrograph Design

The line densities typically available for echelle gratings are 31.6 and 79 lines/mm. The question now is whether it is better to work with a higher line density in low order, or vice versa. We have discussed the length and number of orders already in Sect. 5.6. The constants K for the two gratings are (see Eq. 5.10):

$$K_{79} = 2 \cdot 1/79{,}0\,\text{mm} \cdot 10^6\,\text{nm/mm} \cdot \cos(6) \cdot \sin(63, 4) = 2{,}251 \cdot 10^4\,\text{nm}$$

$$K_{31,6} = 2 \cdot 1/31{,}6\,\text{mm} \cdot 10^6\,\text{nm/mm} \cdot \cos(6) \cdot \sin(63, 4) = 5{,}628 \cdot 10^4\,\text{nm}.$$

We assume that the longest wavelength is 700 nm and the shortest is 400 nm. The corresponding order numbers are then

$$n_{79}(700\,\text{nm}) = \frac{K_{79}}{\lambda_c} = \frac{2,251 \cdot 10^4\,\text{nm}}{700\,\text{nm}} = 32,2$$

$$n_{31,6}(700\,\text{nm}) = \frac{K_{31,6}}{\lambda_c} = \frac{5,628 \cdot 10^4\,\text{nm}}{700\,\text{nm}} = 80,4$$

or

$$n_{79}(400\,\text{nm}) = \frac{K_{79}}{\lambda_c} = \frac{2,251 \cdot 10^4\,\text{nm}}{700\,\text{nm}} = 56,3$$

$$n_{31,6}(400\,\text{nm}) = \frac{K_{31,6}}{\lambda_c} = \frac{5,628 \cdot 10^4\,\text{nm}}{700\,\text{nm}} = 140,7.$$

The grating with 79 lines/mm operates within the required wavelengths between the orders 32 and 56; the grating with 31.6 lines/mm between the orders 80 and 140, i.e. 24 orders for the first and 60 for the second case. For the same wavelength range the 31.6 grating requires 2.5 times more orders, which also means that the orders must be shorter. Therefor, for the 79 grating a long and narrow CCD chip is required. For both gratings we now calculate the order lengths in radians for the longest orders in the red wavelength range $n_{79} = 32$ and $n_{31.6} = 80$:

$$\Delta\beta_{32} = \frac{2 \cdot \sin(63.4°) \cdot \cos(6°)}{32 \cdot \cos(63.4° - 6°)} = 0.1032 \tag{6.15}$$

or

$$\Delta\beta_{80} = \frac{2 \cdot \sin(63.4°) \cdot \cos(6°)}{80 \cdot \cos(63.4° - 6°)} = 0.0413 \tag{6.16}$$

and in the second step, the length of the two orders ΔL_{32} and ΔL_{80} for a given focal length f_{cam} by simple multiplication. The extent of the highest order of the grating with 79 lines/mm is precisely the factor $0.1032/0.0413 = 79/31.6 = 2.5$ times greater than that of the grating with 31.6 lines/mm. Therefore the second grating needs 2.5 times more orders for the same wavelength interval.

We now examine the extent of the echelle spectrum perpendicular to the echelle dispersion direction, because we need to know the geometrical relation of the necessary CCD chip. For this we need to know the dispersion of the cross-disperser and the angular extent to compare this with the angular extent of the orders in the echelle dispersion direction, as done in the Eqs. 6.15 and 6.16. When separating the orders the main goal is not to achieve the highest possible resolving power, but to sufficiently separate the orders. Therefore, when using a grating cross-disperser a line density of 150–300 lines/mm is sufficient. For the estimation of the CCD dimensions we assume an average value of 200 lines/mm.[9] With Eq. 2.81 we have

---

[9] We will deliver a detailed analysis of the required line density in Sect. 6.4.1.

already derived an expression for the dispersion of a grating. Accordingly, we obtain $2.01 \cdot 10^{-4}$ rad/nm for a grating of 200 lines/mm and a central wavelength of 550 nm. For the wavelength range 400–700 nm we obtain therefore an angular extent of $2.01 \cdot 10^{-4}$ rad/nm $\cdot$ 300 nm $= 0.0604$ rad. The width-to-height ratios for the two gratings are 0.0604/0.1032 (1:1.71) for the grating of 79 l/mm and 0.0604/0.0413 (1.46:1) for the grating of 31.6 l/mm, regardless of the camera focal length. The KAF 1602 CCD has $1,536 \times 1,024$ pixels of 9 μm size and thus a width-to-height ratio of 1.5:1. Thus, the width-to-height ratio of the KAF 1602 is better suited to the 31.6 grating. In addition, the chip has a length of 13.85 mm and thus a maximum focal length of 13.85 mm/0.1032 $= 134$ mm is possible for the 79 grating. On the other hand, the 31.6 grating accepts a maximum focal length of 13.85 mm/0.0604 $= 229$ mm, which additionally has a positive effect on the spectral resolving power. Hence, the 31.6 l/mm grating delivers better conditions for the spectrometer. If another CCD camera is used, the parameters need to be re-calculated, of course, but the width-to-height ratios of the commercially available cameras are very similar in most cases, so that a 31.6 grating is almost always the better choice. To what extent the significantly larger number of orders affect the data analysis, will be examined later.

Note: The FEROS spectrometer of Landessternwarte Heidelberg (see Sect. 8.4.4) uses a 79 l/mm echelle grating. For the CCD camera of the spectrograph, a chip with $2,048 \times 4,096$ pixels was selected due to the rectangular dimensions of the echellogramms.

6.3.2 Effects of the Angle of Incidence on the Order Length

Beside the line density, the angle of incidence has an effect on the spectrometer design. When changing Θ the order length changes, too. We examine this quantitatively by calculating the derivative of the function $\Delta\beta$ (Eq. 5.8) to the angle of incidence Θ:

$$\frac{d}{d\Theta}(\Delta\beta) = \frac{d}{d\Theta}\left(\frac{2 \cdot \sin\Theta_B \cdot \cos\Theta}{n \cdot \cos(\Theta_B - \Theta)}\right).$$

We obtain

$$\frac{d}{d\Theta}(\Delta\beta) = \frac{2\sin\Theta_B}{n} \cdot \frac{\cos\Theta\sin(\Theta - \Theta_B) - \sin\Theta\cos(\Theta - \Theta_B)}{\cos^2(\Theta_B - \Theta)}$$

or summarized

$$\frac{d}{d\Theta}(\Delta\beta) = -\frac{2\sin^2\Theta_B}{n \cdot \cos^2(\Theta_B - \Theta)}.$$

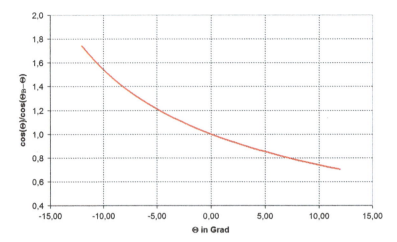

Fig. 6.13 Relative influence of the angle of incidence Θ on the spectral length of an echelle order. The ratio of the two angular functions is normalized to $\Theta = 0$

Especially for $\Theta = 0$ we obtain

$$\frac{d}{d\Theta}(\Delta\beta) = -\frac{2\sin^2\Theta_B}{n\cdot\cos^2\Theta_B} = -\frac{2}{n}\cdot\tan^2\Theta_B$$

and in particular for a 31.6 grating and $\Theta = 0$

$$\frac{d}{d\Theta}(\Delta\beta) = -\frac{2}{n}\cdot 2^2 = -\frac{8}{n}$$

that is, for an average example order n = 40, we obtain $\Delta(\Delta\beta) = -0.2\cdot\Delta\Theta$. For an angle change of $+1°$, the angular extent of order 40 is therefore reduced by $-0.2°$. For $\Theta = +10°$, this value is almost halved and for $\Theta = -10°$ the lengths is doubled. Figure 6.13 shows the ratio $\cos\Theta/(\cos\Theta_B - \Theta)$ from Eq. 5.8 for the order length as a function of Θ in a 31.6 grating. In particular, for negative angles of incidence the angular extent of the orders increases. Thus, the orders become long and correspondingly large CCD chips are needed.

6.3.3 The Influence of the Angle of Incidence on the Echelle Efficiency

With the ansatz for the diffraction theory of an echelle in Sect. 5.13 we are able to estimate the dependence of the efficiency on the angle of incidence Θ. According to Eq. 2.55 we have for the grating angles α, β and Θ_B

$$I(\alpha, \beta, \Theta_B) = I_0\cdot B(\alpha, \beta, \Theta_B)\cdot G(\alpha, \beta)$$

with B the Blaze function again and G the grating interference function:

$$G(\alpha, \beta) = \left[\frac{\sin \left(\frac{1}{2} \cdot N \cdot k \cdot d \cdot (\sin \alpha + \sin \beta) \right)}{\sin \left(\frac{1}{2} \cdot k \cdot d \cdot (\sin \alpha + \sin \beta) \right)} \right]^2 .$$

The interference function G shows maxima at the diffraction angles β_n defined by the grating equation (Fig. 2.22):

$$\sin \alpha + \sin \beta_n = \frac{n \cdot \lambda}{d \cdot \cos \gamma} .$$

The blaze function B represents the envelope of the diffraction order maxima (see Fig. 2.22). The maximum intensities at the diffraction maxima are therefore determined to be (Eq. 2.56)

$$I_n(\alpha, \beta_n, \Theta_B) = N^2 \cdot B(\alpha, \beta_n, \Theta_B).$$

We now calculate the total intensity I_n in the diffraction order n by integrating the interference function $G(\alpha, \beta)$ in the range of the diffraction angle β_n. Using the main theorem of calculus:

$$\int_{\beta_n - \frac{\Delta \beta_n}{2}}^{\beta_n + \frac{\Delta \beta_n}{2}} G(\alpha, \beta) d\beta = G(\alpha, \beta_n) \cdot \Delta \beta_n$$

with $\Delta \beta_n$ the integration range around the diffraction angle β_n. The applicable range of integration $\Delta \beta_n$ is known from diffraction theory and can be identified with the order width σ, which has already been estimated by Eq. 2.69 in Sect. 2.5.1:

$$\sigma \approx 0.85 \frac{\lambda}{d \cdot N} .$$

For the general case of an arbitrary diffraction angle β_n one has to replace the grating constant d by the effective width $d \cos \beta_n$ and we obtain for the integration interval $\Delta \beta_n$:

$$\Delta \beta_n \approx 0.85 \frac{\lambda}{N d \cos \beta_n} .$$

The equation can be considered as follows: The spectral order width $\Delta \beta_n$ in radians corresponds to the grating facet of the effective width $d \cos \beta_n$ and becomes smaller with increasing number of facets N. Again, one can see that the theoretical resolution of the grating increases with its dimension.

For the integration the average of the function $G(\alpha, \beta)$ in the domain of integration is required. This average is just the value of the blaze function at the diffraction maximum of order n, i.e., $G(\alpha, \beta_n) = N^2 \cdot B(\alpha, \beta_n, \Theta_B)$. Thus we obtain for the intensity integral in the range of the order peak

$$I_n(\alpha, \beta_n) = N^2 \cdot B(\alpha, \beta_n, \Theta_B) \cdot \Delta\beta_n = N^2 \cdot B(\alpha, \beta_n, \Theta_B) \cdot 0.85 \frac{\lambda}{Nd \cos \beta_n}.$$

To obtain information on the efficiency ϵ of the grating, we replace the echelle artificially by a perfect mirror of the same reflectivity. Because for a mirror the angles of incidence and reflection are identical, α, we obtain by analogy to the above calculation the order intensity:

$$I(\alpha) = I_0 \cdot \Delta\alpha = I_0 \cdot 0.85 \frac{\lambda}{Nd \cos \alpha}.$$

This result represents simple diffraction at a rectangular aperture of the width $Nd \cos \alpha$, with $\Delta\alpha$ the *FWHM* of the sinc function and I_0 its peak intensity. Comparing these to intensities via their ratio one then obtains the efficiency ϵ

$$\epsilon(\alpha, \beta_n) = \frac{I_n(\alpha, \beta_n)}{I(\alpha)} = \frac{N^2 \cdot B(\alpha, \beta_n, \Theta_B)}{I_0} \cdot \frac{\cos \alpha}{\cos \beta_n}. \tag{6.17}$$

According to Eq. 5.31 we have[10]

$$B(\alpha, \beta_n, \Theta_B) = B(\alpha, \beta_0, \Theta_B) \cdot \mathrm{sinc}^2 \left(\frac{1}{2} \cdot k \cdot d \cdot \cos \alpha \cdot (\sin(\alpha - \Theta_B) + \sin(\beta_n - \Theta_B)) \right).$$

The angle $\beta_0 = 2\Theta_B - \alpha$ represents the diffraction angle of the maximum Blaze intensity according to Eq. 2.74. One can show that

$$I_0 = N^2 \cdot B(\alpha, \beta_0, \Theta_B), \tag{6.18}$$

i.e., in the direction of the blaze the grating acts like a mirror. This can be made clear by "shifting" all grating facets into one single plane. The individual grating facets thereby act like a plane mirror. Diffraction at the small step structures then disappears and thus also the blaze function. Diffraction is only present for the large structures, i.e., the grating dimensions. We then only measure the grating interference function in zero order, i.e., mirror reflection.

---

[10]Note that the facet width s in the Blaze function is again replaced by the effective width $s^* = d \cos \alpha$.

In Fig. 6.14 we now consider the two possible cases $\alpha > \Theta_B > \beta$ (a) and $\alpha < \Theta_B < \beta$ (b) for operating the echelle grating (Fig. 6.15). In case (a) the illuminated facet $s^* = d \cos \alpha$ for $\Theta > 0$ is smaller than the complete facet $s = d \cos \Theta_B$. Substituting Eq. 6.18 in 6.17 we obtain for the efficiency:

$$\epsilon(\alpha, \beta_n) = \frac{\cos \alpha}{\cos \beta_n} \cdot \text{sinc}^2 \left(\frac{1}{2} \cdot k \cdot d \cdot \cos \alpha \cdot (\sin(\alpha - \Theta_B) + \sin(\beta_n - \Theta_B)) \right). \tag{6.19}$$

The result of Eq. 6.19 is depicted in Fig. 6.16. The peak intensity at $\Theta = 5°$ is about 70 % and for $\Theta = 10°$ it is already only 40 %. If large efficiencies for the blaze intensity are required one should opt for a possibly small angle Θ for the construction of a spectrometer. Note that the intensity might be reduced but the blaze function is broadened because of narrow facets. For large wavelength regions it becomes broader.

For case (b) the facet is fully illuminated. However, part of the incoming light is reflected back towards the light source. Vignetting is present. The illuminated part of the facet contributing to the diffraction intensity is then $s^* = d \cos \beta$, i.e., we again observe reduced peak intensity. In addition, however, vignetting is present which can be geometrically estimated with the factor $1 - \frac{\cos \beta}{\cos \alpha}$. The intensity reaching the camera is, hence, significantly smaller than for case (a) which potentially reduces the final S/N. Operating an echelle grating with negative angle of incidence is generally possible. However, for faint targets a γ configuration is preferable instead of a Θ configuration because of the potential loss of light. If the intensity is of

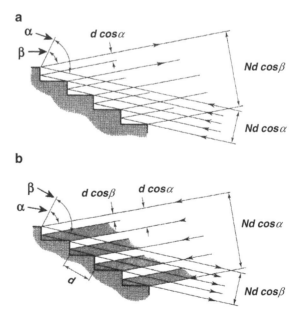

Fig. 6.14 Positive ($\alpha > \Theta_B$) and negative $\alpha < \Theta_B$ aberration of an echelle grating. (**a**) The illuminated part of a facet is smaller but the whole light flux is diffracted towards the camera. (**b**) The facet is fully illuminated but part of the incoming flux is reflected towards the source (vignetting) potentially introducing stray light in the spectrometer (Hutley 1982)

Fig. 6.15 Microscopic image of a Thorlabs 31.6 l/mm R2 grating (*Courtesy*: Bruker Nano GmbH)

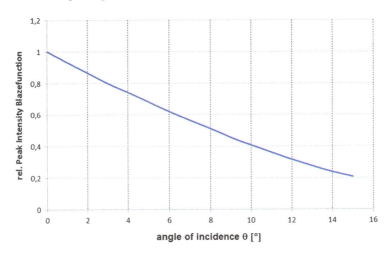

Fig. 6.16 Relative blaze peak intensity vs. angle of incidence Θ in degrees

minor importance the negative angle operation has the advantage of a smaller beam size behind the echelle grating (Fig. 6.14, bottom) and hence a larger f-ratio of the camera lens. This in turn has a positive impact on the necessary lens corrections (Fig. 6.17).

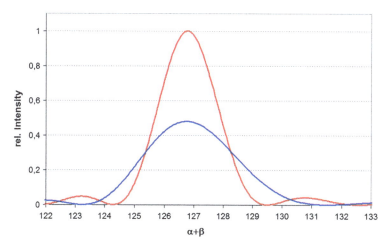

Fig. 6.17 Blaze functions for $\Theta = 0°$ (red) and $\Theta = 10°$ blue

6.4   The Choice of the Cross-Disperser

6.4.1   Grating

In the following we will discuss which grating ruling densities are necessary or useful for a grating cross-disperser. With Eq. 5.12 we already had calculated the wavelength difference between two consecutive orders:

$$\Delta\lambda_c = \lambda_c(n) - \lambda_c(n+1) = \frac{K}{n} - \frac{K}{n+1} = K \cdot \frac{(n+1) - n}{n(n+1)} \approx \frac{K}{n^2}.$$

The order separation is smallest at higher orders, i.e., at short wavelengths. If we choose the grating with 31.6 l/mm the smallest spacing can be found at order $n = 140$. We obtain

$$\Delta\lambda_c = \approx \frac{K}{n^2} = \frac{5.628 \cdot 10^4\,\mathrm{nm}}{140^2} = 2.87\,\mathrm{nm} = 28.7\,\mathring{A}.$$

To geometrically separate this spectral distance sufficiently in the focal plane the cross-disperser, apparently must have a certain linear dispersion which must be of the order of a few Å/pixel. For the linear dispersion in pixels/Å we found with Eq. 4.3

$$\frac{dx}{d\lambda} = \frac{d\beta}{d\lambda} \cdot \frac{f_{cam}}{p} = \frac{g \cdot n}{\cos\beta} \cdot \frac{f_{cam}}{p} = \frac{g \cdot n}{\sqrt{1 - (\sin\beta)^2}} \cdot \frac{f_{cam}}{p}.$$

We can replace the parameter $\sin \beta$ by the grating equation. Initially we assume for the angle of incidence $\alpha = 0$[11]:

$$\sin \beta = \frac{n \cdot \lambda}{d} = n \cdot \lambda \cdot g.$$

Thus we obtain for the linear dispersion in the first diffraction order

$$\frac{dx}{d\lambda} = \frac{g}{\sqrt{1 - (\lambda \cdot g)^2}} \cdot \frac{f_{cam}}{p}.$$

If we multiply the linear dispersion with the wavelength difference between two orders $\Delta\lambda$, we obtain the separation $\Delta\lambda$ in pixels

$$\Delta x = \frac{dx}{d\lambda} \cdot \Delta\lambda = \frac{g}{\sqrt{1 - (\lambda \cdot g)^2}} \cdot \frac{f_{cam}}{p} \cdot \Delta\lambda.$$

By using the pixel size p again and the allowed minimum distance Δx, we can re-arrange the equation to the grating constant g:

$$g = \frac{1}{\sqrt{\left(\frac{f_{cam}\cdot\Delta\lambda}{\Delta x\cdot p}\right)^2 + \lambda^2}}.$$

If we choose the mean wavelength $\lambda = 550$ nm, the minimum distance $\Delta x = 10$ pixels and the pixel size $p = 9\,\mu$m, we obtain

$$g = \frac{1}{\sqrt{\left(\frac{229\text{mm}\cdot 2.87\,\text{nm}}{10\text{Pixel}\cdot 0.009\text{mm/Pixel}}\right)^2 + 550\,\text{nm}^2}} = 137 \,\text{lines/mm}.$$

If the pixels are binned 2-fold and the distance is also held to 10 super-pixels (after binning), we get a grating constant of 271 lines/mm. A grating of 300 lines/mm would thus allow an order distance of 7.4 super pixels at 400 nm for 3-fold binning, a value that is still tolerable. In summary, grating constants of 150–300 lines/mm are sufficient to separate the orders. Of course, the choice should be a blazed grating, regardless of whether it is a reflection or transmission grating. High performance blaze gratings have an efficiency of 70–85 % at their peak, so that light losses by cross-dispersion can be kept relatively small.

---

[11]A non-zero angle of incidence can be structurally important. In this case, we use $\sin \beta = n \cdot \lambda \cdot g - \sin \alpha$.

6.4.2 Prism

Compared to a grating, a prism delivers only "one order", so that the light loss can theoretically be limited to the dielectric reflection losses. However, the significantly pronounced non-linearity of the dispersion causes a strong curvature of the orders, which can be disadvantageous in the data analysis. Because of the nonlinear wavelength behavior of the refractive index the dispersion of a prism can only be calculated numerically in detail. We already discussed this in Chap. 2. In particular, if a variable angle of incidence is to be realized, the general equation 2.14 must be used for the deflection angle δ of a prism. The angle of α_2 is given by Eq. 2.13 for a known angle of incidence α_1

$$\alpha_2 = \delta - \alpha_1 + \gamma$$

with γ the angle of refraction of the prism. The deflection angle is determined by the wavelength-dependent refractive index $n(\lambda)$ with Eq. 2.14

$$\delta(n, \gamma, \alpha_1) = \alpha_1 + \arcsin(\sin \gamma \cdot \sqrt{n^2 - \sin^2 \alpha_1} - \sin \alpha_1 \cdot \cos \gamma) - \gamma$$

where $n(\lambda)$ can again be determinded by the Sellmeier equation

$$n(\lambda) = \sqrt{1 + \frac{B_1 \lambda^2}{\lambda^2 - C_1} + \frac{B_2 \lambda^2}{\lambda^2 - C_2} + \frac{B_3 \lambda^2}{\lambda^2 - C_3}}.$$

The inverse linear dispersion D in Å/pixel is calculated again using the angular dispersion, the camera focal length f_{cam} and the binned pixel size $p \cdot B$

$$D = \frac{d\lambda}{d\delta} \cdot \frac{p \cdot B}{f_{cam}}. \tag{6.20}$$

All parameters for determining D are analytically given but not the dispersion angle. Here it is most appropriate to calculate the derivative by numerical calculation using, e.g., a spreadsheet. With the above equations one can calculate the refractive index $n(\lambda)$ and the corresponding deflection angle $\delta(n, \gamma, \alpha_1)$ in the range of 3,000–7,000 Å for wavelength steps of, e.g., 1 Å. From these data one can now calculate the approximate angular dispersion with the difference quotient

$$\frac{d\lambda}{d\delta} \approx \frac{\Delta\lambda}{\Delta\delta} = \frac{\lambda_{i+1} - \lambda_i}{\delta_{i+1} - \delta_i}.$$

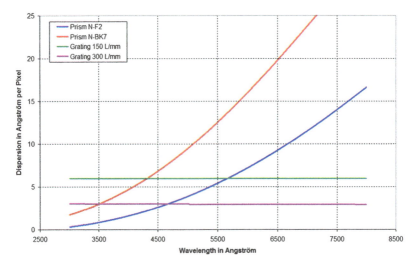

Fig. 6.18 Comparison of inverse linear dispersions for gratings and prisms in Å/pixel versus wavelength

Overall, we therefore obtain

$$D_i \approx \frac{\lambda_{i+1} - \lambda_i}{\delta_{i+1} - \delta_i} \cdot \frac{p \cdot B}{f_{cam}}$$

with i the index of the wavelength.

Figure 6.18 shows a comparison of the linear dispersion for various prisms and gratings, which were calculated for a camera focal length of $f_{cam} = 100$ mm and a pixel size of 9 μm. For the comparison, the popular prism materials N-BK7 and N-F2 were chosen and the gratings were calculated for the angle of incidence $\alpha = 0°$. The two gratings show an almost linear behavior in the entire visible wavelength range. In contrast, the prisms have a non-linear dispersion in the same interval, which exhibits itself in echelle mode by curvature of the orders. Also noticeable is the low linear dispersion of N-BK7 in the red wavelength range. This can lead to the already mentioned cross-talk of the orders. However, if the camera focal length is doubled, the linear dispersion is halved.[12] Another way to sufficiently separate the orders and to increase the total dispersion is positioning two simple BK7 prisms

---

[12]When increasing the pixel size or pixel binning, the linear dispersion is increased according to Eq. 6.20 and N-BK7 can perhaps again be used as prism material.

directly behind the other using them as a single cross-disperser unit. However, this also increases the total deflection angle, which influences the structural design of the spectrometer.

It should be noted that prisms have maximum efficiencies at specific angles of incident. In addition, one should always choose coated prisms, otherwise the advantage of the prism compared to the grating is reduced.

6.5 "SimEchelle": A Simple Echelle Simulation Program

We now will discuss two sample echelle designs and determine their main parameters—the individual order length and their distance from each other in Å, the inverse dispersion in Å/pixels per order, and in particular the necessary dimensions of the CCD chip. In principle, the calculations are not complicated but relatively extensive for the calculation of the entire spectral range. Therefore, it is helpful to perform all calculations in the form of, e.g., a spreadsheet program. By means of such a program, it is more effective to change certain parameters specifically and to optimize the spectrometer in single steps. For this purpose, we developed an Excel sheet named "SimEchelle" based on previous work by Buil (2003) in which all important equations calculate the above output parameters when entering all the important input variables. Table 6.1 lists all input and output parameters of the program. In addition to the above specified parameters, the following variables also appear as output for each order:

- Order number n
- Central wavelength λ_c in Å
- Order length in Å
- Order length in pixels
- λ_{min} in Å
- λ_{max} in Å
- Free spectral range $\Delta\lambda = \lambda_{max} - \lambda_{min}$ in Å
- Inverse Dispersion in Å/Pixel
- Order length in mm
- Order overlap in Å
- Spectral resolving power R

The program "SimEchelle" can be found at http://www.stsci.de/simechelle.xls.

Table 6.1 Input and output parameters of the simulation program "SimEchelle"

Input parameter	Output parameter
Pixel dimension of the CCD p in microns	Diffraction angle $\beta = \Theta_B - \Theta$ in degrees
Horizontal pixel number N_x	Minimum and maximum sampling factor
Vertical pixel number N_y	F/D (f-ratio) of the telescope
Telescope aperture D in mm	Seeing disk in microns
Telescope focal length F in mm	Minimum collimator diameter in mm
Slit width in μm	Maximum collimator depth of focus in mm
Binning in dispersion direction in number of pixels	Maximum camera depth of focus in mm
Collimator focal length f_{col} in mm	Minimum CCD dimension (x-y direction) in mm
Camera focal length f_{cam} in mm	Minimum grating width in mm
Seeing S in arc seconds	Minimum grating length in mm
Grating constant echelle in lines/mm	Minimum cross-disperser diameter in mm
Angle of incidence Θ in degree	Minimum camera f-ratio
Off-axis angle γ in degrees	Camera image field in degree (x-direction)
Blaze angle Θ_B in degree	Camera image field in degree (y-direction)
Grating constant cross-disperser in l/mm	Maximum slit width in mm
Distance echelle to cross-disperser in mm	
Distance cross-disperser to camera in mm	
Geometrical length of the camera lens	
Total angle at cross-disperser	
Minimum and maximum grating order	
Cross-Disperser type	Prism or grating
Glass material for prism	According to internal list
Prism angle at prism	In degrees
Angle of incidence at prism	In degrees

6.6 Project "Mini-Echelle" an Echelle Spectrograph for f/10 Telescopes

For our first example project we want to construct a relatively inexpensive spectrometer for an f/10 telescope that works as efficiently as possible. We choose a simple classical echelle concept as described in the introduction. The sensor is a relatively small, but cooled CCD camera with a KAF-400 chip and 582×752 pixels of $8.6\,\mu$m size.[13] This has the advantage of a small angular field of view with small off-axis

---

[13]Alternatively, an ICX285AL of 6.60 mm $\times$ 8.77 mm size.

aberrations. Despite the small sensor, the spectral resolving power should be as large as possible. For the echelle grating we want to use the usual devices available at the major manufacturers with dimensions 12.5 mm × 25 mm or 25 mm × 50 mm. The grating constants available are 31.6 or 79 l/mm. Because of the format advantages already outlined, we will select the grating with 31.6 lines/mm for our project and an angle of incidence $\Theta = +5°$. As a camera lens a simple achromat should be used. We already know that the correction of an achromatic doublet is limited; however, this simple lens has the advantage of a smaller dimension, which we want to take advantage of for our situation. Of course one has to make certain compromises in terms of spectral resolving power in this application. The distance behind the grating must be kept as small as possible and therefore the entire echelle spectrometer is designed as short as possible to keep the camera f-ratio as large and its aperture as small as possible. Only under these conditions can the spherical aberration be kept within acceptable limits. In addition, the object angle should be kept below 2° to also keep the off-axis aberrations small, as shown in Sect. 3.12.4. The small sensor KAF400 meets this requirement. Its size of 5.01 mm × 6.47 mm delivers directly the maximum selectable camera focal length f_{cam}. First, we calculate the order with the largest angular extension (n = 80):

$$\Delta\beta_{80} = \frac{2 \cdot \sin(63.4°) \cdot \cos(5°)}{80 \cdot \cos(63.4° - 5°)} = 0.0426 \, \text{rad}.$$

From the CCD dimensions and the focal length, we calculate the (half) image angle 1.20° × 1.54°. However, we must consider that the echelle does not only image one 2D spectral image as a standard long-slit spectrograph does but the entire CCD array. The field angle in the sensor diagonal is therefore

$$\arctan\left(\frac{\sqrt{(5.01 \, \text{mm}/2)^2 + (6.47 \, \text{mm}/2)^2}}{120 \, \text{mm}}\right) = 1.95°.$$

As we will show below, this field angle is acceptable. We align the CCD camera so that the dispersion direction of the echelle is arranged parallel to the short chip side, as we had already called for in Sect. 6.3.1. For the maximum camera focal length with a central wavelength of 7,000 Å in order 80 we obtain

$$f_{cam} = 5.01 \, \text{mm}/0.0426 \, \text{rad} = 118 \, \text{mm}.$$

We choose a camera focal length $f_{cam} = 120$ mm as it is again readily available from optics manufacturers, and we increase the minimum order to n = 82. The constant K for the grating with 31.6 lines/mm is

$$K_{31.6} = 2 \cdot 1/31.6\text{mm} \cdot 10^6 \, \text{nm/mm} \cdot \cos(5°) \cdot \sin(63.4°) = 5.6396 \cdot 10^4 \, \text{nm}.$$

We therefore find the largest central wavelength at $\lambda_C (n = 82) = 5.6396 \cdot 10^4 \, \text{nm}/82 = 6{,}878 \, \text{Å}$. As cross-disperser, we select a grating with 200 lines/mm with a dispersion of about $2.01 \cdot 10^{-4}$ rad/nm at a central wavelength of 550 nm. For the wavelength range of 400–700 nm, we obtain an angular extent of

$$2.01 \cdot 10^{-4} \, \text{rad/nm} \cdot 300 \, \text{nm} = 0.0604 \text{rad}.$$

Using a calculated camera focal length of $f_{cam} = 120$ mm, we would thus require a minimum CCD extension of 7.248 mm. Therefore, it is necessary to somewhat limit the imaged wavelength range. From the CCD dimensions of 6.47 mm, we obtain an angular extent of 6.47 mm/120 mm = 0.539 rad corresponding to a wavelength range of $0.539 \text{rad}/2.01 \cdot 10^{-4} \, \text{rad/nm} = 2{,}682 \text{Å}$. The smallest observable central wavelength on the CCD is then $6{,}878 - 2{,}682 \, \text{Å} = 4{,}196 \, \text{Å}$, in order $n = 5.6396 \cdot 10^4 \, \text{nm}/419.6 \, \text{nm} = 134.$[14] Table 6.2 shows all calculated data for each single order.

We now discuss the necessary aperture of the camera doublet. For the minimum diameter of the monochromatic beam, we obtain

$$d_{min} = d_{col} \cdot \cos(63.4° - 5°)/\cos(63.4° + 5°) = \frac{f_{col}}{F\#} \cdot 1.424.$$

For the necessary aperture of the achromat, this results in

$$d_{cam} = \frac{f_{col}}{F\#} \cdot 1.424 + \Delta l \cdot \frac{6.47 \, \text{mm}}{120 \, \text{mm}}.$$

Note: So far we have identified ΔCCD with the length of the CCD chip to keep at least the central CCD pixel row free of vignetting. However, since the entire chip will be used, we must also consider the entire chip diagonal of the camera image field (Fig. 6.19) and need to design d_{cam} accordingly.

In our example the ratio 6.47 mm/120 mm represents the (double) field angle $2 \cdot 1.54°$ of the CCD in radians (one radian = 180/pi degrees). But for echelle spectroscopy the entire chip is used. Hence, we need to use the angular field of view of the (double) diagonal for the determination of d_{cam} to avoid vignetting of the image field:

$$d_{cam} = \frac{f_{col}}{F\#} \cdot 1.424 + \Delta l \cdot 2 \cdot 1.95° \cdot \frac{\pi}{180°} = f_{col} \cdot 0.1424 + \Delta l \cdot 0.068.$$

For a given collimator focal length f_{col} and a given distance Δl between echelle and camera it is, hence, possible to determine the aperture or f-ratio of the camera lens. Due to the small field angle, the distance Δl has only half the influence of

---

[14]If one also considers the non-linearity of the angular dispersion of the cross-disper one finds a minimal central wavelength of 4,272 Å in the highest, still usable order 132.

Table 6.2 Modeling the "Mini-Echelle"

Order #	λ_c Å	Order separation Å	Order Separation Pixel	λ_{Min} Å	λ_{Max} Å	spectral range Å	Dispersion Å/Pixel	Order length mm	Over lap Å
82	6878	85	24	6835	6920	84	0.145	5.01	00
83	6795	83	23	6753	6836	83	0.143	4.96	01
84	6714	81	23	6673	6755	82	0.141	4.89	02
85	6635	79	22	6594	6675	81	0.140	4.83	03
86	6558	77	22	6518	6598	80	0.138	4.78	04
87	6482	75	21	6443	6522	79	0.136	4.72	04
88	6409	74	21	6369	6448	79	0.135	4.67	05
89	6337	72	20	6298	6375	78	0.133	4.61	06
90	6266	70	20	6228	6305	77	0.132	4.56	07
91	6197	69	19	6159	6235	76	0.130	4.51	07
92	6130	67	19	6092	6168	75	0.129	4.46	08
93	6064	66	18	6027	6101	74	0.128	4.41	09
94	6000	65	18	5963	6036	73	0.126	4.37	09
95	5936	63	18	5900	5973	73	0.125	4.32	10
96	5875	62	17	5839	5911	72	0.124	4.27	11
97	5814	61	17	5778	5850	71	0.122	4.23	11
98	5755	59	17	5720	5790	70	0.121	4.18	12
99	5697	58	16	5662	5732	70	0.120	4.15	12
100	5640	57	16	5605	5674	69	0.119	4.10	12
101	5584	56	16	5550	5618	68	0.118	4.06	13
102	5529	55	15	5495	5563	68	0.116	4.02	13
103	5475	54	15	5442	5509	67	0.115	3.98	14
104	5423	53	15	5390	5456	66	0.114	3.94	14
105	5371	52	14	5338	5404	66	0.113	3.90	14
106	5320	51	14	5288	5353	65	0.112	3.87	15
107	5271	50	14	5238	5303	65	0.111	3.83	15
108	5222	49	14	5190	5254	64	0.110	3.79	15
109	5174	48	13	5142	5206	63	0.109	3.76	16
110	5127	47	13	5096	5158	63	0.108	3.73	16
111	5081	46	13	5050	5112	62	0.107	3.69	16
112	5035	45	13	5005	5066	62	0.106	3.66	17
113	4991	45	12	4960	5021	61	0.105	3.63	17
114	4947	44	12	4917	4977	61	0.104	3.59	17
115	4904	43	12	4874	4934	60	0.103	3.56	17
116	4862	42	12	4832	4892	60	0.102	3.53	18
117	4820	42	12	4791	4850	59	0.101	3.50	18
118	4779	41	11	4750	4809	59	0.101	3.47	18
119	4739	40	11	4710	4768	58	0.100	3.44	18
120	4700	39	11	4671	4728	58	0.099	3.41	18

(continued)

Table 6.2 (continued)

Order #	λ_c Å	Order separation Å	Order Separation Pixel	λ_{Min} Å	λ_{Max} Å	spectral range Å	Dispersion Å/Pixel	Order length mm	Over lap Å
121	4661	39	11	4632	4689	57	0.098	3.38	18
122	4623	38	11	4594	4651	57	0.097	3.36	19
123	4585	38	10	4557	4613	56	0.097	3.33	19
124	4548	37	10	4520	4576	56	0.096	3.30	19
125	4512	36	10	4484	4539	55	0.095	3.27	19
126	4476	36	10	4449	4503	55	0.094	3.25	19
127	4441	35	10	4413	4468	54	0.093	3.22	19
128	4406	35	10	4379	4433	54	0.093	3.20	19
129	4372	34	10	4345	4399	54	0.092	3.17	20
130	4338	34	09	4312	4365	53	0.091	3.15	20
131	4305	33	09	4279	4331	53	0.091	3.12	20
132	4272	33	09	4246	4299	52	0.090	3.10	20
133	4240	32	09	4214	4266	52	0.089	3.08	20

By multiplying the order number n by λ_c one obtains the echelle constant $K = 5.6396 \cdot 10^4$ nm for each order

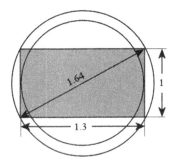

Fig. 6.19 Geometry at the CCD chip with respect to the camera aperture normalized to the short CCD side. To avoid vignetting of the entire chip, as necessary for echelle spectroscopy, the camera image field must match the chip diagonal and not the length of the chip. In the present example, the diagonal corresponds to twice the angle for the field of view $2 \cdot 1.95°$

the collimator focal length. The maximum collimator focal length is determined directly by the short grating. For the smaller 12.5 mm × 25 mm grating, we obtain for the selected f/10 telescope a maximum focal length of $f_{col} = 12.5$ mm $\cdot 10 = 125$ mm. The required grating length for full cross-illumination of the grating is

$$W = 12.5 \, \text{mm} / \cos(63.4° + 5°) = 34 \, \text{mm}.$$

Table 6.3 The necessary camera aperture d_{cam} and its f-ratio and camera blur spot size vs. the echelle—camera distance Δl

Δl [mm]	d_{cam} [mm]	f-ratio	max. spot [μm]
80	10.7	5.6	33
100	11.3	5.3	36
120	11.8	5.1	40
140	12.3	4.9	44
160	12.9	4.7	49
180	13.4	4.5	54
200	14.0	4.3	60

Thus, the smaller grating with 25 mm length is at least 9 mm too short for the maximum collimator focal length of 125 mm. In order not to degrade the resolving power of the spectrometer,[15] we choose for the spectrometer design the larger grating with the dimensions 25 mm × 50 mm which is slightly oversized. Thus, the collimator focal length can theoretically be increased to 50 mm · cos(63.4° + 5°) · 10 = 183.7 mm if the CCD dimensions are also enlarged. We will examine in the next chapter, if this opportunity is meaningful.

In the last step we need to specify the distance of the camera Δl to the echelle grating. This distance is based on the mechanical design of the echelle spectrometer. We must remember that the cross-disperser grating must be mounted directly in front of the camera lens (with a slight misalignment due to the blaze angle). For geometric reasons, the distance between the echelle and cross-disperser Δl_{CD} is at least $d_{min}/\sin(2\Theta)$. We use this measure and take a small distance of 10 mm between cross-disperser and camera lens additionally into account. In Table 6.3 the impact on the required f-ratio of the camera depending on the distance Δl is listed.

6.6.1 The Limit of an Achromatic Lens as a Camera for the Mini-Echelle

The advantage of an achromatic lens is its compact size. We have found in the last chapters that the distances between the optical components and thus indirectly their size, plays a major role in the choice of the apertures. This raises the question whether it is possible to use a simple achromat as camera for an echelle. For a collimator we have already found that this is quite possible with a sufficiently large telescope f-ratio. For the camera, we need to clarify in detail whether it is possible (with the focal lengths mentioned in the last chapter) to find an achromat, whose spot is sufficiently small not to degrade the spectral resolving power.

---

[15]Recall that the spectral resolving power increases with the collimator focal length.

Since the entire CCD camera is filled with orders, field errors—i.e. astigmatism, coma and field curvature—occur beside spherical aberrations. Additionally we have to deal with chromatic aberration. The question then arises whether the aberrations of the doublet can still be tolerated under the given conditions. Figure 6.20 displays the corresponding spot diagrams for the two edge wavelengths of the detectable spectral range from 4,272 Å (blue) to 6,878 Å (red) and the mean wavelength 5,575 Å (green). The illuminated diameter of the lens is approximately 20 mm. All spots show different diameters, defining the respective spectral resolving power.

In principle, it is now necessary to evaluate the corresponding spot diameter for each wavelength, in order to determine the resulting full-width-at-half-maximum (FWHM) of the slit image. In addition to the wavelength-dependent aberrations the focus position is also an important parameter which has extreme impact on the spot

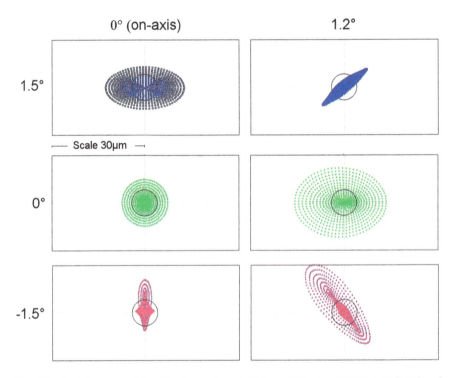

Fig. 6.20 Spot diagrams of the Qioptics achromat 31.5 mm/120 mm for the wavelengths of 4,272Å (*blue order* n = 132), 5,575Å (*green*) and 6,878Å (*red order* n = 81). The focus position is 0.0 mm (Gaussian image plane, paraxial focus). *Left column, middle picture*: on-axis position (CCD sensor center). *Right column*: Symmetrical focus positions at the order ends (1.2° from center). The largest spot of the central wavelength at the outer order edge shows a spot diameter of approximately 30 μm in the horizontal (sagittal) dispersion direction. The spectral order length has a maximum extension of ±45 Å and can therefore be considered as monochromatic for the spot diagrams. The *black circle* represents the diffraction-limited Airy disk

diagram. The best focus must therefore be determined simultaneously by means of an optimization process. The focus optimized and wavelength-dependent FWHM can then be converted to spectral resolving power using the dispersion. However, this method is very time-consuming and not effective for a simple estimate of the resolving power. We therefore choose a simplified approach and use the diameter of the largest spots at the (green) central wavelength at the edge of the order (off axis angle 1.2°). The spectral resolving power following from the blurring circle thus defines a lower limit.

We insert the maximum blurring circle σ_{cam} from the spot diagrams in Fig. 6.20 into Eq. 2.98

$$d_D^2 = \left(A \cdot \frac{f_{cam}}{f_{col}} \right)^2 \cdot \left(S^2 + \sigma_{col}^2 \right) + \sigma_{grating}^2 + \sigma_{cam}^2 + p^2$$

and then estimate the spectral resolving power with Eq. 5.21

$$R = \frac{f_{cam}}{d_D} \cdot \frac{2 \cdot \sin \Theta_B \cdot \cos \Theta}{\cos(\Theta_B - \Theta)}.$$

To maintain the optimum collimator focal length we successively increased it and determined the illuminated diameter of the camera and the resulting maximum spot size. The result is given in Table 6.4. The maximum spot diameter increases abruptly with the illuminated camera aperture. The reason for this increase is the increasing field error. Plotting the maximum blurring circles against the camera diameter in third-order d_{cam}^3 (spherical aberration increases in third order with the aperture diameter) delivers an almost linear relationship in Fig. 6.21. The dominating image error is thus spherical aberration. Using the maximum spot diameters in Table 3.1 and the two Eqs. 2.98 and 5.21 one can estimate the corresponding spectral resolving power. We then obtain the spectral resolving power depending on the collimator focal length f_{col} for different slit widths (Figs. 6.22 and 6.23).

Table 6.4 Illumination of the camera aperture with increasing collimator focal length f_{col}

f_{col} [mm]	d_{cam} [mm]	f-ratio	max. spot [μm]
60	5.5	10.9	12
80	7.2	8.3	18
100	9.0	6.7	24
120	10.7	5.6	33
140	12.4	4.8	42
160	14.2	4.2	57
180	15.9	3.8	89

The spots increase abruptly due to the increasing spherical aberration

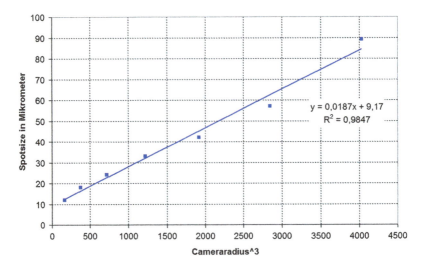

Fig. 6.21 Increase of the maximum spot diameter with σ_{col}^3. The relation is nearly linear, so the spherical aberration is the dominating image errors

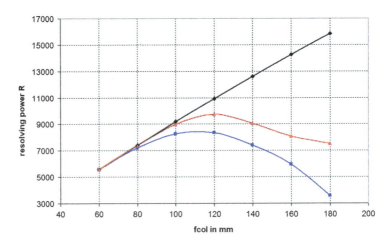

Fig. 6.22 Spectral resolving power vs. collimator focal length f_{col}. The slit width is 50 μm. The *black curve* is the resulting resolving power for the case of diffraction-limited camera spots. The *blue curve* represents the spectral resolving power when using the largest blurring circles and thus gives a lower limit. The spots at long wavelengths are slightly smaller on average, so that larger resolving power in the long wavelength region can be expected. The spectral resolving powers are therefore set within the two limiting curves

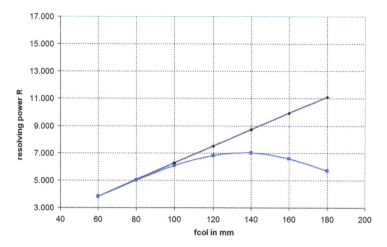

Fig. 6.23 As Fig. 6.22 but with a slit width 75 μm. For both cases, the optimum collimator focal length should not be greater than 120 mm. For larger focal lengths the illuminated aperture of the collimator and thus the camera rises and spherical aberrations increase sharply

6.6.2 Compensation of Longitudinal Chromatic Aberration by Camera Tilt

Some lenses do not show the typical parabolic behavior for longitudinal chromatic aberration, but an almost linear one. Figure 6.24 shows the longitudinal chromatic aberration of a 40 mm/120 mm doublet made of N-BAK1 + N-F2 glass. The longitudinal chromatic aberration increases almost linearly in the long wavelength range from 450 nm. Below 450 nm there is a negative gradient. If one uses the achromat only in the wavelength range 450–700 nm and additionally tilts the sensor surface of the CCD camera appropriately, one can compensate for paraxial chromatic aberration. A mechanical device with appropriate fine-thread screws or micrometers for adjustments in the μm domain is then required (Fig. 6.25).

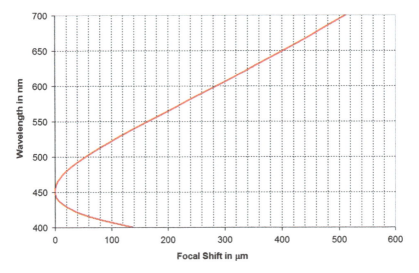

Fig. 6.24 Paraxial chromatic aberration of an achromat (Qioptics 322388000). The branch on the long wavelength side shows an almost linear curve

Fig. 6.25 The mechanical camera adjustment of the research spectrograph

6.7 Projekt "Research-Echelle": First Tests

We performed our first practical tests with a fiber-fed Perkin-Elmer OPTIMA echelle spectrograph. Figure 6.26 shows the structure on the bench. It operates in off-axis configuration ($\Theta = 0$), as illustrated in Fig. 5.5. The instrument operates with a parabolic off-axis collimator of 500 mm focal length, an 80×160 mm grating of 79 lines/mm and a 70×70 mm prism cross-disperser with a refraction

Fig. 6.26 First setup of the OPTIMA echelle spectrograph. The fiber optics are visible on the left side of the echelle grating. The prism cross-disperser is visible in the image center

angle of 60°. In contrast to the design in Fig. 6.1 the light falls centrally onto the collimator and then via the off-axis angle $\gamma = 7°$ onto the echelle grating. The angle of incidence α corresponds exactly to the blaze angle $\Theta_B = 63.4°$. A normal paraboloid would cause aberrations in off-axis illumination. The slit image and the spectral lines are tilted according to Eq. 5.16 (see Fig. 5.7). Figure 6.27 shows the setup for the determination of the grating constant of the echelle grating and Fig. 6.28 shows the echelle spectrum behind the cross-disperser prism.

Since the material of the prism was unknown, the beam deflection was measured with a laser pointer of 635 nm wavelength. For a known angle of incidence on the prism one can estimate the prism exit angle with the laser wavelength and the equations from Sect. 6.4.2. For the angles $\alpha_1 = 40°$, $50°$ and $60°$ against the prism normal one finds the exit angles $\alpha_2 = 73°$, $57°$ and $47°$. This matches with the theoretical values for N-F2 glass within $\pm 2°$ For the following considerations we assume this to be the correct glass material.

The first tests for finding the required camera were carried out with a simple f1.8/50 mm Nikon lens and a KAF400 CCD chip. The fiber optics for input of sunlight can be seen in Fig. 6.26 on the left side next to the echelle grating. Since the fiber at the output has a small f-ratio, both the 70 mm collimator and the Nikon lens were fully illuminated. Figure 6.29 shows the image of the solar spectrum in this configuration. Since the lens introduces chromatic aberration, which appears as blurring in the short- and long-wavelength regions, the focal length of the camera was enlarged and further tests with a 300 mm achromat (and with a Pentacon Orestegor f4.0/300 mm telephoto lens, for comparison) were carried out. As a preliminary sensor we used a CANON 5D with a 24×36 mm chip, which was completely illuminated because of the large camera focal length.

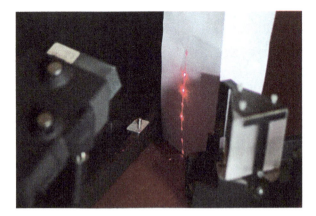

Fig. 6.27 Simple set-up for determining the grating constant of the echelle grating. The grating is illuminated by a simple laser pointer for determining the orders with respect to the diffraction angle. If the angle of incidence α is known as are the diffraction angles β_1, β_2 of two adjacent orders, the orders n and the grating constant g can be estimated by repeated application of the grating equation and then respective division ($n = \frac{\sin\alpha + \sin\beta_1}{\sin\beta_2 - \sin\beta_1}$ and $g = \frac{\lambda_{Laser}}{\sin\beta_1 - \sin\beta_2}$)

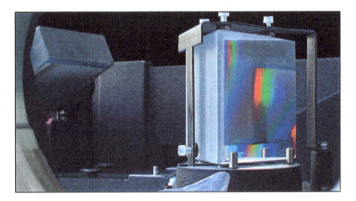

Fig. 6.28 Photo of the prism cross-disperser with depicted echelle spectrum

To check the results, the spectrometer was modeled with "Simechelle". Table 6.5 shows the results of the calculation for the orders 33–53, which can be identified in the corresponding solar spectrum in Fig. 6.33. For a cross-disperser, the refractive index of N-F2 was used and the angle of incidence to the prism normal $\alpha_1 = 56.2°$ was adopted. To test the calculated parameters a continuous light source (white LED) was fed into the fiber spectrometer with a $100\,\mu$m fiber (Fig. 6.30). The corresponding flat image is shown in Fig. 6.31. Since the anamorphic magnification factor is 1 in the γ configuration, the image of the fiber core in the CCD plane should have the size $100\,\mu$m $\cdot$ $300\,$mm$/500\,$mm $= 60\,\mu$m. With a pixel size of $6.4\,\mu$m these are 9.4 pixels accordingly. In the case of the PENTACON f4.0/300 mm all orders have almost the same width. By measuring the order widths, values are

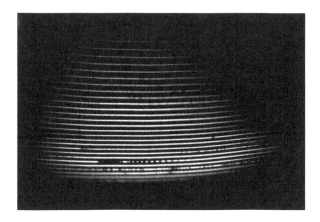

Fig. 6.29 First attempt with a 1.8/50 mm Nikon lens (Series E) with a KAF400 CCD camera. The spectrum is focused only in the green and red wavelength ranges up to about Hα. Particularly in the oxygen bands extreme blurring is obvious

Table 6.5 Modeling of the modified OPTIMA with "SimEchelle"

Order #	λ_c Å	Order Distance Å	Order Distance Pixel	λ_{Min} Å	λ_{Max} Å	Spectral Range Å	Dispersion Å/Pixel	Order Length mm	Over lap Å
33	6811	213	089.0	6709	6913	204	0.0363	36.95	−06
34	6611	200	091.7	6512	6709	198	0.0352	35.82	+01
35	6422	189	094.6	6326	6518	192	0.0342	34.77	+06
36	6243	178	097.7	6150	6337	187	0.0333	33.78	+11
37	6075	169	100.9	5984	6165	182	0.0324	32.84	+16
38	5915	160	104.2	5826	6003	177	0.0315	31.95	+20
39	5763	152	107.6	5677	5849	173	0.0307	31.10	+23
40	5619	144	111.3	5535	5703	168	0.0300	30.32	+26
41	5482	137	115.1	5400	5564	164	0.0292	29.56	+29
42	5351	131	118.9	5271	5432	160	0.0285	28.84	+32
43	5227	124	123.1	5149	5305	157	0.0279	28.15	+34
44	5108	119	127.3	5032	5185	153	0.0272	27.49	+36
45	4995	114	131.8	4920	5069	150	0.0266	26.87	+38
46	4886	109	136.4	4813	4959	146	0.0261	26.27	+39
47	4782	104	141.2	4710	4854	143	0.0255	25.70	+41
48	4682	100	146.4	4612	4753	140	0.0250	25.15	+42
49	4587	096	151.7	4518	4656	137	0.0245	24.63	+43
50	4495	092	157.2	4428	4562	135	0.0240	24.13	+44
51	4407	088	163.0	4341	4473	132	0.0235	23.64	+45
52	4322	085	169.3	4258	4387	129	0.0230	23.19	+46
53	4241	082	175.8	4177	4304	127	0.0226	22.74	+47

Multiplying the order number n with λ_c delivers the constant $K = 2.2475 \cdot 10^4$ nm for each order

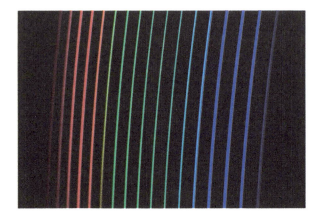

Fig. 6.30 White LED echelle spectrum obtained with an achromat of 300 mm focal length focussed on the central wavelengths. The *FWHM* of the blue and red orders are about twice as large as for green wavelengths. The detector is a Canon 5D with a 24 × 36 mm chip. The fiber diameter is 100 μm

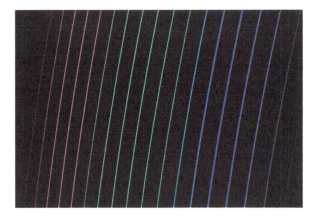

Fig. 6.31 As in Fig. 6.30 now with a f = 300 mm f/4 medium format lens (analogue). With this lens all orders are well focused

obtained in the range of 9.7–11 pixels or 103–117 μm, i.e., in the entire wavelength range the aberrations are probably smaller than the fiber diameter. However, for the achromat the situation is different. One can clearly see the different order widths over the entire spectral range. The focus was at the central wavelengths and the typical parabolic behavior of the longitudinal chromatic aberration of an achromatic lens is expressed in red and blue wavelengths by broader orders. For comparison, one can consider the spot diagrams in Fig. 6.32 for an achromatic lens with 300 mm focal length. The off-axis spots on the right side clearly show astigmatic blurring. This could be remedied by a custom field flattener (field lens). Figure 6.33 shows a section of the solar spectrum.

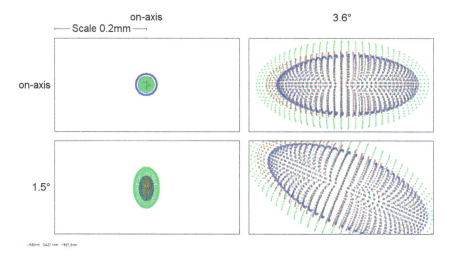

Fig. 6.32   Aberrations of a Qioptics achromat (322305000) with f = 300 mm at best on-axis focus for the three wavelengths 427, 585 and 667 nm. The off-axis spots on the right side show significant field curvature and astigmatism

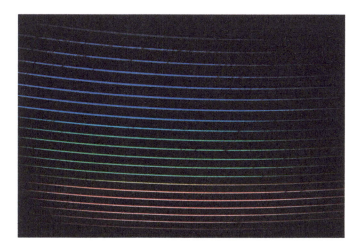

Fig. 6.33   Solar spectrum taken with the Pentacon 300 mm via a fiber of 200 μm diameter. Order 34 contains the Hα line and order 38 the sodium doublet

Order 34 contains the Hα line at 656.3 nm and order 38 the sodium doublet at 589 nm. Both sodium lines can be used for the estimation of the dispersion. The two doublet lines are about 202 pixels apart which corresponds to a dispersion of 0.296 Å/pixel. By estimating the distance between the orders via an intensity plot of the orders (Fig. 6.34) one obtains 92 pixels between the orders 34 and 35, and 167 pixels between the orders 51 and 52 (for comparison see calculation in Table 6.5). For comparing the spectral resolving power, the light of a neon lamp was fed into the spectrometer together with a white LED. Figure 6.35 shows some of the emission lines. The inset shows the profile of the line appearing on the extreme right side. Its wavelength is in the range of 5,400 Å in order 42. The FWHM of the line is approximately 10.5 pixels. Thus we obtain a spectral resolving power of

$$R = 5400\text{Å}/(10.5\text{Pixel} \cdot 0.0285\text{Å/Pixel}) = 18000.$$

If one calculates the resolving power with a mean blurring circle of about 30 μm one obtains $R = 16,700$. In the red region, the FWHM increases from 13 to 15 pixels delivering a resolving power of about 13,000. In summary it can be stated that the calculation of the spectral parameters can be relatively reliably performed.

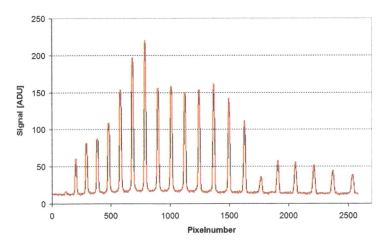

Fig. 6.34 Order cross-section at the central wavelengths. Starting on the left side with the very weak order 32 ($\lambda_c = 7,024$ Å) to order 54 ($\lambda_c = 4,162$ Å), one can clearly see the increasing order separation

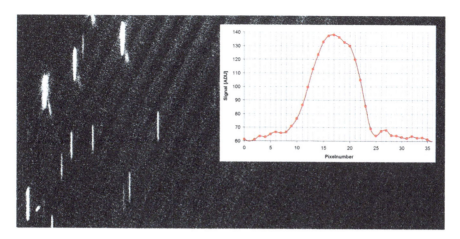

Fig. 6.35 Section of a neon calibration spectrum and the half-width of the extreme right line (*inset*). The weak strip under the neon emission lines is the continuum (orders) caused by additional irradiation of a white LED

6.8 Specific Echelle Design Constraints

In contrast to the standard long-slit spectrographs there are interesting design options for the echelle that we want to address briefly.

6.8.1 The "White Pupil" Concept

In complex echelle spectrographs a parabolic mirror in front of the cross-disperser might produce an intermediate focus between grating and cross-disperser (see, e.g., the FEROS echelle in Fig. 8.49). With this intermediate focus the divergence of the beam path is limited. Figure 6.36 illustrates the principle of this white-pupil concept. The intermediate image is positioned at a point where the orders are not yet separated by the cross-disperser, i.e., they lie above each other. Therefore, at the location of this intermediate image a white slit image as a superposition of all wavelengths is formed (Fig. 6.37). Because of this intermediate focus, all subsequent apertures can be made smaller, which is an advantage with respect to the correction of the camera lens. Another advantage is reduced stray light by a so-called spatial filter. This is done with a baffle at the intermediate focus position matching its diameter. This baffle prevents scattered light potentially appearing in the spectrometer.

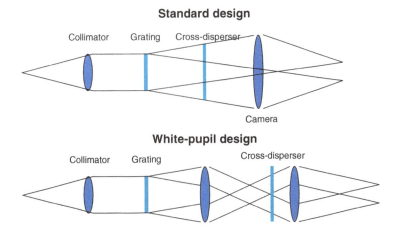

Fig. 6.36 Schematic sketch of the "White Pupil" concept

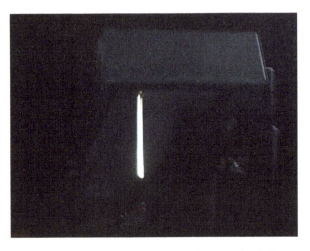

Fig. 6.37 All superimposed echelle orders behind the echelle grating before separation by the cross-disperser

6.8.2 Data Reduction

The concept of echelle spectroscopy has become a standard technology in astronomy. Professional data reduction tools are MIDAS (MUNICH IMAGE DATA ANALYSIS SYSTEM) of the European Organisation for Astronomical Research in the Southern Hemisphere (ESO) and IRAF (IMAGE REDUCTION AND ANALYSIS FACILITY) of the National Optical Astronomy Observatories (NOAO) in the United States. A detailed description can be found in Chap. 12. Because of the static mechanical design (no moving parts) the geometric output of an echelle spectrograph at the detector is stable. Unlike standard spectrographs, this has the advantage that automated reduction routines can be applied and thus the amount of analysis work is reduced.[16] In the best case, all echelle orders are always mapped at the same geometrical positions on the chip. The orders of the target and calibration spectra are localized by the software and corresponding offset against each other is automatically taken into account (see Sects. 12.3.1–12.3.9). Typical effects such as tilted lines (Sect. 5.7) and curved orders (Sect. 5.8) must be processed adequately. However, it is then important to ensure that external influences do not cause geometric shifts in the system. Mechanical stress in the spectrograph as well as fluctuations in the ambient temperature should be avoided since even small changes in geometry cause significant spectral shifts at the detector. Therefore echelle spectrographs are almost always fed by fiber optics to provide mechanical and thermal de-coupling between the telescope and the instrument.

6.8.3   Design Implications by Fiber Optics

If a spectrograph should not be coupled directly to the telescope but instead is fed via optical fibers, the physical properties of fibers must be considered. We will discuss optical fibers and their internal effects in detail in Sect. 11. At this point we already mention that the output f-ratio at the fiber end is between f/2.5 and f/4.5 depending on the fiber coupling. This has implications for the necessary f-ratio of the camera, since it must be now even smaller than the f-ratio at the entrance of the spectrometer. Focal ratios of f/2 or smaller become necessary when using optical fibers. As already highlighted, the requirements for the lens corrections then increase dramatically. For example, the commercial Shelyak "Eshel" spectrograph[17] uses a CANON f1.8/85 mm as camera optics. The lens is just barely able to map the visual spectral region with a sufficiently small spot. Figure 6.38 shows a section through the orders of an "Eshel" example spectrum. Increasing order widths at the order edges can already be seen in the raw spectrum. As a result the order FWHMs at the edges are about two times larger than the FWHMs at the order centers (Fig. 6.39).

---

[16]For commercial spectrographs, reduction software is usually included that offers automated algorithms.

[17]A system description can be found in Sect. 8.4.3.

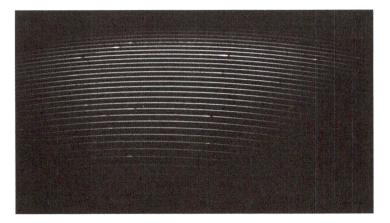

Fig. 6.38 Raw spectrum of P Cygni obtained with the Shelyak "Eshel". In the wavelength range 4,500 Å–7,500 Å, the orders show a sufficient sharpness (*Courtesy*: B. Stober)

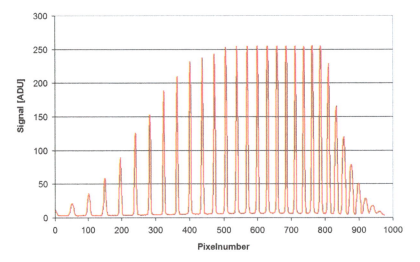

Fig. 6.39 Section through the orders of the P-Cygni spectrum of Fig. 6.38

6.9 Prospect

From the above considerations one should realize that an echelle spectrograph is an attractive system even for small telescope. According to our experience one can obtain spectra of $S/N \approx 100$ for stars of $V \approx 8$ mag within about 30 minutes having $R \approx 20,000$ with a telescope of about 35 cm aperture at a seeing of 1–2 arc seconds. All necessary optical elements are available off-the-shelf. The most popular manufacturers offer different echelle and standard gratings as well as all other optics. Standard apertures of 50 mm are already sufficient for high resolving power as long as they are correctly adapted. The camera lens must be corrected, though, to properly cover the entire visual wavelength range. For wavelength calibration one needs an appropriate calibration lamp with many lines over the entire spectrum (see Sect. 4.1.3). Most such lamps contain standard elements, such as Thorium-Argon or Iron-Argon, but others are also available on the market.

As potential applications we want to mention two examples:

- A relatively high spectral resolving power combined with a very large wavelength range can enable a very precise estimation of radial velocities for components of binary stars. Radial velocities are normally estimated by the wavelength shift of photospheric absorption lines of the system components via the Doppler effect. Then one can estimate certain orbital parameters and the stellar masses. The accuracy of this method depends on the number of available lines and hence on the detected wavelength range. If for example, 100 lines instead of only four lines are available, the measurement accuracy is increased by a factor of five. By analyzing all possible line velocities in the optical spectrum one can significantly increase the precision of the radial velocities.
- By simultaneously detection of different emission lines of massive stars one can investigate dynamic correlations in their winds mainly via the different ionization energies. The ionization levels of atomic lines depend mainly on the temperatures in the plasma which emits the line radiation. For instance, fourfold ionized nitrogen (NV) requires significantly higher temperatures than double ionized Carbon (CIII). And Hα emission in massive stars is produced by recombination in relatively cool environments. Different lines are, hence, an indicator of the plasma temperature. Analyzing all possible line species in the optical wavelength range, one therefore can acquire an indicator of radial temperature stratification in the environment of those stars because of the average temperature drop with the distance from the surface.

Each optical element of a spectrograph should be as efficient as possible. This is especially true when two dispersing elements are used. However, an echelle does not compete with a normal grating spectrograph, but complements it. The echelle delivers high dispersion over large wavelength ranges, while the standard long-slit spectrograph provides low and medium resolution without the problem of having to splice the orders together as in an echelle system.

Recommended Readings for All Spectroscopy Chapters

- A.C. Oliveira, L.S. de Oliveira, J.B. dos Santos, *Studying Focal Ratio Degradation of Optical Fibres with a Core Size of 50 μm for Astronomy*, Monthly Notices of the Royal Astronomical Society, 2005, 356, 1079–1087
- C.R. Kitchin, *Optical Astronomical Spectroscopy*, Taylor & Francis, 1995
- G.C. de Strobel, M. Spite (Editors), *The Impact of Very High S/N Spectroscopy on Stellar Physics*, International Astronomical Union Symposia, Springer 1988

Chapter 7
Reflecting Spectrographs

A Short Story

One day we found a "30 cm concave mirror" in an Internet auction on which no one had yet bid. We did not know if it was a spherical or parabolic mirror. If cheap, the latter, of course, would have been great. So we placed a bid and because we did not know the mirror shape, we limited it to 100 euros. You never know! Who needs spherical mirrors, at all? To our surprise, the auction ended at 30 euros and after a few days we received our prey. To our great regret, it was a spherical mirror and, of course, we were absolutely certain that one cannot do anything with spherical mirrors...

7.1 Basic Design Considerations

So far, only refractive optics were treated as imaging elements of the spectrometer. An alternative to lens optics are concave mirrors. The problem of chromatic aberration does not exist in the case of reflectors and for parabolic mirrors spherical aberration is corrected, although only on-axis.[1] On the other hand mirrors in a spectrograph must always be operated in an off-axis configuration to avoid respective vignetting and this causes coma and astigmatism. However, regarding the calculation of the image errors, the case of only a single reflecting surface is a special case that should be considered separately because of its simplicity. Moreover, in the scope of this book it is possible to fully derive the Seidel aberrations for a single surface. We now provide the reader access to the general procedure for the derivation of image errors of a mirror spectrometer. In the following chapter we therefore consider the main spectrometer mirror design variants and their interpretation.

---

[1] In this context, the expression "on-axis" describes all axis-parallel beams reaching the mirror.

© Springer-Verlag Berlin Heidelberg 2015
T. Eversberg, K. Vollmann, *Spectroscopic Instrumentation*, Springer Praxis Books,
DOI 10.1007/978-3-662-44535-8_7

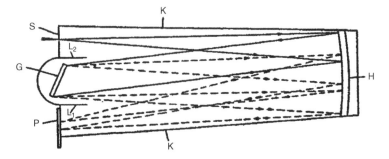

Fig. 7.1 Original sketch of the Ebert spectrograph (Ebert 1889)

7.1.1   Ebert-Fastie Configuration

In 1952 William G. Fastie introduced a very simple high-resolution slit spectrometer consisting of only one spherical mirror as imaging element plus a plane grating (Fastie 1952). However, the original idea stems from a much older consideration by Ebert (1889). Figure 7.1 shows Ebert's spectrograph configuration with a single spherical mirror and the corresponding beam path. In contrast to Ebert, though, Fastie described the optical setup in sufficient detail. An Ebert-Fastie spectrograph consists of a single spherical mirror and a grating centered on the optical axis of the spherical mirror.[2] The input and output slit may be positioned in the focal plane of the mirror. The opening angle is an important parameter of the Ebert-Fastie. It is defined by the distance between the input and output slit with respect to the radius of the mirror. The larger this angle, the more the spherical mirror is used in off-axis mode. This in turn deteriorates the quality of the optical system. Because of the single mirror design, Ebert-Fastie spectrographs can be realized with minimal effort. However, the spherical mirror causes image aberrations such as spherical aberration, coma, astigmatism, and a curved image field.

7.1.2   Czerny-Turner Configuration

The Czerny-Turner configuration is an improvement to the Ebert-Fastie system. The difference is that two separate mirrors are used for collimation and imaging instead of only one mirror. We have introduced the Czerny-Turner design already in Chap. 2. It is often used as a so-called monochromator. The spectrograph isolates individual wavelengths that can be detected by a scanning process. For this purpose the system has not only an input slit but also an output slit at the corresponding position behind the camera lens in order to select individual wavelengths. Figure 7.2 shows the basic design.

---

[2]In his paper, Fastie already mentioned the possible use of two off-axis parabola mirrors for good imaging, which directly hints towards a Czerny-Turner system.

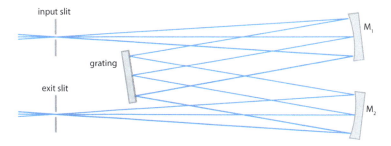

Fig. 7.2 Czerny-Turner spectrograph as monochromator (Source: R. Paschotta, Encyclopedia of Laser Physics and Technology)

Fig. 7.3 Czerny-Turner design. The light coming from the fiber optics is projected through a slit onto a spherical collimating mirror (80 mm, $R/2 \approx 2{,}000$ mm). Then it reaches the grating (50×50 mm, 2,604 lines/mm), the camera mirror (80 mm, $R/2 \approx 2{,}000$ mm) and finally the CCD camera. With this construction, a spectral resolving power of R $= 125{,}000$ is achieved (*Courtesy:* D. Goretzki)

The Czerny-Turner design corresponds to an Ebert-Fastie, but is much more flexible in its correction of aberrations. By selecting the right mirrors, a flat spectral field and a corresponding coma correction can be achieved. However, other aberrations (spherical aberration, astigmatism and field curvature) remain present at all wavelengths. Due to the design with two mirrors the handling of the Czerny-Turner is more complex than that of an Ebert-Fastie but it can be better corrected. As for the Ebert-Fastie the advantage of mirror optics is the complete lack of chromatic aberration. The focus is wavelength-independent. And such devices have a high efficiency. The laboratory setup of a Czerny-Turner spectrograph for solar observations is shown in Figs. 7.3 and 7.4.

If collimator mirror and camera mirror are concave, the off-axis nature of this system produces astigmatism and coma. The monochromatic image of a straight entrance slit has a parabolic shape. This results from the geometry of the oblique light incidence onto the grating[3] and is even visible if other aberrations are negligible (Fastie 1952). To compensate for this effect, the entrance slit should have the shape of a parabola (Fig. 7.5). Another problem is the sensitive response to inaccurate alignment between the slit and its image. In this case, the transfer function of the spectrograph deteriorates and reduces the spectral resolving

---

[3]In Sect. 5.7 and we 5.8 have already treated the problem of spectral line curvature due to an echelle γ-configuration and the resulting oblique light incidence at the grating

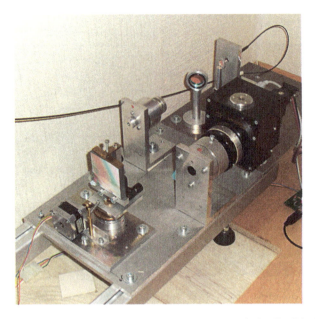

Fig. 7.4 Slit and grating unit of the Czerny-Turner spectrograph in Fig. 7.3 (*Courtesy*: D. Goretzki)

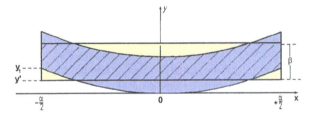

Fig. 7.5 Entrance slit (*yellow*) and its image (*blue*) for a Czerny-Turner spectrograph (Kay and Shepherd 1982)

power and efficiency. The combination of mirror optics makes a Czerny-Turner spectrograph consequently relatively large and heavy. If smaller telescopes are used, the spectrograph can therefore only be fed via optical fibers, which in turn reduces the efficiency (see Chap. 11).

7.2 The Imaging Equation of a Spherical Mirror

To illuminate the physics of mirror spectrometers, we now deal with the equations of a concave mirror and its aberrations. For this purpose Fig. 7.6 represents the geometric relationships with respect to the spherical mirror with an image point on

Fig. 7.6 The geometrical
conditions at the spherical
mirror with an image point on
the optical axis

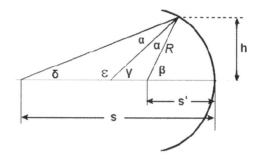

the optical axis. Thus, for the paraxial approximation (small angles) the following
geometric relations are applicable: $\frac{h}{s'} = \tan\beta \approx \beta, \frac{h}{s} = \tan\delta \approx \delta, \frac{h}{R} = \tan\gamma \approx \gamma$.
In addition, the following angular relations are valid: $2\gamma = \delta + \beta$ and hence
$2\frac{h}{R} = \frac{h}{s'} + \frac{h}{s}$. This results in the so-called imaging equation of the spherical mirror

$$\frac{2}{R} = \frac{1}{s'} + \frac{1}{s}. \tag{7.1}$$

For $s = \infty$ we have $s' = f$ and hence $f := \frac{R}{2}$ as the focal length of the mirror.
With this result we also obtain for the concave mirror the already known Gaussian
lens equation for paraxial conditions[4]:

$$\frac{1}{f} = \frac{1}{s'} + \frac{1}{s}.$$

Therefore, if the focal length f is identified as half the radius R of a spherical
mirror, one obtains an expression for determining the distances of object and image
for spherical mirrors in the paraxial case. This is completely analogous to the lens
equation for refractive lenses or parabolic mirrors.

7.3 Aberrations of a Concave Mirror

The calculation of the image error is simpler for a mirror with just one reflective
surface than for all involved surfaces of a complex lens system. The problem of a
single reflecting surface is equivalent to the calculation of a single refracting surface.
For this reason, we will deal in the following with the explicit calculation of the main
imaging errors of a single refracting surface. As for the case of lenses, we first deal
with an estimation of the spherical aberration and we then examine the condition in
which such optics can be used in a spectrograph.

---

[4]If we consider the direction of light propagation with a positive sign as for the case of lenses,
the intersection length s' for the concave mirror will have a negative sign and our equation is now
$\frac{1}{f} = \frac{1}{s} - \frac{1}{s'}$.

7.3.1 Spherical Aberration

For spherical mirrors, spherical aberration significantly limits the applicability in imaging and spectroscopic systems. The crucial question is: To what diameter does a spherical mirror still provide a useful circle of confusion (smaller than the resolution element supplied by the system in the focal plane) for given boundary conditions? Or more specifically: To what diameter can the mirror still work being diffraction limited? For this purpose we first deal with the corresponding geometric derivations of the longitudinal and transverse spherical aberrations of a concave mirror of arbitrary curvature with parallel incident rays.

Longitudinal Spherical Aberration

For the consideration of the longitudinal spherical aberration Fig. 7.7 shows the geometrical situation of the beams. One first obtains the trigonometric relation $\tan 2\Phi = \frac{h}{z_0}$. Using $\Phi_1 = \Phi_2 =: \Phi$ and the addition theorem for $\tan 2\Phi$[5] we then obtain $\tan 2\Phi = \frac{2\tan \Phi}{1-\tan^2 \Phi}$ and from that the parameter z_0

$$z_0 = \frac{h(1 - \tan^2 \Phi)}{2 \tan \Phi}. \tag{7.2}$$

To determine the longitudinal spherical aberration, we need to estimate the focal length depending on the height of incidence $f = f(h)$ for arbitrary mirror shapes. Different mirror types are represented by conic sections and can generally be described by the implicit function[6] $\varphi(h, z) = 0$ (see for example D.J. Schroeder,

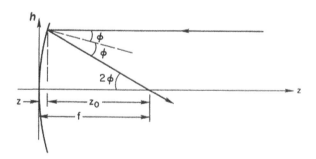

Fig. 7.7 The geometric conditions at a concave mirror for incidence of a collimated light beam (Schroeder 1987)

---

[5]$\tan(\Phi_1 + \Phi_2) = \frac{\tan \Phi_1 + \tan \Phi_2}{1 - \tan \Phi_1 \tan \Phi_2}$.

[6]In contrast to an explicit representation of a function $y = f(x)$ one speaks about an implicit representation when all variables are on one side of the equation $F(x, y) = 0$ and if it is not

Astronomical Optics, Academic Press Inc., 1987):

$$\varphi(h, z) = h^2 - 2Rz + (1 + K)z^2 = 0 \tag{7.3}$$

with $h^2 = x^2 + y^2$. The parameter K defines the type of conic section. In addition to the simple spherical shape, a concave mirror can also have other shapes (paraboloid, hyperboloid, ellipsoid) and for the individual cases K is:

$$K = 0 \qquad \text{Sphere}$$

$$K = -1 \qquad \text{Paraboloid}$$

$$K < -1 \qquad \text{Hyperboloid}$$

$$K > -1 \qquad \text{Ellipsoid}$$

The shape of the mirror surface is described by the function $h = h(z)$. The tangent to this curve (which defines the mirror shape) is $\frac{dh}{dz}$ and the normal (which defines the angle of incidence) is consequently its negative reciprocal $-\frac{dz}{dh}$. To calculate the tangent we now can follow two different paths:

1. Calculating the derivative of the explicit functional representation $h = h(z)$.
2. Estimation of the normal $\vec{n}$ by applying the gradient in the directions h and z to the implicit representation $\vec{n} = \nabla\varphi$.

For the first approach, we first have to re-arrange the conic parameterization $\varphi(h, z)$ for the determination of the normal to z. As a result we obtain a square root function because of Eq. 7.3, which results in a complicated derivation. In particular for non-linear curves, the use of the implicit representation is often a simpler method to determine the normal. When applying the gradient to Eq. 7.3 we obtain the vector

$$\nabla\varphi = \left(\frac{\partial\varphi}{\partial h}, \frac{\partial\varphi}{\partial z}\right) = (2h, -2R + 2(1 + K)z)$$

and hence for the inverse derivation of the function $h = h(z)$

$$-\frac{dz}{dh} = -\frac{\partial\varphi}{\partial h} \cdot \frac{\partial z}{\partial\varphi} = \frac{h}{R - (1 + K)z}.$$

Geometrically, the derivative $-\frac{dz}{dh}$ is the slope $\tan\Phi$. Hence

$$-\frac{dz}{dh} = \tan\Phi = \frac{h}{R - (1 + K)z}. \tag{7.4}$$

---

immediately obvious what is the independent variable. Example of a circle: $F(x, y) = x^2 + y^2 - R^2 = 0$.

We now use Eq. 7.2 for the derivation of the spherical aberration of the concave mirror and calculate for the value z_0

$$z_0 = \frac{h}{2} \cdot \frac{1 - \tan^2 \Phi}{\tan \Phi} = \frac{h}{2} \cdot \left(\frac{1}{\tan \Phi} - \tan \Phi \right).$$

With our result for $\tan \Phi$ from Eq. 7.4 we then obtain

$$z_0 = \frac{h}{2} \cdot \left(\frac{R - (1 + K)z}{h} - \frac{h}{R - (1 + K)z} \right).$$

For the focal length of the concave mirror we have $f = z + z_0$. Hence

$$f = z + \frac{h}{2} \cdot \left(\frac{R - (1 + K)z}{h} - \frac{h}{R - (1 + K)z} \right).$$

By multiplication and combination we obtain the intermediate result

$$f(h, z) = \frac{R}{2} + \frac{(1 - K)}{2} z - \frac{h^2}{2(R - (1 + K)z)}. \tag{7.5}$$

Beside the variable z, the z-dependent incidence height $h(z)$ of the conic section occurs in this equation. Both parameters are hence interdependent. In order to obtain an expression for the longitudinal spherical aberration we must now determine the focal-length $f(h)$ dependence on the incidence height. This means we will eliminate z. We first re-arrange the cone paramerization from Eq. 7.3 to a quadratic equation of the variable z:

$$z^2 - \frac{2R}{1 + K} z + \frac{h^2}{1 + K} = 0.$$

The solutions of this quadratic equation are

$$z = \frac{R}{1 + K} \pm \sqrt{\left(\frac{R}{1 + K} \right)^2 - \frac{h^2}{1 + K}}$$

or

$$z = \frac{R}{1 + K} \left[1 \pm \sqrt{\left(1 - \frac{h^2}{R^2}(1 + K) \right)} \right]. \tag{7.6}$$

To simplify the relatively complex expression for z, we develop the inner root function in a Taylor series around $\zeta = 0$ (with $\zeta = \frac{h^2}{R^2}(1 + K)$) and truncate the development after the 2nd order:

$$\sqrt{1 - \zeta} \approx 1 - \frac{1}{2}\zeta - \frac{1}{8}\zeta^2 - \ldots \tag{7.7}$$

Even for relatively very large distances from the optical axis this approximation introduces only very small errors.[7] For example, we consider the case $D = f = R/2$ ($F\# = 1$!) for the sphere, where D is the diameter of the mirror. With these conditions $\zeta = h^2/R^2 = D^2/4R^2 = R^2/16R^2 = 0.0625$ and thus the quadratic approximation is accurate to 0.0016%. Re-substitution of ζ provides

$$\sqrt{1 - \zeta} \approx 1 - \frac{1}{2}\frac{h^2}{R^2}(1 + K) - \frac{1}{8}\left(\frac{h^2}{R^2}(1 + K)\right)^2 - \ldots$$

and for z we finally obtain[8]

$$z \approx \frac{h^2}{2R} + \frac{h^4}{8R^3}(1 + K). \tag{7.8}$$

To determine the focal length $f(h)$ one can neglect the z-term in the denominator of the 3rd summand in Eq. 7.5.[9] We thus first obtain the focal length

$$f(h) = \frac{R}{2} + \frac{(1 - K)}{2}z - \frac{h^2}{2R}$$

and after substitution of z

$$f(h) = \frac{R}{2} - \frac{h^2}{4R}(1 + K) + \frac{h^4}{16R^3}(1 + K)(1 - K).$$

---

[7]The accuracy of the square root series decreases with increasing ζ. However, this has only a significant effect for very large ζ. For instance, $\zeta = 0.5$ is equivalent to an F-number of 0.35. For this value the error is only 1.65%.

[8]There are two solutions for z, which are symmetrical and completely equivalent. We use the solution with the negative root, since the 1 in the root of Eq. 7.6 then vanishes when implementing the approximation.

[9]For this we consider the denominator $2R - 2(1 + K)z$ of the 2nd summand of Eq. 7.5. If we substitute z by the quadratic term $h^2/2R = D^2/8R$ from Eq. 7.8 we obtain, for example, for the sphere ($K = 0$) the denominator $2R - D^2/4R$. If we choose for example the case $D = f = R/2$ (F# = 1!), we obtain $2R - R/16$ for the denominator. For this challenging case the correction term $R/16$ contributes only 3%, which decreases quadratically with increasing f-ratio. Thus, it is sufficient to consider only the term $2R$ in the denominator

As expected, for $h = 0$ we obtain the paraxial focal length $f(0) = R/2$. Thus the deviation from the paraxial case Δf_{SA} becomes

$$\Delta f_{SA}(h) = -\frac{h^2}{4R}(1 + K) + \frac{h^4}{16R^3}(1 + K)(1 - K). \qquad (7.9)$$

The focal length difference Δf_{SA} is the longitudinal spherical aberration of the concave mirror for the development of the root function to the 2nd order. It results from the parameters of the mirror shape K, the mirror radius R and the height of incidence of the beam h. For example, if we presume a parabolic concave mirror, then $K = -1$ and we obtain for the focal length of all incidence heights h the constant result $f(h) = \frac{R}{2}$, as expected. For the case of a sphere $K = 0$ the longitudinal spherical aberration is of the order $-\frac{h^2}{4R}$. The intersection lengths for collimated beams with $h > 0$ are thus smaller than the paraxial focal length (under-corrected spherical aberration).

Transverse Spherical Aberration

Completely equivalent to the lens (see Fig. 3.6) we define the transverse spherical aberration ρ, which again reflects the circle of confusion (blurring circle) of an object point projected into the image plane. In Fig. 7.8 we show the geometric situation for transverse spherical aberration. We calculate the transverse deviation from the paraxial image location by deriving the following relation from Fig. 7.8:

$$\frac{\rho}{\Delta f_{SA}} = \frac{h}{f - z}.$$

If we now consider in a first approximation for Δf only terms that include h^2 (see Eq. 7.9), we obtain for the transverse spherical aberration ρ

$$\rho \approx -\frac{h^2}{4R}(1 + K) \cdot \frac{h}{f - z}.$$

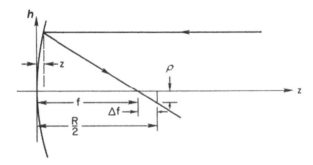

Fig. 7.8 Definition of the transverse spherical aberration. ρ defines the half circle of confusion in the Gaussian plane. f indicates the focal length for the incident height h and Δf the deviation from the paraxial focal length (Schroeder 1987)

To good approximation we have $f - z \approx R/2$, and hence we obtain for ρ

$$\rho \approx \frac{h^3}{2R^2}(1 + K).$$ (7.10)

The parameter ρ is the transverse spherical aberration of a concave mirror for developing elements of the series expansion of the focal length up to 1st-order terms, i.e., terms that include h^2. Analogous to the longitudinal aberration the transverse spherical aberration results from the mirror shape K, the mirror radius R and the height of incidence of the beam h.

The Circle of Least Confusion of a Spherical Mirror

We use Eq. 7.10 to calculate the circle of confusion for a spherical mirror. Using the mirror diameter D and its focal length f we re-arrange the equation and obtain with $R = 2f$ and $h = D/2$ for the diameter of the circle of confusion 2ρ in the Gaussian plane

$$2\rho \approx \frac{D^3}{32f^2}.$$

One now can show,[10] that the radius of the <u>smalles</u> circle of confusion ρ' is $\rho/4$. Hence, we obtain for the desired diameter of the circle of confusion

$$2\rho' = \frac{2\rho}{4} \approx \frac{D^3}{128f^2} = \frac{D}{128 \cdot (F\#)^2}.$$ (7.11)

An Example

As an example we consider an f/10 spherical mirror with 150 mm diameter and thus obtain for the diameter of the smallest circle of confusion $2\rho' = 11{,}7\,\mu$m. Comparing this value with the diameter of the diffraction disk one obtains $2.44 \cdot 0.55\,\mu$m$\cdot F\# = 13.4\,\mu$m. Hence, this spherical mirror functions just at the limit of diffraction. Doubling the diameter of the mirror, while maintaining the f-ratio ($F\#$),

---

[10]The desired radius of the smallest constriction results from the following consideration: The smallest beam constriction is located where the light reflected from an arbitrary incidence height h of the mirror beam intersects with the beam reflected from the mid-height $\frac{h}{2}$ on the opposite side of the mirror (see Fig. 7.9). The intersection of the two beams must be positioned on the caustic (the envelope of all reflected/refracted light rays), exactly where the $D/2$ beam touches the caustic. The parameterisations of the caustic in polar coordinates are $x = R(\frac{3}{2}\cos\alpha - \cos^3\alpha)$ and $y = R\sin^3\alpha$ with $\sin\alpha = \frac{h}{R}$. The desired radius of the narrowest constriction is the coordinate y at the point $\frac{h}{2}$, hence $\rho' = y(\frac{h}{2}) = \frac{h^3}{8R^2}$. Comparing this result with Eq. 7.10, we obtain $\rho' = \rho/4$.

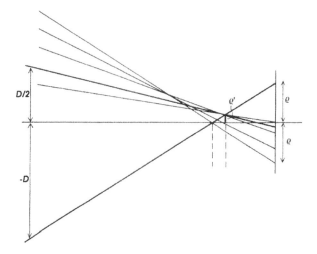

Fig. 7.9 The emergence of the "circle of least confusion". The intersection of the two rays, starting from D (lens or mirror diameter) and $D/2$ defines the circle of least confusion $2\rho'$, which is smaller than the circle of confusion 2ρ in the Gaussian image plane by a factor of 4

the circle of confusion doubles to 23.4 μm. For this case the blurring circle is larger than the Airy disk and the mirror is inappropriate by definition. However, if one takes the seeing disk as boundary condition instead of the diffraction, a $3''$ seeing disk with a diameter of $3'' \cdot 3{,}000\text{mm}/206{,}265 = 43.6\,\mu$ would not be exceeded. A 300 mm telescope mirror with f/10 might therefore be appropriate for an average seeing of $2'' - 3''$ without significant loss of quality. We now can specify a simple relationship for the maximum diameter of a spherical mirror with a given f-ratio:

$$D_{max} \leq 128 \cdot 2.44 \cdot \lambda \cdot (F\#)^3.$$

If this condition is fulfilled the mirror is diffraction limited.

The performance of a sphere is often underestimated. The dependence of the smallest circle of confusion on the mirror diameter in Eq. 7.11 now plays an essential role in the feasibility of using spherical mirrors in spectrometers with respect to spherical aberration. The diameters of the imaging elements are normally much smaller than the primary telescope mirror. For this reason, the circle of confusion $2\rho'$ is correspondingly smaller for the same f-ratio, and it may well be useful to apply simple spherical mirrors for spectroscopic purposes. This is especially true if spectroscopic broadband (Echelle) measurements are applied and if chromatic aberration limits the spectral resolution. To what extent the necessary off-axis angle will have a negative impact during operation of the mirror will be discussed further below.

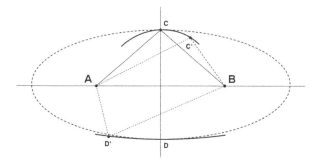

Fig. 7.10 The optical path for a concave mirror can assume a maximum or minimum, depending on the curvature of the mirror

7.4   Fermat's Principle

In 1657, Pierre de Fermat formulated the following assertion: Between two points, light propagates in the path that requires the shortest time. This extremum principle is a result of diffraction theory and is known as Fermat's principle. In Chap. 2 we already briefly mentioned it. However, this formulation of the principle leads in some cases to incorrect results and therefore requires a modification. As an example we consider Fig. 7.10 in which the reflection on a highly curved concave mirror is shown (top) and the path the light will choose is not directly known. According to Fermat's principle, a beam coming from point A will take a path to point B for which it requires the shortest time or shortest path (at constant speed of light). For comparison, we consider an ellipse whose foci are just the points A and B. We know that all the locations of an ellipse are defined by the totality of all possible paths with $[ACB] = constant$. The reflection path $[ACB]$ (solid line) chosen by the light beam[11] is therefore longer than any other theoretically possible reflection $AC'B$ at the concave mirror (dashed line). In this case, the light chooses the longest path. Let us now consider a concave mirror whose deflection is weaker than that of the ellipse (bottom). Then the path $[AD'B]$ is obviously longer than the real path $[ADB]$. In this case the light apparently chooses the shortest path. In summary, the chosen light path in curved mirrors is therefore not necessarily a minimum, but rather an extremum. For a quantitative understanding of these issue, we first analyze a simple case. We consider a beam of light running from point Q to point Q' while the two paths s_1 and s_2 travel in the media with the refractive indices n_1 and n_2. For the entire path the light needs the total time

$$T = \frac{s_1}{v_1} + \frac{s_2}{v_2}$$

---

[11]In the next chapter we will derive the necessary law of reflection with the generalized Fermat's principle.

with v_1 and v_2 the light propagation velocities in the two media. From optics we know that the propagation velocity in a medium is the ratio of the speed of light to the index of refraction c/n. Thus we can write

$$T = \frac{1}{c} \left(n_1 s_1 + n_2 s_2 \right).$$

The sum $n_1 s_1 + n_2 s_2$ is called the optical path L. Thus we can now rephrase Fermat's assertion: A light beam chooses the smallest optical path length. If the refractive index n changes along the path s, i.e. $n = n(s)$, we can express the optical path length as a path integral over the path C:

$$L = \int_C n(s)ds \rightarrow \text{minimum}.$$

Since it is a extremal principle, one can express another and still more general formulation. For that, we recall the properties of extreme values of any function $f(x)$: A function $f(x)$ has an extremum at the point $x = x'$ if and when the derivative $\frac{df}{dx}$ at the point $x = x'$ vanishes. The function $f(x)$ then only weakly changes in the vicinity of the extremum of $f(x')$, i.e. $f(x) \approx f(x')$ for $x \approx x'$. This means that the extremal principle of Fermat can now generally be formulated in the following way:

For light traveling between two points the derivative of the optical path length disappears with respect to the geometrical location.

The extremum principle immediately says that adjacent paths have almost the same optical path length. If the optical path length L is represented by a Taylor series and if we then set the first derivative at the point $s = s'$ to zero, we obtain to first approximation[12] $L(s) \approx L(s')$.

7.4.1 The Law of Reflection

As the simplest application we now consider the derivation of the reflection law with Fermat's principle. The general geometry of the reflected beam is shown in Fig. 7.11. A light beam is emitted at point A with the coordinates $(0, y)$ and is received at point B with the coordinates (x_r, y) after reflection at point P in a mirror plane. The point P is on the x-axis with the coordinates $(x, 0)$. The refractive index above the mirror plane is $n = 1$. Thus, the optical path L defined by $L = l_i + l_r$.

---

[12]In this context the first approximation means that higher order derivatives are not considered but do not necessarily disappear.

Fig. 7.11 Geometrical conditions for the reflection law

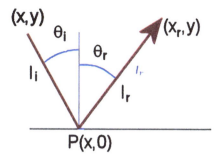

The optical path becomes a minimum when this path L is stationary. The point P is moved on the x-axis until

$$\frac{dL}{dx} = 0.$$

We first determine the optical path L

$$L(x) = \sqrt{x^2 + y^2} + \sqrt{(x_r - x)^2 + y^2}.$$

For the derivative to the spatial coordinate x, we obtain

$$\frac{dL(x)}{dx} = \frac{x}{\sqrt{x^2 + y^2}} - \frac{x_r - x}{\sqrt{(x_r - x)^2 + y^2}} = 0.$$

Together with the y-axis the rays AP and PB span the angles Θ_i and Θ_r. Thus we can easily derive the trigonometric relations

$$\sin \Theta_i = \frac{x}{\sqrt{x^2 + y^2}}$$

$$\sin \Theta_r = \frac{x_r - x}{\sqrt{(x_r - x)^2 + y^2}}$$

and we obtain

$$\sin \Theta_i = \sin \Theta_r$$

or

$$\Theta_i = \Theta_r.$$

Apparently, the incident angle Θ_i and the reflection angle Θ_r are the same under the requirement of the Fermat principle.

7.4.2 Snell's Law of Refraction

In the second example, we deal with the derivation of Snell's law (see Sect. 2.1). For this purpose, we consider two different media of refractive indices n_1 and n_2. As for reflection, we consider a beam of light emitted at point A with the coordinates $(0, y_1)$ received at point P with the coordinates $(x, 0)$. The angle of incidence is again Θ_1 and the individual optical paths l_1 and l_2. The beam is deflected behind the boundary layer and passes through the point B of the coordinates (x_2, y_2). Our task is the estimation of the refraction angle Θ_2 with Fermat's principle. For this purpose, we first consider the complete optical path L:

$$L(x) = n_1 \sqrt{x^2 + y_1^2} + n_2 \sqrt{(x_2 - x)^2 + y^2}.$$

Apparently, the optical path only differs from the reflection case by the two occurring refractive indices. Putting to zero the derivative with respect to the spatial variable x again yields the smallest optical path L:

$$\frac{dL(x)}{dx} = \frac{n_1 x}{\sqrt{x^2 + y_1^2}} - \frac{n_2(x_2 - x)}{\sqrt{(x_2 - x)^2 + y^2}} = 0.$$

As for the reflection case, the rays AP and PB and the y-axis span the angles Θ_1 and Θ_2. From Fig. 7.12 we can obtain the relations

$$\sin \Theta_1 = \frac{n_1 x}{\sqrt{x^2 + y_1^2}}$$

$$\sin \Theta_2 = \frac{n_2(x_2 - x)}{\sqrt{(x_2 - x)^2 + y^2}}$$

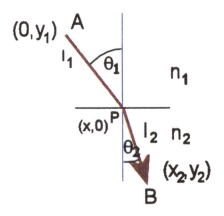

Fig. 7.12 Geometric conditions for Snell's law of refraction

and thus obtain by inserting the already known law of refraction

$$n_1 \sin \Theta_1 = n_2 \sin \Theta_2.$$

Hence, both laws can be derived very elegantly by Fermat's principle. If we choose other paths for the reflection and refraction law than those determined by the Fermat principle, they are longer in any case. For both laws, the actual optical path is therefore minimal.

Let us consider a further application of the reflection law; a curved surface, such as the elliptical mirror in a homogeneous medium with refractive index n. For an ellipsoidal mirror all light rays emitted by the focal point are focused in the other focal point. The optical path through each reflection point on the ellipse to the other focus is constant. For this case, obviously there is no specific path. This situation is given in a similar way for the case of a parabolic mirror. In Sect. 7.3.1 we have already calculated that for a collimated incident bundle all beams are focused in the focal point of the paraboloid. For the purposes of Fermat's principle, this means that not only a specific path exists, but all paths are equal. Comparing two adjacent optical paths for the parabolic mirror, the difference is therefore nil. This obviously generally applies to all ideally focusing systems. As we already know however, the situation for the sphere is fundamentally different. Because of spherical aberration, different intersection points for paraxial beams of different incident heights h exist. Hence, the optical path lengths differ, as well.

From the stationarity of the optical paths (the spatial derivatives vanish) one can thus derive a more simple conclusion for the calculation of aberrations:

If one determines the difference between two adjacent optical paths that already satisfy Fermat's principle[13] (e.g., the principle ray with the optical path length $L(s)$ and in addition a general ray with the optical path length $L(s')$) and calculate their optical path difference $\Delta L = L(s) - L(s')$, no aberrations are present if $\Delta L = 0$. Conversely, if $\Delta L \neq 0$ aberrations are present, with Fermat's principle satisfied for each partial beam.

The calculation of aberrations thus can easily be traced back to the calculation of various optical paths in the Fermat sense.

7.5 Seidel Aberrations of a Single Refracting Surface

We will now derive the Seidel aberrations of <u>one</u> refracting surface, which is designed as a conic. As a reminder: The derivative is, as might be expected, completely equivalent to the reflective surface. The special case of a single surface treated here is a supplement to Sect. 3.7 where we specified Seidel aberrations for

---

[13]"Satisfy the Fermat principle" in this context means the application of the physical laws that follow from Fermat's law. If we consider, for example, an obliquely incident principle ray on the refracting surface, Snell's law of refraction must be applied.

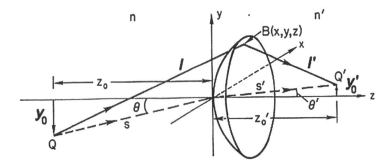

Fig. 7.13 The general geometric conditions for a refracting surface with off-axis angle Θ (Schroeder 1987)

any number of surfaces of refractive imaging optics. Starting from the optical path and applying Fermat's principle we will now show the complete derivation of the Seidel aberrations for a refracting surface. As already mentioned in Sect. 3.7, there are several ways to derive the aberration coefficients.

Here we essentially follow the procedure of D.J. Schroeder ("Astronomical Optics", see reference at the end of this chapter), which is without symmetry assumptions[14] somewhat complex but straight-forward, and hence, immediately understandable for the beginner. For an alternative calculation the work of, e.g., Born and Wolf (1959) is recommended.

The procedure for this problem (i.e. the determination of the Seidel errors of a single refracting surface under general conditions) is similar to the derivation of the refraction law.[15] In this case, the algebraic calculation is more complicated only because of the curved surface. Figure 7.13 graphically represents the problem to be solved. The optical path L starting at point Q via point B of the refracting surface to point Q' can be derived from the sum of all products of the geometrical paths and their respective diffraction indices:

$$L = n[QB] + n'[BQ'] = nl + n'l'.$$

Geometrically, the paths $[QB] = l$ and $[BQ'] = l'$ in the y-z plane are determined to be

$$l = \sqrt{(y - y_0)^2 + (z_0 - z^2) + x^2}$$

$$l' = \sqrt{(y - y_0')^2 + (z_0' - z^2) + x^2}.$$

---

[14]Assuming rotational symmetry of the optics and introducing polar coordinates in the object and image space, the occurring imaging errors can be directly obtained with general polynomials. To do so, one needs to find the linear combinations of the coordinates that are rotationally invariant. Thus, the calculation method presented here is shortened considerably.

[15]We therefore expect again the derivation of Snell's law, because essentially nothing has changed.

Furthermore, we use the following geometric and trigonometric relationships

$$y_0 = s \sin \Theta \qquad z_0 = s \cos \Theta \qquad s^2 = z_0^2 + y_0^2$$

$$y_0' = s' \sin \Theta' \qquad z_0' = s' \cos \Theta' \qquad s'^2 = z_0'^2 + y_0'^2$$

and also the identity $h^2 = x^2 + y^2$. The coordinates in the object space s, y_0, z_0 are negative and those in the image space s', y_0', z_0', Θ' are also negative. Hence, the angle Θ has a positive sign.

We obtain for the two different paths in the media with the refractive indices n and n'

$$l = \sqrt{h^2 - 2ys \sin \Theta + s^2 - 2zs \cos \Theta + z^2} \qquad (7.12)$$

$$l' = \sqrt{h^2 - 2ys' \sin \Theta' + s'^2 - 2zs' \cos \Theta' + z^2}. \qquad (7.13)$$

Since both equations do not differ in their structure, we only use Eq. 7.12 for the further calculations and replace the corresponding variables for Eq. 7.13. To simplify the square root, we develop it into a Taylor series. We factor out s and obtain after re-arrangement

$$l = s\sqrt{1 - \frac{2y}{s} \sin \Theta + \frac{h^2}{s^2} - \frac{2z \cos \Theta}{s} + \frac{z^2}{s^2}}.$$

The variable $z(x, y)$ still occurs in the square root, describing the sphere of the refracting surface. We replace $z = z(x, y) = z(h)$ by the already found approximation equation 7.8

$$z(h) = \frac{h^2}{2R} + \frac{(1 + K)h^4}{8R^3}$$

and obtain

$$l = s\sqrt{1 - \frac{2y}{s} \sin \Theta + \frac{h^2}{s^2} - \frac{2 \cos \Theta}{s}\left(\frac{h^2}{2R} + \frac{(1 + K)h^4}{8R^3}\right) + \frac{z^2}{s^2}}$$

or after factoring out and re-arrangement

$$l = s\sqrt{1 - \frac{2y}{s} \sin \Theta + \frac{h^2}{s}\left(\frac{1}{s} - \frac{2 \cos \Theta}{R}\right) - \frac{2(1+K)h^4 \cos \Theta}{8sR^3} + \frac{1}{s^2}\left(\frac{h^2}{2R} + \frac{(1+K)h^4}{8R^3}\right)^2}.$$

Because of the already known accuracy of the square root series 7.7 we take only terms up to h^4 into account and finally obtain

$$l = s\sqrt{1 - \frac{2y}{s}\sin\Theta + \frac{h^2}{s}\left(\frac{1}{s} - \frac{2\cos\Theta}{R}\right) - \frac{2(1+K)h^4\cos\Theta}{8sR^3} + \frac{h^4}{4s^2}\left(\frac{1}{R^2} - \frac{(1+K)}{R^3}s\cos\Theta\right)}.$$

We now use again the Taylor series of the function $\sqrt{1 - \zeta}$ with the variable $\zeta = \frac{2y}{s}\sin\Theta + \frac{h^2}{s}\left(\frac{1}{s} - \frac{2\cos\Theta}{R}\right) - \frac{2(1+K)h^4\cos\Theta}{8sR^3} + \frac{h^4}{4s^2}\left(\frac{1}{R^2} - \frac{(1+K)}{R^3}s\cos\Theta\right)$ and take into account all terms up to quadratic order

$$\sqrt{1 - \zeta} \approx 1 - \frac{1}{2}\zeta - \frac{1}{8}\zeta^2.$$

We determine the three terms of the Taylor series, take again only h-terms up to 4th order into account and get with $h^2 = x^2 + y^2$ for the root function

$$\sqrt{1-\zeta} = 1 - \left[\frac{y}{s}\sin\Theta - \frac{y^2}{2s}\left(\frac{1}{s} - \frac{\cos\Theta}{R}\right) - \frac{x^2}{2s}\left(\frac{1}{s} - \frac{\cos\Theta}{R}\right)\right] - \cdots$$

$$\left[\frac{y^2}{2s^2}\sin^2\Theta - \frac{y^3}{2s^2}\sin\Theta\left(\frac{1}{s} - \frac{\cos\Theta}{R}\right) - \frac{x^2 y}{2s^2}\sin\Theta\left(\frac{1}{s} - \frac{\cos\Theta}{R}\right)\right] - \cdots$$

$$\left[\frac{h^4}{8s^2}\left(\frac{1}{R^2} - \frac{(1+K)}{R^3}s\cos\Theta\right) - \frac{h^4}{8s^2}\left(\frac{1}{s} - \frac{\cos\Theta}{R}\right)^2\right].$$

We replace $\sin^2\Theta = 1 - \cos^2\Theta$, multiply and re-arrange to powers of x, y and h

$$\sqrt{1-\zeta} = 1 - \frac{y}{s}\sin\Theta + \frac{y^2}{2s^2}\left(\cos^2\Theta - \frac{\cos\Theta}{R}\right) + \frac{x^2}{2s}\left(\frac{1}{s} - \frac{\cos\Theta}{R}\right) - \cdots$$

$$\frac{y^3}{2s^2}\sin\Theta\left(\frac{1}{s} - \frac{\cos\Theta}{R}\right) + \frac{x^2 y}{2s^2}\sin\Theta\left(\frac{1}{s} - \frac{\cos\Theta}{R}\right) + \frac{h^4}{8s^2}\left(\frac{1}{R^2} - \frac{(1+K)}{R^3}s\cos\Theta\right) + \cdots$$

$$\frac{h^4}{8s^2}\left(\frac{1}{s} - \frac{\cos\Theta}{R}\right)^2.$$

We finally obtain the first partial path l (in the medium with refractive index n) of the complete path as a function of x, y:

$$l = -s + y\sin\Theta - \frac{y^2}{2}\left(\frac{\cos^2\Theta}{s} - \frac{\cos\Theta}{R}\right) - \frac{x^2}{2}\left(\frac{1}{s} - \frac{\cos\Theta}{R}\right) - \cdots$$

$$- \frac{y^3}{2s}\sin\Theta\left(\frac{\cos^2\Theta}{s} - \frac{\cos\Theta}{R}\right) - \frac{x^2 y}{2s}\sin\Theta\left(\frac{1}{s} - \frac{\cos\Theta}{R}\right) - \cdots$$

$$- \frac{h^4}{8}\left[\frac{1}{R^2}\left(\frac{1}{s} - \frac{(1+K)\cos\Theta}{R}\right) - \frac{1}{s}\left(\frac{1}{s} - \frac{\cos\Theta}{R}\right)^2\right]$$

and respectively for the 2nd partial path l'

$$l' = s' - y \sin \Theta' + \frac{y^2}{2} \left(\frac{\cos^2 \Theta'}{s'} - \frac{\cos \Theta'}{R} \right) + \frac{x^2}{2} \left(\frac{1}{s'} - \frac{\cos \Theta'}{R} \right) - \cdots$$

$$- \frac{y^3}{2s} \sin \Theta' \left(\frac{\cos^2 \Theta'}{s'} - \frac{\cos \Theta'}{R} \right) + \frac{x^2 y}{2s} \sin \Theta' \left(\frac{1}{s'} - \frac{\cos \Theta'}{R} \right) + \cdots$$

$$+ \frac{h^4}{8} \left[\frac{1}{R^2} \left(\frac{1}{s'} - \frac{(1+K)\cos \Theta'}{R} \right) - \frac{1}{s'} \left(\frac{1}{s'} - \frac{\cos \Theta'}{R} \right)^2 \right].$$

The opposite signs of the summands in l and l' are introduced by the negative definition of s. We are now in a position to calculate the total optical path L of any ray and then the Seidel image errors by searching for the extremum.

$$L = nl + n'l' =$$

$$+ [-ns + n's']$$

$$- y \left[n \sin \Theta - n' \sin \Theta' \right]$$

$$+ \frac{y^2}{2} \left[\frac{n' \cos^2 \Theta'}{s'} - \frac{n \cos^2 \Theta}{s} - \frac{n' \cos \Theta - n \cos \Theta}{R} \right]$$

$$+ \frac{x^2}{2} \left[\frac{n'}{s'} - \frac{n}{s} - \frac{n' \cos \Theta - n \cos \Theta}{R} \right]$$

$$+ \frac{y^3}{2} \left[\frac{n \sin \Theta}{s} \left(\frac{\cos^2 \Theta}{s} - \frac{\cos \Theta}{R} \right) - \frac{n' \sin \Theta'}{s'} \left(\frac{\cos^2 \Theta'}{s'} - \frac{\cos \Theta'}{R} \right) \right]$$

$$+ \frac{x^2 y}{2} \left[\frac{n \sin \Theta}{s} \left(\frac{1}{s} - \frac{\cos \Theta}{R} \right) - \frac{n' \sin \Theta'}{s'} \left(\frac{1}{s'} - \frac{\cos \Theta'}{R} \right) \right]$$

$$+ \frac{h^4}{8} \left[\frac{1}{R^2} \left(\frac{n'}{s'} - \frac{n}{s} - \frac{(1+K)}{R} \right) (n' \cos \Theta' - n \cos \Theta) + \frac{n}{s} \left(\frac{1}{s} - \frac{\cos \Theta}{R} \right)^2 \right]$$

$$- \frac{h^4}{8} \left[\frac{n'}{s'} \left(\frac{1}{s'} - \frac{\cos \Theta'}{R} \right)^2 \right].$$

$$(7.14)$$

In the decisive equation 7.14 of the entire optical path, apparently various powers and products of the surface coordinates x and y occur. We can summarize the polynomial for L using the following coefficients

$$L = L_0 + A_0 y + A_1 y^2 + A_1' x^2 + A_2 y^3 + A_2' x^2 y + A_3 h^4.$$

As we will see in the next chapter each of these terms corresponds to a Seidel aberration contribution, except $L_0 = (-ns + n's')$ and A_0y. The term L_0 corresponds precisely to the chief ray, a value independent from the surface coordinates x, y. It here appears because of the chosen coordinates and does not contribute to the search for extreme values. Therefore, we define the so-called optical path difference ΔL

$$\Delta L = L - L_0 = A_0y + A_1y^2 + A_1'x^2 + A_2y^3 + A_2'x^2y + A_3h^4 \qquad (7.15)$$

which is the difference of the optical paths between the general and the principle ray. For the case that in the further application of Fermat's principle, all coefficients disappear by selected conditions, all beams converge in point Q' and aberrations do not appear.

Before we turn to the detailed determination of the individual coefficients, we first try to understand the relationship between the calculated optical path differences and transverse aberration already discussed in Sect. 3.7 .

7.5.1 The Connection Between Wave Aberration and Longitudinal and Transverse Aberration

The so-called wave front is defined by all locations of a periodic wave having the same phase. This front moves with the propagation speed of the wave in the direction of their surface normal. Light rays are therefore basically perpendicularly oriented to the wavefront. For example, the wave fronts of a collimated beam bundle are planes perpendicular to the propagation direction. In this case one refers to a plane wave. Locations of identical propagation-time can also be interpreted as a wave front. In the sense of geometric optics a wavefront is thus the envelope of all beams having the same phase. Let us now consider two light beams taking different paths to a common location. They may arrive at different times. Using the speed of light, the resulting time difference can be converted to the so-called optical path difference. Optical path differences and differences of wavefronts are therefore equivalent.

If a spherical wavefront is emanating from a point source and is then imaged onto the Gaussian plane at point I by an ideal optical system, then the Gaussian wave front is convergent and also spherical.[16] Non-ideal imaging systems lead to deformations of the Gaussian wave front, causing aberrations. The so-called wave aberration is defined as the difference between the Gaussian and the distorted wavefront. Figure 7.14 shows the situation. A beam of the ideal convergent wavefront W_1 starting from point A crosses the optical axis at point I of the paraxial plane. If the optical system introduces aberrations, the exiting wavefront w_2 is

---

[16]The Gaussian wave front is also referred to as the reference wavefront and corresponds to the paraxial solution.

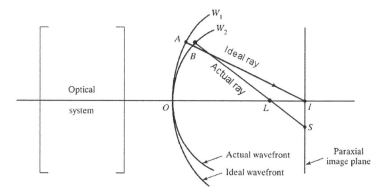

Fig. 7.14 The effects of the distorted wavefront W_2

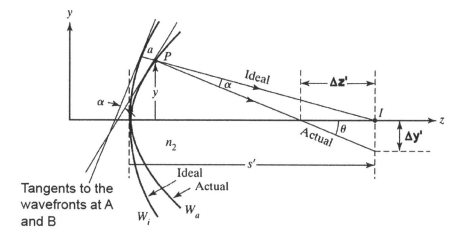

Fig. 7.15 Wavefront, transverse and longitudinal error

deformed, and a wavefront beam starting from point B penetrates the optical axis at point L and the Gaussian plane at point S.

As we already know, the distance $[LI]$ defines the longitudinal aberration and the distance $[IS]$, the transverse aberration.[17] In our example, the corresponding wave aberration is defined by the distance $[AB]$, or by the optical path difference $n'[AB]$, with n' the refractive index in the image space.

We now consider the situation in Fig. 7.15 in more detail. We are particularly interested in the wave aberration, which we now want to express in the form of the optical path difference ΔL between the ideal and the distorted wavefront with the coordinates W_i and W_a, respectively. If the wave fronts differ, then also the

---

[17]In Sect. 3.7 meridional and sagittal aberrations were indicated as $\Delta(y'_k)_{mer}$ and $\Delta(x'_k)_{sag}$.

tangents in a selected point P differ. The normals to the wavefronts (and therefore on the tangents) represent the propagation direction of the beams. If the tangents in Fig. 7.15 differ by the angle α, this leads to different propagation directions. The slopes of the two tangents are $\frac{dW_i}{dy}$ and $\frac{dW_a}{dy}$. If we denote the coordinate difference of the wave fronts with $\Delta W = W_i - W_a$ we can estimate the difference of the slopes: $\frac{dW_i}{dy} - \frac{dW_a}{dy} = \frac{d\Delta W}{dy} = \tan\alpha_i - \tan\alpha_a$. For small angle differences α of the tangents we can apply the addition theoreme[18] and write $\tan\alpha_i - \tan\alpha_a \approx \tan\alpha \approx \alpha \approx \frac{d\Delta W}{dy}$. Since the product of the wavefront difference ΔW and refractive index n' is just the optical path difference, i.e. $\Delta L = n'\Delta W$, we have $\frac{d\Delta L}{dy} = \alpha n'$. The transverse (or lateral) beam aberration $\Delta y'$ is $\Delta y' \approx \alpha s' = \frac{s'}{n'}\frac{d\Delta L}{dy}$, where $s' = [OI]$ represents the paraxial image distance.[19] For the longitudinal ray aberration $\Delta z'$ we obtain $\Delta z' = \frac{\Delta y'}{\tan\Theta} = \frac{s'\Delta y'}{y+\Delta y'} \approx \frac{s'\Delta y'}{y}$. Overall, we obtain the following expressions for the lateral and longitudinal aberrations.

$$\Delta x' = \frac{s'}{n'}\frac{d\Delta L}{dx}$$

$$\Delta y' = \frac{s'}{n'}\frac{d\Delta L}{dy} \tag{7.16}$$

$$\Delta z' = \frac{s'^2}{n'y}\frac{d\Delta L}{dy}$$

In the last step, we consider the transverse aberrations $\Delta x'$ and $\Delta y'$ by using the relationships following from the general equation 7.15 and compare them with the results obtained for a general lens system in Sect. 3.7. By differentiating and consideration of $A_2 = A'_2$ for $\Delta x'$ and $\Delta y'$ we obtain:

$$\Delta x' = \frac{s'}{n'}\frac{d\Delta L}{dx} = 2A'_1 x + 2A'_2 xy + 4A_3(x^2 + y^2)x$$

$$\Delta y' = \frac{s'}{n'}\frac{d\Delta L}{dy} = 2A_1 y + A_2(x^2 + 3y^2) + 4A_3(x^2 + y^2)y.$$

If we also consider the angular dependence of the coefficients A_1, A'_1, A_2, A'_2 we obtain with the new coefficients C_1, C'_1, C_2, C'_2 for the example of infinite object distance $s = -\infty$ and $s' = f$

$$\frac{\Delta x'}{f} = -\frac{1}{2}(x^2 + y^2)xC_3 + \frac{1}{2}\Theta xy C'_2 - \frac{1}{2}x\Theta^2 C'_2$$

$$\frac{\Delta y'}{f} = -\frac{1}{2}(x^2 + y^2)yC_3 + \frac{1}{2}(x^2 + 3y^2)\Theta C_2 - \frac{1}{2}y\Theta^2 C_1.$$

---

[18] $\tan(\Phi_1 - \Phi_2) = \frac{\tan\Phi_1 - \tan\Phi_2}{1+\tan\Phi_1\tan\Phi_2}$.

[19] The same is true for the x-direction, of course.

If we compare these results with the Eqs. 3.23 and 3.24 (see Sect. 3.7) we obtain a structural consistency when identifying the coefficients occurring here with the Seidel coefficients.[20] Hence, the coefficient A_3 represents the spherical aberration, A_2, A_2' the meridional and sagittal coma and A_1, A_1' the meridional and sagittal field curvature.

7.5.2 The Estimation of the Aberration Coefficients

We now consider the aberration coefficients A_1, A_1', A_2, A_2' in detail. Fermat's principle demands that the spatial derivatives of the optical path difference ΔL disappear:

$$\frac{\partial \Delta L}{\partial x} = 0 \qquad \frac{\partial \Delta L}{\partial y} = 0$$

Let us recall once more what these terms mean. The partial derivatives represent gradients, i.e. angles in zero approximation. In Sect. 7.5.1 we have already shown that the derivative $\frac{\partial \Delta L}{\partial y}$ is just the product of the deviation α from the Gaussian reference beam (see Fig. 7.15) and the refractive index[21] n':

$$\frac{\partial \Delta L}{\partial y} = \alpha n'.$$

We recall also the fact that the optical path QBQ' was determined. For the case that the partial derivatives vanish, this means that a light beam proceeding from point Q is refracted at point B and then crosses point Q'. Any deviation of the derivatives from zero means that the beam is not guided towards Q and thus the resulting image is not point-like.

We consider now the various coefficients of the given polynomial and try to find out under what conditions they disappear.

Coefficient A_0 of y

As already expected, the factor A_0 of y satisfies Snell's law of refraction at $A_0 = 0$, here for the principle ray due to the occurring angles Θ and Θ'.

$$A_0 = n' \sin \Theta' - n \sin \Theta. \tag{7.17}$$

---

[20]The term proportional to Θ^3 occurring in Eq. 3.23 (Distortion) was not taken into account in the development and is, hence, missing in this derivation.

[21]The same is true for the x-direction, of course: $\frac{\partial \Delta L}{\partial x} = \beta n'$ with β the deviation angle in this direction.

Coefficient A_1 of y^2

The coefficient of y^2 is

$$A_1 = \frac{1}{2}\left(\frac{n'\cos^2\Theta'}{s'} - \frac{n\cos^2\Theta}{s} - \frac{n'\cos\Theta' - n\cos\Theta}{R}\right)$$

and vanishes if

$$\frac{n'\cos^2\Theta'}{s'} - \frac{n\cos^2\Theta}{s} = \frac{n'\cos\Theta' - n\cos\Theta}{R}.$$

We first consider the paraxial case. Here Θ and Θ' are small. This yields for $\cos\Theta = \cos\Theta' \approx 1$

$$\frac{n'}{s'} - \frac{n}{s} = \frac{n' - n}{R}$$

which corresponds to the already known Abbe invariants in Eqs. 3.2 and thus the coefficient A_1 vanishes in the sense of Fermat's principle. This means all the rays meet on the image side in a point on the intersection length $l' \approx s'$. If we develop the cos function up to the next order, i.e. $\cos\alpha \approx 1 - \frac{\alpha^2}{2}$, we obtain for A_1:

$$A_1 = \frac{n'}{s'}(1 - \Theta'^2) - \frac{n}{s}(1 - \Theta^2) - \frac{n'(1 - \frac{\Theta'^2}{2}) - n(1 - \frac{\Theta^2}{2})}{R}.$$

Due to the Abbe invariants one can replace the last fraction[22]

$$A_1 = \frac{n'}{s'}(1 - \Theta'^2) - \frac{n}{s}(1 - \Theta^2) - \frac{n'}{s'}(1 - \frac{\Theta'^2}{2}) + \frac{n}{s}(1 - \frac{\Theta^2}{2})$$

and hence

$$A_1 = -\frac{1}{2}\left(\frac{n'}{s'}\Theta'^2 - \frac{n}{s}\Theta^2\right).$$

If we now replace $\Theta'^2 = \frac{n^2\Theta^2}{n'^2}$ due to the squared Snell's law for small angles we immediately obtain

$$A_1 = -\frac{n^2\Theta^2}{2}\left(\frac{1}{n's'} - \frac{1}{ns}\right) \tag{7.18}$$

with A_1 the coefficient of the <u>meridional</u> field curvature.

---

[22]The different factors $(1 - \Theta'^2)$ and $(1 - \Theta^2)$ act like a change of the refractive indices, which are consistently present on both sides of the Abbe equation.

Coefficient A_1' of x^2

The coefficient of x^2 is

$$A_1' = \frac{1}{2} \left(\frac{n'}{s'} - \frac{n}{s} - \frac{n' \cos \Theta' - n \cos \Theta}{R} \right)$$

and vanishes if

$$\frac{n'}{s'} - \frac{n}{s} = \frac{n' \cos \Theta' - n \cos \Theta}{R}.$$

For A_1' we obtain in a similar manner as in the derivation of A_1

$$A_1' = \frac{n^2 \Theta^2}{2} \left(\frac{1}{n's'} - \frac{1}{ns} \right) \tag{7.19}$$

with A_1' the coefficient of the <u>sagittal</u> field curvature.

Coefficient A_2 of y^3

The coefficient of y^3 is A_2

$$A_2 = \frac{1}{2} \left[\frac{n \sin \Theta}{s} \left(\frac{\cos^2 \Theta}{s} - \frac{\cos \Theta}{R} \right) - \frac{n' \sin \Theta'}{s'} \left(\frac{\cos^2 \Theta'}{s'} - \frac{\cos \Theta'}{R} \right) \right].$$

For small Θ and Θ' we consequently have $\cos \Theta = \cos \Theta' \approx 1$ and because of $n\Theta \approx n'\Theta'$

$$A_2 = \frac{\Theta}{2} \left[\frac{n}{s} \left(\frac{1}{s} - \frac{1}{R} \right) - \frac{n}{s'} \left(\frac{1}{s'} - \frac{1}{R} \right) \right].$$

According to Eq. 3.2 we generally have for a refracting surface

$$\frac{n'}{s'} - \frac{n}{s} = \frac{n' - n}{R}.$$

This can be re-arranged to

$$n' \left(\frac{1}{s'} - \frac{1}{R} \right) = n \left(\frac{1}{s} - \frac{1}{R} \right).$$

With these auxiliary equations we can write A_2 as

$$A_2 = \frac{n^2 \Theta}{2} \left[\frac{1}{ns} \left(\frac{1}{s} - \frac{1}{R} \right) - \frac{1}{n's'} \left(\frac{1}{s} - \frac{1}{R} \right) \right]$$

and we obtain the result for the coefficient A_2

$$A_2 = \frac{n^2 \Theta}{2} \left[\left(\frac{1}{s} - \frac{1}{R} \right) \left(\frac{1}{ns} - \frac{1}{n's'} \right) \right] \tag{7.20}$$

with A_2 the coefficient of the meridionalen coma.

Coefficient A'_2 of $x^2 y$

The coefficient of $x^2 y$ is

$$A'_2 = \frac{1}{2} \left[\frac{n \sin \Theta}{s} \left(\frac{1}{s} - \frac{\cos \Theta}{R} \right) - \frac{n' \sin \Theta'}{s'} \left(\frac{1}{s'} - \frac{\cos \Theta'}{R} \right) \right].$$

Again, we have for small Θ and $\Theta' \cos \Theta = \cos \Theta' \approx 1$ or $n\Theta \approx n'\Theta'$ and thus

$$A'_2 = \frac{\Theta}{2} \left[\frac{n}{s} \left(\frac{1}{s} - \frac{1}{R} \right) - \frac{n}{s'} \left(\frac{1}{s'} - \frac{1}{R} \right) \right].$$

Thus, the coefficient of $x^2 y$ is just $A'_2 = A_2$

$$A'_2 = \frac{n^2 \Theta}{2} \left[\left(\frac{1}{s} - \frac{1}{R} \right) \left(\frac{1}{ns} - \frac{1}{n's'} \right) \right] \tag{7.21}$$

with A'_2 representing the coefficients of the sagittal coma.

Coefficient A_3 of h^4

The coefficient of $h^4 = (x^2 + y^2)^2$ is

$$A_3 = \frac{1}{8} \left[\frac{1}{R^2} \left(\frac{n'}{s'} - \frac{n}{s} - \frac{(1+K)}{R} (n' \cos \Theta' - n \cos \Theta) \right) + \frac{n}{s} \left(\frac{1}{s} - \frac{\cos \Theta}{R} \right)^2 \right] - \frac{1}{8} \left[\frac{n'}{s'} \left(\frac{1}{s'} - \frac{\cos \Theta'}{R} \right)^2 \right].$$

For small Θ and Θ' we consequently have $\cos \Theta = \cos \Theta' \approx 1$ and thus

$$A_3 = \frac{1}{8} \left[\frac{1}{R^2} \left(\frac{n'}{s'} - \frac{n}{s} - \frac{(1+K)}{R} (n' - n) \right) + \frac{n^2}{ns} \left(\frac{1}{s} - \frac{1}{R} \right)^2 - \frac{n'^2}{n's'} \left(\frac{1}{s'} - \frac{1}{R} \right)^2 \right].$$

With

$$n'^2 \left(\frac{1}{s'} - \frac{1}{R} \right)^2 = n^2 \left(\frac{1}{s} - \frac{1}{R} \right)^2$$

follows

$$A_3 = \frac{1}{8} \left[\frac{1}{R^2} \left(\frac{n'}{s'} - \frac{n}{s} - \frac{(1+K)}{R}(n'-n) \right) + \frac{n^2}{ns} \left(\frac{1}{s} - \frac{1}{R} \right)^2 - \frac{n^2}{n's'} \left(\frac{1}{s} - \frac{1}{R} \right)^2 \right]$$

or

$$A_3 = \frac{1}{8} \left[\frac{1}{R^2} \left(\frac{(n'-n)}{R} - \frac{(1+K)}{R}(n'-n) \right) + n^2 \left(\frac{1}{ns} - \frac{1}{n's'} \right) \left(\frac{1}{s} - \frac{1}{R} \right)^2 \right]$$

or as result for the coefficient of spherical aberration A_3

$$A_3 = -\frac{1}{8} \left(\frac{K(n'-n)}{R^3} \right) - \frac{n^2}{8} \left(\frac{1}{n's'} - \frac{1}{ns} \right) \left(\frac{1}{s} - \frac{1}{R} \right)^2 . \qquad (7.22)$$

Apparently, the coefficients A_1, A_1', A_2, A_2', A_3 can vanish only under certain conditions. In the image space of the refracting surface, aberrations thus do occur.

The Meaning of the Coefficients A_1 and A_1': Astigmatism

From the two coefficients A_1 and A_1' of the two field curvatures follow the specific solutions for the disappearance of the coefficients:

$$\frac{n'}{s_m'} = \frac{1}{\cos^2 \Theta'} \left(\frac{n \cos^2 \Theta}{s} + \frac{n' \cos \Theta' - n \cos \Theta}{R} \right)$$

for the meridional direction y and

$$\frac{n'}{s_s'} = \frac{n}{s} + \frac{n' \cos \Theta' - n \cos \Theta}{R}$$

for the sagittal x-direction. s_m' represents the focal line of the meridional and s_s' of the sagittal image surface.

As mentioned in Sect. 3.7.2 we introduce the astigmatism again by subtraction of the two image spheres. In addition, we use the trigonometric identity $\frac{1}{\cos^2 \Theta'} = 1 + \tan^2 \Theta'$ and obtain

$$\frac{n'}{s'_m} - \frac{n'}{s'_s} = n'\frac{(s'_m - s'_s)}{s'_m s'_s} = \ldots$$

$$n'\frac{\Delta s'}{s'_m s'_s} = (1 + \tan^2 \Theta') \left(\frac{n \cos^2 \Theta}{s} + \frac{n' \cos \Theta' - n \cos \Theta}{R} \right) - \frac{n}{s} - \frac{n' \cos \Theta' - n \cos \Theta}{R}$$

with $\Delta s' = s'_m - s'_s$ the difference between the two image spheres. By multiplication we obtain

$$n'\frac{\Delta s'}{s'_s s'_m} = \tan^2 \Theta' \left(\frac{n \cos^2 \Theta}{s} + \frac{n' \cos \Theta' - n \cos \Theta}{R} \right) + \frac{n \cos^2 \Theta}{s} - \frac{n}{s}$$

or because of $\cos^2 \Theta - 1 = -\sin^2 \Theta$

$$n'\frac{\Delta s'}{s'_m s'_s} = \tan^2 \Theta' \left(\frac{n \cos^2 \Theta}{s} + \frac{n' \cos \Theta' - n \cos \Theta}{R} \right) - \frac{n}{s} \cos^2 \Theta \tan^2 \Theta.$$

For small angles Θ and Θ' we have again $n^2 \Theta^2 = n'^2 \Theta'^2$ and thus

$$n'\frac{\Delta s'}{s'_m s'_s} = \Theta^2 \frac{n^2}{n'^2} \left(\frac{n}{s} + \frac{n' - n}{R} \right) - \frac{n}{s} \Theta'^2.$$

We use again the Abbe invariant and obtain

$$n'\frac{\Delta s'}{s'_m s'_s} = \Theta^2 \frac{n^2}{n'^2} \left(\frac{n}{s} + \frac{n'}{s'} - \frac{n}{s} \right) - \frac{n}{s} \Theta^2 = \Theta^2 \frac{n^2}{n'^2} \frac{n'}{s'} - \frac{n}{s} \Theta^2$$

and thus finally

$$\frac{\Delta s'}{s'_m s'_s} = \frac{\Theta^2 n^2}{n'} \left(\frac{1}{n' s'} - \frac{1}{ns} \right).$$

If we compare this result with the results for the coefficients A_1 and A'_1 (Eqs. 7.18 and 7.20) we can obviously express the astigmatism by using the coefficients $A1$ or A'_1:

$$\frac{\Delta s'}{s'_s s'_m} = \frac{-2A_1}{n'}. \tag{7.23}$$

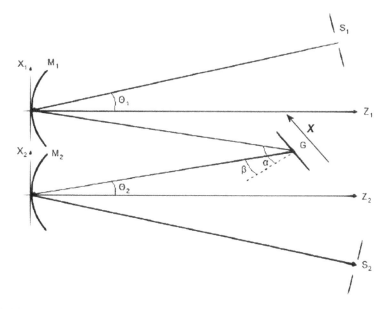

Fig. 7.16 The geometry of the Czerny-Turner spectrometer (*Courtesy*: D. Sablowski)

7.6   Calculation of the Czerny-Turner Spectrometer

The Czerny-Turner spectrograph consists of two concave imaging mirrors and a dispersive grating. Since the mirrors naturally have to work in an off-axis arrangement, corresponding image errors arise which depend on the geometric design of the system. Figure 7.16 shows the geometry of the Czerny-Turner with the collimator M_1, the camera M_2, the grating G and the corresponding angles. The mirror vertices lie at the origins of the two coordinate systems x_1, z_1 and x_2, z_2. As usual, the angle at the grating is denoted by the angle of incidence α and the diffraction angle β and the off-axis angles Θ_1 and Θ_2. For the determination of the aberrations we need to consider two reflective surfaces, i.e. the Seidel aberrations for both mirrors add to yield an overall error.

For the spherical surface, we first obtained for the coefficient of astigmatism the Eq. 7.19

$$A_1' = \frac{\Theta^2 n^2}{2} \left(\frac{1}{n's'} - \frac{1}{ns} \right).$$

Since this general equation is valid for each of the two mirrors of the Czerny-Turner spectrograph, we find for the total coefficient

$$A_1' = \frac{\Theta_1^2 n_1^2}{2} \left(\frac{1}{n_1's_1'} - \frac{1}{n_1s_1} \right) + \frac{\Theta_2^2 n_2^2}{2} \left(\frac{1}{n_2's_2'} - \frac{1}{n_2s_2} \right).$$

For a mirror, the refractive indices $n_1 = n_2 = n'_1 = n'_2 = -1$ in the coefficients must be negative.[23] We obtain for A'_1

$$A'_1 = \frac{\Theta_1^2}{2}\left(\frac{1}{s_1} - \frac{1}{s'_1}\right) + \frac{\Theta_2^2}{2}\left(\frac{1}{s_2} - \frac{1}{s'_2}\right).$$

According to the Gaussian imaging equation of the mirror 7.1 we have

$$\frac{1}{s_2} - \frac{1}{s'_2} = \frac{1}{f} = \frac{2}{R}.$$

Both mirrors have the same y-coordinate. Thus, the term of the total astigmatism is

$$A'_1 = \frac{\Theta_1^2}{R_1} + \frac{\Theta_2^2}{R_2}.$$

Furthermore, for the coefficient of the coma A_2, Eq. 7.20 is valid

$$A_2 = \frac{\Theta n^2}{2}\left(\frac{1}{s} - \frac{1}{R}\right)\left(\frac{1}{n's'} - \frac{1}{ns}\right).$$

We now consider the two mirrors separately. For the collimator M_1 one has $s' = \infty$ and $s = f_1$ and again $n' = n = -1$ for the refractive indices. Substituting we obtain

$$A_{2,Col} = \frac{\Theta_1}{2}\left(\frac{1}{f_1} - \frac{1}{R_1}\right)\left(\frac{1}{\infty} - \frac{1}{f_1}\right) = \frac{\Theta_1}{2}\left(-\frac{1}{f_1^2} + \frac{1}{f_1 R_1}\right)$$

$$= \frac{\Theta_1}{2}\left(\frac{4}{R_1^2} - \frac{2}{R_1^2}\right) = \frac{\Theta_1}{R_1^2}.$$

For the camera M_2 we obtain in accordance with $s' = f_2$ and $s = \infty$

$$A_{2,Cam} = \frac{\Theta_2}{2}\left(\frac{1}{\infty} - \frac{1}{R_2}\right)\left(\frac{1}{f_2} - \frac{1}{\infty}\right) = -\frac{\Theta_2}{2}\frac{1}{R_2}\frac{2}{R_2} = -\frac{\Theta_2}{R_2^2}.$$

---

[23]This rule can be derived from the already discussed lens-maker equation 3.12. This applies to spherical surfaces and thus consequently also to mirrors. However, if $n = 1$ for the refractive index, we do not obtain a meaningful result for f. For reflections there is a change of direction, so one must formally set $n = -1$. For $r_1 = \infty$ (the concave mirror has only one curvature radius) and $r_2 = R$ one obtains the relation between focal length and curvature radius of the mirror using the lens-maker equation $\frac{1}{f} = (n-1)\left(\frac{1}{r_1} - \frac{1}{r_2}\right)$: $\frac{1}{f} = (-2)\left(\frac{1}{\infty} - \frac{1}{R}\right) = \frac{2}{R}$. Insertion into the lens equation again provides the imaging equation 7.1 of the mirror.

Overall, we obtain for the total coefficient of the coma A_2

$$A_2 = \frac{\Theta_1}{R_1^2} - \frac{\Theta_2}{R_2^2}. \tag{7.24}$$

Since the coordinates of two separated mirrors must be taken into account, the coma of the overall system of the Czerny-Turner is

$$A_{2,1}x_1^3 + A_{2,2}x_2^3 = \frac{\Theta_1}{R_1^2}x_1^3 - \frac{\Theta_2}{R_2^2}x_2^3.$$

The coma of the system can obviously be compensated by requiring:

$$\frac{\Theta_1}{R_1^2}x_1^3 - \frac{\Theta_2}{R_2^2}x_2^3 = 0.$$

Thus

$$\frac{\Theta_2}{\Theta_1} = \left(\frac{R_2}{R_1}\right)^2 \left(\frac{x_1}{x_2}\right)^3. \tag{7.25}$$

One can obtain from Fig. 7.16 the following trigonometric identities

$$x_1 \cos \Theta_1 = x \cos \alpha$$

$$x_2 \cos \Theta_2 = x \cos \beta$$

with x the coordinate along the grating. For a given wavelength λ_0 we specifically have

$$x_1 \cos \Theta_1 = x \cos \alpha_0$$

$$x_2 \cos \Theta_2 = x \cos \beta_0.$$

Putting this into the condition 7.25 we obtain the so-called Shafer condition 7.26

$$\frac{\Theta_2}{\Theta_1} = \left(\frac{R_2}{R_1}\right)^2 \left(\frac{\cos \Theta_2 \cos \alpha_0}{\cos \Theta_1 \cos \beta_0}\right)^3. \tag{7.26}$$

The equation indicates a condition in which the coma of the Czerny-Turner disappears. The two angles Θ_1 and Θ_2 need to have the specified relation. Under these conditions, one of the off-axis aberrations is therefore compensated.[24] Assuming

---

[24]The grating angles α_0 and β_0 determine the wavelength. Thus, the coma is indeed exactly compensated only for a single wavelength λ_0, but is also negligible for a wide range in wavelength.

spherical aberration plays a minor role, only the astigmatism and field curvature limit the spectral resolution of the Czerny-Turner system. The off-axis angles Θ_1 and Θ_2 should be kept as small as possible because of the resulting astigmatism. Under these conditions, one can set $\cos \Theta_i \approx 1$ and we have

$$\frac{\Theta_2}{\Theta_1} = \left(\frac{R_2}{R_1}\right)^2 \left(\frac{\cos \alpha_0}{\cos \beta_0}\right)^3 .$$

Substituting this condition in Eq. 7.6 we obtain with Eq. 7.23 an expression for the total astigmatism. Then only the angle Θ_1 of the two original angles remains.

$$A_1' = \frac{\Theta_1^2}{R_1} + \frac{\Theta_1^2}{R_2}\left(\frac{R_1}{R_2}\right)^4 \left(\frac{\cos \alpha_0}{\cos \beta_0}\right)^6 = \frac{\Theta_1^2}{R_1}\left[1 + \left(\frac{R_1}{R_2}\right)^3 \left(\frac{\cos \alpha_0}{\cos \beta_0}\right)^6\right].$$

In the equations for coma (7.20 and 7.21) and field curvature/astigmatism (7.5.2 and 7.5.2) the radii of the mirrors M_1 and M_2 appear in the denominator. The greater these radii are, the smaller the off-axis aberrations. The same is true for the influence of the focal length and f-ratios of the mirrors. In addition, we should remember that in the spectrometer only the dispersion direction determines the spectral resolution. If one places the astigmatic foci onto the slit or CCD camera, the other focal line has no effect on the spectral resolution. The pixels perpendicular to the dispersion direction will be integrated during the subsequent reduction process. Therefore, the astigmatic defocus direction plays only a minor role. Another possibility to compensate or suppress the resulting residual astigmatism is to have, e.g., cylindrical lenses, which are also used for the correction of astigmatic telescope systems (Schiefspiegler). As an example, Fig. 7.17 shows the spot diagrams at the exit slit for different spectrograph configurations from Shafer et al. (1964). (a) Ebert-Fastie configuration. (b) Czerny-Turner configuration with $R_1 = R_2 = 2,000$ mm (Coma correction done by increased angle α'). (c) Czerny-Turner configuration with $R_1 = 2,000$ mm, $R_2 = 1,800$ mm. The coma in (a) is reduced in (b). On the other hand, astigmatism is increased. In both patterns a curvature of the astigmatism pattern is visible. Both coma and astigmatism are reduced in the asymmetric Czerny-Turner configuration. The scale is in microns.

7.7 Focusing Gratings

One can even go a step further by giving the grating the shape of a concave mirror. Focusing gratings are a useful way to dispense with collimator and camera optics. This is important if maximum efficiency is the goal of spectroscopic measurements or for measurements in ultraviolet light, for which glass opacities increase dramatically (see Appendix D). In addition, the reflectivity of optical gratings decrease from approximately 90 % in the visual domain to a maximum of

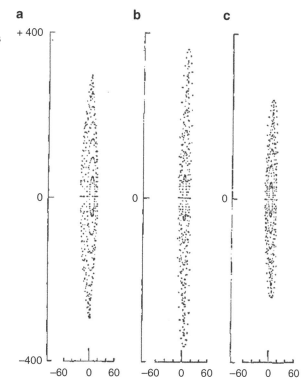

Fig. 7.17 Spot diagrams for Ebert Fastie and Czerny-Turner configurations (Shafer et al. 1964)

20 % in the far UV. However, measurements in the UV are today an integral part of astronomical observations and so concave gratings are used in various space-based telescopes (see Sect. 8.5).

At the end of the nineteenth century Henry Rowland examined focusing gratings. He had to rely on optical elements made of metal instead of glass, whose reflectivity was only about 70 %. Each additional optical element reduced the efficiency dramatically. Rowland therefore manufactured a spherical concave grating so that the distances and angles of the grating lines are constant with respect to the chord of the curved grating, but not against the grating surface (Fig. 7.18). Rowland showed the following: If a point source is positioned on a circle having the half radius of the grating radius and its light passes through its apex, then its diffraction image is also positioned on this circle. Then the grating focus must also be located on this circle. When a point source is on this "Rowland circle" the curved grating diffracts this source into a plane which is a tangent to the Rowland circle (Fig. 7.19).

Figure 7.20 depicts the situation. The point C is the center of the grating curvature. At P is the light source of wavelength λ and Q is the grating focus. The diffraction conditions at point A again follow the grating equation.

$$d(sin\alpha + sin\beta) = n\lambda$$

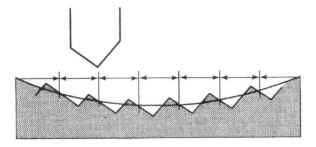

Fig. 7.18 Sketch of a concave diffraction grating (Hutley 1982)

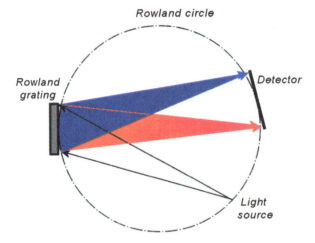

Fig. 7.19 Rowland circle principle. The radius of the curved Rowland grating is twice as large as the radius of the Rowland circle

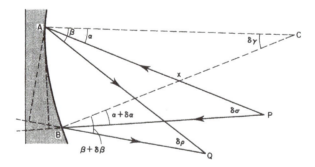

Fig. 7.20 Calculation of the Rowland circle (Hutley 1982)

Fig. 7.21 Construction of the
Rowland circle (Hutley 1982)

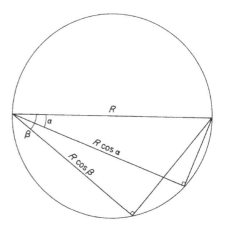

In order to investigate what exactly happens at spherical gratings, we first consider
only a small area around the grating apex. We neglect the groove length and consider
the grating curvature as part of the Rowland circle (in reality the grating only touches
the circle). The beam is emitted at point P and reaches the grating surface at point
A with an angle α with respect to the grating normal $\overline{AC}$. It is then diffracted in the
direction β towards point Q. The same happens for the angles $\alpha + \delta\alpha$ and $\beta + \delta\beta$
at point B, again with respect to the grating normal $\overline{BC}$. If we now differentiate the
grating equation, we obtain

$$\cos\alpha \cdot \partial\alpha + \cos\beta \cdot \partial\beta = 0. \tag{7.27}$$

From the triangles ACX and BPX, the angle $\delta\sigma$ between the two beams emanating
from P and the angle $\delta\rho$ between the two beams converging at Q we obtain $\alpha + \delta\gamma = \alpha + \delta\alpha + \delta\sigma$ and therefore $\delta\alpha = \delta\gamma - \delta\sigma$ and $\delta\beta = \delta\gamma - \delta\rho$. Because of the observed
small apex region at very small angles $\delta\alpha$, $\delta\beta$, $\delta\gamma$ and $\delta\sigma$ we obtain with the grating
radius R = $\overline{AC}$ and the distances $\overline{AP} = r$ and $\overline{AQ} = r_1$

$$\delta\gamma = \frac{\overline{BC}}{R}, \delta\sigma = \frac{\overline{AB}\cos\alpha}{r}, \delta\rho = \frac{\overline{AB}\cos\beta}{r_1}.$$

Substituting $\partial\alpha$ and $\partial\beta$ from Eq. 7.27, we obtain

$$\frac{\cos\alpha}{R} - \frac{\cos^2\alpha}{r} + \frac{\cos\beta}{R} - \frac{\cos^2\beta}{r_1} = 0. \tag{7.28}$$

This is the focus equation for the spherical concave grating. A solution can be found
if the respective terms for α and β vanish separately, so that $\frac{\cos\alpha}{R} = \frac{\cos^2\alpha}{r}$ and
$\frac{\cos\beta}{R} = \frac{\cos^2\alpha}{r_1}$. Thus we have $r = R\cos\alpha$ and $r_1 = R\cos\beta$. This in turn means that
both P and Q are located on a circle of diameter R (Fig. 7.21).

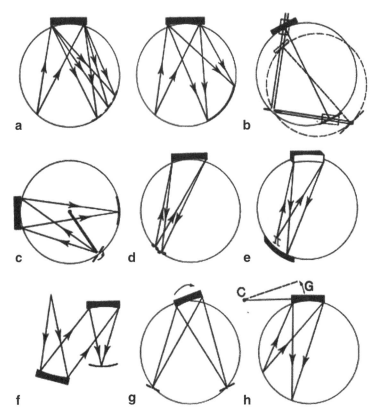

Fig. 7.22 Examples of spectrographs with concave gratings. (**a**) Paschen-Runge, (**b**) Rowland, (**c**) Abney, (**d** and **e**) Eagle, (**f**) Wadsworth, (**g**) Seya-Namioka, (**h**) Johnson-Onaka (Hutley 1982)

Spherical concave gratings show massive astigmatism when used on the Rowland circle. Therefore, a point source is imaged as a line. However, the foci of the meridional plane and the sagittal plane are different (Fig. 3.8). Thus, choosing the correct plane with respect to the dispersion direction, astigmatism can be completely eliminated. A common variant is the Wadsworth system (Fig. 7.22f). In this configuration, the incident light is collimated, so that r in the focus equation 7.28 is taken as infinity. This is then

$$r_1 = \frac{R \cos^2 \beta}{\cos \alpha + \cos \beta}$$

The design is characterized by the fact that astigmatism disappears if the image is generated in the middle of the grating ($\beta = 0$). Wadsworth systems can be advantageously used where the incident light is naturally collimated (e.g., orbital observatories or spectro-helioscopes). In addition, they also offer the advantage of compact size.

Suggested Readings

- H.G. Beutler, *The Theory of the Concave Grating*, Journal of the Optical Society of America, Vol. 35, Number 5, 1945
- M. Czerny, A.F. Turner, *Über den Astigmatismus bei Spiegelspektrometern*, Z. Physik, Vol. 61(11–12), 1930
- H. Ebert, *Zwei Formen von Spectrografen*, Annalen der Physik, Vol. 274, Number 11, p. 489–493, 1889
- W.G. Fastie, *A Small Plane Grating Monochromator*, Journal of the Optical Society of America, Vol. 42, Number 9, 1952
- M.C. Hutley, *Diffraction Gratings*, Academic Press, 1982
- A. Reader, *Optimization Czerny-Turner Spectrographs: A Comparison between Analytical Theory and Ray Tracing*, Journal of the Optical Society of America, Vol. 59, Number 9, 1969
- D.J. Schroeder, *Astronomical Optics*, Academic Press, 1987
- W.T. Welford, *Aberration Theory of Gratings and Grating Mountings*, Progress in Optics, Volume IV, North-Holland Publishing Company, 1964

Chapter 8
Practical Examples

A Short Story

When Peter Stoffer in Switzerland started the development of a spectrograph for the 60 cm telescope in Bern-Zimmerwald, he contacted Thomas and Klaus to discuss the design. Fortunately, Peter had already begun his studies in physics and all three spoke the same technical language. The exchange quickly went forward with new insights. The discussion took place in the online discussion forum of the German VdS amateur spectroscopy section. One Saturday night, Klaus and Thomas were sitting in front of their computers in Germany and Peter in Switzerland—LOG-IN—and the arguments flew through the network. Various thoughts and designs were developed and discarded, and the evening was getting longer. Soon some books and catalogues of optics manufacturers piled up on the kitchen table. Focal lengths and apertures were checked and sent to Switzerland. Peter's private library was obviously well equipped; the corresponding product numbers and answers quickly came back. Klaus placed the first beer bottles beside the catalogues, the first design concepts took shape and excitement grew. Thomas uncorked the whisky bottle and surprisingly, the design drafts were getting better. Early in the morning the whisky was half empty, the heads dizzy and the first practical design was finalized. LOG-OUT!

8.1 From Unique Instruments to Mass Production

Spectroscopic measurements in astronomy were an almost exclusive domain of professionals until about 25 years ago. This has changed dramatically through technical developments. Today one can buy spectrographs of various designs and customized system parameters in the mass market and equip telescopes down to about 20 cm aperture. Thus one can successfully deliver data for professional

© Springer-Verlag Berlin Heidelberg 2015
T. Eversberg, K. Vollmann, *Spectroscopic Instrumentation*, Springer Praxis Books,
DOI 10.1007/978-3-662-44535-8__8

research. The spectroscopic principles, as we have shown in the first Chaps. 2 and 5, remain valid, of course, regardless of the spectrograph size.

Whether a spectrograph is developed or bought ready-to-go depends on the system size, the required parameters and the available financial resources. For large telescopes (D $\geq$ 4 m) a mass-market does not exist, given the different telescope parameters at such facilities. Therefore appropriate spectrographs on such telescopes are all unique. Since these large systems are very expensive, development errors must be avoided. For that reason they are designed, developed and built by modern management methods. These methods include management phases, which ensure full control over design and construction. First, from a scientific perspective a science case is defined, which includes all scientific issues, goals and the necessary technical parameters (e.g. spectral resolution, signal-to-noise ratio, exposure time, limiting magnitude). Usually scientists from around the world or from those countries who finance the project define the science case. In a feasibility study (Phase A—Feasibility) the development team shows that the minimum technical requirements can (or cannot) be achieved. Once Phase A is passed, a subsequent definition phase (Phase B—Definition) ends with the preliminary design (Preliminary Design Review). Then, the final instrument design (Phase C—Detailed Design) is completed. Only after phase C will the construction of the instrument start (Phase D—Production). All phases are reviewed by external experts invited to scrutinize the draft and put it in context to the requirements. All deviations from the requirements and open points are recorded in Review Item Discrepancies (RID). All RIDs must be adequately answered by the development team and resolved before phase D and thus the construction of the device can begin.

Only for telescopes of moderate size and common focal lengths has a mass market, albeit modest, been developed. The company Boller & Chivens has developed some systems for focal lengths of 10–20 m found in many observatories (Sect. 8.3.3). These spectrographs are robust, easy to handle and have standardized connections for a CCD cryostat. For many years they were the workhorses of modern astrophysics. For even shorter focal lengths and smaller apertures the several-dozen kilograms of the Boller & Chivens instruments are too much to carry, though. Therefore, spectrographs for telescopes under one meter aperture were also custome-made until only a few years ago. Examples are the Garrison spectrograph of the 60 cm University of Toronto Southern Observatory (UTSO) at Las Campanas in Chile (today Leoncito in Argentina) and the one at Danish 0.5 m telescope at La Silla in Chile (Fig. 8.1). For some years now there have been "off-the-shelf" spectrographs available for very short focal-length telescopes. These are mainly devices for the widespread class of Schmidt-Cassegrain Telescopes (e.g., Meade, Celestron) optimized for focal lengths of 2–4 m. These devices can as well be adapted to telescope focal lengths of up to 8 m (with compromises in spectral resolving power) by using focal reducers (e.g. simple Shapley lenses). All these systems are slit spectrographs. Commercial spectrographs for short focal length telescopes can in principle be divided into three categories.

Fig. 8.1 Spectrograph at the Danish 0.5 m telescope at La Silla (*Courtesy*: ESO)

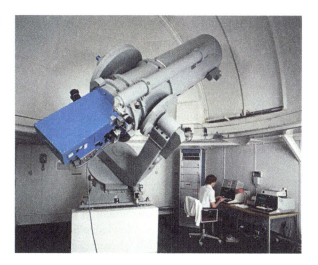

1. spectroscopes of low resolving power ($R \sim 1,000\text{–}3,000$),
2. spectrographs of medium to high resolving power ($R \sim 6,000\text{–}14,000$) and
3. Echelle systems of high resolving power ($R \geq 10,000$).

The spectral resolving power, of course, depends on the telescope focal length used and the slit width with respect to the target and seeing conditions. The first such devices were introduced by the company SANTA BARBARA IMAGE GROUP (SBIG) in California with their DEEP SPACE SPECTROGRAPH (typical resolving power of $R \sim 1,200$) and the SELF-GUIDING SPECTROGRAPH (typical resolving power of $R \sim 3,000$). These low-resolution systems were supplemented in 2009 by the DADOS spectrograph of the German company BAADER-PLANETARIUM. This last apparatus delivers a resolving power of $R \sim 400\text{–}3,000$ depending on the grating used. This is sufficient for broadband spectroscopy and/or faint-object observations and appropriate for educational purposes. For advanced line-profile analysis, however, a spectral resolving power of at least 4,000 is required (see Chap. 14). When writing this text, we realized that this domain has only been reached with the Littrow High Resolution Spectrograph III (Lhires III) by the French company SHELYAK INSTRUMENTS. This is a Littrow spectrograph optimized for f/10 systems of 8–11 inch aperture telescopes (i.e. for telescopes of the commercially available Schmidt-Cassegrain types). Hence, a mass-produced product for small telescopes was available that meets professional standards for the first time. Since then it has even been successfully used in some professional campaigns. Already in the mid 1990s the American company OPTO MECHANICS RESEARCH introduced a first echelle spectrograph for very small telescopes. Their SE200 system is fed by fiber optics and provides a spectral resolving power of about 6,000. This unit was later supplemented by the less expensive SE100 system with a resolving power of about 3,000. In the world of echelle spectroscopy this was a breakthrough, and again in 2008, SHELYAK INSTRUMENTS introduced an alternative system. Their eShel is also fed by a fiber optical system for various telescope focal lengths with a resolving power of about 10,000. Christian Buil detected the extrasolar planet of τ Bootis with this device and reached a radial velocity accuracy of a few dozen meters per second. In our view, only such devices with medium to high resolution are suitable for scientific objectives in the small-telescope domain, and all off-the-shelf spectrographs have their advantages and disadvantages. One should bear in mind that the price determines the instrument quality; mechanical problems are not completely eliminated even at the high end. We waive an extensive analysis at this point, but we recommend carefully assessing the pros and cons and examining each system on the merits of its mechanical and optical stability as well as the suitability for the particular task at hand before buying the device.

The requirements for working with small telescopes trigger the decision between choosing a self-constructed instrument or an off-the-shelf device. However, it is obvious that a vast amount of knowledge about optics and mechanics is acquired when developing a spectrograph. Therefore we always recommend to try developing a spectrograph optimally adapted to the telescope and CCD camera. In the following we present a selection of specially adapted systems of all sizes and ingenuity. These are Littrow spectrographs which are preferably used at small telescopes because of their compact design, as well as standard and echelle systems for telescopes of 20 centimeter to 8 meter aperture.

8.2   Littrow Systems

Littrow spectrographs are compact measurement devices, mainly for small telescopes. For this reason one preferably finds them in the amateur domain (a professional example can be found in Sect. 8.2.4) delivering professional data.

8.2.1   Keyhole Littrow: Good Data for Little Money

Norbert Reinecke is a physicist with a minimalistic approach driven by technical efficiency. His aim is to use parts "off-the-shelf" which should be available for any astronomer. His approach provides the basis for spectroscopic use of telescopes by any enthusiast. He pursues the Littrow principle with a blazed 30×30 mm grating of 1,200 lines/mm and an MX916 (Starlight) or ST-6 (SBIG) CCD camera. In contrast to a difficult design with mirror or prism units he uses a low-cost Celestron off-axis guider for light injection, which is quite appropriate for mechanical reasons. Usually the most difficult part of any Littrow design is mounting of the adjustable folding mirror or prism in front of the CCD chip. Its construction usually requires special tools and material. The commercial Celestron off-axis guider was used to avoid this expense. Reinecke performed additional mechanical modifications so that the guider can now be used for light injection. The design is shown in Fig. 8.2 and the

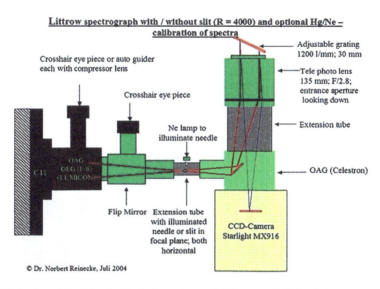

Fig. 8.2 Design of the Reinecke Keyhole spectrograph (*Courtesy*: N. Reinecke)

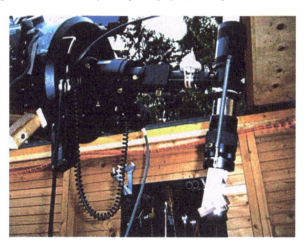

Fig. 8.3 The Keyhole spectrograph at the telescope (*Courtesy*: N. Reinecke)

set-up is shown in Fig. 8.3. A special innovation is the calibration unit whose core element is a sewing needle. Instead of a mechanically complicated slit, this needle may be positioned in the focal plane and may be illuminated by a spectral lamp perpendicular to the optical axis. The spectrum of this thin reflection can then be used for calibration. The needle curvature only reflects a very narrow strip, thus simulating an optical slit. Thus, the calibration reflection is positioned exactly at the location of the imaged target star.

Basic spectrograph parameters	
Design	Low-cost spectrograph made from off-the-shelf components
Wavelengths	5,000–7,000 Å
Field of View	6 × 6 arcminutes
CCD Detectors	ST-6 375 × 242, 23 × 27 μm Pixels
Resolving power	$R \sim 4,000$
Slit	Neon illuminated needle reflection

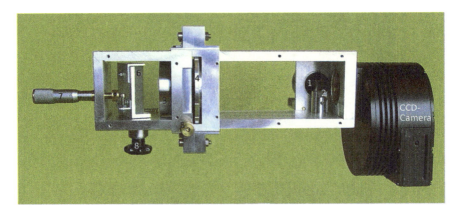

Fig. 8.4 Details of the Mahlmann Littrow. (*1*) Optical slit. (*2*) Feeding prism. (*3*) Collimator/camera lens. (*4*) Focuser ring. (*5*) Lens fixation. (*6*) Optical grating. (*7*) Adjustment screw. (*8*) Grating fixation (*Courtesy*: W. Mahlmann)

8.2.2   Mahlmann Littrow: Solid Mechanics

An example of a very solid small spectrograph is provided by Wolfgang Mahlmann. Within one year he realized a practical design for his 190 mm/760 mm-Lichtenknecker Flat-Field-Camera (Slevogt). He waived expensive custom optical components but only used off-the-shelf components (Fig. 8.4). The spectrograph is designed for small telescopes with a focal ratio of f/4 to f/6 and focal lengths from 760 to 2,000 mm. A f/4.5 Tessar lens of 210 mm focal length matches the spectrograph requirements best, and the focal length of 210 mm offers sufficient space for the mechanical design. For maximum dispersion with respect to the 46 mm collimator/camera aperture a commercially available 50 mm grating of 1,200 lines/mm has been selected. The optical slit has been adjusted to the above telescope focal lengths and seeing conditions of up to 4 arcsec. The spectrograph is housed by a rectangular aluminum tube of 82 × 80 mm cross section and 270 mm length. Particular attention was paid to the holders of the lens and the grating, as these components must be adjusted to the spectral range of 4,000–7,000 Å. After respective adjustment both components can be fixed mechanically. The lens is mounted in a threaded sleeve. The sleeve itself is guided by a threaded bushing. In part the bushing is axially split which can be clamped by a tangentially arranged

Fig. 8.5 Details of the
Mahlmann Littrow. (*1*)
Optical slit. (*2*) Knife edge
for adjustments (*Courtesy*:
W. Mahlmann)

threaded bar to fix the lens. The grating holder is mounted in a socket and can
be adjusted and fixed by means of a micrometer screw. The compact housing
and the fixation of all moving parts ensure a maximum one pixel shift or 0.25
Å, respectively, at any spatial location of the spectrograph . The mechanical slit,
the slit image and the lens axis are in the same plane (mechanical center plane
of the spectrograph). The grating rotation axis is perpendicular to this plane. This
arrangement ensures that the inclination of the spectral lines does not change
during the grating adjustment. The lens axis is perpendicular to the detector surface,
hence, light reflections of the slit on the lens rear side do not strike the detector
surface.

An interesting detail is a unit consisting of prism and a knife blade that
can be swiveled in front of the slit (Fig. 8.5). Stellar light coming from the
telescope is directed onto this knife edge. Applying the usual Foucault knife-
edge method the whole system can be focused. The knife edge is adjusted so
that it corresponds optically with the slit. Hence, the stellar image is focused and
centered after removing the knife edge. For better visual accessibility the knife edge
can be projected out of the casing by a lens of 20 mm focal length (picture not
shown).

Basic parameters	
Design	Solid amateur made Littrow-spectrograph
Wavelength range	4,000–7,000 Å
Dispersion	0.25 Å/pixel
Resolving power	$R \sim 6,000$

8.2.3   Lhires III: A Littrow for All

A technically oriented team in the Astronomical Ring for Access to Spectroscopy (ARAS), an internationally active group of amateur and professional astronomers, developed the first series of identical high resolution spectrographs, the Lhires III (Littrow High Resolution Spectrograph—now in third version). This device is optimized for widespread 8- and 11-inch Schmidt-Cassegraig (SC) telescopes. The spectrograph is commercialized by the company SHELYAK INSTRUMENTS. The main use of Lhires III is for observations of the region around Hα because of this line's general prominence in many sources, including Be stars (see Sect. 14.6). For wavelength calibration one then can use very inexpensive and permanently installed neon lamps that go with almost every baby lamp. The design of the Lhires III stems from the desire to reach high spectral resolving power of the order $R = 10,000$. In order to minimize mechanical stress, the casing of the Lhires III was designed from strengthened iron sheets. A 200 mm collimator/camera doublet keeps the final price within reasonable limits. Lhires can be used with different gratings of 150–2,400 lines per millimeter and has a revolving slit to deliver maximum efficiency for a given seeing. Autoguiding is done by a webcam on a polished slit. The optical design is shown in Fig. 8.3. Thus a design was introduced which can easily be copied by anyone. All drawings are freely available. The spectrograph is now widely used. For instance, it was successfully used for the recent professional-amateur campaign on the Wolf-Rayet + O colliding-wind binary WR 140 at the Observatorio del Teide.

In principle one can also use this spectrograph at telescopes of other focal ratios than f/10 by applying respective focal optics. However, because of the internal collimator/camera lens one might accept reduced resolving power. This can be circumnavigated by closing the slit jaws for the price of reduced telescope-spectrograph efficiency, hence, applicable only for bright targets (Figs. 8.6, 8.7, 8.8 and 8.9).

Basic spectrograph parameters	
Design	Commercial Littrow spectrograph for standard SC telescopes
Wavelengths	450–800 nm
Resolving power	R up to 18,000
Slit	15–35 μm mirror, reflective for guiding
Calibration	Neon lamp

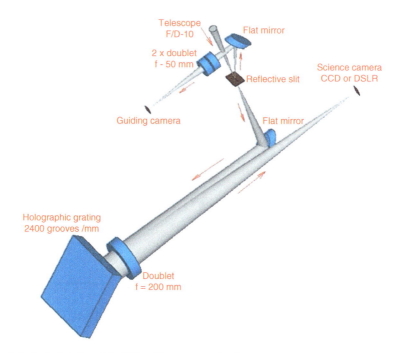

Fig. 8.6 Optical design of the Lhires III spectrograph (*Courtesy*: O. Thizy)

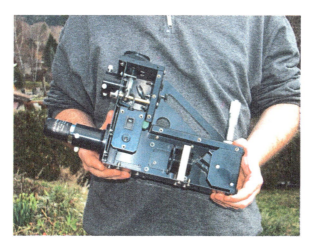

Fig. 8.7 Inside the Lhires III spectrograph (*Courtesy*: O. Thizy)

Fig. 8.8 Lhires III at the Celestron C14 focus (*Courtesy*: O. Thizy)

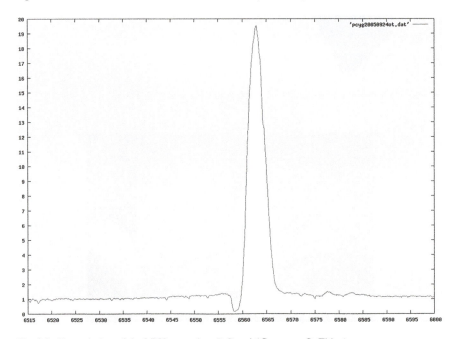

Fig. 8.9 Hα emission of the LBV supergiant P Cygni (*Courtesy*: O. Thizy)

8.2.4   SPIRAL: *A Littrow for Large Telescopes*

Professional telescopes with large apertures usually have long focal lengths, despite small f-ratios down to f/1, thus generating a correspondingly large seeing disk. With telescope focal length f_t a stellar image of angular diameter α has a geometric diameter $d_1 = \tan\alpha f_t \approx \alpha f_t$ at its focus. As we have already discussed in Sect. 2.6.3, the spectrograph slit width in the spectrograph camera image plane d_1 is calculated by $d_D = f_{col}/f_{cam} \cdot d_1$, with the collimator and camera focal lengths f_{col} and f_{cam} (see Fig. 2.30). The collimator f-ratio must not exceed that of the telescope. Otherwise light is lost. Two groups at the Anglo-Australian Observatory and the Institute of Astronomy at Cambridge University have built a Littrow spectrograph that is fed through a fiber bundle. The SPIRAL fiber bundle of segmented pupil/image reformatting array lenses images the entire telescope aperture into the focus and the bunch is then positioned in a row, so that all fibers work together as a virtual optical fiber slit (Figs. 8.10 and 8.11).

SPIRAL works in two modi: (A) Integral Field Spectroscopy—The focal image field is projected onto the micro-lenses via magnification and collimator optics

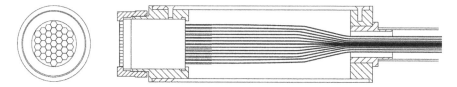

Fig. 8.10 The SPIRAL lens array and the fiber connection (Kenworthy et al. 2001)

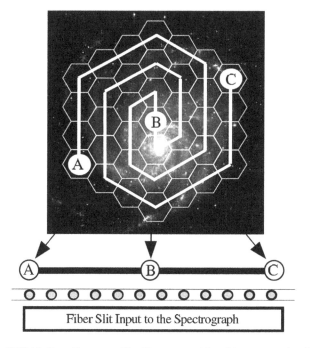

Fig. 8.11 The SPIRAL fiber slit pattern. The fibers are positioned in a row so that they can act as an optical slit (Kenworthy et al. 2001)

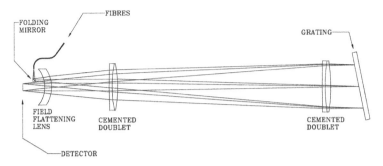

Fig. 8.12 Optical setup of the SPIRAL Littrow spectrograph (Kenworthy et al. 2001)

so that a spectral mapping for extended objects (e.g., galaxies) can be provided. (B) Spectroscopy of point sources—For point source spectroscopy a focusing lens images the target onto all micro-lenses.

All fibers have a diameter of 50 μm and act as a virtual slit of the same width. The micro-lenses are fed by an f/5 beam. The collimator/camera optics of 150 mm diameter and 720 mm focal length has been designed for f/4.8 to reduce light loss by focal ratio degradation (Fig. 8.12). The Littrow configuration has the advantage that it can be used at different large telescopes of up to 8 m aperture with appropriate imaging optics. SPIRAL uses so-called Petzval optics out of two doublets to reduce chromatic aberration and to increase the spectral field of view. The field flattener corrects the necessarily occurring image curvature.

Basic spectrograph parameters	
Design	Fiber-fed multi- and single-object Littrow spectrograph
Collimated beam diameter	150 mm
Coll./Cam. focal length	720 mm
Wavelength range	1,200 l/mm grating: 500 Å
Spectral dispersion	1,200 l/mm grating: 0.314 Å/pixel

8.3 Classical Systems

Classical spectrographs use two different lenses for collimator and camera. This means they can easily be adapted to the appropriate telescope and CCD parameters. They are used at telescopes of all sizes.

8.3.1   The Mice Mansion: A Classical Grating Spectrograph

In 2006 the German chemist Lothar Schanne initially built a classical spectrograph for his 5 inch f/8 Maksutov-Newton in order to gain experience in spectroscopy. His CCD camera was an AUDINE type with a KAF 401E chip and 9 μm pixels. Lothar Schanne had no machine shop and the mechanical implementation was a significant problem. By chance, he discovered a small plywood house for hamsters (!) in a pet shop. This casing provided an opening of the right size for the collimator and camera optics. An opening for the installation of the CCD camera was also available and he installed all optics in this box. He called it "The Mice Mansion". Both lenses, auctioned on the internet, fit exactly and the first measurements were carried out. Figures 8.13 and 8.14 show this most original design.

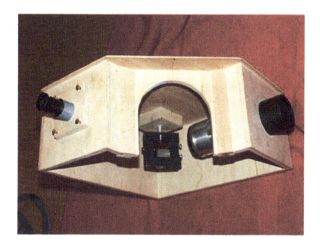

Fig. 8.13 The MICE MANSION from a pet shop (*Courtesy*: L. Schanne)

Fig. 8.14 The MICE MANSION with internal optics (*Courtesy*: L. Schanne)

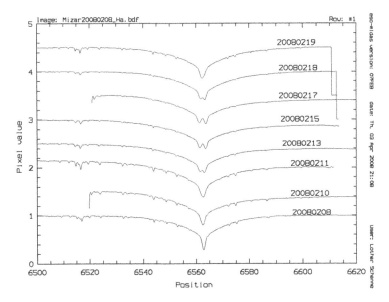

Fig. 8.15 A spectroscopic binary: Double peak of the Hα and other absorption lines in the spectrum of Mizar due to the stellar binary nature (*Courtesy*: L. Schanne)

The spectrograph for about 100 Euro has a symmetrical structure with a collimator focal length of 135 mm, a 25 × 25 mm reflection grating and a 135 mm camera lens; it provides a spectral resolving power $R = 4,600$ at Hα. The imaged wavelength interval is 290 Å. In zeroth order (grating used as a mirror) one can position the desired object in the sky onto the optical axis with the viewfinder. Over time, the data quality was improved gradually and now binary stars are regularly measured and time series of variable stars recorded. Two examples are demonstrated in Fig. 8.15 for Mizar and in Fig. 8.16 for ε Aurigae.

Basic spectrograph parameters	
Design	Standard design in wooden box
Wavelengths	500–750 nm
Resolving power	$R \sim 4,600$
Slits	No slit

8.3.2   Spectrashift: A Czerny–Turner for Exoplanets

Spectrashift is an international campaign team which dedicates its work to the observations of extrasolar planets and has developed a number of spectrographs,

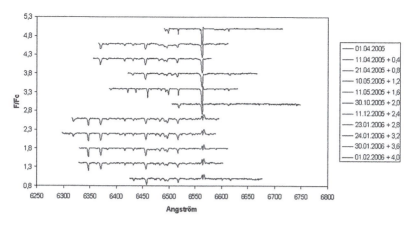

Fig. 8.16 Time series of ϵ Aurigae in the interval around the Hα line (*Courtesy*: L. Schanne)

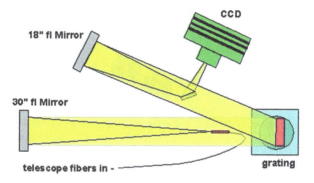

Fig. 8.17 The Spectrashift Czerny-Turner installation (*Courtesy*: Spectrashift)

including a Czerny-Turner. Figure 8.17 shows the layout of the system. The device has a thermally stabilized wooden cabinet, in which there is a base plate made of granite carrying the optical elements. Figure 8.18 shows the collimator and camera mirror, the diffraction grating and the CCD camera. Both mirrors have an aperture of 150 mm, 750 mm focal length for the collimator and 450 mm for the camera. They are also both mounted on granite holders to avoid disturbing temperature drifts. The spectrograph is fed through a bundle of seven individual optical fibers each of 120 microns diameter (Fig. 8.19). They are mounted in a Teflon casing of a bicycle brake(!). Wavelength calibration is carried out via two additional fibers, which inject the light from the Thorium-Argon calibration source into the spectrograph. Both fiber ends are mirror-polished for maximum transmission (Fig. 8.20).

Fig. 8.18 View of the parabola mirrors (*Courtesy*: Spectrashift)

Fig. 8.19 Schematc picture of the bundle of seven single fibers (*Courtesy*: Spectrashift)

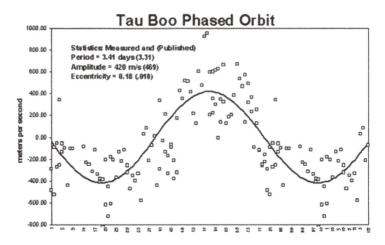

Fig. 8.20 Combined phase data points for a single orbit of τ Bootis (*Courtesy*: Spectrashift)

Basic spectrograph parameters	
Design	Thermally controlled on granite plate
Wavelengths	450–700 nm
Connection	7-fold fiber bundle
Resolving power	$R \sim 7{,}000$
Pixel drift per night	< 0.1 pixel

Fig. 8.21 Large Boller & Chivens spectrograph in the focus of the Observatoire du Mont Mégantic telescope (*Courtesy*: Robert Lamontagne)

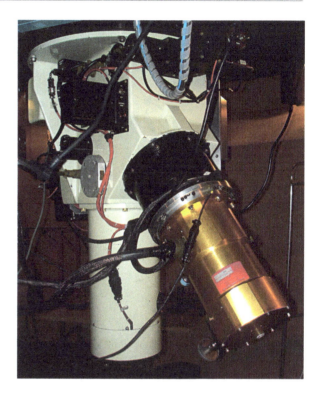

8.3.3   Boller & Chivens: Work-Horse Spectrographs for Midsize Telescopes

The company Boller & Chivens supplied different spectrographs virtually off-the-shelf for telescopes of intermediate size. Almost every observational professional astronomer knows them or has worked with them at some time. The extremely robust standard spectrographs were adapted to different telescopes via appropriate collimator and camera lenses. Basically these spectrographs were the blueprint for today's small off-the-shelf spectrographs. In Figs. 8.21 and 8.22 we show typical adaptations and adjustments to the telescope at Observatoire du Mont Mégantic (OMM) in Canada. The spectrograph at the 1.6 m OMM is a unit of the series 31523 which uses a parabolic f/8 mirror collimator of 720 mm focal length adapted

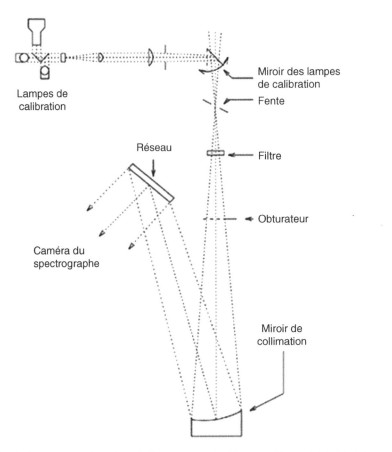

Fig. 8.22 Optical configuration of the OMM spectrograph (*Courtesy*: Université de Montréal)

to the Ritchey-Chrétien telescope. This mirror needs to be focused by hand before each observation according to the external temperature. In front of the optical slit is a switch mirror unit for feeding the calibration light. The spectrograph provides a dispersion from 6.4 Å/pixel (150 l/mm grating) to 0.2 Å/pixel (3,600 l/mm grating) with grating sizes of 102 mm × 128 mm. Richardson gratings from Newport Spectra-Physics are used. The 8 element f/1.55 camera lens from Institute National d'Optique (INO) provides a transmissivity of 98 % over the entire visible spectrum from 3,700 to 9,500 Å. The 2.5-m du Pont at Las Campanas Observatory is also a Ritchey-Chrétien telescope, of f/7.5. The spectrograph uses a collimator of 675 mm focal length. The gratings with 300–1,200 l/mm deliver a dispersion of 3.0–0.8 Å/pixel via a camera of 140 mm focal length.

Basic spectrograph parameters	
Design	Maximum rigidity for mid-size telescopes
Wavelengths	400–800 nm

Fig. 8.23 Bonette of Hectospec in the focal plane. Each single robotic arm in the imaged focal plane of 0.6 m diameter carries an optical fiber which feeds Hectospec. The positions of all arms are defined by the observer before his observation run. For each campaign the Bonette is newly adjusted (*Courtesy*: MMT)

8.3.4   Hectospec: Multi-Object Spectroscopy at the MMT

The MMT supplies a usable spectroscopic field-of-view of 1°. To use this field, a team at the Smithsonian Center for Astrophysics developed a robotic unit that can simultaneously position 300 fiber optics in the focal plane according to the individual observation program by freely moving mechanical arms (Fig. 8.23). These arms (Bonette) catch the focal object light via fiber optics and feed this light into two spectrographs of medium and high dispersion. These two multi-object spectrographs are HECTOSPEC and HECTOECHELLE.

HECTOSPEC is a medium resolution spectrograph of $R = 1,000$–$2,500$, which works from 3,650 to 9,200 Å. All optical elements are positioned on an optical bench made of Invar (Fig. 8.24). Core elements are three interchangeable Richardson gratings. A camera mirror of 397 mm focal length images the dispersed light onto a CCD with $3,400 \times 3,400$ pixels (Fabricant et al. 1994). The key to multi-object spectroscopy at the MMT is the freely positioned fiber optics which also feeds Hectoechelle. All 300 optical fibers are positioned within 300 arc-seconds to any place in the 1° Cassegrain focal plane. This is done by six-axis robots delivering a positional accuracy of $\sim 25\,\mu$m. The optical fibers have a length of 25 m each. Each fiber of $250\,\mu$m diameter covers a diameter of 1.5 arcsec on the sky. The distance between the nearest fibers is at least 20 arcsec. The optical fibers are the

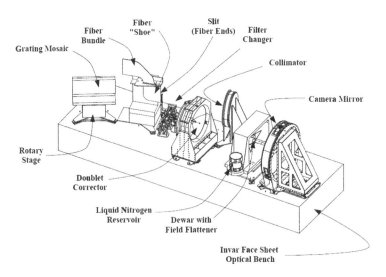

Fig. 8.24 Isometric view of Hectospec (*Courtesy*: MMT)

core elements of Hectospec and Hectoechelle. Given the known technical problems
of these components (see Chap. 11), utmost attention has been devoted to mitigate
these problems. To achieve maximum optical efficiency, the company Heraeus-
Amersil developed a new broadband fiber of high transparency and low intrinsic
focal ratio degradation (FRD). By avoiding external mechanical stress and by using
an f/6 input ratio the enormous fraction of about 95 % of the injected light passes
through the 25-m optics and remains within an f/6 cone at the output.

A mask is installed in front of the grating that geometrically limits the f-ratio of
the output beam at the fiber. Thus the price for avoiding FRD by mechanical stress
is a reduced efficiency. Based on the developer's experience, V-shaped grooves in
which the fibers are fitted and glued by epoxy, reduce stress significantly. Stress
caused by friction at the fiber cladding was reduced through a Teflon coating. In
addition, measures to reduce the thermal stress were taken. To reduce stress by
bending on the path to the spectrograph, all fibers are carried by cable chains,
which are also used for guiding electrical cables (Fig. 8.25). To reduce thermal
stress by shrinking Teflon mantles, so-called Thermal Breaks are implemented. This
is a simple break in the Teflon mantle to compensate for differential stretching.
The Teflon ends have been carefully deburred and kept in line by outer metal
tubes. These units have been mounted at each point of significant bending. The
Thermal-Brake concept for two different temperatures is shown in Fig. 8.26. A
critical element is the path from the focal plane edge along the telescope where
image rotation due to the azimuth telescope mount had to be taken into account.
However, various measures compensate for potential stress sufficiently well.

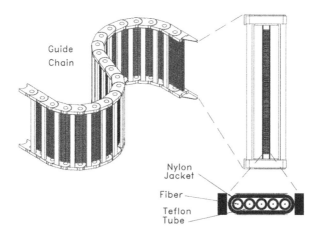

Fig. 8.25 The Hectospec fibers in the guide chain (*Courtesy*: MMT)

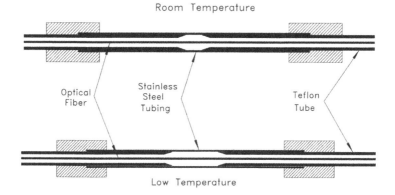

Fig. 8.26 Thermal-Brake concept for Hectospec (*Courtesy*: MMT)

Basic spectrograph parameters	
Design	Fiber-fed standard multi-object spectrograph
Collimated beam diameter	397 mm
Camera focal length	259 mm
Wavelengths	270 l/mm grating: 4,488–8,664 Å
	600 l/mm grating: 5,609–7,522 Å
	1,200 l/mm grating: 6,084–7,038 Å
Spectral resolution	270 l/mm grating: 6.2 Å
	600 l/mm grating: 2.6 Å
	1,200 l/mm grating: 1.1 Å

8.3.5   MODS: A Multi-Object Double Spectrograph
for the LBT

The two Multi-Object Double Spectrographs at the Large Binocular Telescope on Mount Graham in Arizona are two identical systems for low and medium resolution spectroscopy. They were built by the Department of Astronomy at Ohio State University (Pogge et al. 2006, 2010; Osmer et al. 2000; Byard and O'Brien 2000). Both devices are always working together in the f/15 Gregorian focus of one of the 8.4 m mirrors to use both mirrors simultaneously. Unlike Hectospec, multi-object spectroscopy with MODS does not use robotic arms but specially made metal slit masks. Standard masks for long-slit spectroscopy are used in the focal planes of the two telescopes. On the other hand, tailored laser-cut masks with many optical slits can be applied so that MODS can obtain spectra for many target objects simultaneously in the image field. A mask cartridge at the instrument can hold 24 masks. Fifteen mask slots are available for the observer which are virtually designed with a special software in advance of the campaign. The production of the respective campaign masks is done by the LBT observatory. The remaining nine masks consist of segmented long slits of different widths, which remain as default in the cartridges. The two seeing-limited spectrographs provide low to medium resolution spectra in the wavelength range from 3,200 to 10,000 Angstroms in a field of 6 × 6 arc minutes. Reflection gratings provide an average resolving power of $R \sim 2,000$ whereas double prisms are used for low resolving power ($R = 150–500$). A dichroic filter separates the light into two channels behind the focal mask, each of which is optimized for red or blue light. The separation of the two light channels is done at 5,650 Å. The collimators, dispersers, camera optics and detectors in the two channels are appropriately optimized in order to achieve maximum efficiency. To ensure high efficiency of the CCD cameras, highly sensitive 3k × 8k E2V chips have been selected. For both channels, the corresponding coating layers were adjusted individually. Each of the two MODS devices contains two different optical paths for the red and blue spectral regions, which are separated directly behind the telescope focus at 5,500 Å by dichroic filters with an overlap of about 500 Å. Both systems use mirror collimators and cameras. The design is motivated by optimized gratings for the two channels, minimizing the number of optical surfaces (high efficiency) as well as the adjustment of surface coatings to the two channels.

The optical path of MODS first passes a unit to compensate for atmospheric dispersion in front of the telescope focus. The multi-object slit array is positioned in the convex focal surface of the Gregory type LBT. This is to perfectly adapt the slit widths to the telescope image quality. A field lens behind the focal plane along with the collimator positions the exit pupil of the telescope on the position of the three available optical gratings. For observations at highest resolving power (> 10,000), the gratings work in high orders and order sorting filters are applied. Alternatively, a grism can be positioned directly in front of the camera to select individual orders. In both channels, cameras of 700 mm focal length work in Maksutov-Schmidt mode. A special feature of the two cameras is their off-axis position with respect to the

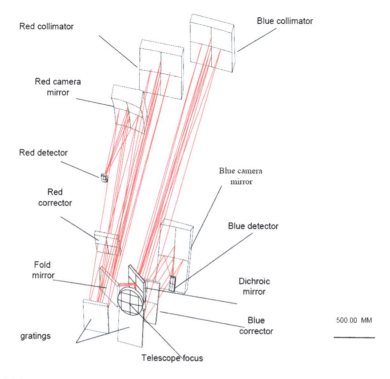

Fig. 8.27 Beams in the red and blue MODS channels (*Courtesy*: Ohio State University)

dispersed beam. At the camera, field lens and detectors are outside the incident beam. On the one hand, vignetting is prevented. On the other hand, scattered light cannot be reflected back from the CCD camera in the grating direction (Figs. 8.27, 8.28, 8.29, and 8.30).

Basic spectrograph parameters	
Design	Seeing-Limited Optical Double Spectrograph
Wavelengths	3,200–10,000 Å
Field of View	6 × 6-arcminutes
CCD Detectors	E2V CCD231-68, 15 μm Pixels
Pixel Scale	0.13 arcsec/pixel
Grating Spectroscopy	$R \sim 2,000$
Prism Spectroscopy	$R = 500$–150
Slits	Laser-cut spherical slit masks (up to 24)
Dichroic	Blue-transmit/Red-reflect, 565 nm cross-over wavelength

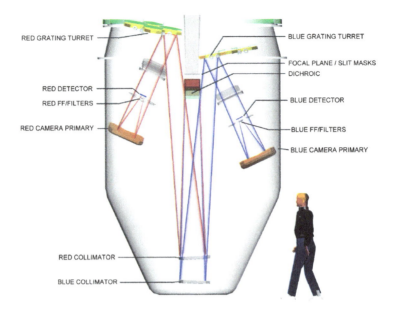

Fig. 8.28 MODS optical layout (*Courtesy*: Ohio State University)

Fig. 8.29 First Multi-Object Spectrum: Galaxy Cluster Abell 1689 2011 March 17 (*Courtesy*: Ohio State University)

Fig. 8.30 Images and Long-Slit Spectra of the Crab Nebula 2010 November 19 (*Courtesy*: Ohio State University)

8.3.6 COMICS: Ground Based Thermal IR Spectroscopy

In order to observe regions and states at lower energies one has to detect radiation in the thermal infrared at wavelengths between 3 and 50 μm. However, this requires a significant technical input in the form of highly cooled instruments, so that their own thermal radiation does not interfere with the target signals. And one has to use the appropriate atmospheric windows at lowest opacity. High observational altitudes and dry atmospheric conditions are therefore essential.

The Cooled Mid-Infrared Camera and Spectrometer (COMICS) at the Cassegrain focus of the 8.2-m Subaru telescope on 4,200 m Mauna Kea provides long-slit spectra between wavelengths of 8 and 26 μm. Between 8 and 13 μm (N-band), the spectral resolving power is between 250 and 10,000, depending on the chosen grating. For the range of 16–26 μm (Q-band) it is about 2,500. The cooled vacuum chamber (cryostat) that contains the entire spectrograph (Figs. 8.31, 8.32, and 8.33) cools the optics and the five Raytheon (320 × 240 Si:AS IBC) detectors to 35 K and 4 K, respectively. To reach these temperatures the cooling unit needs about 6 days. The optical elements are installed below a cooled base plate. Of all the Subaru instruments COMICS operates at the longest wavelengths. A special feature of the COMICS camera is a spherical mirror whose radius of curvature is located close to the grating. It focuses the spectra into a spherical surface. The five detectors are positioned in a row on this focal sphere and completely cover the wavelenth range 8–13 μm at high resolution ($R = 2,500$) with two grating positions. The system efficiency including all optics is about 1 % (Kataza et al. 2000).

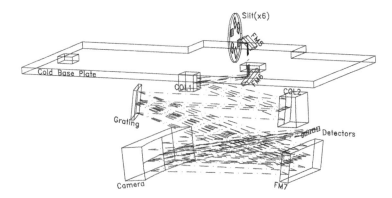

Fig. 8.31 Optical layout of the COMICS spectrograph. The three gratings are mounted on a holder which is controlled from the outside. The five detectors lie in the imaging sphere of the camera (*Courtesy*: National Astronomical Observatory of Japan, Kataza et al. 2000)

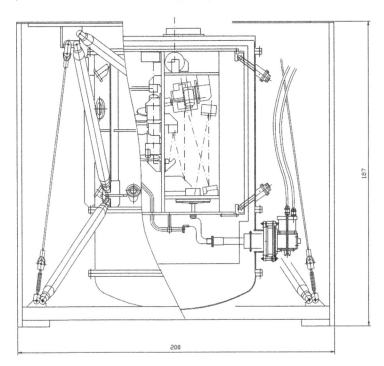

Fig. 8.32 Sketch of the cryo vacuum chamber of 2 m × 2 m × 2 m volume and weight of two tons (*Courtesy*: National Astronomical Observatory of Japan, Kataza et al. 2000)

Fig. 8.33 COMICS attached to the Cassegrain focus (*Courtesy*: National Astronomical Observatory of Japan)

Basic spectrograph parameters	
Design	Mid-infrared long-slit spectrograph
Detector	5 × 320 × 240 element Si:As IBC
Pixel size	50 μm (0.165")
Focal plane scale	0.165 arcsec/pix
Slit width	0.33, 0.50, 0.66 or 1.00 arcsec
Slit length	40 arcsec
Resolving power	250 (8–13 μm)
	2,500 (8–13 μm)
	2,500 (16–26 μm)
	10,000 (8–13 μm) (only NeII 12.8 μm)

8.4 Echelle Systems

Today the spectroscopic work-horses of professional astronomy deliver high spectral resolution and efficiency. In the last few years work-horse spectrographs have also reached the domain of small-telescopes in the form of self-constructed devices or off-the-shelf equipment.

8.4.1   Stober Echelle: A Physician on New Tracks

Trained as a specialist in general medicine, Berthold Stober wanders with his interests in steam engines and his activities in astronomical spectroscopy between very contrasting disciplines. His first experience in spectroscopy was the construction of a Littrow system with a grating in a wooden chamber. He mainly studied Hα emission lines of massive stars and performed corresponding monitoring campaigns to investigate the temporal variability of these lines in close binary stars with rather short orbital periods. Later he focused on echelle spectroscopy and developed his own instrument. Stober's pragmatic approach is guided by his machining capabilities in his own workshop. For his first echelle spectrograph, he made the base plate and the optical mounting out of wood. This served as a 1-to-1 model for finalizing all important parameters (Fig. 8.34). After all the parameters were fixed, a more stable and lighter version with metal reinforcement was manufactured (Fig. 8.35).

After several improvements and a telescope coupling via fiber optics the system provides a spectral resolving power of about 12,000 over the spectral range 4,800–7,000 Å (Figs. 8.36 and 8.37). At shorter wavelengths the resolving power decreases significantly. The cause of this is the relatively large longitudinal chromatic

Fig. 8.34 Wooden layout of the Stober echelle (*Courtesy*: B. Stober)

Fig. 8.35 Realization of the Stober spectrograph with a rigid casing (*Courtesy*: B. Stober)

Fig. 8.36 Echelle spectrum of the Hα line of ε Aurigae (*Courtesy*: B. Stober)

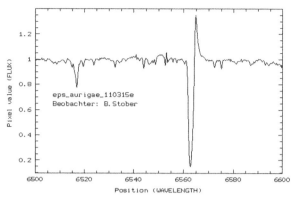

Fig. 8.37 Echelle spectrum of ε Aurigae (*Courtesy*: B. Stober)

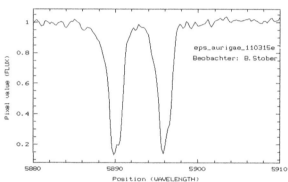

aberration of the camera lens in this range. A Thorium-Argon lamp is used for wavelength calibration. One should keep in mind the exceptional performance of this physician ranging from spectroscopist to surgical assistant. Berthold Stober is probably the first amateur astronomer in Germany who built a functional and routinely usable echelle spectrograph.

Basic spectrograph parameters	
Design	Echelle with transmission grating cross-disperser
Wavelengths	420–750 nm
CCD Detectors	KAF 3200 type
Camera	1.4/85 Zeiss Planar
Resolving power	$R \sim 11{,}000$

8.4.2 Feger Echelle: From Mechatronics to Optics

Feger I

When we talk about professional spectrographs we preferably talk about devices that are developed by physical institutions. The case of Tobias Feger is different. His proposal to develop an echelle spectrograph as part of his master thesis in mechatronics was received with skepticism. But the student prevailed. He then succeeded in building a very slim echelle spectrograph for which he received the Innovation Award of the University of Ulm (Fig. 8.38). The collimator is an 85 mm achromatic lens. Behind the standard echelle grating with 79 l/mm and a blaze angle of 63.4° he uses a transmission grating with 207 l/mm as a cross-disperser. Using an aspheric Summicron-R 2/90 apochromat as camera (Fig. 8.39) he achieves a resolving power of about 17,000. The reference light source is fed by a fiber via a slit revolver with four slit widths (Fig. 8.40) and guided automatically into the beam path. The materials used are aluminum, Polycetal (POM), carbon sandwich and brass. The spectrograph overall weight is about 5 kg.

Fig. 8.38 Principle sketch of the Feger echelle spectrograph (*Courtesy*: T. Feger)

Fig. 8.39 Grating cross-disperser and camera optics of the Feger echelle spectrograph (*Courtesy*: T. Feger)

Fig. 8.40 The Feger echelle with the focus feeding mimic (*Courtesy*: T. Feger)

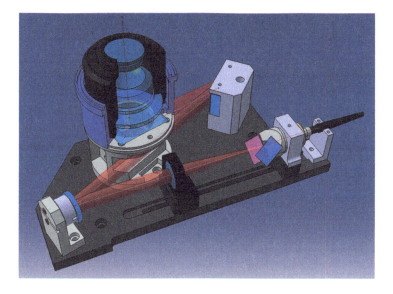

Fig. 8.41 Components of the Feger spectrograph mounted on a 10 mm aluminium bench (*Courtesy*: T. Feger)

Feger II

With this previous experience and taking a CAOS image slicer into account (Sect. 9.7) Feger designed an improved fiber-fed echelle spectrograph with a home-made image slicer. The collimator is a 85 mm doublet. A standard 300 l/mm transmission grating separates the orders behind the echelle grating. A LEICA APO-Summicron images the orders onto the CCD. The cross-disperser is housed in a solid casing, which also acts as a camera flange. A special feature is a mirror behind the echelle grating in the casing of the cross-disperser. The mirror directs the light perpendicular to the optical plane towards the camera. This means that the camera is

Fig. 8.42 Camera and cross-disperser mount of the Feger echelle (*Courtesy*: T. Feger)

taken out of the optical path and the design hence becomes more compact (Figs. 8.41 and 8.42). The mechanical design of aluminum, brass and plastic is driven by high mechanical and thermal stability. The spectrograph is fed by a 50 μm fiber. The image slicer narrows the virtual slit to 25 μm. In addition, a modular construction allows quick and easy exchange of components for various telescopes and CCD pixel sizes.

Basic spectrograph parameters	
Design	Echelle with transmission grating cross-disperser
Wavelengths	420–750 nm
Fiber	50 μm
Echelle grating	79 l/mm, 50 mm × 25 mm
Image slicer	40 μm width, 160 μm length
Cross-disperser	300 l/mm, 50 mm × 50 mm
Camera	LEICA APO-Summicron-R 1:2/90 ASPH
CCD Detectors	KAF 1603 type
Resolving power	$R \sim 9{,}000$

8.4.3 eShel: A Stable Off-the-Shelf Fiber Echelle

After the great success of Lhires III Littrow in the amateur and professional community, the team around SHELYAK INSTRUMENTS developed and commercialized a ready-to-go echelle spectrograph, the ESHEL. Eshel is a compact standalone unit with a prism cross-disperser and fed by fiber optics (Figs. 8.43 and 8.44). Figure 8.45

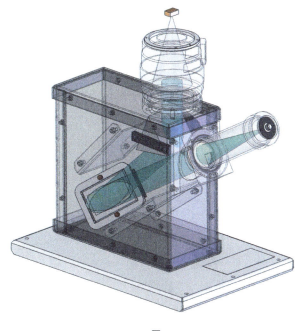

Fig. 8.43 3D schematics of the eShel spectrograph (*Courtesy*: Shelyak Instruments)

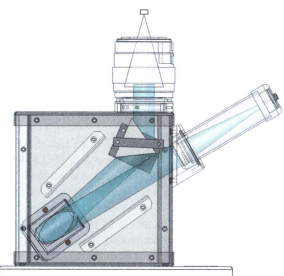

Fig. 8.44 3D side schematics of the eShel spectrograph (*Courtesy*: Shelyak Instruments)

Fig. 8.45 The eShel fiber spectrograph used at 80 cm Teide IAC80 telescope. The device is fed by an orange 50 micron data fiber

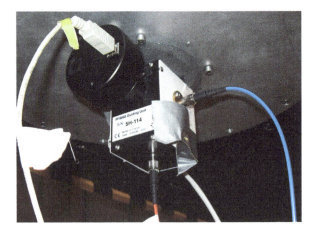

Fig. 8.46 The eShel guiding unit in the focus of the Teide IAC80 telescope. Note that the telescope focal length had to be reduced by a Shapley lens focal reducer to match the seeing disk and the necessary fiber f-ratio for maximum transmission. Front: Guiding CCD with USB connection. Orange data and blue calibration fiber. Back: Power connection for the flip mirror

shows the device working at the Teide IAC80 telescope. In the telescope focus the starlight is taken by a guiding unit (Fig. 8.46). This unit provides accurate tracking using a standard video camera and an efficient light transmission to the spectrograph. The light is fed into the polished 50 μm fiber by an f/6 beam. A second optical fiber of 200 μm diameter feeds Thorium-Argon calibration light coming from a corresponding source unit into the spectrograph. Switching between star and calibration light is automated via a computer controlled RS232 interface and corresponding software. The spectrograph uses a commercial Canon 85 mm f/1.8

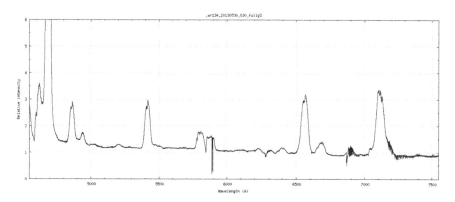

Fig. 8.47 Spectrum of the 8 mag star WR 134 obtained with the eShel spectrograph at IAC80 telescope after combining all echelle orders (no final rectification)

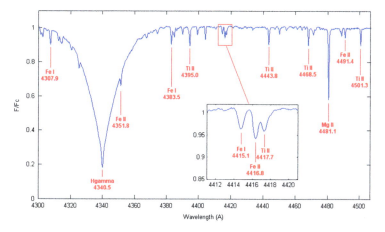

Fig. 8.48 Spectrum of Vega around Hβ (*Courtesy*: C. Buil)

lens camera imaging the entire visible spectral region of about 4,000–8,000 Å. A data reduction software package is included, based on the MIDAS echelle task. The reduction procedure (darks, biases, flats, calibration) is fully automated. The typical resolving power is $R = 12,000$ (Fig. 8.47).

One should keep in mind that the f/6 fiber input is mandatory not to lose light by focal ratio degradation (see Sect. 11.5). That means, larger telescope f-ratios need to be reduced by a focal reducer. In addition, the seeing and focal length dependent focal seeing disk needs to match the 50 micron fiber aperture. This has to be taken into account for good results. In practice, the eShel is highly sensitive for mechanical and CCD camera distortions. Touching might be fatal for the entire data reduction loop. This is also valid for the Shelyak fibers of unknown efficiency. For professional use it is recommended to use tailored fibers not to further reduce the overall system efficiency of already below 10 % (Fig. 8.48).

Basic spectrograph parameters	
Design	Commercial echelle spectrograph
Cross-disperser	Coated prism
Telescope connection	50 μ fiber
Wavelengths	400–800 nm
CCD Detector	KAF 1603 based CCD
Resolving power	$R \sim 12.000$
Calibration	Thorium-Argon lamp
Data reduction	Fully automated

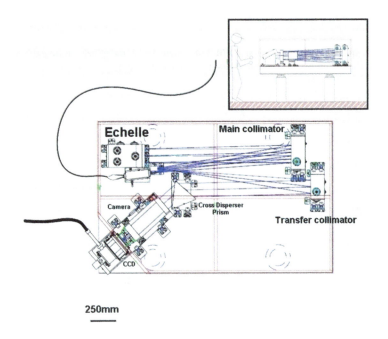

Fig. 8.49 The FEROS design (Kaufer and Pasquini 1998)

8.4.4 FEROS: An Echelle for Chile

A high-level example for a flexible system and different purposes is FEROS (Fiber-fed Extended Range Optical Spectrograph), which was developed and built by an international consortium led by Landessternwarte Heidelberg (Kaufer and Pasquini 1998). Initially for the ESO La Silla 1.52 m CAT Observatory it is now used at the 2.2 m Planck telescope. The basic design of FEROS has been copied for the "Bochum Echelle Spectrograph for the Optical" (BESO) at the Hexapod Telescope at Cerro Armazones in Chile. FEROS is fed by fiber optics and uses prism cross-dispersers (Fig. 8.49).

Efficient light feeding is provided by micro-lenses at the telescope focus in front of the two fiber apertures used in FEROS. After passing the fibers the light is degraded to low f-ratios due to focal-ratio degradation (see Sect. 11.5). Then the two beams are adjusted by a lens to the f/11 spectrograph collimator. Fiber optics have the advantage that the spectrograph can be operated under mechanically stable conditions and at a controlled temperature. A disadvantage, however, is their low efficiency (see Chap. 11). At FEROS the latter is compensated for by an image slicer that we introduce in Sect. 9.4. Thus, the slit width is designed to be narrower and one can reach a higher spectral resolution with the same efficiency, in FEROS $R = 48,000$. Moreover FEROS uses a so-called "white pupil configuration". After dispersion by the echelle grating, the light passes an intermediate focus between collimator and a transfer collimator towards the cross-disperser, thus minimizing stray-light from the echelle grating. Figure 9.9 is a drawing of the FEROS image slicer. We note that a 3-dimensional understanding of this relatively complicated device is not easy and we further illuminate the working principle in Chap. 9. The present FEROS arrangement provides a typical S/N of about 100 for stars of 12 mag over the entire visual spectrum. Figure 8.50 shows an example spectrum of P Cygni.

Basic spectrograph parameters	
Design	Echelle with prism cross-disperser
Wavelengths	360–920 nm
Entrance aperture	2.7 arcsec (circular)
Efficiency	1 %@360 nm, 16 %@440 nm, 17 %@550 nm
	16 %@640 nm, 11 %@790 nm
Resolving power	$R \sim 48.000$

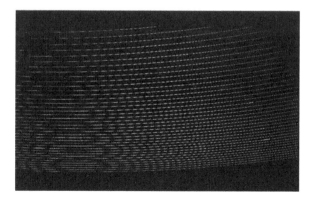

Fig. 8.50 FEROS raw spectrum of P Cygni (Kaufer and Pasquini 1998)

8.4.5 HDS: Highest Resolution at the Nasmyth Focus

There are two options for an astronomical spectrograph. Either the device is permanently installed at and moves with the telescope (usually at the Cassegrain focus) or the light of the targets is transmitted by fiber optics from the focus to a fixed spectrograph. Both options have been described here. Advantages and disadvantages of each must be weighed. For instance, the direct installation at the telescope can cause mechanical problems (e.g., bending, temperature drift). On the other hand, fiber optics can introduce optical problems (e.g., Focal Ratio Degradation FRD, efficiency). In all cases, these side effects must be sufficiently analyzed and eliminated. Alternatively, measuring devices can be placed in stationary Coudé- or Nasmyth foci (normally they are not used entirely stationarily but often they are rotated to conserve the orientation on the sky). Apart from the disadvantage of non-axisymmetric optical geometries which make polarization measurements almost impossible (see Chap. 15), very heavy spectrographs can be installed there, which should be adapted to the necessarily large focal lengths, of course.

One such device is the "High Dispersion Spectrograph" (HDS), which is working at the f/12.6 Nasmyth focus of the 8.2 m Subaru telescope on Mauna Kea / Hawaii. HDS is a colossal instrument of approximately $6 \times 6 \times 3$ m in size and weighs 6 tons (Fig. 8.51). It is located in one of the two temperature-stabilized Nasmyth rooms next to the telescope. HDS is a conventional echelle spectrograph with an echelle grating of 300 mm × 840 mm usable area and a catadioptric camera with three correction lenses. The instrument provides spectra from 3,000 to 10,000 Å. In standard configuration, HDS provides a spectral resolving power of 90,000 with a 0.4 arcsec slit. With an even narrower slit, a resolving power of 160,000 can be achieved (Noguchi et al. 2002).

Figure 8.52 shows the principle light path for HDS. The incident light from the telescope behind the slit (tilted by 15° for tracking the reflected target) is reflected onto the echelle grating by one of the two possible parabolic off-axis mirror collimators. The two collimators have different coatings for the red and blue spectral ranges. Behind the slit is installed a filter wheel, whose filters can block individual orders of the echelle spectrum in order to perform the usual long-slit spectroscopy. The collimated beam has a diameter of 272 mm. The echelle grating with 31.6 l/mm and a blaze angle of 71.3° was manufactured from two of the largest commercially available Richardson gratings. The off-axis angle γ at the echelle grating between incident and reflected beams towards the cross-disperser is 12°. In HDS, either of two cross-dispersers can be selected: either a 400 l/mm grating for the blue spectral range or a 250 l/mm grating for the red spectral range. Alternatively, a flat mirror for long-slit spectroscopy can be used. The camera consists of a field lens for image correction and a parabolic mirror at whose focus the CCD chip is positioned. To cover the entire CCD area of 55 mm × 55 mm the corrector and the f/0.96 camera mirror have apertures of 610 or 800 mm. Thus, a resolution of 0.12″/pixel is achieved.

Fig. 8.51 HDS in the lab. Note the size of the computer monitor in front (*Courtesy*: Subaru Telescope)

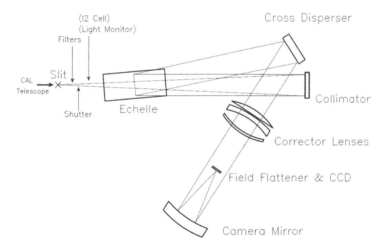

Fig. 8.52 The optical layout of HDS (Noguchi et al. 2002)

Basic parameters	
Design	Echelle spectrograph for highest resolution
Echelle grating	31.6 l/mm, R = 2.8
	Blaze angle = 71.5 °
Cross-disperser	Red: 250 l/mm
	Blue: 400 l/mm
Wavelength range	300–1,000 nm
CCD Detectors	Two 2K × 4K CCDs (E2V CCD42-80)
Pixel size	13.5 µm
Resolving Power	R up to 165,000
Slit width	0.2–4″
Image slicer	Bowen-Walraven (5 slices)
Efficiency	13 %@500–600 nm, 8 %@4,000 nm

8.4.6   X-Shooter: 20,000 Å in a Single Shot

The instruments of the Very Large Telescope (VLT) are at the present limit of technical feasibility. One of these instruments, representing the second generation VLT instruments, is X-shooter. This system consists of three highly efficient echelle spectrographs with prism cross-dispersers of medium resolution for the measurement of individual objects at the Cassegrain focus. The salient feature of X-shooter is its simultaneously covered spectral range of 3,000–25,000 Å with a resolving power of R = 4,000–17,000, depending on the wavelength. The unit was built with a budget of 5.3 million Euro by a consortium of four countries under the authority of ESO (Vernet et al. 2011).

To simultaneously accommodate the large spectral range the light first passes through two dichroic filters behind the optical slit. The first filter reflects 98 % of the incident light between 3,500 and 5,430 Å towards the direction of the UVB spectrograph and transmits 95 % of the light between 6,000 and 23,000 Å. The second dichroic mirror again reflects 98 % of the light from 5,350 to 9,850 Å, and transmits 95 % of the light between 10,450 and 23,000 Å. The reflected or transmitted light is then fed into the three spectrograph units accordingly. The functional diagram is shown in Fig. 8.53.

The Spectrographs for the UV and blue light (UVB) and the one for the visual range (VIS) barely differ. Essentially, these are just different prism materials, echelle gratings and camera lenses. The layout for the UVB spectrograph is shown in Fig. 8.54. Both devices use spherical collimators plus correction lenses in Maksutov mode. Behind the corrector the light passes through the cross-disperser, is diffracted at the echelle grating (180 l/mm for the UVB spectrograph and 99.4 l/mm for the VIS spectrograph) and passes through the cross-disperser for a second time (Littrow mode) to sufficiently separate the orders. The diffracted light is then again reflected by the collimator towards a field mirror (analogous to a field lens) which then feeds a camera out of four lens groups, which depicts the spectrum.

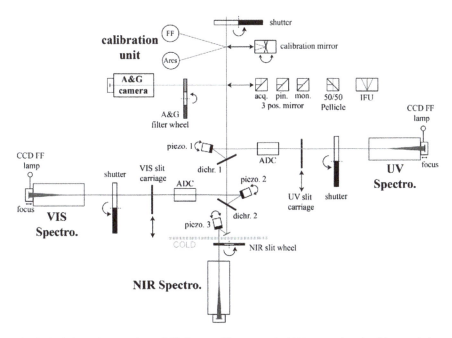

Fig. 8.53 Schematic overview of X-shooter (Vernet et al. 2011, reproduced with permission ©ESO)

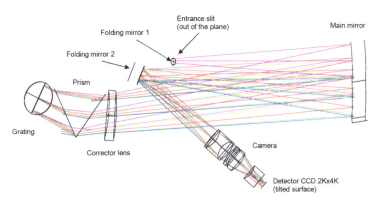

Fig. 8.54 The X-shooter UVB spectrograph optical layout (Vernet et al. 2011, reproduced with permission ©ESO)

The Near Infrared (NIR) spectrograph, as well, uses a setup similar to the other two instruments (Fig. 8.55). Here, however, the collimator consists of two spherical mirrors plus a corrector lens. The collimated light passes through two different prism cross-dispersers arranged in tandem, is diffracted at the echelle with 55 l/mm and passes through the two prisms again. Analogous to the UVB and VIS spectrographs, the light is then directed towards the camera above the

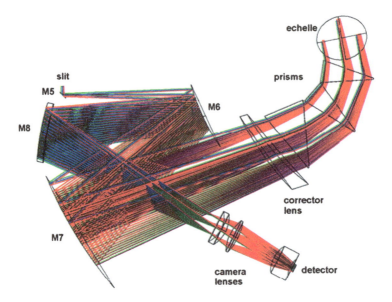

Fig. 8.55 The X-shooter NIR spectrograph optical layout (Vernet et al. 2011, reproduced with permission ©ESO)

collimator. Unlike the CCD detectors of the first two spectrographs operating at 153 and 135 K, the NIR spectrograph is a HgCdTe detector, cooled by a solid copper bar which is directly immersed in a nitrogen bath. The working temperature is 81 K.

X-shooter is working at the Cassegrain focus and is therefore subject to all potentially relevant deflection effects. To avoid these serious problems, piezo-controlled correction mirrors direct the light into the spectrograph unit after passing the dichroic filter mirrors, and thus sufficiently compensate for all bending effects. Figure 8.56 shows X-shooter in the Cassegrain focus of the VLT UT2. In the center is the cryostat of the NIR spectrograph. Above is the UVB spectrograph and below the VIS spectrograph. The two boxes on the left and right contain electronics.

The dichroic filters for beam selection reduce the overall efficiency. However, an otherwise necessary correction of atmospheric dispersion over the entire spectral range of X-shooter is impossible. Instead, this is done by two correctors, one in each of the UVB and VIS spectrographs. In addition, the instrumental interfaces had to be carefully defined to manage the different work packages of different working groups and to make them not too complex. With a joint slit for all instruments that would have been impossible. Given the single-target design, a corresponding mirror image slicer was also developed (Sect. 9.5).

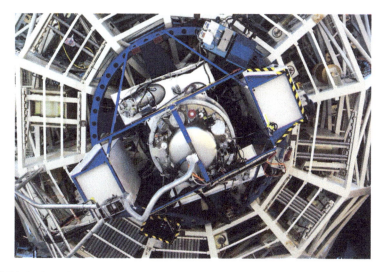

Fig. 8.56 A view of X-shooter at the Cassegrain focus below the primary mirror cell of the VLT UT2. In this view from below the instrument one can see the UVB and VIS spectrographs at the top and bottom, respectively. The NIR cryostat is visible in the center. The two boxes on the left and on the right are electronic cabinets (Vernet et al. 2011, reproduced with permission ©ESO)

Basic spectrograph parameters	
Design	High efficiency, single-target, intermediate-resolution spectrograph
Wavelengths	UVB: 3,000–5,595 Å
	VIS: 5,595–10,240 Å
	NIR: 10,240–24,800 Å
CCD Detectors	CCD (UVB/VIS) & HgCdTe (NIR)
Resolving power	$R \sim 4,000$–17,000
Slit	Image slicer

8.5 Spectrographs with Spherical Convex Gratings

8.5.1 FUSE: The Far Ultraviolet Spectroscopic Explorer

The FUSE satellite observatory was a spectroscopic mission for the far ultraviolet wavelength range between 905 and 1,195 Å with a spectral resolving power of $R = 24,000 - 30,000$. The system was designed to preferentially deliver answers on fundamental questions about the origin of the Universe. The main driver for the FUSE design, however, was the cost (as for all space missions). This in turn had an impact on the maximum satellite size (large payload = large rocket = high cost) as well as the appropriate manufacturability. This involved interfaces between the satellite bus and the instrument which had to be designed as simply as possible, which in turn has a direct impact on the management scope.

Fig. 8.57 **Fig. 8.57** The FUSE UV
channels: the optical path in
the instrument (*Courtesy*:
STScI)

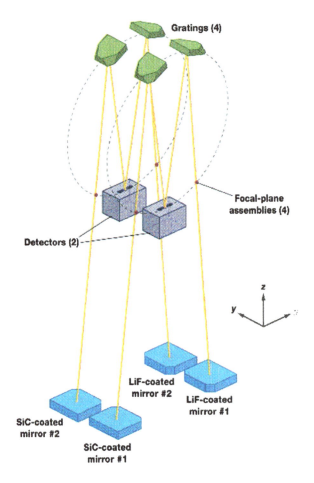

The on-board instrument of Rowland cycle type (Fig. 8.57) consists of 2×2 spherical telescope mirrors with an aperture of 390×350 mm and 2,245 mm focal length. Each pair of mirrors covers two optical channels. Two channels with SiC coating conduct light between 905 and 1,100 Å and two other channels with LiF coating cover the range 1,000–1,195 Å. The collimated light of these four mirrors passes through either a narrow slit of 1.25 arcsec width for highest spectral resolution or a slit of 4 arcsec size providing low resolution for maximum efficiency. For very bright objects one can use a pinhole of 0.6 arcsec diameter, so that the detector is not dazzled. Behind the slit/pinhole the light is directed towards two aberration-corrected, curved holographic gratings. The dispersed light is then received by two MCP detectors. The mirror system is actively controlled by actuators so that 90 % of the incident light is concentrated in a circle of 1.5 arcsec (Sahnow et al. 1996). The solid design allows the maximization of the effective telescope aperture at reasonable grating sizes and a strictly limited size of the launcher fairing (Fig. 8.58).

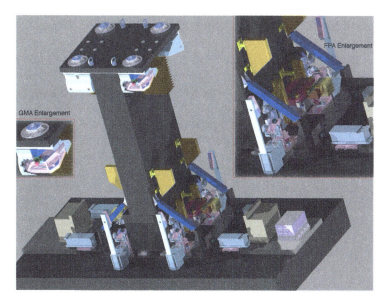

Fig. 8.58 The FUSE instrument (*Courtesy*: STScI)

One of the scientific drivers of FUSE was the measurement of the cosmic deuterium abundance. Since it is known that baryons represent only a relatively small fraction of all matter in the Universe, the question arises how to measure it. According to the Big Bang standard model deuterium can be used as a direct measure for the determination of the baryon density, a critical cosmological parameter. Deuterium is quickly destroyed in stellar nuclear processes because of its low binding energy. In addition, it is generated only in small amounts (typically one part in 10^5). Therefore, the measurement of this isotope in physically relatively quiet and undisturbed volumes of the universe is needed to determine its primordial distribution. This is done by monitoring the absorption spectra of clouds against bright background objects such as Quasars. With the FUSE spectrograph it was possible to extend previous satellite measurements to large distances.

Basic spectrograph parameters	
Design	High efficiency, single-target, intermediate-resolution spectrograph
Mirrors	Four off-axis parabolae
Wavelengths	905–1,195 Å
Grating	Holographic, spherical, aberration-corrected
Grating size	260 × 259 mm clear aperture
Detectors	MCP with double delay-line anode
	Curved to match Rowland circle
Resolving Power	$R \sim 24{,}000{-}30{,}000$

8.5.2   COS: The Cosmic Origins Spectrograph for HST

The Cosmic Origins Spectrograph (COS) is the most sensitive spectrograph for ultraviolet light ever built for space applications. It works at the focus of the Hubble Space Telescope (HST). Contractor for the design and production was Ball Aerospace & Technologies Corporation in Boulder, Colorado.

COS is a system of moderate spectral resolution for measuring point sources. It works with two different channels. The channel for the far ultraviolet (FUV) operates between 1,150 and 2,050 Å while the other channel covers the near UV (NUV) between 1,700 and 3,200 Å. In the FUV three different gratings are available and four in the NUV. A key problem in the far UV is the efficiency of the optical elements. Therefore their number had to be minimized leading to the choice of an optical Wadsworth configuration with curved gratings. Transmission optics were omitted completely, because they (a) have a very low UV transmissivity (see Appendix D) and (b) CCD cover glasses are potential sources of scattered light due to internal phosphorescence of UV photons. Beside the correction elements for spherical aberration and astigmatism of the telescope, the gratings are hence the only optical elements in COS. The aberration corrections for the telescope are carried out in the spectrograph. Point sources in the telescope focus are, hence, distorted by the telescope optics and an accordingly classic narrow slit would lead to light loss despite the lack of the Earth's atmosphere in space. On the other hand, a wider slit adjusted to the distorted image would further reduce the instrument resolution (Fig. 8.59). To avoid these problems COS just focuses the sky background so that the spectral width of a monochromatic point source is only as great as the imaging function of HST plus the internal aberrations in the spectrograph (Green et al. 2012).

Basic spectrograph parameters	
Design	High efficiency, single-target, intermediate-resolution spectrograph
Mirrors	Four off axis parabolas
Wavelengths	905–1,195 Å
Grating	Holographic, spherical, aberration-corrected
Grating size	260 × 259 mm clear aperture
Detectors	MCP with double delay-line anode
	Curved to match Rowland circle
Resolving Power	$R \sim 1,500$–41,000

Fig. 8.59 Optical layout for COS (*Courtesy*: STScI)

Suggested Readings

- Website SANTA BARBARA INSTRUMENT GROUP (SBIG): http://www.sbig.com
- Blog of the CAOS group—http://spectroscopy.wordpress.com
- Website SHELYAK INSTRUMENTS: http://www.shelyak.com
- Website BAADER PLANETARIUM: http://www.baader-planetarium.de
- Website OPTOMECHANICS RESEARCH: http://www.echellespectrographs.com
- Website LARGE BINOCULAR TELESCOPE OBSERVATORY: http://www.lbto.org/index.htm
- Website MULTI MIRROR TELESCOPE OBSERVATORY: http://mmto.org
- Website W. M. KECK OBSERVATORY: http://keckobservatory.org
- Website EUROPEAN SOUTHERN OBSERVATORY: http://www.eso.org
- Website MODS-WEBSITE: http://www.astronomy.ohio-state.edu/MODS/
- Website INSTRUMENTS AT THE LBT: http://www.lbto.org/instruments.htm
- Website INSTRUMENTS AT THE MMT: http://mmto.org/instruments
- Website X-SHOOTER: http://www.eso.org/sci/facilities/paranal/instruments/xshooter/overview.html
- Website FEROS: \http://www.eso.org/sci/facilities/lasilla/instruments/feros/index.html

Chapter 9
Image Slicer

A Short Story

Several years ago we visited a colleague at the Astronomical Institute of the University of Leuven in Belgium. During the day he showed us their instrument laboratory. We came up with the need of an image slicer for the large focal length of our Cassegrain telescope. A laboratory employee then showed us one specially designed slicer for the Institute telescope. He said the price driver would be the very thin glass plate between the two prisms and the minimum price would be around 10,000 Euros. This was not very motivating for many years until ESO colleagues introduced a mirror slicer for around 300 Euros.

9.1 Basic Remarks

In Sect. 4.1.2 we described the optical slit as the resolution-defining optical element. It suppresses scintillation effects by cutting the wings off the seeing disk and accepting only the central part as the resolution-determining portion (Fig. 4.4). However, a substantial portion of the light flux is lost for data acquisition, even more so if the slit is narrowed in relation to the seeing disk for very high resolution. For improved efficiency, the image diameter in the telescope focus should be adapted so that its width matches with the slit width. In Sect. 2.6.3 however, we have already determined that this image size depends directly on the telescope focal length ($d_D = f_{col}/f_{cam} \cdot f_t \tan \alpha$) and that the optical slit width for high resolution would reduce the efficiency if the seeing disk is significantly larger.

By increasing the telescope focal length one inevitably reaches certain limits. On the one hand one would like to ensure as high a spectral resolution as possible by choosing a slit width which meets the Nyquist criterion for a given pixel size (Sect. 2.6.6). On the other hand the seeing disk at the telescope focus increases with

© Springer-Verlag Berlin Heidelberg 2015
T. Eversberg, K. Vollmann, *Spectroscopic Instrumentation*, Springer Praxis Books,
DOI 10.1007/978-3-662-44535-8_9

the telescope focal length. To overcome this contradiction, one can only "compress" the seeing disk so that it can entirely pass through the slit (we already introduced a variant in Sect. 8.2.4). Modern large telescopes however, provide very large seeing disks with their often long focal lengths. For instance, a 0.5″ seeing disk of the VLT telescopes has a diameter of at least 265 μm at the focus.

For adequate sampling (Nyquist criterion) we would need CCD pixels of at least about 100 μm in size. Hence, the telescope in-focus image diameter in the dispersion direction must therefore ideally be adapted to the slit width, so that most of the stellar light can pass through the narrow slit, and hence, lead to reasonably high spectral resolution. The specific optical element here is called an "Image Slicer". Such a device "cuts" the seeing disk into individual strips and repositions these strips in a single row to feed the narrow slit. The optical tools required are either mirrors or plane-parallel plates, together with prisms. The basic procedures are fully equivalent, though. To understand the slicer principle described in the following sections we first introduce the fundamental slicer principle: We position the telescope input (here a fiber aperture for simplicity) on a plane-parallel plate, so that one edge of the plate lies exactly in the center of the fiber output as is shown in Fig. 9.1.

The fiber aperture coming from the telescope is divided into two half-pieces. One half is guided unobstructed along the plane-parallel plate, the other half hits the plate. As long as the plate and the beam paths are orientated perpendicular to each other, the plate does not alter the seeing disk position.

Only when the beam hits the plate surface under a certain angle (tilted glass plate) is the beam displaced by an amount which can be calculated by the equations for the offset in a plane-parallel plate (Chap. 2). This displacement is

$$s = d \sin\alpha \left(1 - \frac{\cos\alpha}{\sqrt{n^2 - \cos^2\alpha}} \right)$$

with the plate thickness d, the angle of incidence α and the refractive index n. The offset concerns only the half of the seeing disk which passes through the glass plate. The disk is thus cut in two pieces which are shifted against each other. The result after the slicer is shown in Fig. 9.2 for a fiber-fed Echelle spectrograph.

Feeding this focal image to a spectrometer, we obtain two closely adjacent spectral strips that can be analyzed separately. The spectral resolution, however, is

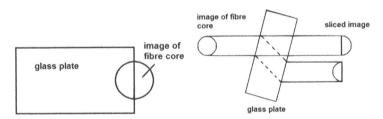

Fig. 9.1 Depiction of the slicer principle by light refraction. *Left*: Front view. *Right*: Side view (Sablowski 2012)

Fig. 9.2 The two shifted slicer input parts after being sliced by the above configuration (Sablowski 2012)

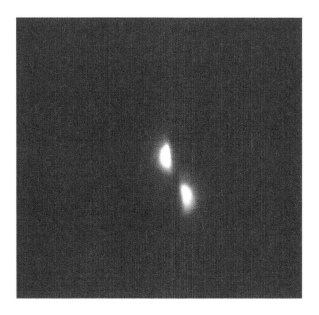

now reduced due to the displacement between the two half images in the dispersion direction. It would be better to obtain two images directly above each other (with respect to the dispersion direction) so that the two halves can act as one narrow slit of half the seeing-disk diameter. In that case both spectral strips would not show a wavelength shift against each other and could be considered as one single slit. In the following sections we describe various methods solving this very problem.

9.2   The Bowen Slicer

With respect to increasing telescope sizes, Ira Bowen at the California Institute of Technology discussed the possibility of how one might achieve maximum spectral resolution without potential light loss at the slit. In a groundbreaking article he presented the design of a so-called image slicer (Bowen 1938).[1] To change the image form, Bowen used a set of mirrors that visually "cut" the stellar image into individual strips and then in turn imaged all these strips together in a row. They are then fed to a narrow optical slit. Figure 9.3 shows Bowen's original concept.

The incident mirror MN reflects the light from one side towards the individual slit mirrors. At first glance, a holder for the mirrors QP to WX of 0.5–2 mm in size seems quite complicated. But the light from mirror MN does pass through the entire

---

[1] The light path inside an images slicer is relatively complex and requires a good three-dimensional imagination to understand it. This fact was probably the decisive step in the design of the first image slicer.

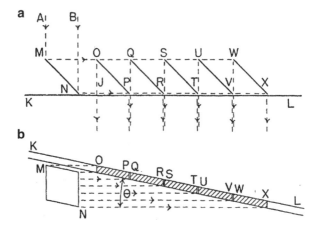

Fig. 9.3 The original Bowen concept for an image slicer (Bowen 1938). (**a**) Side view of all 45°
mirrors (OP, QR, ST, UV, WX) across the slit (KL) and the incidence mirror MN beside the slit.
(**b**) Top view onto the slit. The extrafocal incidence mirror MN, which is positioned beside the slit
with a slight offset, reflects an area towards the slit mirrors QP to WX. The area properly matches
the seeing disk (not the entire telescope field-of-view should be sliced but rather the seeing disk).
Due to the lateral slit-mirror offset with respect to the incident beam, all slit mirrors reflect only a
narrow portion of the beam towards the slit coming from mirror MN. Note that mirror MN is not
necessary if the beam configuration is rotated by 90° with respect to the telescope optical axis

slit, so that a relatively simple solution for the mirror mount exists. Bowen used
aluminized glass plates whose thickness corresponded exactly to the distance of the
individual mirrors to each other. These platelets can be shifted over the slit from one
of its sides. The principle is illustrated in Fig. 9.4.

With Fig. 9.5 Bowen defined the exact conditions for his design.

- The thickness of the individual glass plates T has to be $\sqrt{2}$ times smaller than
  the diameter of the seeing disk a on the sky. This arises from the requirement that
  the slices should "cut" the disk in its full length.
- The number of platelets is a divided by the slit width b since each plate provides
  only one slice.
- The angle between the mirror plane DPE and slit plane EPK is 45°.
- The platelet angle KPE against the slit should be $\pi/2 - \Theta$. $\sin \Theta = b/\sqrt{2}T = b/a$.
- The angle between the planes DPE and CPE is $\pi/2 - \Phi$ with $\tan \Phi = \tan \Theta/\sqrt{2}$,
  and must not exceed 45°.
- The angle between the mirror plane DPE and the plane CPD must not exceed
  $\pi/2 - \Phi$.
- The planes MN and DPE should be parallel.

The Bowen slicer has the obvious drawback of different optical path lengths
for the different light portions (slices), so that they cannot be perfectly focused all
together. This is not a problem as long as the path length differences are smaller
than the range in optical depth (see 3.47) of the telescope. Today, image slicers are

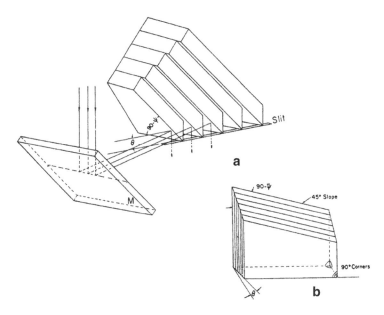

Fig. 9.4 (**a**) Positioning of silvered glass platelets for a Bowen image slicer above the optical slit. (**b**) Stacked platelets for production in the workshop (Pierce 1965). The thickness of the plates corresponds to the distance of the mirrors perpendicular to their surfaces in Fig. 9.3

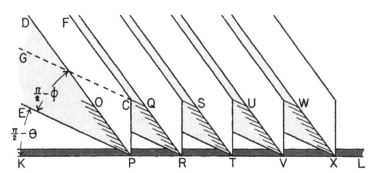

Fig. 9.5 Side perspective view of the slicer construction. The white areas like DPCF represent the narrow side surfaces of the slicer mirrors which all end at the slit edge. The gray areas represent the reflective surfaces

standard components for spectroscopy and are built in different variants. Given the seeing-disk size imaged at the telescope focus, these components are very small and highly delicate to manufacture. For example, the smallest slicer for Kitt Peak Observatory (Fig. 9.4) of 1 mm side length slices the image field into 20 parts! The critical production parameters, which determine the quality of the slicer, are the exact platelet thickness, their parallelism and their angular accuracy.

9.3   Bowen–Walraven Slicer

The efficiency of mirror reflections is approximately 92 %. In contrast, in fused silica almost all of the light is transmitted if the angle of incidence at which the light hits the boundary surface in the interior of the glass is greater than the critical angle for total reflection. This critical angle can be obtained from Snell's law (Eq. 2.2) for the exit angle $\beta_1 = 90°$. For different glasses (quartz, crown and flint glass) this critical angle $\beta_{1,c} = \arcsin(n_2/n_1)$ is smaller than 43.6°, when air surrounds the glass. In contrast, the critical angle does not exist when the surrounding medium has the same refractive index. So, we can guide light through glass under multiple light reflections if it is surrounded by air (in Chap. 11 we will discuss this in more detail). On the other hand, the light can be decoupled and then further guided if the ambient air is replaced by identical glass material. Because of the high reflection efficiency of nearly 100 % this technique can be used for a slicer, which uses refractions rather than reflections. This is accomplished by a Bowen-Walraven slicer (Walraven and Walraven 1972). The elegant and tricky optical element consists of two glass prisms and a very thin glass sheet between them. This unit slices the seeing disk or fiber input aperture in sections that are superimposed and fed to the slit (Fig. 9.7). The working principle is shown in Fig. 9.6. The Bowen-Walraven slicer makes use of the fact that light is completely reflected at boundary layers of different refractive indices as long as the critical angle between the incident beam and the surface normal is not reached or exceeded.

Figure 9.6a shows the core element of the slicer from the side. It is a glass plate which is tilted by 45° against the incident beam. The beam hits the lower left polished plate surface which in turn is cut by 45° for better light entry (real Bowen-Walraven slicers use a prism glued onto the plate). The illustrated but simplified parallel beam (in reality it converges depending on the telescope F-number) is then totally reflected back and forth inside the plate because it never reaches the critical angle. Figure 9.6b again shows a side view but now rotated by 90°. The circles are the "footprints" of the reflected beam on the lower glass side corresponding to 9.6a (in this case 1, 2 and 3). The light beam falls into the plate end-surface and hits the lower side three times, at 1 without reflection, and at 2 and 3 after two and four internal reflections, respectively. If we now apply a glass body to the lower glass side, for example a prism, the light is no longer reflected but decoupled into this

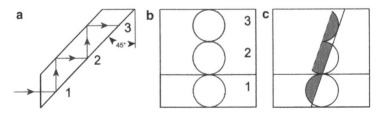

Fig. 9.6   The working principle of a Bowen-Walraven image slicer (Avila and Guirao 2009)

prism because of the identical material refractive index. In order not to decouple all
the light at once but rather only in successive individual beam portions, the prism
edge can be diagonally positioned over the beam footprints so that only one edge of
the seeing disk (here in gray) is decoupled, while the remaining part (here in white)
is further reflected (Fig. 9.6c). Consecutive reflections thus lead to continuous exits
of individual beam portions, which are orientated in a row towards the lower prism
and along its edge. There the optical slit can be positioned, which is for the present
example three times smaller than the incident beam. Due to the quartz material,
light loss is almost non existent. At the entry and exit surfaces of the two prisms, the
critical "Brewster" angle is never reached.

   A real device is shown in Fig. 9.7. This Bowen-Walraven slicer has been constructed for the high-resolution spectrometer of the 1.52 m f/27.6 Coudé telescope
at Haute Provence Observatory (OHP) (Gillet et al. 1994).

   The entrance window of the slicer comprises a prism (1) on which the incident
light falls vertically. The incident beam (here represented as a circle) passes through
the prism and enters the underlying slicer platelet at an angle of 45°. There, the beam
is reflected and cut into five sections (2), which are supplied to the slit (3). The input
lens has a diameter of 600 µm, which corresponds to 3″ on the sky. The input disk
is sliced into five parts, each 0.6″. The slicer efficiency is about 90 %. To facilitate
the slicer manufacturing, the entrance pupil is increased by a factor of 3.62 and the
telescope then works effectively at f/100.[2] An even smaller Bowen-Walraven slicer

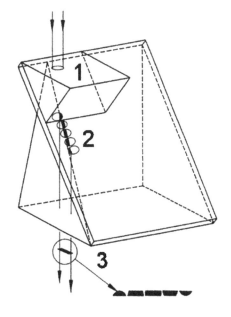

Fig. 9.7 Bowen-Walraven
image slicer for the
spectrometer of the 1.52 m
OHP Coudé Telescope (Gillet
et al. 1994, reproduced with
permission ©ESO). The
large and small prisms have a
base length of 25 and 12 mm,
respectively. The
plane-parallel plate has a
thickness of 1.58 mm

---

[2]The slicer is already very small. By virtually increasing the entrance focal length the entrance
pupil becomes larger and manufacturing becomes easier.

Fig. 9.8 The
Bowen-Walraven image slicer
for ESPaDOnS (Laboratoire
Astrophysique de Toulouse)

is used at the Canada-France-Hawaii Telescope for the Echelle spectropolarimeter
ESPaDOnS (Fig. 9.8). This 15 mm slicer simultaneously processes the input of three
optical fibers.

The two advantages of the Bowen-Walraven slicer—high efficiency and a significant increase in spectral resolution—are contrasted however by non-negligible
drawbacks. They are:

1. The equivalent slicer slit is tilted against the optical collimator axis. The slit
   image is sharp only in the slit center because of the limited chromatic field
   depth. Towards the slit wings perpendicular to the dispersion the image is blurred,
   thereby reducing the resolution somewhat. One can compensate for this effect by
   appropriate exit-prism adaptation. This would make the slicer more complex,
   though.
2. Quartz glass has a relatively low refractive index. Therefore one needs large F
   numbers at the entrance of at least 11 because of the internal reflection angle.
3. Slicers are expensive. The platelet is very thin. Particularly for small image fields
   the platelets and the prisms have to be cut and polished very precisely in terms
   of good slicer mapping quality.
4. All slicer parts must be assembled via molecular contacts, which is not a standard
   technique for manufacturers of optical components.

9.4   FEROS: A Modified Bowen–Walraven Slicer

For highly accurate measurements it is often necessary to remove the sky background carefully. If background and object are to be measured simultaneously while
the resolution has to be maximized, one therefore would need two slicers working in
parallel. For the Fiber-Fed Extended Range Optical Spectrograph (FEROS), which

measures both the sky background and the target object through fiber optics, a
Bowen-Walraven slicer has been modified so that it can accommodate two apertures
simultaneously and slices both once (Fig. 9.9).

Thus the FEROS slicer works like two separate and independent slicers and it
achieves a spectral resolving power of 40,000 (Kaufer 1998).

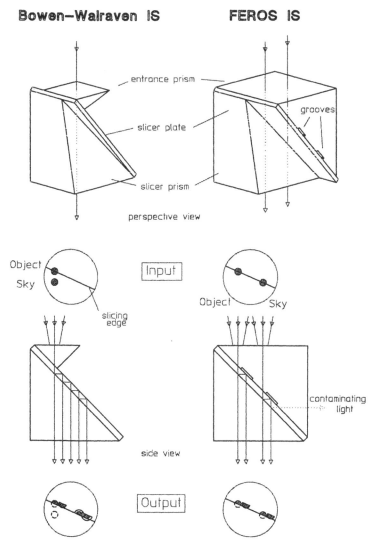

Fig. 9.9 Comparison of the Bowen-Walraven and FEROS image slicers (Kaufer 1998)

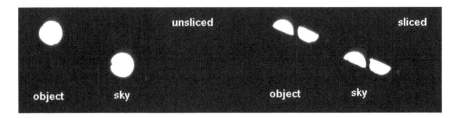

Fig. 9.10 Pictures of the sliced seeing disk in the FEROS slicer prototype (Kaufer 1998)

The slicer was designed to minimize defocusing due to the different optical path lengths. For this task it is positioned in the intermediate f/11 focus of the fiber optics, which is generated by intermediate optics. Two air-grooves in the quartz entrance prism (Homosil) guarantee the required total reflections for object and sky. In addition, the material allows a more accurate surface polishing, which in turn makes adhesive material unnecessary. This has the following advantages: (a) Two prisms of identical edge length facilitate the slicer manufacturing, (b) a high-quality surface finish makes bonding unnecessary, which in turn ensures sharp slicer edges and clean air grooves, (c) the air grooves control the maximum number of slices. A potentially bad alignment of the base prism or tolerances in the slicer plate can introduce undesirable slices to be recognized as straylight. This scattered light is faded out by the well-defined groove width and disappears vertically from the slicer. Figure 9.10 shows the input and output of the FEROS slicer prototype.

Again, the seeing disks in telescope foci and fiber apertures have very small diameters and thus the glass plate necessary for light separation must be designed to be accordingly thin. It is extremely delicate to handle such plates of quartz glass— and they can break easily. Therefore, the plate is the true price driver of all Bowen-Walraven image slicers.

9.5   X-Shooter Mirror Slicer

An entirely different concept pursues the image slicer of X-shooter at the VLT (Sect. 8.4.6) that was developed at the Observatoire de Paris. The so-called integral-field unit (IFU) system with reflective optical slits and four spherical mirrors re-arranges a $4'' \times 1.8''$ image field to a $12'' \times 0.6''$ field (Fig. 9.11). The central region of the seeing disk runs through the slit without loss. The wings, however, are reflected back from the front side of the two slit wings, which are designed as mirrors, towards two spherical mirrors. The latter mirrors reflect the two beams again along the slit towards a second mirror pair. The light is then reflected onto the backside of the slit and then towards the image of the central beam part so that all three image areas are arranged next to each other (Fig. 9.12). The entire IFU is positioned within a cube volume of 80 mm edge length.

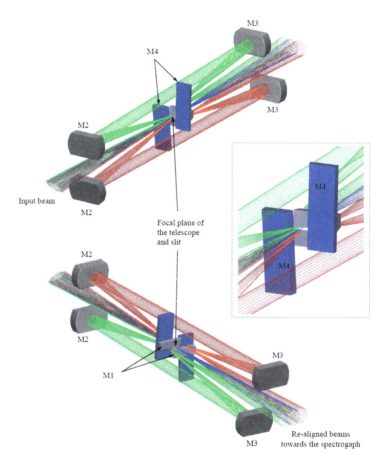

Fig. 9.11 Sketch of the X-shooter IFU. The blue central region passes undisturbed through the slit while the wings (*green* and *red*) are reflected towards the mirrors and positioned adjacent to the central beam (Vernet et al. 2011)

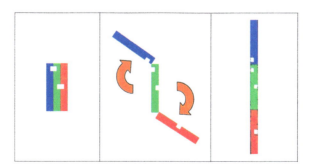

Fig. 9.12 Representation of the re-positioning of the image field by the X-shooter slicer. The original field is sliced twice (Vernet et al. 2011)

Particular attention was paid to suppress slicer straylight (Guinouard et al. 2006). A field stop concept has been developed, which limits the incident light to a fixed frame by a diaphragm. Moreover, the slit is defined by two flat mirror pairs, whose edge phases act as baffles or reflect light towards the housing inner wall, depending on the geometrical position of an incident beam. The four reflections at the slit and at the mirrors reduce the overall efficiency to about 85 % of the central beam flux. Because of limited coating efficiency in the blue this value falls to around 50 % for wavelengths below 400 nm.

9.6   The Waveguide

Because of the optical path and, hence, focus differences, Bowen slicers can only be used for fast telescopes.[3] Bowen-Walraven slicers, however, reflect the light several times, so that in fast telescopes light loss occurs through many reflections. Therefore, they are better suited for slow telescopes. A completely different option is delivered by waveguide image slicers (Suto and Takami 1997). This slicer must be fed by fiber optics (Fig. 9.13). The very simple design (although not easy to build) uses the numerical aperture of a fiber. It avoids optical path differences and is particularly suitable for telescopes as fast as f/1.

The slicer has 10 stacked glass plates of 50 μm thickness and 500 μm width each. It is stacked and bonded with an adhesive of suitable refractive index. The result is a long, square-faced glass block, which acts as entrance window of 0.5 mm × 0.5 mm edge length (the adhesive layers have a thickness of about 2 μm and limit the

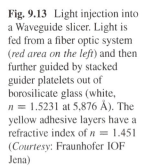 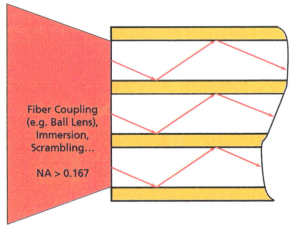

Fig. 9.13 Light injection into a Waveguide slicer. Light is fed from a fiber optic system (*red area on the left*) and then further guided by stacked guider platelets out of borosilicate glass (white, $n = 1.5231$ at 5,876 Å). The yellow adhesive layers have a refractive index of $n = 1.451$ (*Courtesy*: Fraunhofer IOF Jena)

Fiber Coupling (e.g. Ball Lens), Immersion, Scrambling…

NA > 0.167

---

[3]Astronomers refer to small and large telescope F-numbers as "fast" and "slow", respectively.

light loss at f/1 to 0.1 %. Since the height of each tile is identical to the width of the slit to be fed, and thus they are a measure of the spectral resolution, the latter may be influenced by the slice thickness (Fig. 9.14). However, 50 μm represent a lower limit in the production process and the glass polishing.

The incident light coming from the telescope hits the slicer face, is transmitted by the individual platelets and reaches a mirrored and by 45° inclined end surface at the other waveguide end. This mirror reflects the individual beam vertically out of the slicer (Fig. 9.15). The exit windows are positioned stepwise to each other and have an offset in dispersion direction. The resulting pixel shift between the adjacent imaged spectra is compensated by appropriate wavelength calibration before merging them. Since each exit window represents a single slit, there is no path difference between individual sliced portions and they are all in focus. This is an important advantage over the Bowen Slicer.

Waveguide technology for spectroscopy is used at the 8.4 m Large Binocular Telescope (LBT). The Potsdam Echelle Polarimetric and Spectroscopic Instrument (PEPSI) delivers high-resolution spectroscopic data with a spectral resolving power of up to $R = 300{,}000$. In order to achieve an acceptable combination of fiber efficiency and resolution, the spectrograph can be fed by 100, 200 and 300 μm

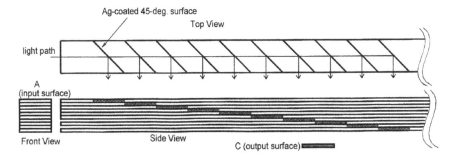

Fig. 9.14   Front, side and top view of a Waveguide image slicer by Suto and Takami (1997). The length of each tile is between 1.5 and 6.45 mm. The 0.5 mm × 0.5 mm entrance window is large enough to compensate for optical aberrations and seeing effects of the object

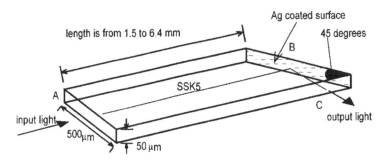

Fig. 9.15   Schematic view of a single Waveguide platelet (Suto and Takami 1997)

fibers via interchangeable Waveguide image slicers (Fig. 9.16). The system consists of seven stacked glass layers of uniform thickness. The resolution limiting factors are the 30 μm minimum layer glass blank thickness and the apparent focal ratio degradation in the fiber optics (Sect. 11.5). To ensure total internal reflection at the appropriate numerical fiber aperture, the supplied platelets have to be surrounded by a medium of refractive index n < 1.47. This is ensured by an adhesive, which also delivers a stable configuration. At the top end opposite to the input window, each platelet ends with a coated 45° mirror that reflects the light towards the exit. Figure 9.17 shows the output of a prototype with 210 μm glass layers. No FRD is detected—the input f/13.9 beam leaves the slicer with f/13.9.

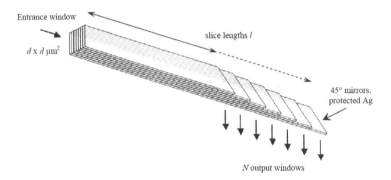

Fig. 9.16 Layout of the LBT Waveguide image slicer (Beckert et al. 2008)

Fig. 9.17 Waveguide prototype for PEPSI under a microscope (*Courtesy*: Fraunhofer IOF Jena)

9.7 CAOS Low Cost Slicer

The Club of Amateurs in Optical Spectroscopy (CAOS) is a group of aficionados at the European Southern Observatory (ESO) and the Max Planck Institute for Extraterrestrial Physics (MPE) in Garching. The group works on all spectroscopic aspects and devotes its knowledge to astronomical amateur spectroscopy.[4] CAOS has developed and built a mirror image slicer for a 50 μm fiber feed whose principle element is the Bowen-Walraven slicer and which is completely analogous to it (Sect. 9.3).

Figure 9.18a shows a side view of the two mirrors which are tilted by 45° with respect to the incident beam reflecting it back and forth. Figure 9.18b shows the "footprints" of the 50 μm beam delivered by the fiber optics. The beam hits the bottom mirror, leaves a "footprint" at 1 and is reflected to the upper mirror to leave a second "footprint" at 2 and a third at 3. If we now rotate the lower mirror against the upper one (Fig. 9.18c), the light is no longer partially reflected but can pass through the lower mirror (right) and leaves the slicer (gray). After the first reflection only the reflected beam part remains (white) and is further reflected. Consecutive reflections thus lead to a continuous suppression of individual beam portions which are positioned in a row along the mirror. The optical slit can now be positioned at this point. In the present example the slit is three times smaller than the incident beam.

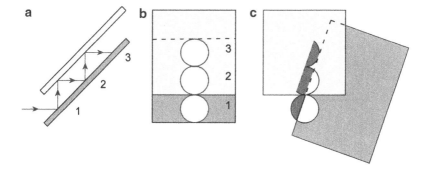

Fig. 9.18 Working principle of the CAOS slicer (Avila and Guirao 2009). (**a**) Side view with the multiply reflected beam between the two mirrors. (**b**) Top view in the incident beam direction with the corresponding "footprints" (*white*). (**c**) Same as (**b**) but now with a tilted back mirror. The white areas of the "footprints" are further reflected; the gray areas leave the slicer

---

[4]In fact, it is highly professional work applied to small telescopes. This includes investigations on fiber optics, slicer design and complete spectrographs.

However, the mirror slicer has a lower efficiency than the Bowen-Walraven slicer. This is because of light loss during reflections and, hence, only a single slice should be performed. Figure 9.19 describes the geometric facts for the slicer construction.

After a number of tests, the CAOS group lists the following advantages of their slicer:

- Compared to the Bowen-Walraven Slicer the mirror slicer is easy to construct, compact and cheap. The slicer can be used with a slit width down to $50\,\mu$m. This makes it interesting even for very small telescopes and appropriately short focal lengths.
- The slicer delivers an adequate quality for sharp slit edges.

However, the slicer has also two non-negligible drawbacks.

First, the slit tilt along the optical axis is larger than for the Bowen-Walraven Slicer. The reason is the lower reflectivity. And the defocusing is $\sqrt{2}$ times larger than the beam diameter because of longer reflection paths. Second, the efficiency falls rapidly with the number of slices. The concept is, therefore, best suited for a maximum of two slices (three sections), if the mirror coating is of good quality.

Tobias Feger used the CAOS slicer design for his Echelle spectrograph (Sect. 8.4.2). The connection with the telescope was realized with a $50\,\mu$m fiber (SMA connector), which provides a f/4.5 beam at the output. The fiber aperture is imaged onto the slicer by a small achromatic doublet of 8 mm focal length. The 13 mm distance between fiber end and doublet produces a beam of $80\,\mu$m

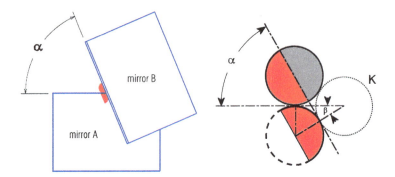

Fig. 9.19 Geometry determination for a two-slice image slicer (left graph from Feger 2012). To position the two beam slices next to each other and in series, the two mirrors are rotated against each other by the angle α. The angle β (*right*) can be derived with the Thales theorem and the external circle K which has the same diameter as the fiber aperture. From β we then obtain $\alpha = 60°$. For manufacturing the slicer, we need to derive the true rotating angle γ which delivers the projected angle α in the observation direction. The real tilt angle must obviously be larger to provide α. This true angle $\gamma = \arctan(\sqrt{2}\tan\alpha)$ required by construction can be determined by elementary geometry and is 67.8° for two slices. The distance s between the two mirrors depends on the fiber beam diameter d and is $s = d/\sqrt{2}$. In the same way we can determine the angles for three slices (see Fig. 9.18). The external circle K must then be twice as large as the fiber beam. In this case, $\alpha = 70.53°$ and $\gamma = 75.96°$

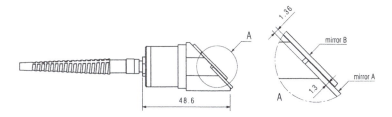

Fig. 9.20 Detailed view of the mirror separation (Feger 2012)

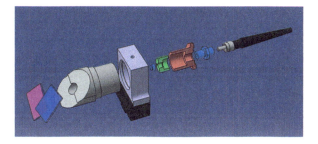

Fig. 9.21 3D drawing (Feger 2012)

Fig. 9.22 Laboratory setup (Feger 2012)

diameter (f/7.2) in the doublet image plane. Figure 9.20 shows the side view of the stepped mirror holder according to the instructions by the CAOS group and Fig. 9.21 shows the mechanical realization.

The stepped mirror holder guarantees an air gap of 60 μm which is necessary for the slice through the stellar or fiber image. The two sliced pieces are then imaged at 45°. Figure 9.22 shows the corresponding laboratory setup.

The mirror material are two front-surface mirrors of 1.2–1.3 mm thickness. The edges were ground with sandpaper of 600–1,200 grits per square centimeter and then polished with fibre polishing paper to obtain sharp edges. Figure 9.23 shows the two spot images on the slicer. The first test spectrum is shown in Fig. 9.24.

Fig. 9.23 Input and output slicer image (Feger 2012)

Fig. 9.24 Echelle spectrum obtained with the Feger slicer (Feger 2012). Note the weaker spectrum due to reflection loses

Chapter 10
Some Remarks on CCD Detectors

A Short Story

Our first CCD camera was water cooled. More than 300 watts heating power had to be removed via 12 volts—Who has such a power supply? For cooling the water we bought an old car radiator and screwed it to the outer wall of the observatory. This beast made quite a noise. And to avoid breaking the camera body by freezing water in the rather cool winters, we also used a coolant. What a mess when the camera had to be disconnected! The whole construction was spectacular and visitors were quite impressed—until a lightning strike finished the whole rigmarole and the camera broke.

10.1 High Quantum Efficiencies

Today scientific data are mainly recorded digitally by electronic means. Photographic emulsions are no longer applied in professional astronomy due to their low sensitivity and precision. Among the many different techniques (photomultipliers, photodiodes), the CCD camera (Charge Coupled Device) is by far the most common detector system in astronomy. A CCD is a semiconductor-based sensor. This detector usually produced in chip form consists of microscopically small, matrix arranged silicon elements (pixels) that convert incident photons into electrons via the inner photoelectric effect (photoconduction). For reading-out, these electrons are electronically shifted in rows or columns towards the edge of the pixel array and then transferred to an analog-to-digital converter and sequentially counted. Each pixel is thus allocated to a gray level associated to the number of free electrons in the pixel, i.e., white sections of the CCD are areas where many photons and black where few have been collected. The unit in which the grayscale value is measured is called Analog to Digital Unit or abbreviated to ADU. The maximum grayscale

© Springer-Verlag Berlin Heidelberg 2015
T. Eversberg, K. Vollmann, *Spectroscopic Instrumentation*, Springer Praxis Books,
DOI 10.1007/978-3-662-44535-8_10

value corresponds to the maximum digital output value of the AD converter. For a 16 bit converter we obtain, for example, a maximum signal level of $S_{max} = 2^{16}$ ADU = 65.563 ADU. Since the pixels have a limited capacity, there may be an over-exposure of the CCD. In this case, excess electrons in a given pixel are transferred into adjacent ones. The pixel shows "overflow", much like a bucket of water. For the image one refers in this case to "blooming". The maximum number of electrons that can be accommodated by a pixel is called the "quantum full-well capacity". This parameter is strongly dependent on the pixel size and thus on the particular chip. The "full-well capacity" also determines the "dynamics" of the CCD, i.e., the number of distinguishable gray levels. Without a deeper discussion of CCD characteristics, there are two decisive factors:

1. high quantum efficiency and
2. high sensitivity over a wide wavelength range.

The quantum efficiency η is the ratio between incident photons N_p and registered charge carriers N_E (electrons).

$$\eta(\lambda) = \frac{N_e}{N_p}. \tag{10.1}$$

The quantum efficiency is wavelength dependent. For commercially available CCD chips it is of the order of 50–80 % in the visible spectral range and for specially designed research chips is even significantly higher. The human eye and photo emulsions with an efficiency of about 1–5 % are far inferior. For low-energy radiation the quantum efficiency cannot be greater than unity, as one absorbed photon can at most produce one electron. If not all photons are absorbed the quantum efficiency is less than one. Figure 10.1 shows the efficiency curves of a Tektronix TK 1024 chip and Fig. 10.2 that of a KODAK KAF 1603ME. The red line near the zero level in Fig. 10.1 indicates the efficiency and the corresponding

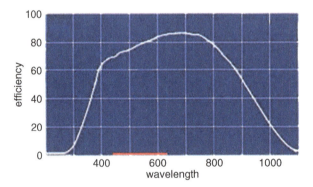

Fig. 10.1 Quantum efficiency of a Tektronix TK 1024 CCD chip. The *red line* indicates the sensitivity of the human eye

Fig. 10.2 Quantum
efficiency of a Kodak
1603ME CCD chip

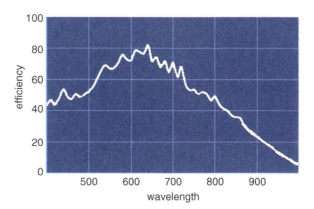

sensitivity wavelength-interval of the human eye and of photo emulsions. We have
used a linear scale rather than a usual logarithmic scale to illustrate this dramatic
situation. It is obvious that CCD technology is a unique innovation enabling highly
improved results compared to photographic techniques. We already have met an
impressive comparison in the introduction (Fig. 1.4). Some inherent disadvantages
are associated with the CCD technology, though, which can be eliminated by
specific corrections.

10.2   Linear Response: The Gain

If a CCD is read-out, the photon generated charges (pixel electrons) must be converted to a voltage signal via a capacitor. An analog-to-digital converter transforms
this voltage into a digital signal in analog-to-digital units (ADU), which can be
easily read thereafter.[1] The ratio between the number of generated electrons per
pixel and the signal in ADU is referred to as a gain G and is a property of the read-
out amplifier (measured in e^-/ADU). This parameter indicates how many electrons
are combined into one count. The gain must be adjusted to the AD converter,
otherwise for example, a bright object provides high voltage values that can then
lead to signal overflow or noise dominates when there is not enough gain. One of the
main positive attributes of the CCD is its linear response with respect to the arriving
photon signal over large intensity intervals (from a few electrons to the full well
capacity of up to several hundred thousand electrons per pixel). The ADU interval
in which a CCD can provide signals is defined by the AD converter. Therefore,
a 12 bit camera delivers theoretically (!) $2^{12} = 4,096$ different intensity values.
However, the process is generally degraded by the ever-present amplifier noise. For

---

[1]The incident signal is not a digital one because the current induced by single photons is transferred
into a "continuous" signal by an internal capacity.

a non-illuminated detector the noisy dark value may not exhibit negative values introduced by noise, otherwise there will be distortions of the measured value.[2] This is similar for the maximum signal values, i.e., the useful value range is smaller than the dynamical range of the AD converter. The number of photons necessary for the maximum intensity value defines the full-well capacity. A CCD camera behaves linearly over almost the entire range of all intensity values (ADU). Just below the maximum intensity value (typically at about 70–80 % of the maximum value), however, a CCD reaches a saturation region and does not provide more ADU despite more incident photons (Fig. 10.3). To remain in the linear signal region one should, hence, avoid exposure times close to this saturation region.

The same is true for very low signal levels: For example, in order keep the noise symmetrically around a positive mean value, most CCD are equipped with a positive offset (the so-called bias, Chap. 12). In addition, residual charges always remain on the chip. This means, even for an exposure time of 0 s, each CCD has a small positive offset signal in ADU. In reality, the gradient of the curve in Fig. 10.3 more or less deviates from linear behavior. This variation is called "integral nonlinearity" and is generated by imperfect analog-to-digital converters. As all imperfect parameters this difference has negative influence on the quality of the spectral signal. Let us consider a camera with a gain of 4 electrons/ADU. In this case, a linearity error of 10 ADU provides a deviation of 40 electrons at the amplifier output, potentially unknown to the observer. For a 16-bit camera these 10 electrons mean a deviation of about 0.015 %. In contrast, for a 12-bit camera this would be about 0.25 % deviation. The linearity deviation is in each case therefore significantly lower for a large dynamic range than for a small dynamic range. Cameras with more bits thus provide more accurate results.

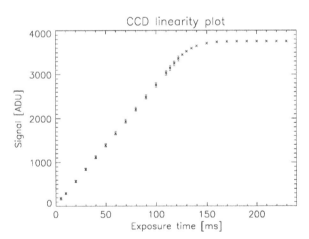

Fig. 10.3 ADU dependence on the exposure time for a 12-bit CCD. Just below the ADU maximum value the chip saturates and no additional ADU are delivered with increasing exposure time (more photons). Note also the offset signal for the exposure time $t = 0$, which is caused by the bias (Bristol University, https://wikis.bris.ac.uk/display/astro/CCD+Linearity)

---

[2]Signals which would be limited to positive values would introduce a shift of the mean value towards a positive signal.

Almost every CCD camera can also operate in a so-called "binning mode" where
several pixels are combined to <u>one</u> "superpixel". This binning mode is, for example,
used for objects of very low intensity in order to reduce read-out noise and to
increase the sensitivity or to avoid undersampling. In the case of 2×2 binning the
signal of 4 pixels is co-added. Thus, a superpixel can collect a much greater number
of electrons than in the 1×1 mode. However, the binning mode does not change the
maximum number of counts by the AD converter. As a result, the gain in binning
mode differs from that of the 1×1 mode. The user must take this into account when
comparing images taken with different binnings.

10.3 Noise

Of course, the quantum efficiency is a critical parameter for the sensitivity of any
CCD. However, one should not jump to the conclusion that the quantum efficiency
is the sole factor for the image quality. It is merely a measure of the ability to
convert photons into electrical signals, and thus an intrinsic property of the chip.
By specifying the quantum efficiency, other parameters at least equally important
are not taken into account. These other parameters determine the noise performance
and thus the accuracy of signal reproduction. The main noise sources are the photons
themselves, the noise of the silicon chip and the noise of the read-out amplifier.

10.3.1 Photon Noise: The Number Is It

Despite the above advantages, interference sources cannot be totally eliminated even
in CCDs. Already in Sect. 2.6.6 we have considered deviations from the incident
signal at the telescope. This is in the form of interpolation errors in the sampling of
a continuous signal by discrete pixels. In the following, we consider real noise in
the detector.

A fundamental source of noise is found in the nature of light. Since photons
are considered as particles in CCD data acquisition, they are subject to the corresponding statistics for countable ensembles occurring stochastically (randomly).
The respective statistics are the Poisson distribution (Appendix B.2). This means
that the standard deviation of all events (the generated electrons) alone defines the
noise. This noise thus depends on the number of incident photons N_p (the light
flux) and is $\sqrt{N_p}$ (The actual noise per pixel will not be in the number of incident
photons, but rather the number of detected photons converted into electrons). Since
the signal-to-noise ratio S/N is crucial for the data quality, there is an interesting
consequence for the observer. The signal is proportional to the number of detected
photons with respect to the quantum efficiency η (Eq. 10.1), $S \sim \eta N_p$. However,
according to Poisson statistics the noise is proportional to the square root of the

measured number of photons. Thus we obtain the ratio $S/N \sim \sqrt{\eta N_p}$. The number of photons depends on the aperture surface, which in turn is proportional to the square of the mirror diameter, and thus $S/N \sim D$. The signal-to-noise ratio thus depends linearly on the mirror diameter (telescope aperture) and increases with the square root of the exposure time (see also Sect. 10.7).

The quantum full-well capacity limits the S/N. For the exposure of a single image (frame) we expect for the electrons $(S/N)_{max} = \sqrt{N_{fullwell}}$, but only if the detection resolution of the analog-to-digital converter is sufficient to detect single electrons. If the full-well capacity is, for example, $N_{fullwell} = 350{,}000 e^-$ and the AD converter has a dynamic range of 16 bit, only packages of $5.3 e^-$ can be distinguished per count. It is therefore convenient to calculate directly in the units of the AD converter. We obtain for a 16-bit converter $(S/N)_{max} = \sqrt{2^{16}} = \sqrt{65536} = 256$ in ADU. If for some reason a $S/N > 256$ is required, this can only be achieved by a larger number of frames which are then averaged. For a number of N_{frames} individually recorded frames the corresponding noise of the average frame decreases with $1/\sqrt{N_{Frames}}$.

10.3.2 Dark Noise: Bad Vibrations

The generation of electrons in the chip is always subject to statistical fluctuations, which depend on the square-root of the number of electrons $\sqrt{N_e}$, analogous to photon noise (even without a light source). These fluctuations appear as background noise in every image and can, as for the photon noise, not be eliminated. This is called thermal noise. CCD chips are made of silicon crystals, whose atoms oscillate with temperature, and thereby, transport electric charges from the valence band into the conduction band. These charges are taken from the potential well of the pixel and are read-out like the charges generated by photons. This is an intrinsic property of any CCD and the additional thermal charges affect the data generated by the signal photons. Thus, the sensor also produces a signal, even when no radiation illuminates it. This current of randomly generated charge carriers is called "dark current" and its interferences are called "dark noise". It is correlated with both the exposure time and the temperature. The number of thermally excited electrons follows a Boltzmann distribution

$$N_{therm} \propto exp\left(-\frac{E_g}{kT}\right)$$

with E_g the band gap of silicon, T the absolute temperature, and $k = 8.62 \cdot 10^{-5} eV/K$ the Boltzmann constant. Silicon has a band gap of about $1.1\,eV$. Thus we can now easily derive a temperature difference, for which the number of charge carriers is reduced by a factor of two: We calculate the above Boltzmann factors for two different temperatures T and $T + \Delta T$ and then determine the relation corresponding to $1/2$.

$$\frac{N_{T+\Delta T}}{N_T} = exp\left(\frac{E_g}{kT} - \frac{E_g}{k(T+\Delta T)}\right) = \frac{1}{2}.$$

As a good approximation for the temperature difference we obtain

$$\Delta T \approx \frac{\ln(0.5) * 8.62 \cdot 10^{-5} \, \text{eV}/K}{1.1 \, \text{eV}} \cdot T^2.$$

Assuming room temperature $T = 300 \, \text{K}$ we obtain for the necessary temperature difference $\Delta T = 4.88 \, \text{K}$. For halving the number of thermal charge carriers the temperature must be lowered from room temperature of $300 \, \text{K}$ to about $295 \, \text{K}$. The dependence of the dark current on temperature can be specified using the following empirical relationship (Kodak 2003):

$$I_D = 2.5 \cdot 10^{15} \cdot A \cdot I_{D0} \cdot T^{1.5} \cdot e^{-\frac{E_g}{2kT}}$$

with I_D the dark current in $e^-/\text{Pixel}/s$, A the pixel area in cm^2 and I_{D0} the measured dark current in nA/cm^2 at $300 \, \text{K}$. The band gap E_g of the silicon is also temperature-dependent and is

$$E_g(T) = 1.1557 - \frac{7.02 \cdot 10^{-4} \cdot T^2}{1108 + T}.$$

If the dark current density of a CCD is given for example as $I_D/A = 1 \, \text{pA/cm}^2$ at $300 \, \text{K}$, one can calculate $I_D = 6.24 \cdot 10^{-2} e^-/\text{s}/\mu\text{m}^2$ in the unit $e^-/\text{s}/\mu\text{m}^2$. For sensor pixels of $9 \, \mu\text{m} \times 9 \, \mu\text{m}$ (=$81 \, \mu\text{m}^2$) we obtain a surface dark current at room temperature of about $5e^-/\text{s}$. For longer exposure times, this leads to problems in the average signal and the noise level. One can now influence the dark current by varying the temperature. Measurements show that reducing the detector temperature by about $60 \, ^\circ\text{C}$ can reduce the dark current by about four orders of magnitude (Fig. 10.4).[3] To minimize the disturbing charges, CCD chips must thus generally be cooled. Professional systems are cooled with liquid nitrogen via a so-called cryostat (dewar) to $-186 \, ^\circ\text{C}$ while cost-effective systems use electrical (Peltier element) cooling down to typically $30 \, ^\circ\text{C}$ below ambient temperature. The price plays a secondary role, since liquid nitrogen is relatively inexpensive. Since the dark noise increases with the square-root of the exposure time, it plays a minor role for fast video signals. It must be taken into consideration, though, via reducing the science images by using dark fields if the science exposure times are longer than a few seconds (see Chap. 12).

On average, thermal electrons are created at a constant rate (in units of counts/s), independent of the incident intensity. Therefore, the dark current is proportional

---

[3]For a detailed analysis of the temperature dependence of dark noise, we refer to Widenhorn et al. (2002).

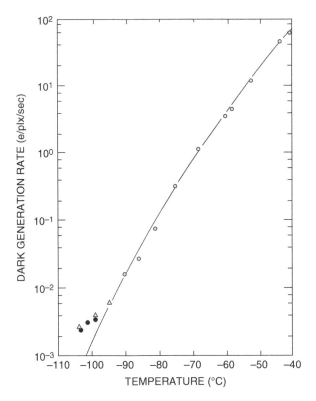

Fig. 10.4 Dark current in electrons per pixel and second versus temperature (Robinson 1988)

to the exposure time. Besides the chip temperature, the exposure time also plays an important role, in particular if the dark current is not sufficiently minimized by corresponding cooling. Figure 10.5 shows the dark noise of an "AlphaMaxi" camera from the company OES for exposure times between 0.01 and 20 s. Dark frames were taken for all exposure times for each recording, with the standard deviation also being determined. Plotting the determined dark noise against exposure time we find a linear relationship with a gradient of about 44 ADU/Pixel/s. The KAF400 has a full well capacity of about $85ke^-$, i.e., for a 16 bit converter the gain can be approximated by $g \approx 850,00e^-/65,536 \text{ADU} = 1.29$. Thus the dark noise in the units of electron numbers is $57e^-$/pixel/s at maximum. The axis intercept of approximately 46 ADU/pixel gives an indication of the read-out noise and is with the estimated gain about $59e^-$/pixel. A more accurate method for the determination of the gain and read-out noise of the Alpha Maxi is described in Sect. 10.6.

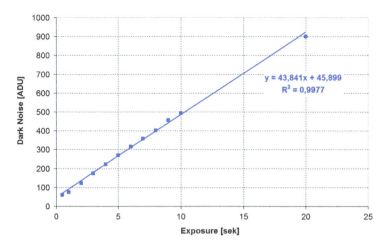

Fig. 10.5 Dark Noise in ADU versus exposure time in seconds obtained for an AlphaMaxi camera (KAF400)

10.3.3   Read-Out Noise: Electronic Influences

A significant noise source is the read-out electronics, which is needed by all the CCD cameras to read the signals and transfer them as computable data. First, the analog-to-digital converter provides only a statistical distribution that varies around a mean value when transmitting an analog pixel value to a digital signal at the amplifier output. Pixel values are in principle not exactly repeatable but noisy. In addition, the internal amplifier causes additional noise during the read-out process, generating secondary electrons which are superimposed on the original signal. This read-out noise is independent of the temperature and essentially depends on the circuit layout of the read-out amplifier. Practice shows that significant development experience and many experimental setup trials are needed to find the optimal design of the electronics and to minimize the read-out noise. In this context the read-out speed plays an important role. One has to keep in mind that large frequencies in the read-out process may cause unwanted resonant circuits on the circuit board which can have undesired consequences. For example, if a chip of $1,000 \times 1,000 = 1,000,000$ pixels is read in $10\,\mathrm{s}$, we are in a frequency domain that requires careful electronic design in order to suppress the generation of additional noise signals. Lower read-out speeds are preferable for the measurement. However, the read-out times must be set in an appropriate relation to the duration of the entire measurement process within an observation period.

10.3.4   Additional Noise: Pixel-to-Pixel Variations

Sensor pixels generally have slightly different sensitivities. Thus, a noise source is artificially introduced. Even a completely homogeneous light source with no brightness fluctuations will therefore lead to some noise in the resulting frame.[4] The sensitivity variations (standard deviation) over the sensor are usually in the percentage domain of the averaged signal. This has the consequence that the absolute noise caused by pixel-to-pixel sensitivity variations increases linearly with the signal. Mathematically, one can express the noise by the following relation:

$$\sigma_{PtP} = k \cdot S \tag{10.2}$$

with S the signal and k the proportionality constant which is derived from the standard deviation of the sensitivity variations. Figure 10.6 shows the situation for three different k-values compared to the photon noise.

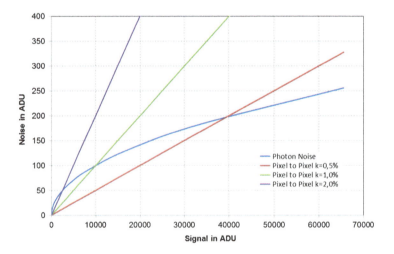

Fig. 10.6 Noise caused by pixel-to-pixel variations compared to photon noise for three different standard deviations of the chip sensitivities. To keep the pixel-to-pixel noise below the photon noise, one needs a high-quality CCD with $k < 0.5\,\%$

---

[4]In the literature this is sometimes noted as "flat field effect".

10.4   The Combination Is Crucial

The efficiency curves of the Tektronix TK 1024 (Fig. 10.1) and the Kodak KAF 1603ME (Fig. 10.2) lead to a preference for the Tektronix chip due to better efficiency. However, one should keep in mind that the read-out noise of the camera can significantly affect the signal quality. For clarity, spectra of the star γCas were recorded at 60 s exposure time with these two chips in two different cameras. The two spectra around Hα are shown in Figs. 10.7 and 10.8. It is immediately apparent that the Megatek delivers significantly poorer data than the Sigma 1603, despite significantly higher quantum efficiency of the chip. The first spectrum has a S/N of about 20, whereas the second spectrum provides a S/N of about 240. To find the cause, one should analyze a dark image in the shortest possible exposure time, a so-called bias. Figures 10.9 and 10.10 show the corresponding standard

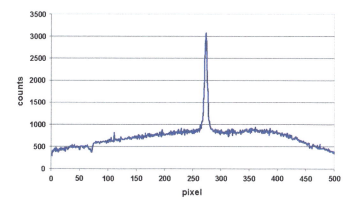

Fig. 10.7   Uncalibrated 60s raw spectrum of the Be star γCas at Hα, taken with a Tektronix TK 1024 chip in a Megatek camera

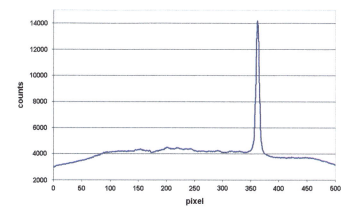

Fig. 10.8   As in Fig. 10.7 but for a Kodak KAF 1603ME Chip in a Sigma 1603 camera

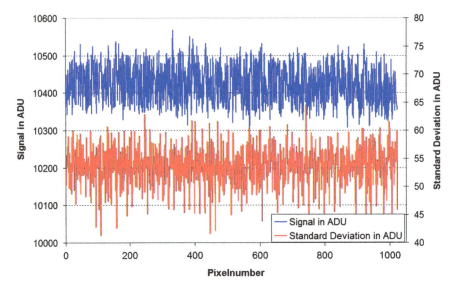

Fig. 10.9 Signal and standard deviation of the bias of a Megatek camera

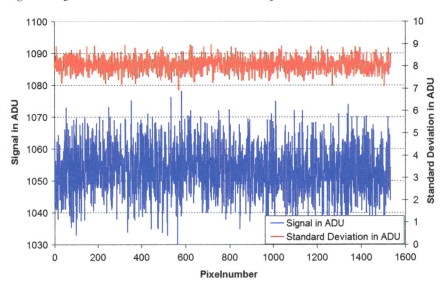

Fig. 10.10 Signal and standard deviation of the bias of a Sigma camera

deviations of the two biases. Now we can make a statement about the average of all pixels, the corresponding standard deviations and the manufacturer's specified gains (Table 10.1).

Table 10.1 Gain, average
and standard deviation for a
Megatek and a Sigma camera

	Megatek	SIGMA
Gain	$2.3e^-$ /ADU	$1.8e^-$ /ADU
Average value Σ	10,500 ADU	1,050 ADU
Standard deviation σ	55 ADU	8 ADU

Obviously, the Megatek provides an approximately seven times larger dispersion of the pixel value σ. With the gain we can now calculate the standard deviation per pixel via $N_E = \text{Gain} \cdot \sigma$. The chip scattering of the Megatek camera converted into electrons is thus nine times larger ($55 \cdot 2.3/8 \cdot 1.8$) than that of the Sigma-camera. Thus we can conclude that the Sigma shows a better noise performance than the Megatek. The read-out electronics of the Megatek introduces additional noise which explains the lower S/N at the same conditions. Before one decides for a specific camera, one should request a bias field from the manufacturer for analysis. The excellent deep-sky images in the advertising brochures of different CCD cameras are not sufficient for an evaluation.

10.5 A Simple Sensor Model

CCD sensors may be parameterized with a simple linear model. To define all necessary parameters, we first specify them in Table 10.2. Using a CCD one can directly measure S_{ADU}, S_{bias}, N_{ADU} and R_{ADU}. The parameters S_{e^-} and N_{e^-} are initially unknown but can be determined from the directly measurable signal variables. We begin with the relation of the total number of charge carriers

$$S_{e^-} = g \cdot S_{\text{ADU}} \qquad (10.3)$$

with g the conversion factor of counts into charges in units of e^-/ADU, which must be known. To illustrate the relationship in more detail, we now investigate the relation between the accessible measurement signal S_{ADU} and the charges generated by photons. n_P photons illuminate one CCD pixel during the exposure time t. On average, $\eta \cdot n_p = n_{e^-}$ of these photons are transferred to electrons when assuming the quantum efficiency η. In addition, the afore-mentioned electronics-related signal offset still exists. In units of ADU we obtain

$$S_{ADU} = S_{\text{bias}} + \frac{1}{g} \cdot n_{e^-}. \qquad (10.4)$$

The model thus assumes a linear amplification with amplification factor $1/g$ in ADU/e^-. The number of electrons is subject to the physical laws of statistical fluctuations and is Poisson distributed. This allows the estimation of the overall noise N_{e^-} from the variance of the number of all electrons S_{e^-}:

$$N_{e^-}^2 = var(S_{e^-}) = \sigma^2(S_{e^-}).$$

Table 10.2 Parameters for a simple sensor model

Parameter	Description	Unit
S_{ADU}	Measured overall signal	ADU
S_{bias}	Measured signal offset (bias)	ADU
S_{e-}	Overall signal	Electrons
N_{ADU}	Overall noise	ADU
N_{e-}	Overall noise	Electrons
R_{e-}	Read-out noise	Electrons/pixel
R_{ADU}	Read-out noise	Electrons/pixel
g	Gain	e^-/ADU
n_p	Number of incoming photons	
n_{e-}	Number of electrons converted from photons	
η	Quantum efficiency	$e^-/photon$
t	Exposure time t	Seconds

Multiplying Eq. 10.4 with g, we obtain the total number of electrons. By substituting into the variance operator we obtain

$$N_{e-}^2 = var(g \cdot S_{bias}) + var(n_{e-}).$$

The variance $var(g \cdot S_{bias})$ corresponds exactly to the variance of the bias field, and thus the square of the read-out noise R_{e-}^2. Due to Poisson statistics the variance of the electrons generated per photons is $var(n_{e-}) = n_{e-}$. Thus we obtain for the total noise in units of "electrons" (charge carriers)

$$N_{e-}^2 = R_{e-}^2 + n_{e-}. \qquad (10.5)$$

10.6   Measuring the Read-Out Noise and the CCD Gain

The signal comparison of two different CCD cameras can be carried out exclusively in the comparison of the generated electrons. To do so, however, the "conversion factor" or "system gain" must be known. Usually this is specified in the data sheet of a camera, but the value may vary individually. Therefore, we give a simple method to determine the main CCD parameters "read-out noise" R and "gain" g by measurement. The gain is the ratio of the number of electrons generated and the signal of the AD converter in ADU, i.e., if one knows the number of incident photons and multiplies this number with the quantum efficiency η one can directly determine the gain from the measured signal. However, the number of incident photons is difficult to access. Here photon statistics helps. Accordingly, we first use Eq. 10.5 and then the conversion equations $N_{e^-} = g \cdot N_{\mathrm{ADU}}$ for the total noise and $R_{e^-} = g \cdot R_{\mathrm{ADU}}$ for the read-out noise. With Eq. 10.4 we replace also n_{e^-} and find

$$g^2 \cdot N_{\mathrm{ADU}}^2 = g^2 \cdot R_{\mathrm{ADU}}^2 + g \cdot (S_{\mathrm{ADU}} - S_{\mathrm{bias}})$$

or

$$N_{\mathrm{ADU}}^2 = \frac{1}{g} \cdot (S_{\mathrm{ADU}} - S_{\mathrm{bias}}) + R_{\mathrm{ADU}}^2. \qquad (10.6)$$

This is a linear equation in which the variance of the noise is plotted against the bias-corrected signal. A fit to the data points provides the slope and the intercept and thus the gain factor and the square of the read-out noise. Figure 10.11 shows the parameters for the Megatek. The reciprocal of the slope provides a gain of $g = 5.8e^-/\mathrm{ADU}$ and the square-root of the intercept provides the read-out noise of $R_{\mathrm{ADU}} = 36\mathrm{ADU/pixel}$. Converted to electrons we obtain $R = 5.8e^-/\mathrm{ADU} \cdot 36\mathrm{ADU/pixel} = 209e^-/\mathrm{pixel}$.

According to the data sheet of the Tektronix CCD (Table 10.3) the maximum read-out noise is around $10e^-/\mathrm{pixel}$. The CCD camera and its electronics exceed this value by a factor of 20. The data sheet also provides information about the CCD full-well capacity. Typically, a maximum of $350{,}000e^-/\mathrm{pixel}$ can be generated. With a gain of $5.8e^-/\mathrm{ADU}$ the camera can thus be driven to approximately 60,000 ADU before the signal goes into saturation.

Note: When determining the parameter slope for increasing exposure times it may be difficult to accurately estimate the variance. For the determination of the

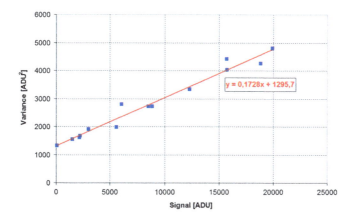

Fig. 10.11 Determination of the read-out noise and the conversion factor for the Megatek camera. The signal level is increased by increasing the exposure time. The scatter of the data points indicates unfavorable illumination conditions; however, the correlation coefficient of the fit is 98 %. Thus, the linear function predicted with Eq. 10.6 is estimated with great certainty

Table 10.3 Device specifications from the SITe Tk 1024 × 1024 CCD data sheet

	Minimum	Typical	Maximum
Format		1,024 × 1,024 pixels	
Pixel size		24 μm × 24 μm	
Imaging area		24.6 mm × 24.6 mm	
Dark current (MPP), 20 °C equiv.		50 pA/cm^2	70 pA/cm^2
read-out noise front		5 electrons	9 electrons
read-out noise back		7 electrons	10 electrons
Full well signal	300,000 electrons	350,000 electrons	
Output gain	1.0 μV/electron	1.5 μV/electron	
CTE per pixel	0.99995	0.99999	

Measured at −45 °C unless otherwise indicated, 45 kpixels/s and standard voltages using a dual-slope CDS circuit (8 μs integration time)

slope the sensor should be illuminated as homogeneously as possible.[5] However, this can be difficult to achieve. There are, for example, dust particles in the vicinity of the CCD plane, which can lead to undesirable signal variations (see Fig. 10.13). In this case the variance is overestimated. The same situation exists in the case of very short exposure times. "Hot" or "dead" pixels[6] can also increase the Gaussian noise of the bias fields (Fig. 10.12). Under these circumstances one estimates the so-called "fixed pattern noise" of the CCD.[7] To circumvent this technical problem one can take two flat fields of identical exposure time directly after the other and subtract them from each other. The difference frame only contains a noise component increased by a factor of $\sqrt{2}$, which can be corrected by correspondingly halving the variance. Newberry (2000) presents a robust method for determining the gain and the system noise of a CCD camera through the following procedure:

"For each intensity level, do the following:

1. *Obtain two images in succession at the same light level. Call these images A and B.*
2. *Subtract the bias level from both images. Keep the exposure short so that the dark current is negligibly small. If the dark current is large, you should also remove it from both frames.*
3. *Measure the mean signal level S in a region of pixels on images A and B. Call these mean signals S_A and S_B. It is best if the bounds of the region change as little as possible from one image to the next. The region might be as small as 50 × 50 to 100 × 100 pixels but should not contain obvious defects such as cosmic ray hits, dead pixels etc.*

---

[5]One therefore also refers to a so-called "flat field".

[6]Hot pixels have a very high dark current and dead pixels have a greatly reduced light sensitivity.

[7]The term fixed pattern noise is misleading because the pixel errors are spatially constant, thus having no temporal variation as noise and thus are correctable.

Fig. 10.12 The bias field of an Apogee AP8P camera with a Tektronix 1024 × 1024 CCD. In the upper right corner one can see a number of hot pixels. The standard deviation of the whole flat frame (fixed pattern noise) is 5.4 ADU/pixel. Analysis of an undisturbed area delivers a noise of 3.7 ADU/pixel. For a given gain of about $4.5e^-$/ADU one obtains a read-out noise of $16.6e^-$/pixel, which corresponds well to the specified system noise $15e^-$/pixel given in the data sheet of the AP8 system (Source: Magnus Schneide, Hamburger Sternwarte)

4. *Calculate the ratio of the mean signal levels as* $r = S_A/S_B$.
5. *Multiply image B by the number r. This corrects image B to the same signal level as image A without affecting its noise structure or flat field variations.*
6. *Subtract image B from image A. The flat field effects present in both images should be cancelled to within the random errors.*
7. *Measure the standard deviation over the same pixel region you used in step 3. Square this number to get the Variance. In addition, divide the resulting variance by 2.0 to correct for the fact that the variance is doubled when you subtract one similar image from another.*
8. *Use the Signal from step 3 and the Variance from step 7 to add a data point to your Signal-Variance plot.*
9. *Change the light intensity and repeat steps 1 through 8."*

In a somewhat modified form this procedure has been applied to data obtained with an OES AlphaMaxi. Instead of increasing the light source intensity the exposure time has been increased successively at constant illumination from 0.01 to 20 s. Despite cooling, the CCD still showed a relatively strong dark current, which increased with increasing exposure times (see Fig. 10.5). The image frames show the typical flat-field structures already after 0.5 s exposure time (Fig. 10.13). Take two such flat field images, one after the other; subtract the dark frame and form the difference of the resulting flats; one then obtains the frame in Fig. 10.14, which can

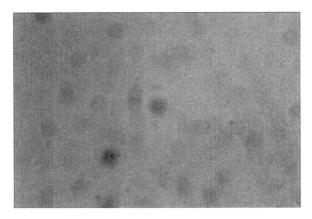

Fig. 10.13 Five-second flat field image obtained with an AlphaMaxi (KAF 400) camera. The standard deviation of the flat is 399 ADU

Fig. 10.14 Difference of two flat fields obtained with the same CCD camera and 5 s exposure time. The structures are completely eliminated. The standard deviation is 173 ADU. The difference between the two flats causes an increased noise level by a factor of $\sqrt{2}$ that has to be subsequently scaled

now be analyzed for noise. The resulting frame has an increased noise by a factor of $\sqrt{2}$ which can be corrected by halving the variance. When plotting the estimated variance per exposure time against the averaged signal, as described in the above procedure, one obtains the plot in Fig. 10.15. From the parameters one estimates a gain of $g = 1e^-/\text{ADU}$ and a read-out noise of $R_{e^-} = \sqrt{3333} = 58e^-/\text{pixel}$.

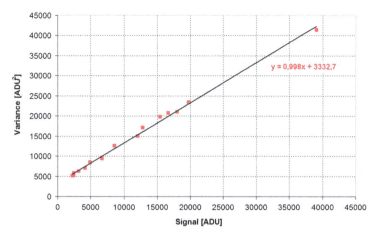

Fig. 10.15 Gain and read-out noise of an AlphaMaxi camera

10.7   The Signal-to-Noise Ratio and Detection Threshold

Modern astronomical detectors provide a signal approximately proportional to the incident photon flux. Their behavior is almost linear. However, since the actual source signal of the target object and background effects are both recorded, the source signal must be extracted from this signal combination. This is done by the data reduction, as we discuss in Sect. 12.3. To detect source signals, one would think that these signals should only be stronger than all the noise together. However, this is not the case because the mean of the background can be eliminated by simple subtraction of bias and dark and is thus irrelevant.[8] The decisive question is whether and how much the actual signal should be stronger than the noise fluctuations around this mean (see Fig. 10.16).

The background noise is composed of signals of photons which independently occur at a constant rate. Therefore, the noise is described by Poisson statistics (see Appendix B.2). When collecting N photons with a detector in the time interval t, the probability to produce n photoelectrons within the detector area within this time is

$$p(n,t) = \frac{(Nt)^n e^{-Nt}}{n}.$$

In Poisson statistics the standard deviation is simply the square root of the mean number of events.

$$\sigma = \sqrt{Nt}. \tag{10.7}$$

---

[8] At this point we neglect the sky background.

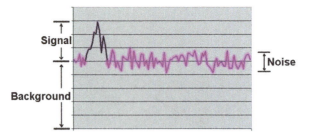

Fig. 10.16 Synthetic emission-line with background noise

To illustrate the accuracy of a measurement or to express the probability that a detected signal is actually real, the signal S is compared with the background noise N by the signal-to-noise ratio S/N. It is the signal strength in units of its standard deviation. S/N is therefore unitless and is expressed as a number. The noise level of any instrument normally follows a standard Gaussian distribution. If one wants to detect any signal, it must be stronger than the noise level (here, by definition, 1σ, often also 3σ and 5σ). This corresponds to a signal-to-noise ratio of 1 (the signal level is just at the 1σ value of the Gaussian distribution). For 68 % of all cases the noise level is then smaller than the signal and is then detectable. For 32 % of all cases, the signal is not detected because it is obscured by noise (wings of the Gauss function). Thus, from three signals, about two are detected in average. For a $S/N = 2$ the signal level is at 2σ. For that case, 95.4 % of all noise signals are smaller and 4.6 % are greater than the signal. Thus, only about one of twenty signals is not detected. Analogously about 0.3 % of all signals are not significantly measured for a $S/N = 3$. Statistically only one of 300 signals is lost. Therefore, a $S/N = 3$ is usually regarded as a reliable detection level. However, unexpected and non-quantified errors often distort a signal and even highly accurate measurements carried out with the utmost care are disturbed in their quality. Therefore, only $S/N = 5$ is often considered reliable. In this case, the probability that a signal is spurious is about one in a hundred thousand. The reason to increase the S/N above 3 or 5 is the dynamic range of the corresponding measurement. If, for example, an object is detected with a $S/N = 100$, one can separate 100 different intensity values and the weakest signals should reach at least an $S/N = 1$. If we want to transfer this idea to spectroscopy, we must keep in mind that a spectrum provides average information per unit exposure time. The spectral pixel-to-pixel noise is the result of different pixel values. These values fluctuate stochastically in time around a mean value. Therefore, the above percentage distribution also applies to an entire pixel ensemble. If for example an emission line of $S/N = 1$ is exactly covered by 100 pixels, only 68 pixels in average significantly record the line flux above the 1σ level. If we now consider the typical noise sources and calculate the background fluctuations according to Eq. 10.7 we can then use the signal-to-noise ratio of a source, which covers n pixels on the detector perpendicular to the dispersion direction[9]:

$$S/N = \frac{I \cdot t}{\sqrt{(I + B \cdot n + D \cdot n)t + R^2 \cdot n + var(B \cdot n \cdot t)}} \tag{10.8}$$

with I the total number of photo-electrons measured from the target source during the integration time t, $B \cdot t$ the total number of photo-electrons per pixel and time obtained from the background, $D \cdot t$ the dark current of the detector in electrons per pixel and time, R the read-out noise per pixel, and $var(B \cdot n \cdot t)$ the variance of the *total* background per pixel and time. The last term represents the uncertainty in the determination of the background, which is not shown on the photon statistics, e.g., true variations in the background or inaccurate measurements. Since the source signal shows statistical fluctuations as all other sources, which are indistinguishable from other noise sources, I is also in the denominator. In addition, the background can usually be determined by a large number of pixels or a long time, so that the term $var(B \cdot n \cdot t)$ can often be neglected.[10] The above description of S/N is valid for all astronomical observations and contains the entire instrumental behavior. This includes the efficiency of the detector and its noise performance as well as the efficiency of the telescope and instrument optics. This is because I and D are determined by the number of photo-electrons on the detector and not by photons collected by the telescope aperture. Only this term and the exposure time required for a particular measurement define the limiting magnitude of an observation. Based on Eq. 10.8 we now can distinguish three different cases, which define their own monitoring strategies.

1. **Limitation by photon noise**. If the observed object is bright and its photon noise dominates over other noise sources, S/N simplifies to

$$S/N = \sqrt{I \cdot t}$$

and a higher system sensitivity is achieved by increasing the exposure time.
2. **Limitation by detector noise**. If the target as well as the background signal are weak and if the detector noise dominates, S/N simplifies to

$$S/N = \frac{I \cdot t}{\sqrt{D \cdot n \cdot t + R^2 \cdot n}}.$$

---

[9]This implies that the spectral noise is due to fluctuations of a single pixel.

[10]This is usually not valid for only a few pixel elements and if the background varies (e.g., infrared measurements). Because $var(B \cdot n \cdot t)$ is often difficult to determine, one should invest corresponding effort to stabilize all background signals.

This is applicable for spectroscopy of stars in medium and high spectral resolution because the noise per spectral element decreases by correspondingly increasing the spectral resolution. In this case (and also for the following case, the limitation by background noise) the S/N increases proportional to the number of detected target photons. In practice, individual spectral images should then be exposed as long as possible. The corresponding S/N grows linearly with $\sqrt{t}$ until the respective background signal is large enough so that its fluctuations are larger than the read-out noise. Furthermore, now the maximum sensitivity of the telescope depends on its aperture (mirror surface), and the time to perform a measurement with the same quality, decreases with the fourth power of the mirror diameter.

3. **Limitation by background noise**. This case occurs for ground-based observations, when the target is weak and the natural background (e.g., zodiacal light) dominates the noise. Then we have

$$S/N = \frac{I\sqrt{t}}{\sqrt{B \cdot n}}.$$

Thus, the measurement accuracy is directly proportional to the square root of the exposure time, and inversely proportional to the square root of the background. This applies to minimal instrumental effects (detector, thermal noise). Image quality enhancements can only be reached by increasing the telescope aperture, improving the image quality (adaptive optics) or the reduction of the background (e.g., reduction of the thermal emission of the optics).

To first approximation, the number of photo-electrons of the background B is given by

$$B \approx \Phi\sigma^2 A \cdot t$$

with Φ the photon flux of the background (sky) per square arcsecond, σ the angular diameter of the source in the sky, A the telescope surface and t the exposure time. Already for telescopes larger than about 20 cm aperture the image of a point source σ is defined by the local seeing, and thus the solid angle within which the signal and background photons occur, remains constant. An increased telescope aperture enlarges both photon fluxes of signal and background with the same rate. Since the signal is linearly dependent on the photon flux and hence on the mirror surface area, the noise, however, by the square root of the photon flux, the S/N increases only with the telescope aperture (see also Sect. 10.3.1).

Suggested Readings

- Berry, R., Kanto, V., Munger, J., *The CCD Camera Cookbook*, Willmann-Bell, 1994
- Howell, S.B., *Handbook of CCD Astronomy*, 2nd ed., Cambridge University Press, 2006
- Martinez, P., Klotz, A., *A Practical Guide to CCD Astronomy*, Cambridge University Press, 2008
- Wodaski, R., *The New CCD Astronomy*, Multimedia Madness, 2002

Chapter 11
Remarks on Fiber Optics

A Short Story

For our first spectrograph we had been thinking about a fiber feed. But we had no idea about that and we could only imagine having problems. Therefore, we installed the spectrograph at the Newtonian focus, fed through a small prism. We did not use fibers. The first measurements delivered nice spectra but strangely enough, the orientation of the spectrum on the chip was not parallel to the CCD pixel rows. And when the telescope was moved along its two coordinates, the spectrum moved somewhere, but not orthogonally to the chip. Everything was somehow wrong. Unfortunately, we had forgotten that the prism had to feed the light parallel to the optical axis of the spectrograph but we had more focused on a good-looking orientation when mounting the device. In any case, the beam was not parallel to the telescope. We were very sure that we had understood the optical geometry and had already completed all mechanics. Who needs fiber optics anyway?

11.1 Basic Remarks

When a spectrograph or any other measuring system is positioned and fixed at the telescope focus we add additional masses, and both the optical and the mechanical behaviour of the optical system is changed. This also applies to the bending behaviour (i.e. flexure) of the spectrograph with respect to the telescope guiding and positioning. In short—additional masses disturb both the static and dynamic behavior of the entire measurement system. This can be compensated for by a mechanical reinforcement of all parts. However, strengthening in turn leads to more fiber-stress and deflection, which may eventually exceed the mechanical system boundaries. Telescope movements and resulting mechanical stress in the spectrograph in turn lead to spectral shifts that potentially affect the measurements.

© Springer-Verlag Berlin Heidelberg 2015
T. Eversberg, K. Vollmann, *Spectroscopic Instrumentation*, Springer Praxis Books,
DOI 10.1007/978-3-662-44535-8__11

In addition, temperature and humidity fluctuations can also cause spectral shifts. Hence, it might be useful to disconnect the instrument from the telescope and run it in a different local environment.

Instead of flanging the instrument to the telescope, fiber optics can transmit light from the telescope focus to the spectrograph, which can be placed in a stable, fixed position away from the telescope. Beside circumnavigating the above problems, fiber optics have even more advantages. The observer can be almost completely independent of the telescope and the spectrograph is not subject to weight or volume limits. This includes using techniques to achieve a very high spectroscopic temperature stability and thus avoiding corresponding spectral shifts. In addition, the observing efficiency can be increased when many target objects can be measured simultaneously (Sect. 8.3.4). This is also possible with slit masks (Sect. 8.3.5), though, but their use is limited with respect to flexible target acquisition. A useful application is the bundling of several very thin fibers with which the star image is taken and which are then arranged in a line, so that they act as a narrow slit (Sect. 8.3.2). In addition, the same fiber bundle can be used for "Integral Field Spectroscopy", in which this bundle entry is arranged to analyze different parts of an extended object in the sky, e.g., nebulae, galaxies (Sect. 8.2.4). Meanwhile, the fiber technique is widely used and is applied to large as well as relatively small telescopes (Sect. 8.4.3).

If fiber optics are well designed and furnished they can be installed without major problems. However, fiber optics have also some disadvantages. These include transmission loss, F-number change of the incident beam, poor subtraction of the sky background in the data reduction (see Chap. 12), intrinsic fiber noise and a sophisticated handling when high sensitivities are to be achieved.

11.2   A Few Words About Fiber Types

Optical fibers are made of glass, quartz or plastic and have typical diameters of 10–2,000 μm. The two relevant astronomical fiber types are multimode and single-mode fibers. Multimode fibers include step-index and gradient-index fibers.

11.2.1   Multimode Fibers

Multimode fibers have a fiber core and a relatively thin cladding. Different possible light paths (modes) are transmitted simultaneously depending on the angle of incidence for different single incoming rays. The so-called primary mode is guided along the optical fiber axis and, therefore, passes the fiber fastest. The larger the fiber diameter the greater the number of transmitted modes. In Sect. 11.6 we will describe fiber modes in detail, in particular the interesting fact that the data quality increases steadily with the number of modes.

Step-Index Fibres In astronomy step-index fibers are usually used. These are fibers having a core and a mantle (cladding) whose refractive index differs from that of the core. Such multimode fibers out of silica are manufactured with core diameters of 50–2,000 μm. This material has high transmissivities of more than 90 % over a large wavelength range even for fiber lengths of 20 m.

Gradient Index Fibers In contrast to step-index fibers with a constant refractive index thereby having straight light paths, gradient index fibers have a decreasing refractive index towards the cladding. The optical path in such fibers is therefore curved. Because of reduced transmissivity at blue wavelengths and very long fibers normally used, these fiber types are not favorable for astronomical applications. Note that for fiber lengths of less than 10 m, the efficiencies of gradient index fibers are similar to those of step-index fibers. However, gradient index fibers show unfavourable behaviour with respect to focal-ratio degradation (FRD, Sect. 11.5). They need to be fed with very small f ratios and have reasonable efficiencies of more than 90 % only for output f-ratios below f/2. In order to transfer a sufficient amount of light, the beams must arrive at the fiber input under a very small angle of incidence. In addition, the photometric distribution at the fiber output depends strongly on that at the input. Hence, such fibers react easily to photometric shifts. This introduces a spectral wavelength shift at the spectrograph depending on the stellar image position at the fiber input (Sect. 11.7).

11.2.2   Single Mode Fibers

As the name suggests, in a single mode fiber only the primary mode is transmitted. This is achieved by a very small fiber diameter of only a few wavelengths of the incident light. The reflection paths for the light within the fiber are so short that the light cannot follow longer paths and runs only along the optical axis. Multimode fibers widen incident pulse signals by different light paths, only allowing single-mode fibers for the measurement of coherent signals in interferometry.

Gradient-index and single-mode fibers have only limited use in spectroscopy and we refer to the literature for details on this. In the following we only discuss step-index fibers.

11.3   Step-Index Fundamentals

First, we consider the light feed into the fiber and its boundary conditions. To avoid vignetting at the fiber aperture the arriving beam must be larger than the focal image size d of the seeing disk ω in the focus. The focal image size depends on the focal length of the telescope f_{tel}.

$$d = f_{tel} \cdot \tan \omega \tag{11.1}$$

Hence, a one arcsec seeing disk in the sky is imaged at the focus of a telescope of 10 m focal length with a diameter of $d \sim 50\,\mu\mathrm{m}$. However, as this is only the half-width of the seeing disk and its wings are not included, one has to choose a fiber diameter of at least $100\,\mu\mathrm{m}$ not to lose too much light. On the other hand, the fiber diameter at the entrance to the spectrograph acts as a virtual aperture slit, thereby determining the spectral resolving power. So, especially for large telescope F-numbers (Cassegrain, Gregory etc.) one has to take different requirements for the fiber feeding and the desired spectral resolving power with respect to the fiber aperture into consideration. For the highest possible resolving power one either has to choose possibly a small fiber diameter or use an image slicer at the fiber output (see Chap. 9). One now might suppose that the simplest solution for both problems is the choice of a short telescope focal-length to keep the seeing disk as small as possible. This is a realistic approach even for large telescopes. As we will see in Sect. 11.5, however, this can cause an angle of incidence penetrating the fiber optics too steeply, resulting in significant light loss. However, if the focal angle of incidence is decreased by a large f-ratio of the telescope, the seeing disk focal image becomes larger again and one potentially loses light already at the fiber input according to Eq. 11.1. In principle, however, one must ensure that the entire incident light is available for transfer through the fiber. Already these simple considerations show that an optical fiber must be considered in its entirety including input and output to ensure optimal light transfer from the telescope to the instrument.

Since the diameter of a fiber is large compared to the wavelength of (optical) light the physics of wave optics can be neglected and the corresponding requirements are presented by considerations of geometrical optics.[1] The meridional incident light falling into the fiber under a certain angle is reflected at the boundary layer between core and cladding, and is further guided along the medium (Fig. 11.1). However, the depiction of the beam path is a simplification. It only shows beams which are lying in the so-called meridional plane. In reality, however, the majority of rays fall into the fiber so that they never intersect the fiber optical axis (Fig. 11.2). This is the case when the surface of the fiber aperture and the incident light plane are not perpendicular to each other. If one then sends a collimated beam (e.g., a laser) at an angle into the fiber (Fig. 11.3) its light leaves the fiber at the same angle but

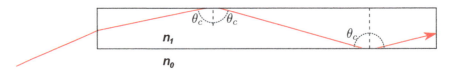

Fig. 11.1 Beam path in fiber optics for meridional rays which cross the optical fiber axis

---

[1]This applies to the present consideration of step-index fibers with discretely different refraction indices for core and cladding. For the so-called mono-mode fibers with core diameters of the order of the light wavelength and for gradiant-index fibers with a continuous refraction index profile, we refer to the corresponding literature.

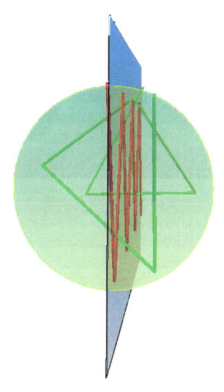

Fig. 11.2 Light path in fiber optics. *Red*: Meridional beams cross only the optical fiber axis. *Green*: Skew rays do not cross the fiber axis (Avila and Guirao 2009)

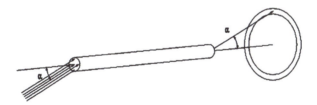

Fig. 11.3 Skew rays from a parallel incoming beam (Avila and Guirao 2009). Because of the finite extent of the laser beam, so-called skew rays are generated in different directions, leaving the fiber into different directions depending on the number of internal reflections

now into all directions. If the exit beam is projected onto a screen, it causes a light ring (Fig. 11.4) whose sharpness depends on the surface roughness of the fiber aperture (Sect. 11.9). However, the internal reflections do not occur without loss. To estimate this loss we consider a fiber which is surrounded only by air, taking the corresponding refractive indices n_1 and n_0 for the fiber and for the surrounding air, respectively, into consideration. The path length of a beam which falls into a fiber depends on the fiber diameter D, its length L and the reflection angle θ_c inside the

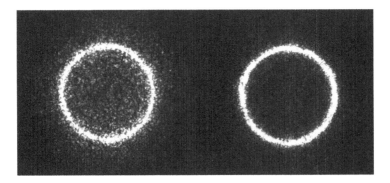

Fig. 11.4 Light distribution according to Fig. 11.3 with respect to focal-ratio degradation for fibers with different surface roughness of 245 nm rms (*left*) and 8 nm rms (*right*). Incoming angle $\theta_i = 8°$, $\lambda = 6{,}330$ Å (Haynes et al. 2008). The sharpness of the light ring depends on the surface roughness of the fiber apertures (Sect. 11.9)

fiber. According to Snell's law the latter depends on the angle of incidence θ_i again. It is

$$l = \frac{L}{\cos \theta_c} = \frac{L}{\sqrt{1 - \sin^2 \theta_c}} = \frac{L}{\sqrt{1 - \left(\frac{n_0}{n_1}\right)^2 \sin^2 \theta_i}}$$

$$l = \frac{n_1 \cdot L}{\sqrt{n_1^2 - n_0^2 \cdot \sin^2 \theta_i}}.$$

This results in the total number of reflections

$$N = \frac{l}{D/\sin \theta_a} \pm 1 = \frac{n_0 \cdot L \cdot \sin \theta_i}{D \cdot \sqrt{n_1^2 - n_0^2 \cdot \sin^2 \theta_i}} \pm 1.$$

The summand ± 1 depends on the location where the light enters the end of the fiber. Since N is usually large, the ± 1 can be neglected. We calculate 6,500 reflections per meter for a fiber of 50 µm diameter with $n_1 = 1.6$ and $n_0 = 1$ (air). One must heed the biggest concern that as little light is lost as possible at each reflection. This is impossible for a reflective coating out of aluminum. The reflectivity of aluminum reaches only up to 95 % over the entire visible spectrum (Fig. 11.5). If the fiber cladding would be coated with aluminum, about 5 % light would be lost at each reflection and for several thousand reflections the fiber optics would be entirely opaque. Even if only 1 per mill of light would be lost at each reflection, the fiber efficiency would thus be about 0.1 % per meter. Since the light leaking out of the fiber is lost for further analysis, it is therefore mandatory to reach total internal reflection as closely as possible at the internal fiber layers, because corresponding reflection effects at the fiber edge drastically reduce the efficiency of every fiber. In principle, one could use air as the fiber core surrounding medium, but

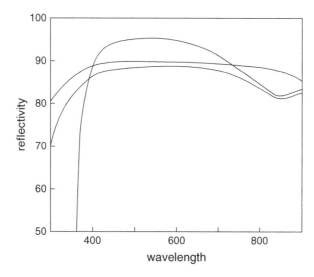

Fig. 11.5 Reflection characteristics for different aluminum coatings (Burle Technical Memorandum)

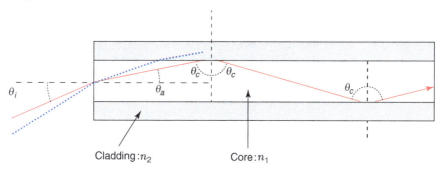

Fig. 11.6 Beam path in fiber optics. Beams inside (*red*) and outside (*blue*) the acceptance cone (from Wikipedia)

then one would have a very thin and mechanically weak fiber. It would break easily under mechanical stress. For this reason, fiber optics are coated with a material whose refractive index is smaller than that of the fiber core (step-index fibers). The cladding makes the fiber stable, keeps the surface clean and avoids light loss due to scattering. The structure of a step-index fiber and the corresponding light path in its interior is shown in Fig. 11.6. The light loss at the interface of two transparent media is less than 0.001 % per reflection. For the above 6,500 reflections, this results in an efficiency of around 94 % per meter. Since the number of reflections decreases linearly with the diameter of the fiber, this value rises to above 98 % for a fiber diameter of 200 μm. Because a fiber length of 1 m is usually not sufficient, the final lengths must be taken into account. Typically the diffraction indices of a fiber and its cladding are $n_1 = 1.4571$ and $n_2 = 1.44$, respectively. For total reflection according to Snell's law we obtain $\sin \theta_c = n_2/n_1$.

There is a maximum angle θ_{max} for the light feed, which is still accepted by fiber optics for the supply from air $(n = 1)$. This will be described by the numerical aperture (*NA*).

$$NA = n_0 \cdot \sin \theta_{max} = \sqrt{n_1^2 - n_2^2}$$

Using the values for n_1 and n_2 we get a numerical aperture NA $= 0.22$. Light with an angle of incidence at the fiber input smaller than θ_{max} will be transferred by the fiber. If this angle is exceeded, light will leak through the fiber cladding. For fast telescope optics with smaller f-ratios, it may then very well be problematic to transfer light by fiber optics of certain diffraction indices. The angle θ_{max} necessarily defines the smallest possible focal-ratio acceptable for the optical fiber. According to Fig. 11.6 we will then obtain $NA = n_0 \cdot \sin \arctan(D/2f) \approx D/2f$ and thus

$$f/D \approx \frac{1}{2NA}.$$

The above typical values $n_1 = 1.4571$ and $n_2 = 1.44$ deliver $f/D|_{krit.} \approx 2.72$. This f number is thus the absolute limit for fiber coupling. If the applied value is smaller, light will leak through the fiber cladding and is lost for further analysis.

11.4   Transmission and Attenuation

The attenuation as a function of the fiber path length is relatively small compared to all other effects mentioned above. The light transmitted through the fiber I_t follows an exponential law and of course depends on the intensity at the fiber input I_0 as well as the fiber length L.

$$I_t = I_0 \cdot e^{-\kappa \cdot L}$$

κ represents the so-called damping constant and is often referred to as fiber loss. Fiber loss is typically specified in decibels per kilometer (!) and follows the relation

$$\kappa_{dB} = -\frac{10}{L} \log \frac{I_t}{I_0}$$

However, attenuation increases with shorter wavelengths. Avila and Guirao (2009) have performed extensive laboratory measurements and determined some fiber transmission values. For instance, a 20 m long silica step-index fiber transmits approximately 96 % of the incident light at 600 nm. This value decreases to about 84 % at 400 nm. Reflection losses within the fiber and losses at the fiber input are not included here. Figures 11.7 and 11.8 show the total transmission for two different fiber optics of 10 and 26 m length. The transmission significantly falls at

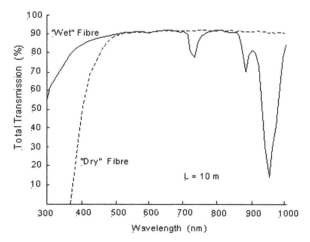

Fig. 11.7 Total transmission of two different 10 m fiber optics (Avila et al. 2007)

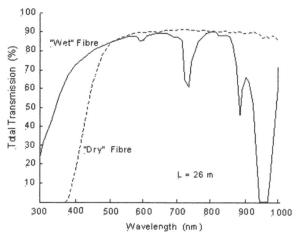

Fig. 11.8 Total transmission of two different 26 m fiber optics (Avila et al. 2007). The transmissivity is reduced due to an increasing number of internal reflections. Light intensities at shorter wavelengths are additionally reduced due to attenuation

wavelengths below about 400 nm and decreases above about 900 nm. When working outside the visible wavelength range one may use very short fiber optics to ensure appropriately high transmission values.

11.5   Focal-Ratio Degradation (FRD)

A significant reduction in efficiency is caused by the sensitivity of optical fibers to any kind of stress as the fiber bends or twists, causing compressive forces and variations in fiber dimensions as a function of fiber length. Such forces will directly affect every reflection angle inside the fiber and lead to increasing numerical apertures at the output, thus degrading the F-number. This so-called focal-ratio degradation (FRD) is decisive in the adjustment process of the spectrometer and

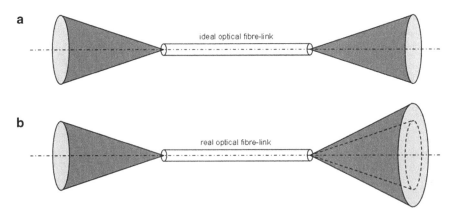

Fig. 11.9 Degradation of the F-number by bending of fiber optics. (**a**) Ideal fiber without FRD loss. (**b**) Real fiber with an expanded output cone (Feger 2012)

telescope. FRD is the main source of light loss in fiber-coupled instruments. Figure 11.9 shows the principle for an optical fiber with and without bending.

The magnification of the beam aperture at the fiber output, inevitable for all fiber optics, has the consequence that a collimator diameter is too small if only designed for the incident beam. This means that light is potentially lost at the collimator behind the fiber output. Enlargement of the beam aperture depends also on the angle of incidence at the fiber input.[2] This efficiency loss is so substantial that it should never be neglected. If a fiber is not bent and perfectly manufactured, the numerical apertures at the input and output are the same; the F-number is not degraded. However, if the fiber is subjected to any stress, the numerical aperture is reduced and the F-number becomes smaller on the way through the fiber. This is because rays almost reaching the entry angle Θ_{max} first leak over the cladding on their way through the stressed fiber (Sect. 11.2). This also takes place in principle with rays entering at a smaller angle of incidence. They are initially further reflected inside the fiber, but with continuous increase of the corresponding reflection angle. If such a ray reaches the critical angle after multiple reflections and after a longer path in the fiber, the ray also leaks out of the fiber. Hence, the closer a ray is to the angle α_{max}, the sooner it leaks out of the fiber because of FRD. The efficiency decreases. In addition, the numerical aperture between telescope and spectrograph fiber must be correctly adjusted. If not, light can be lost at the fiber input if it exceeds the maximum angle α_{max} or due to poor adjustment of the spectrograph collimator at the fiber exit. On the input side "fast" focal-ratios f/3 to f/7 (depending on the fiber diameter) perform the task best (Ramsey 1988). FRD is then primarily generated by micro-bending. Laboratory studies of Avila et al. (2007) quantify the

---

[2]In the following we only consider incident rays inside the acceptance angle. All other beams leak through the fiber cladding.

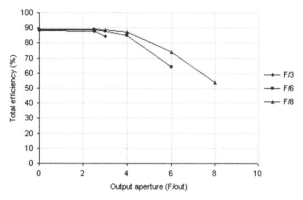

Fig. 11.10 Efficiency for different F-numbers of the incident beam (f/3, f/6 and f/8) plotted against the F-number of the emergent beam (Avila et al. 2007)

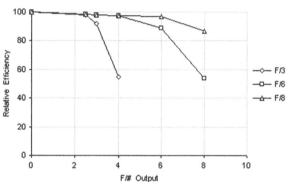

Fig. 11.11 Normalized relative efficiency for a 200 μm fiber (Avila et al. 2007)

light loss caused by FRD. Figure 11.10 shows the fiber efficiency for three different F-numbers of the incident beam plotted against the F-number of the emergent beam. If a beam falls with f/8 into a fiber, its f-number is degraded to about f/3 to f/4. An f/8 collimator at the fiber output would then only collect a fraction of the light cone—in this case about 55 % (the effect increases for even greater F-numbers). To collect more light the collimator should be adapted accordingly, for the present case ideally with f/3. This effect increases dramatically when stress is applied to the fiber. For illustration Fig. 11.11 shows the normalized efficiency for a 200 μm fiber, depending on the incident and emergent beams and Fig. 11.12 shows the same efficiency when strongly bending the fiber.

The measurements show that FRD is significant especially for large F-numbers (slow lenses). For an f/8 collimator at the fiber output the relative light loss increases from about 15 to 55 %. This is an efficiency reduction of about 50 %! However, FRD loss is reduced with decreasing F-number. For an f/6 beam at the input and a corresponding output f/6 collimator the efficiency is reduced from approximately 90 % to about 65 % (reduction by almost 30 %) and for an f/3 beam from about 95 % to about 90 % (reduction by about 6 %). Apparently FRD affects small F-numbers at the input (fast lenses) much less than large F-numbers (slow optics). In order to keep FRD losses small one should feed with small F-numbers. Therefore, for most telescopes additional feeding optics must be used to adjust the F-number at the fiber

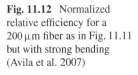

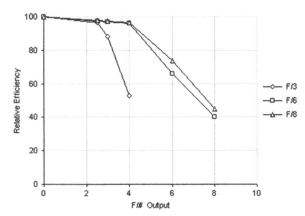

Fig. 11.12 Normalized relative efficiency for a 200 μm fiber as in Fig. 11.11 but with strong bending (Avila et al. 2007)

input accordingly.[3] This should be done considering the respective acceptance cone so that all beams are fed into the fiber input, so that no light leaks out of the fiber cladding.

On the other hand, as much light should be picked up at the end of the fiber through a collimator with a small F-number. With respect to the spectrograph optics, not only the collimator then becomes large, but necessarily also the optical grating and the camera optics. This can significantly affect the financial budget for small telescopes. To counteract this, the F-number in turn could then be increased with intermediate lenses (reduction of the beam diameter). However this only occurs for the price of correspondingly reduced resolving power. Since the fiber aperture regularly serves as a virtual slit, increasing the image of the fiber aperture would affect the virtual slit size. To avoid a larger slit width and hence reduced resolving power, one could in turn use a thinner fiber. Then however, the telescope focal length might be too large (seeing disk diameter). As for the system telescope—spectrograph—CCD pixel, fiber optics with their parameters must therefore be harmoniously integrated into the overall system (we will discuss a special form of optical fibers in Sect. 11.8). For the measurement of FRD we refer to Avila et al. (2007).

So far we have considered only fibers of circular cross-section. Feger (2012) performed comparative FRD studies of fibers with circular, square and octagonal cross-sections (Fig. 11.13). Figure 11.14 shows the F-number at the fiber output for different cross-section geometries as a function of F-number at the fiber input. Obviously the octagonal test fiber suffers significantly lower FRD compared to conventional fibers. Interestingly, this does not apply to the fiber with a square cross-section, which is inferior to all other fiber shapes. However, the square fiber has a very thin cladding and is therefore subject to increased stress (Feger 2012). The successful use of an octagonal fiber has been carried out at the SOPHIE echelle spectrograph at the 1.93 m telescope of the Observatoire de Haute-Provence

---

[3]Fiber optics at the ESO VLT F/13.41 Cassegrain focus without adjusting optics would make any reasonable measurement impossible.

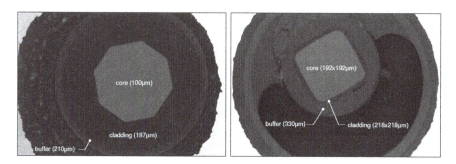

Fig. 11.13 Image of $100\,\mu$m octagonal core fiber and a $200 \times 200\,\mu$m square core fiber from CeramOptec mounted into a stainless steel ferrule (Feger 2012)

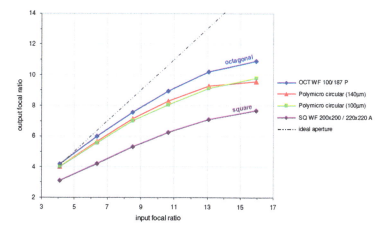

Fig. 11.14 The F-number at the fiber output as a function of the F-number at the input for various fiber cross-section geometries. The *dashed line* represents the ideal case without FRD losses (Feger 2012)

(OHP) performing measurements of stellar radial velocities. With this configuration (SOPHIE plus octagonal fiber = SOPHIE+) the accuracies could be significantly increased from 5–6 to 1–2 m/s (Bouchy et al. submitted).

11.6 Fiber Noise

A well-defined monochromatic beam introduced into a thin fiber is repeatedly reflected on its way through the fiber. Its intensity distribution at the fiber end is not uniform but exhibits "discrete" maxima (speckles), spread stochastically over the fiber aperture (Fig. 11.15). This is due to the many reflections of light rays within the non-perfect fiber as well as scattering at the entrance aperture and the resulting random reflection effects. The resulting thousands of rays introduced by different path lengths are called modes. Modes are produced by interfering light paths. They provide a structured light distribution at the fiber output. Blue light suffers more reflections than red light, thus generating a greater number of modes.[4] The emergence of modes is not an effect of FRD: modes are also formed in fiber optics which are not subjected to any stress. However, their distribution reacts highly sensitively to the smallest deformations or changes in position far below stress levels where FRD occurs. The motion of a telescope that feeds light to a spectrograph via fibers causes time-variable fiber modes. In practice they are inevitable. The reason is the illuminated entrance aperture of the fiber and its small irregularities, unavoidable even for a meticulously polished surface (Fig. 11.16).

Each sub-region of the entrance surface is a source of scattered light rays. Depending on the surface roughness and the wavelength, the phase differences between individual modes can be up to several thousand radians. Therefore, the field intensity at each observation point P is determined by the mutual interference of a large number of wavelets that are coherent but not in phase when leaving the fiber. Stochastic effects from interference minima and maxima are inevitable at the fiber exit. If the observer moves to another place P, the path lengths of the scattered rays change and a new intensity distribution is introduced. The result is a far-field distribution of bright (constructive interference) and dark (destructive interference) spots. It could now be argued that the speckles disappear behind an

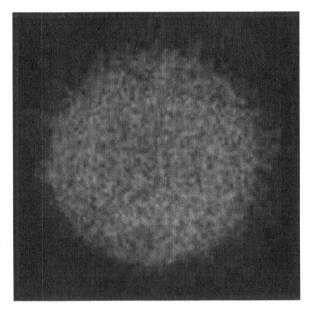

Fig. 11.15 Speckle distribution of monochromatic light of a 660 nm-HeNe laser at the end of an optical fiber (Grupp 2003, reproduced with permission ©ESO)

---

[4]The color-dependent transit time differences additionally introduce a reduction of signal quality and reduced bandwidth (modal dispersion). This can be avoided by the use of gradient-index fibers. Their fiber core has continuous refractive indices decreasing towards the cladding. However, these effects can be neglected in astronomical spectroscopy.

Fig. 11.16 Setup for the observation of fiber speckles in the far field (Sharma et al. 1981)

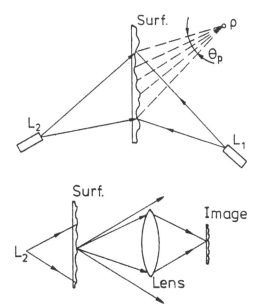

Fig. 11.17 Setup for the observation of fiber speckles by an imaging lens. The lens cannot image all rays of phase space (Sharma et al. 1981)

imaging lens if the rays of a small area can be appropriately mapped. But that is not the case, since we cannot correspondingly image all rays even with a very large lens (Fig. 11.17). Any finite aperture, even if it is perfect, provides only a portion of all phase information and therefore produces abrupt intensity changes in the image.

Because of their stochastic nature in time (telescope movement) fiber modes are an additional source of noise. Completely analogous to the stochastic seeing of the night sky, the light at the fiber end is not uniform, but rather shows an intrinsic fiber noise over the entire aperture. The beams exit the fiber at different geometrical positions in time and are then vignetted stochastically during exposure through the optical slit. Therefore, for a given apparatus the attainable theoretical signal-to-noise ratio is significantly reduced. The number of modes at the fiber end depends on both the F-number at the fiber input and the wavelength. The bigger the input F-number (more reflections) and the shorter the wavelength, the more modes that develop in the fiber. Since the total light flux remains constant, few modes have a stronger effect on the S/N than many modes. As the number of modes at red wavelengths is much smaller than at blue wavelengths, one observes especially for longer wavelengths a significant reduction of the theoretically achievable S/N. In addition, one must keep in mind that the intensity fluctuations that produce the modes together with the vignetting slit, are greater the more the slit is vignetting. That is, especially at high resolution, achieved with a narrow slit, is fiber noise increased. Figure 11.18 shows a measurement result obtained with the FOCES spectrograph at Calar Alto Observatory. At red wavelengths the measured S/N is reduced by a factor of 2.5 compared to theoretical photon statistics. To minimize this massively disturbing effect and achieve maximum S/N in high-resolution spectroscopy mode, it is obvious to artificially generate as many modes as possible,

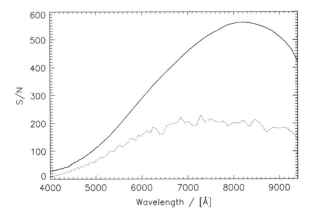

Fig. 11.18 Theoretical S/N according to photon statistics for the FOCES spectrograph (*full line*) and measured S/N (*dotted line*) (Grupp 2003, reproduced with permission ©ESO)

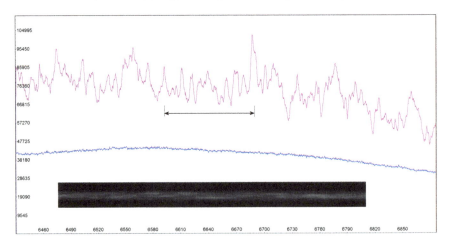

Fig. 11.19 "Shaking out" fiber modes. *Top*: Spectrum of a continuous tungsten lamp fed into a spectrograph with a 50 micron step-index fiber of 2 m length. A 30 micron slit is positioned at the fiber entry to enhance the effect by stopping modes of large input angles. *Bottom*: 2D image of the top spectrum. The respective wavelength interval is indicated. *Middle*: Same as top but now with shacking the fiber during exposure. *Note*: The latter is obtained with a 20 micron slit again enhancing any mode effect. Even with lower entry angles the modes are entirely eliminated (Waldschläger 2014)

which have an adequate distribution during exposure time and thus ensure minimal fluctuations due to slit vignetting. This is done by non-harmonic motions of the fiber during recording. Amazingly, one can "shake out" the modes by hand. Even better, though, is an apparatus that produces non-harmonic motion of the fiber (for example, double-pendulum) to provide a stochastic distribution of the modes. This is shown in Fig. 11.19 for the light of a tungsten lamp.

11.7 Photometric Shift and Scrambling

Grating diffraction depends on the respective angle of incidence α. For that reason, any angle of incidence of stellar light into the telescope aperture in dispersion direction also affects the position of the spectrum in the dispersion direction. Since the seeing disk at the telescope focus is subject to movements by seeing or tracking errors, the stochastic position and thus wavelength changes limit the accuracy of any spectral measurement to a few meters per second. This is a non-negligible effect for high-precision measurements of radial velocities. Given the symmetric light distribution after the passage of a collimated light beam through a fiber (Fig. 11.4) one would assume that the location information is completely lost at the fiber output and photometric offsets are thus fully compensated. However, this is not the case; the effect is true also for observations with fiber optics. On the other hand, optical fibers have the property to minimize this displacement. This is called "photometric scrambling". We consider the displacement of the stellar image d in relation to the aperture of the fiber optic D as well as the relation between the corresponding shift of the PSF in the form of an emission line s and its $FWHM$. The ratio of these two relations is referred to as scrambling gain (Fig. 11.20). It indicates the stability of the spectral resolution element in relation to the geometric variations of the stellar image at the fiber input aperture. The larger G the more accurate are the measurements.

$$G = \frac{d/D}{s/FWHM}$$

Avila et al. (2007) show that the spectral accuracy G can dramatically be increased by simple means. According to their investigations, a non-bent, 3-m-long 60 µm fiber optic cable, which is illuminated by a 50 µm light beam of f/2.5 has a scrambling gain of 120. If this fiber is bent so that the FRD increases by 20 %, the gain increases to 500. The effect is even more dramatic for a non-bent, 3-m-long 600 µm fiber, which is illuminated by a 250 µm beam at f/3. For this case, a continuous bending of 15 mm radius causes a scrambling increase of 150–3,000. Hence, when the fiber is subject of moderate stress, the spectral measurement accuracy increases by a factor of 20! A close examination of scrambling and its gain, respectively, for different fiber cross-sections has been performed by Avila et al.

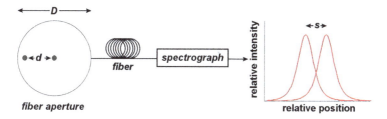

Fig. 11.20 Photometric shift

(2010) (also note the references therein). They point out that the number of fiber modes depends on both the square of the fiber diameter and the numerical aperture $n = 1/2(\pi NA \cdot D/\lambda)^2$. Therefore, thicker fibers with a larger NA provide a more uniform light distribution at the fiber output and the scrambling efficiency should grow. And indeed, the measurements seem to confirm this aspect. However, at least for a round fiber, the F-number at the input fiber has no influence on the scrambling gain.

In Sect. 11.5 we have shown that mechanical stress causes light loss by increased FRD. In Sect. 11.6 we described the possibility to mechanically "shake out" modes and thus improve the achievable spectral S/N. Interestingly, one can also increase the scrambling gain by stress, i.e. by squeezing the fiber (this in turn by increased FRD). In their summary Avila et al. (2010) point out that circular fibers deliver the highest scrambling gain if they have a large NA with large diameters and if they are as long as possible. It is then advantageous to squeeze the fiber as they describe it in their text.

11.8   Tapered Fibers

In Sect. 11.5 we explained that an optimum light throughput despite FRD is only possible through an adequate adjustment of telescope optics, fiber diameter and collimator. Since the fiber normally acts as a virtual slit, its initial diameter should be kept as small as possible. On the other hand, it may be difficult to get a seeing disk of a long focal length telescope without light loss into the fiber aperture. It may be too large. One can increase the fiber diameter and use lenses at the fiber output to limit the collimator aperture and that of all subsequent optical elements. However, this is potentially compensated for by reduced spectral resolution.

To circumvent these problems, tapered fibers can be used. In this special form fiber diameters vary linearly with their length: they are tapered (Fig. 11.21). One then can work without additional optics at the fiber ends for feeding the rays of large focal length and thus minimize FRD. As can be easily seen, tapered fibers can adjust the F-numbers of telescope and spectrograph without additional optics. That means no losses in the matching optics otherwise occur. In addition, tapered fibers are thermally and mechanically very stable. Because of their design tapered Fibers

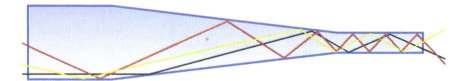

Fig. 11.21 Sketch of a tapered fiber and three light rays which fall into the fiber under different angles (Avila and Guirao 2009). Note that the input F-number is larger than the output F-number

introduce FRD. As a result, further FRD is significantly reduced in the consecutively following fiber optics.

In addition, the spectral resolution can be significantly increased with a smaller fiber aperture in front of the collimator of the spectrograph. This, however, for the price of a larger F-number proportional to the decrease of the diameter (due to conservation of geometrical etendue). The spectrograph collimator must be adjusted accordingly if one does not want to pay for the increase of spectral resolution by loss of light. Hence, a condition for the use of tapered fibers is thus again an exact geometric adaptation to the spectrograph. In addition, the fiber diameter must be perfectly conical in order to avoid light loss. This in turn makes the fiber expensive.

The etendue is conserved in geometrical optics and according to $A \cdot \Omega = A' \cdot \Omega'$ it does not change during the passage of a beam through an optical system. A is the area of the solid angle Ω emitted from the object, and A' is the area of the corresponding image which receives the solid angle Ω'. Therefore, if we consider a tapered fiber whose length is much larger than its diameter, the F-number of the emergent beam $F\#_{out}$ is calculated by

$$F\#_{out} = \frac{d_{in}}{d_{out}} \cdot F\#_{in},$$

with $F\#_{in}$ the F-number of the incident beam and the input and output apertures d_{in} and d_{out}.

When adjusting the telescope to the spectrograph via a tapered fiber one should of course not neglect that any F-number reduction at the fiber end may require an adjustment of the collimator. If this is not done, a small collimator optical unit may introduce vignetting.[5] However, tapered fibers are never perfectly conical and are also subject to FRD, which reduces the efficiency. Tables 11.1 and 11.2 show measurements of Avila and Guirao (2009) for two different tapered fibers, coupled to a 35 m fiber.

The reason for the increased efficiency of the 400–100 μm fiber is its larger diameter and the optimum exit beam ratio with respect to reduced FRD. The authors also found that a shorter fiber length of 20 m only provides little higher efficiency than a 35 m fiber.

Table 11.1 Efficiencies depending of the input and output ratios for a tapered fiber coupled to a standard fiber of 35 m length

$F\#_{in}$–$F\#_{out}$	Total efficiency
$F/15$–$F/5$	35 %
$F/11$–$F/3.7$	53 %
$F/8$–$F/2.7$	65 %

---

[5]Here again, in spectroscopy the overall system and the harmonic tuning of all parameters is essential.

Table 11.2 Efficiencies
depending on the input and
output ratios for a
400–100 μm tapered fiber
coupled to a standard fiber of
35 m length

$F\#_{in}-F\#_{out}$	Total efficiency
$F/15-F/3.75$	53 %
$F/13-F/3.25$	57 %
$F/11-F/2.75$	61 %
$F/8-F/2$	70 % (extrapoliert)

11.9 Lenses for the Telescope Link

To reduce FRD, telescopes of f/8 and higher require a corresponding F-number
reduction. This is typically the case for a feeding at f/4. This adaptation is performed
by lenses of appropriate focal length and can be done in two ways.

11.9.1 Imaging the Star onto the Fiber Aperture

To project a stellar image in the telescope focus onto an optical fiber one needs
appropriate optics. If we choose a single lens, different regions of the stellar image
will fall at different angles into the fiber. This is independent of whether the focus
of the intermediate lens is before or behind the optical fiber (Fig. 11.22). The

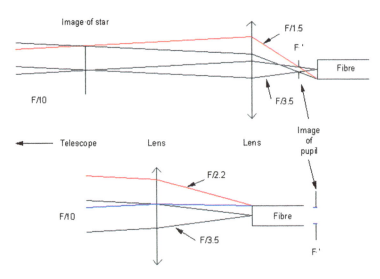

Fig. 11.22 Imaging a star on a fiber with a single lens (Avila and Guirao 2009). *Top*: Real stellar
image. *Bottom*: Virtual stellar image

position of the intermediate optics between the telescopic and fiber aperture may be determined from the lens equation.

$$d_{fiber} = f \left(1 + \frac{F\#_{fiber}}{F\#_{tel}} \right)$$

and

$$d_{tel} = \frac{F\#_{tel}}{F\#_{fiber}} d_{fiber}$$

with $F\#_{tel}$ the telescope F-number, $F\#_{fiber}$ the desired F-number at the fiber input, f the focal length of the lense as well as d_t and d_f the distances between lens and telescope focus and between lens and fiber aperture, respectively.

We copy here the practical example of Avila and Guirao (2009) for a single thin lens with a 10 mm focal length and an f/10 telescope. To reduce the beam to f/4 at the fiber input one has $f = 10$ mm, $F\#_{tel} = 10$ and $F\#_{fiber} = 4$. Then the distances from the lens to the fibre and to the telescope focal plane are $d_{fiber} = 14$ mm and $d_{tel} = 35$ mm, respectively.

However, this approach poses problems. The lens provides different angles of incidence at the fiber aperture for different places of origin in the extended focal stellar image. Rays from the edge reach the fiber with a smaller F-number and are subject to FRD as opposed to the central rays. That is, the edge beams leave the fiber with a substantially lower F-number than the central rays, and can be vignetted by the spectrograph collimator. One can avoid vignetting of slower rays through a larger focal length of the lens, but one must then accept a larger size of the lens unit. Alternatively one can increase the diameter of the collimator, but then one has to enlarge all other optical spectrograph elements as well. The best solution for the problem is a lens system which places the telescope pupil at infinite distance (Fig. 11.23). Hence, one simply uses two lenses whose F-numbers match the F-numbers of the telescope and fiber and whose foci lie in the telescope focus or the fiber aperture, respectively. Then we have $F\#_t / F\#_f = f_{lens1} / f_{lens2}$. With respect to

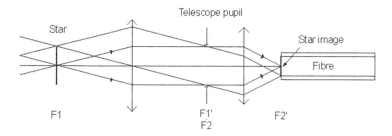

Fig. 11.23 Best coupling between telescope and fiber for a real stellar image (Avila and Guirao 2009)

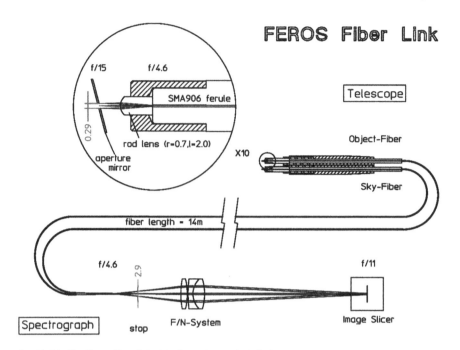

Fig. 11.24 The Feros fiber link (Kaufer and Pasquini 1998)

spherical and chromatic aberration a simple light feeding unit is the combination of two doublets.

A practical application for imaging a virtual stellar image is the FEROS fiber link (Sect. 8.4.4). There one has two pinhole micro- or rod-lense apertures of 0.29 mm diameter fed into two fibers of 100 μm diameter. The input F-number is 4.6 and thus minimizes FRD. The rod lens is only 2 mm long and has been adhered directly to the fiber which in turn is held by a SMA connector. The fiber output on the spectrograph side only was polished. The light at the fiber output is focused by a lens F/N system which converts from f/4.6 to f/11 (Fig. 11.24). There the fiber image of 240 μm diameter is fed into an image slicer (Sect. 9.4).

11.9.2 Imaging the Telescope Pupil onto the Fiber Aperture

Imaging the telescope pupil (i.e. the telescope aperture) on the fiber aperture has the advantage that a single lens is sufficient. The focal plane of the normally used plan-convex lenses often exactly matches the backside of the lens. This has the advantage that the lens can be directly glued onto the fiber aperture and thus two optical surfaces and their respective efficiency losses can be eliminated. However, the diameter of the seeing disk depends on the focal length of the lens and needs

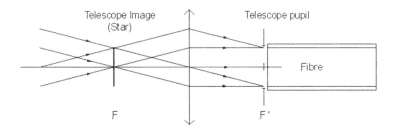

Fig. 11.25 Coupling a telescope to a fibre by projecting the pupil on the fibre input end. The telescope pupil is in the image plane of the lens and the star is in the object focal plane of the lens (Avila and Guirao 2009)

to be adapted to the fiber diameter. As a result, the focal lengths are of the order of only 1 mm and such lenses have a very small size. The principle beam path is shown in Fig. 11.25.

This opto-mechanical solution requires that the focus of the feeding lens coincides with the telescope focus while the telescope pupil is imaged by the lens exactly in the fiber aperture. Only in this configuration, will all seeing disk rays fall parallel into the fiber. That is, the aperture of the beam passing through the fiber is defined by the seeing disk. The focal length of the microlens f and the F-number of the fiber end beam $F\#_{fiber}$ can be calculated by

$$f = d \cdot F\#_{tel} \tag{11.2}$$

and

$$F\#_{fiber} = f/s \tag{11.3}$$

with the focal diameter of the seeing disk s, the F-number of the telescope $F\#_{tel}$ and the fiber diameter d. As an example we calculate for a telescope of 0.8 m aperture and an F-number $F\#$ of 12.5 the necessary fiber diameter in order to transfer a seeing disk of 1 arcsec.

We summarize the Eqs. 11.1–11.3 and we obtain

$$d = \frac{F\#_{tel}}{F\#_{fiber}} \cdot F_{tel} \cdot \tan \omega.$$

If we now achieve an F-number of 4 at the fiber input to reduce FRD ($F\#_{fiber} = 4$), the diameter of the seeing disk on the fiber aperture is calculated to be 150 μm. Again using Eq. 11.2 we then get $f = 1.875$ mm for the focal length of the transfer lens.

It is better to image the telescope pupil onto the fiber aperture than to image the stellar image. This is because one can use a single lense and, hence, avoid additional light-loss at optical surfaces (Fig. 11.25). However, this requires very small optics and higher opto-mechanical complexity. If the use of micro-lenses should be avoided for technical reasons, imaging the star on the fiber (Fig. 11.23) is the better choice.

In addition to an adequate and effective opto-mechanical coupling it must be ensured that no light is lost at the fiber entry and exit. When light passes through a medium, its intensity is reduced by the refractions at the material surfaces for a given wavelength (see Fresnel equations in Appendix B.3). To lose as little light as possible at the transition into the fiber medium, the fiber surface has to be polished. An analogy are scratched lenses. The fewer scratches the better.[6] For the polishing process lapping films are used. These are typically polyester sheets that are coated with well-defined minerals of sizes from 0.01 to 50 µm. After the polishing process they provide uniform and well-defined input and output surfaces to minimize light loss. Two fiber surfaces after the polishing process are shown in Fig. 11.26.

However, if micro-lenses are glued to the fiber aperture, the requirement for the fiber surface quality is reduced. The adhesive used should have a refractive index as close as possible to that of the fiber. Thus, small scratches can be compensated for by the bonding process. However, we point out that all necessary bonding processes should be carried out with great care and accuracy as they can potentially affect the efficiency of the overall system. As illustrated in Sect. 11.5 fiber stress and respective FRD result in a significant reduction of fiber efficiency. Shrinkage during the hardening processes with too much glue between the microlens and fiber can cause so much traction force in the first few millimeters that FRD is greatly enhanced.

Fig. 11.26 Two fiber apertures under the microscope. *Left*: After polishing with 12 µm lapping sheet, peak height = 270 nm rms. *Right*: After polishing with 0.3 µm lapping sheet, peak height = 8 nm rms (Haynes et al. 2008)

---

[6]The internal transmissivity is usually well over 90 % in the visual wavelength range.

11.10 Opto-Mechanical Coupling

To image a star or the entire telescope pupil onto the fiber input an opto-mechanical coupling is needed, which not only performs this task with sufficient precision but also ensures appropriate control options. In addition to the exact mechanical positioning a corresponding optical interface should exist to accomplish this task for both the static (position) and dynamic (tracking) cases. This means in practice that it must be possible to control the positioning of star/pupil on the fiber end, first visually or via detector and then actively maintain this position during dynamic telescope guiding. Thus, a stable fiber coupler at the telescope focus is required. These devices should also be compact, especially for small telescopes. At this point we can only discuss the essential principles and refer to the literature for further considerations.

A basic principle which fulfills the above requirements has become popular for many telescopes. The fiber aperture is imaged on a guiding CCD camera through a lens and mirror combination. Even though the stellar light must completely fall into the aperture for maximum efficiency and thus should be invisible in reflection for optimal transmission, the image wings (often described as "corona") can be used for guiding the telescope with an appropriate software.[7] The corresponding standard optical principle is shown in Fig. 11.27 and a sample implementation is shown in Fig. 11.28. The telescope focus is projected onto the fiber aperture via intermediate optics (Sect. 11.9.1) or via a micro-lens (Sect. 11.9.2), which is typically encased by a SMA connector.

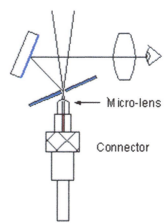

Fig. 11.27 Typical opto-mechanical fiber coupling with guiding optics. Imaging the telescope pupil with a micro lens; see Avila and Guirao (2009)

Micro-lens

Connector

---

[7]At this point we neglect the possibility of guiding the telescope with external guiding optics.

Fig. 11.28 Simple fiber
coupler and guiding unit
(Feger 2012)

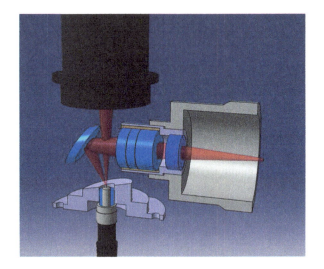

Fig. 11.29 Optomechanical
fiber coupling with beam
splitter and stellar image on
the fiber optics (Avila and
Guirao 2009)

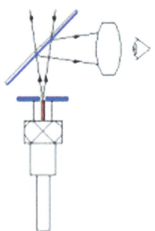

The reflecting surface in which the fiber pinhole is embedded (aperture plate) is
slightly tilted with respect to the optical axis[8] and reflects the image again towards
a complementary tilted mirror. This second mirror again reflects the image towards
the imaging optics (compare with Fig. 11.24). One can basically waive a second
mirror, but then a larger physical size has to be accepted. This principle can be
modified for various mechanical and optical conditions, wherein the above working
principle remains the same.

As an alternative to the two folding mirrors one can also position a 45° beam
splitter in front of the pinhole (Fig. 11.29). The advantage of a visible stellar image

---

[8]Given the very small fiber aperture, the pinholes potentially act as light tunnel introducing
potential vignetting when tilted. The thickness of the pinhole plate should be thinner than the
diameter of the pinhole.

on the fiber aperture, however, is paid by reduced efficiency, depending on the optical beam splitting ratio. As above, one can construct variants that follow the principle of decoupling via beam splitter.

For a more detailed consideration of the various options for opto-mechanical couplings we refer again to Avila and Guirao (2009) who have extensively considered the issue of fiber coupling.

11.11 Resume

If one wants to appreciate the benefits of fiber optics one must take into account the following effects:

1. The seeing disk and thus the telescope focal length must be adapted to the diameter of the optical fiber in order to avoid vignetting of the fiber aperture.
2. Step-index fibers are superior to other fibers for astronomical applications with regard to their optical properties.
3. To avoid leaking light at the fiber cladding, all incoming beams must be within the acceptance cone.
4. To minimize efficiency reduction due to focal-ratio degradation, the F-number at the fiber input should be as small as possible. The F-number of the telescope may then have to be reduced with lenses accordingly.
5. Fiber attenuation is relatively small compared to all other interfering effects, as long as the near UV is not considered.
6. Fiber noise and thus a potential S/N reduction at high resolving power can be minimized by applying non-harmonic movements to the fiber during exposure time.
7. Tapered fibers are helpful for slow telescope optics ($\geq F/4$). For faster telescopes a normal fiber works fine because FRD decreases with the F-number.
8. If mechanically possible, the telescope pupil rather than the stellar image should be imaged onto the fiber aperture. Although mechanically simpler the latter leads to reduced efficiency.
9. To avoid loss due to stray light the fiber input and output surfaces must be polished.
10. To avoid additional FRD the mechanical design must be accurate and should avoid potential forces on the fiber.
11. For very precise radial velocity measurements the photometric shift and its reduction must not be neglected. For this case, mechanical stress (squeezing) can improve the scrambling gain, but this in turn causes FRD.

However, each optical element reduces the overall efficiency. Hence, before one decides in favour of a fiber coupling of the spectrograph to the telescope, the total number of additional lenses, mirrors, and interfaces should be considered and included in the determination of the efficiency of the overall system.

Our general considerations have a practical character as is also expressed in many previous studies of fiber optics. The results depend to a large extent on the material used and are not generally transferable to other work. New tests can be carried out with new results for various spectroscopic instruments. This is unsatisfactory from a physical point of view, pointing out that the optical fiber manufacturing processes are not uniformly reproducible. A fundamental theoretical approach based on the wave equation, however, is provided by the extensive work of Bures (2008), which we also list as further reading at the end of this chapter. We point out expressly once more that a reduced efficiency of the entire system greatly affects the data quality, which is represented in this context by the signal-to-noise ratio (S/N). As already noted above this does not depend on the mirror surface of the telescope from A but only on the diameter D (see Sect. 4.1.1 and Chap. 10). Today maximum fiber efficiencies of more than 70 % can be achieved in specialized laboratories. With respect to the important signal-to-noise ratio, this effectively reduces a 1 m telescope aperture to a 70 cm one. So, before one decides to use fiber optics, one should make clear that the maximum achievable brightness of the star can be significantly reduced if the above effects are not sufficiently allowed for.

Suggested Readings

- J. Bures, *Guided Optics: Optical Fibers and All-fiber Components*, Wiley-VCH, 2008
- A.B. Sharma, S.J. Halme & M.M. Butusow, *Optical Fibre Systems and Their Components*, Springer, 1981
- Blog of the CAOS group http://spectroscopy.wordpress.com/

Chapter 12
Data Reduction

A Short Story

Thomas had already reduced spectroscopic data with both MIDAS and IRAF. It was never in question that his choice would be these professional tools. However, since he had to run Linux, of course a second computer was needed. He installed the other operating system on a second PC and he also bought a switch to use only one keyboard and monitor. This configuration sometimes crashed, though, but there was no better solution. Thomas did not yet know virtual machines which can emulate a LINUX-MIDAS environment...

12.1 Open Tools for Reliable Results

As for all astronomical applications spectroscopic data must be freed from interferences with other light sources. This so-called data reduction made extraordinary progress by the introduction of modern semiconductor detectors and the appropriate computer technology. An important aspect is stable and reliable software whose contents and procedures are understandable for the user. It is often unacceptable for the researcher to work with a "black-box" of unknown algorithms. The appropriate software should be "open source" in order to know what is really done with one's own data. In addition, astronomers usually work in groups. Local Area Networks (LAN) or even Wide Area Networks (WAN) are common techniques. Multitasking is a "must" and therefore appropriate operating systems must be network adapted. If one retains the "open" character, all modern operating systems meet the above requirements. One of the first operating system which was network adapted, was UNIX. It meets all the above requirements and is free of charge. Therefore,

© Springer-Verlag Berlin Heidelberg 2015   439
T. Eversberg, K. Vollmann, *Spectroscopic Instrumentation*, Springer Praxis Books,
DOI 10.1007/978-3-662-44535-8_12

researchers are still working with UNIX-based operating systems such as SUN SOLARIS or LINUX.

Astronomers often develop their own programs, writing them mostly in C, C^{++} or FORTRAN. But the data reduction packages there are also self-developed and freely accessible for everyone. As already mentioned in Chap. 6, the two most powerful tools are MIDAS (Munich Image Data Analysis System) of the European Southern Observatory (ESO) and IRAF (Image Reduction and Analysis Facility) of the National Optical Astronomy Observatories (NOAO) in the United States. Both packages run only on UNIX or LINUX. But we also mentioned that it is possible to work with Linux under Windows. We will now describe the appropriate steps and procedures in more detail.

Meanwhile the programming language PYTHON (see "Astropython" at http://www.astropython.org/resources) is the language-of-choice for upcoming astronomers. Note that MIDAS and IRAF have cross-overs to PYTHON in the form of PYMIDAS (http://www.eso.org/sci/software/esomidas/) and PYRAF (http://www.stsci.edu/institute/software_hardware/pyraf). We also refer to the ESO "Scisoft Collection" package, which includes installations of MIDAS and IRAF for LINUX and MacOSX (https://www.eso.org/sci/software/scisoft/ and http:// scisoftosx.dyndns.org).

12.2   LINUX and Windows

As an alternative to UNIX, one can today also work with Windows. The operating system is widely used, well-developed and very convenient for the average user. Nowadays one can find various programs for all applications on the market. They include catalogues, planetariums, image processing, and programs for the control of telescopes and CCD cameras. However, almost all the sources are not public domain in the Windows world. Meanwhile, there are also a few Windows programs for spectroscopic data reduction, developed exclusively by amateurs. They can perform complete data reduction, including wavelength calibration and rectification, as well as line-profile analysis. However, different programs might be an obstacle for the desirable harmonization of reduction procedures. In contrast to professional research with well-tested tools, the internal algorithms of amateurs are often not sufficiently explained and remain unknown to the user. Therefore, the procedures are not necessarily interchangeable between different observers. For coordinated campaigns it is hence recommended to leave the data reduction to just one person and to consider spectral parameters such as equivalent widths or radial velocities with great care.

MIDAS and IRAF are the most advanced, most powerful and most developed reduction packages, but they only run on UNIX/LINUX. We believe that they represent by far the best solution for any data reduction. In Appendix A we exemplarily describe the main reduction steps under MIDAS.[1]

Working with LINUX and MIDAS opens the most powerful toolbox for any astronomical application—from stellar photometry to the analysis of quasars. But even if one prefers Windows, one can simulate Linux with a so-called emulator and work with MIDAS or IRAF. A Linux emulator mimics a LINUX operating system and runs UNIX/LINUX programs. The program results are identical to the corresponding real LINUX environment. The most popular LINUX emulator for Windows is the free VMware program. It works with a Dynamic Link Library (DLL) for Windows delivering a respective emulation layer. The program provides all functionalities of a LINUX operating system and all necessary tools.

12.3 CCD Reduction

The goal of reducing CCD data is the elimination or minimization of all instrumental effects. In addition, information will be obtained about all noise sources, thus allowing an assessment of data quality and how it is affected by noise. In addition to the original 2D image containing the stellar spectrum we need various correction data for an accurate CCD reduction. They are:

1. A dark image (dark field), which is recorded under the same conditions as the image with the original spectrum (exposure time, temperature etc.) but without being exposed (shutter closed). It includes the internal noise behavior of the chip (read-out noise of the amplifier and thermal noise).
2. A so-called flat field (image) for the global compensation of sensitivity variations on the CCD chip. They are caused by the entire instrument setup (CCD, dust, vignetting etc.). To create a flat field one exposes the chip as evenly as possible, e.g. by a uniformly illuminated area in the observatory or an internal instrument lamp. The exposure time is chosen such that the chip does not saturate and remains in the upper linear range of its sensitivity.
3. A so-called bias field, which is basically a dark field with the shortest possible exposure time (normally zero). It contains all the potential read-out level of the CCD amplifier.

Dark and bias must be subtracted from the raw image and the result is then divided by the net flat. Schematically, the instrument algorithm is therefore

---

[1]IRAF is just as powerful as MIDAS but has a different task structure. Both programs were developed to satisfy all requirements for reliable astronomical data reduction. MIDAS was developed by the Image Processing Group at ESO in Garching whereas IRAF was written by the IRAF programming group of the National Optical Astronomy Observatories (NOAO) in Tucson, Arizona. Both packages use line commands.

$$\text{Reducedimage} = \frac{raw - dark - bias}{flat - dark - bias}.$$

The above reduction scheme is merely the basic procedure for dealing with the data. We explicitly refer to Massey and Hanson (2010) for more details. They say *"The basic premise... is that one should neither observe nor reduce data by rote. Simply subtracting biases because all of one's colleagues subtract biases is an inadequate reason for doing so. One needs to examine the particular data to see if doing so helps or harms. Similarly, unless one is prepared to do a little math, one might do more harm than good by flat-fielding. Software reduction packages, such as IRAF or ESO-MIDAS are extremely useful tools-in the right hands. But, one should never let the software provide a guide to reducing data. Rather, the astronomer should do the guiding. One should strive to understand the steps involved at the level that one could (in principle) reproduce the results with a hand calculator!".*

For a detailed understanding of how the data should be treated, a small detour to the mathematical foundations is inevitable.

12.3.1   General Mathematical Considerations

To understand the above algorithm, one has to keep in mind how a pixel signal is assembled. The signal in ADU^2 of a pixel on the CCD chip can be described as follows:

$$s(x, y, t) = B(x, y) + t \cdot D(x, y) + t \cdot G(x, y) \cdot I(x, y) + noise. \qquad (12.1)$$

$B(x, y)$ is the bias in each pixel at the chip position (x, y) in ADU, $D(x, y)$ is the corresponding net dark current in ADU/sec, $G(x, y)$ is the corresponding pixel sensitivity (local CCD gain) in $ADU/photon$, $I(x, y)$ is the corresponding light flux in $photons/sec$, t is the exposure time and *noise* a statistical noise contribution coming from various sources (see next chapter). We are therefore able to determine the acquired light flux by re-arranging Eq. 12.1 to $I(x, y)$, if we know the bias, the dark and the pixel sensitivity in addition to the signal on the chip. These unknown quantities have to be obtained by various measurements with the complete system "Telescope + Spectrometer + CCD".

---

[2]The unit "ADU" is Analog Digital Unit. For a 16-bit CCD amplifier the CCD dynamics is therefor $2^{16} = 65{,}536$ ADU.

12.3.2   The Bias Field

As mentioned above, the bias image contains the read-out noise of the amplifier, after it has been read out. The bias is obtained by the shortest possible exposure time (normally zero) when the shutter is closed. At best, then $t = 0$, and Eq. 12.1 is given by

$$s(x, y) \equiv b(x, y) = B(x, y) + noise. \tag{12.2}$$

$b(x, y)$ is the measured signal, composed of the true signal and an undefined noise, i.e. we use the measured signal and identify it as a noise-affected overall signal which varies randomly around a mean value. The bias field contains the readout noise of the CCD. D and G are noisy, as well. However, this means that we insert additional noise when combining all data for the estimation of the measured data by offsetting in the determination of $I(x, y)$. We can only reduce the error of the mean with repeated measurement of b. The arithmetic mean of many individual measurements is calculated by

$$\overline{B}_M(x, y) = \frac{1}{n} \cdot \sum_{i=1}^{n} b_i(x, y) \tag{12.3}$$

and its standard deviation (see Chap. 13) by

$$\sigma_{\overline{B}_M}(x, y) = \frac{\sigma(b(x, y))}{\sqrt{n}}. \tag{12.4}$$

The "tilde" over the B means, that we are dealing with a statistical expectation value, which converges to the true value when the number of measurements n[3] is very large. This master bias corresponds best to the true bias the more noisy individual biases are obtained. The error of the master bias decreases according the $\sqrt{n}$ rule (see also Chap. 13). From experience, about 10–20 individual biases are sufficient to obtain a good master bias. Good means that one does not significantly degrade the net quality of the spectrum after its application.

12.3.3   The Dark Field

The dark current increases (normally linearly) with exposure time. A dark field $d(x, y, t_{data})$ is hence obtained with closed camera shutter and the same exposure time as the data image t_{data}. Equation 12.1 becomes

---

[3] This behavior is known as "law of large numbers".

$$s(x, y, t_{data}) \equiv d(x, y, t_{data}) = \overline{B}_M(x, y) + t_{data} \cdot D(x, y) + noise \qquad (12.5)$$

where we used the already estimated $\overline{B}_M(x, y)$ as an estimation for the bias field. Completely analogous to the estimation of the master bias we can now eliminate the scattering of the dark fields by consecutive averaging of many measurements:

$$\overline{D}_M(x, y, t_{data}) = \frac{1}{m} \cdot \sum_{i=1}^{m} d_i(x, y, t_{data}).$$

For each measurement, $d(x, y, t_{data})$ the bias field is always implicitly included. If we subtract the above master bias from the measured dark $d(x, y, t_{data})$ and divide by the exposure time we first obtain a bias-reduced single dark $D(x, y)$:

$$D(x, y) = \frac{d(x, y, t_{data}) - \overline{B}_M(x, y)}{t_{data}} + noise.$$

Analogous to the bias procedure we now average the dark by many individual measurements, reduce the noise and get closer to the true dark. The average dark per time unit is then

$$\overline{D}(x, y) = \frac{1}{m} \cdot \sum_{i=1}^{m} \frac{d_i(x, y, t_{data}) - \overline{B}_M(x, y)}{t_{data}}$$

with the noise given again by the standard deviation. Using the master dark and the master bias we finally can write

$$\overline{D}(x, y) = \frac{1}{t_{data}} \left(\overline{D}_M(x, y, t_{data}) - \overline{B}_M(x, y) \right). \qquad (12.6)$$

Hence, the unknown dark current per time unit is now defined for each pixel (x, y) by the master bias, the master dark and the exposure time.

12.3.4   Saving Time

The above procedure requires darks with the same exposure time as that of the target raw images. However, this might be time consuming especially for very long exposures. On the other hand, if the dark level increases linearly (depending on the CCD) one can also use short exposure darks by scaling them in time.[4] To do so, one creates a bias reduced master dark with the short exposure time t_{dark}. We take

---

[4]The linear behavior should be examined for any CCD camera with real data series.

a master bias, subtract it from each single dark and average by the number of this procedure. Then we obtain the master dark reduced from the average bias:

$$\overline{D}'(x, y, t_{dark}) = \frac{1}{m} \cdot \sum_{i=1}^{m} \left(d_i(x, y, t_{dark}) - \overline{B}_M(x, y) \right). \tag{12.7}$$

For the initially unknown dark field per time unit we then get

$$\overline{D}(x, y) = \frac{1}{t_{dark}} \left(\overline{D}_M(x, y, t_{data}) - \overline{B}_M(x, y) \right).$$

If we now compare the last two expressions for $\overline{D}(x, y)$ and $\overline{D}'(x, y, t_{dark})$ we see that the bias reduced master dark $\overline{D}'(x, y, t_{dark})$ can be expressed over the exposure time of the master dark:

$$\overline{D}'(x, y, t_{dark}) = t_{dark} \cdot \overline{D}(x, y).$$

If the original data are exposed at integration time t_{data}, the two first terms of Eq. 12.1 can be estimated with a master dark $\overline{D}_M$

$$\overline{D}_M(x, y, t_{data}) = \overline{B}_M(x, y) + t_{data} \cdot \overline{D}(x, y) \tag{12.8}$$

$$\overline{D}_M(x, y, t_{data}) = \overline{B}_M(x, y) + \frac{t_{data}}{t_{dark}} \cdot \overline{D}'(x, y) \tag{12.9}$$

and used in Eq. 12.6. If the dark level increases linearly in time and does not lead to saturated pixels, one can save a lot of time with this method.[5]

12.3.5   The Flat Field

To determine the incident light flux, we must now determine the relative sensitivity of each individual pixel. For doing so we expose the chip as uniformly as possible by mapping an evenly radiating surface of intensity $L(x, y) = L_0$ with "Telescope + Spectrometer". Equation 12.1 becomes

$$s(x, y, t_{flat}) \equiv f(x, y, t_{flat}) = B(x, y) + t_{flat} \cdot D(x, y) + t_{flat} \cdot G(x, y) \cdot L_0 + noise.$$

$$\tag{12.10}$$

---

[5]If the time behaviour of the dark level is not linear, it is possible to estimate the appropriate function with many measurements of different exposure times for data reduction.

t_{flat} has to be chosen so that no pixel reaches its full well capacity. In first approximation we assumed L to be constant and independent of the pixel position. This, however, will never be achieved in practice. Since we want to eliminate only local small-scale fluctuations with this procedure, a reasonably evenly illuminated white surface in the observatory is sufficient.

The exposure time for flat-field recordings usually differs from the exposure time of the target data. First, we must therefore determine a master dark $\overline{D}^F_M$ only for the flat fields. One chooses either the same exposure time as for the flats or one scales according to Eqs. 12.8 and 12.9:

$$f'(x, y, t_{flat}) = f(x, y, t_{flat}) - \overline{D}^F_M(x, y, t_{flat}) = t_{flat} \cdot G(x, y) \cdot L_0 + noise.$$

For a master flat and to minimize the amount of noise one again needs many flats which are then averaged:

$$\overline{F}_M(x, y, t_{flat}) = \frac{1}{n} \cdot \sum_{i=1}^{n} f'_i(x, y, t_{flat}) = \frac{1}{n} \cdot \sum_{i=1}^{n} f_i(x, y, t_{flat}) - \overline{D}^F_M(x, y, t_{flat}).$$

$$(12.11)$$

Because the flat $f'(x, y)$ is now known, we could solve the above equation for $G(x, y)$ by using the known detected intensity L_0. However, the absolute value of L_0 is almost always unknown. In addition, it might vary due to aging effects in the lamp. Hence, the absolute value of $G(x, y)$ is also unknown. In most cases the analysis of stellar spectra is a relative comparison of radiation fluxes. Hence, we keep the relative scatter but eliminate the absolute mean level of G as best as possible:

$$G(x, y) = \overline{G} \cdot g(x, y).$$

If we now have chosen $\overline{G}$ such that the average value of $g(x, y)$ is 1, we obtain

$$\overline{F}_M(x, y, t_{flat}) = t_{flat} \cdot \overline{G} \cdot g(x, y) \cdot L_0.$$

$$(12.12)$$

From this we can determine the average over the entire chip

$$\overline{F} = \sum_x \sum_y t_{flat} \cdot \overline{G} \cdot g(x, y) \cdot L_0 = t_{flat} \cdot \overline{G} \cdot L_0$$

and then identify $g(x, y)$ as the normalized master flat:

$$\frac{\overline{F}_M(x, y, t_{flat})}{\overline{F}} = g(x, y).$$

$$(12.13)$$

12.3.6 Why Flat Fielding

The flat field is repeatedly the subject of discussion and it is often not clear what exactly is behind this procedure. To understand the reason for the present considerations one should keep in mind that a reliable rectification of the 1D spectrum in the data reduction process is one of the main goals. For doing so, one simulates the spectral continuum with an artificial fitting function, which also covers the extended spectral lines, in contrast to the real continuum (see Fig. A.32). Within emission and absorption line intervals there is no information available about the local continuum. Hence, one has to rely on a reliable function simulating the continuum where there are no obvious lines. The best choice is a low order function. This has two reasons:

1. A low order (e.g., a 3rd order spline, see Appendix B.4) ensures a relatively smooth variation of the fit and represents the true continuum better than a fit of higher order (e.g., 5th order Legendre function).
2. The higher the order of the continuum fit the more reference points for the fit are needed in the real spectrum. However, since no reference points can be specified within the line intervals (especially for wide stellar wind lines), higher-order functions tend to deviate ("oscillate") from the true continuum.

By choosing a low-order fitting function for the entire spectrum, the spectrum should show a smooth continuum without strong intensity discontinuities. In optimal conditions this is always the case because the spectral continuum is only a convolution (see Sect. 2.6.4) of relatively smooth functions for the star (Planck function) and the transmission curve of the telescope and the spectrograph (including its image aberrations). In reality, however, effects occur that can disrupt this smooth profile. This can be caused by dust particles on the CCD window close to the focus, which create diffuse concentric rings in the image field, introducing strong intensity disturbances. If one would try to sufficiently eliminate such interference by a continuum fit, one would have to select a function of very high order.[6] This, however, is in contradiction to the above requirements of a low-order fit. Therefore,

---

[6]One can in principle approximate any function $f(x) \equiv I(\lambda)$ (here our spectrum), which is sufficiently often differentiable, at each point λ_0 as the sum of a power series: $I(\lambda) = I(\lambda_0) + \frac{I'(\lambda_0)}{1!}(\lambda - \lambda_0) + \frac{I''(\lambda_0)}{2!}(\lambda - \lambda_0)^2 + \ldots + \frac{I^n(\lambda_0)}{n!}(\lambda - \lambda_0)^n + R_n(\lambda)$. This so-called Taylor series is the associated power series $I(\lambda) = \sum_{k=0}^{\infty} \frac{1}{k!} I^k(\lambda_0) \cdot (\lambda - \lambda_0)^k$. The remainder $R_n(\lambda) = \frac{I^{n+1}(\lambda_0)}{(n+1)!}(\lambda - \lambda_0)^{n+1} + \ldots$ becomes smaller and converges to zero with higher orders . $I(\lambda)$ is thus a function of the type $I(\lambda) = a_n \cdot x^n + a_{n-1} \cdot x^{n-1} + \ldots + a_0$. The powers of the individual summands thus represent the "flexibility" of the individual terms and the coefficients defining their appropriate weighting. This means that any function $I(\lambda)$ can be arbitrarily closely approximated by a polynomial series. The more complex is $I(\lambda)$, the more terms need to be selected in increasing order. One might therefore map each spectrum one-to-one by a series expansion, as long as one considers adding infinitely many terms. The remainder $R_n(\lambda)$ would then be zero and we would unnecessarily also fit the spectral noise.

it is necessary to image an area of as few <u>local</u> intensity deviations as possible
(e.g., an extended, uniform white screen) with the entire optical system (telescope,
spectrograph). A modest <u>global</u> brightness gradient over the entire screen does not
interfere with the process, since it can be mapped with a low-order function by
subsequent rectification. This is valid only for the dispersion direction. In spatial
direction, perpendicular to dispersion direction this gradient must be fitted out to
guarantee a good background subtraction[7] during data reduction. Thus, the division
by such a flat eliminates the <u>local</u> brightness variations and a low order fit is
applicable again.

12.3.7 Collapsing the Spectrum

With the above accurately performed procedures we obtain a fully reduced 2-
dimensional (2D) spectrum on the CCD chip without the effects degrading the data
quality (mainly bias, flat-field, dark). To obtain a more useful 1-dimensional (1D)
spectrum in counts versus pixel position and finally intensity versus wavelength (see
Sect. 12.9) all relevant pixels containing information perpendicular to the dispersion
direction (CCD columns) need to be combined (normally co-added with proper
weighting) so that all counts refer to a single pixel. The resulting values of counts
versus pixel number (row) along the dispersion direction then represents the 1D
spectrum. This procedure is called "collapsing" the spectrum. This summation of all
appropriate pixel values within a reasonable window perpendicular to the dispersion
direction will then have a positive impact on the final S/N in the 1D spectrum,
which will be significantly increased over the case of individual pixels even near the
peak intensity in the 2D spectrum.

12.3.8 Flats for Echelle Spectroscopy

The above considerations refer to the respective spectral range to be used. For
good flat field reduction it is therefore necessary to ensure sufficient flat light flux
in all relevant wavelengths. For relatively narrow-band applications of standard
spectroscopy (see Chap. 2) this can be well achieved with appropriate lamps. Much
more efforts are required for echelle spectroscopy (see Chap. 5). Flat light sources
are needed that provide sufficient light flux simultaneously at all wavelengths. In
addition, the source should be free of intrinsic spectral lines that could jeopardize a
proper data reduction. A continuous light source with a Planck spectrum (black
body) is indeed free of lines, but provides dramatically different light fluxes
for different wavelength regions, so that the above requirements for an adequate

---

[7]Professional data reduction tools like MIDAS and IRAF offer this fitting procedure.

flat field are not fulfilled at all wavelengths. The colour temperatures of typical incandescent and halogen lamps lie between 2,300 and 2,900 K, and according to the Wien displacement law $\lambda_{max} \approx \frac{3,000\,\mu m}{T/K}$ their maximum radiation is thus at a wavelength of about 1 μm. According to Planck's law the radiation flux decreases dramatically at shorter visible wavelengths. Therefore such Halogen lamps provide about five times more flux in the red than in the blue light. However, according to the above considerations this would have a corresponding impact on the quality of the spectra, since the achievable S/N in the blue spectral region would be significantly lower than in the red light. Already with only these considerations, it becomes obvious that a single light source cannot provide flat field data needed for high quality spectra, without making an unreasonably large number of short exposures (to avoid saturation in the red) to increase the overall S/N in the blue.

The solution for this problem is the combination of different light sources for different spectral ranges of an echelle spectrograph. A typical example is the FEROS echelle spectrograph of ESO (see Sect. 8.4.4). FEROS uses two different halogen lamps of different strength. The lamp for the blue spectral range is about five times stronger than that for the red range to compensate for the intensity ratio of identical color temperatures in blue and red wavelengths. In order to suppress the red light components, thereby avoiding again a high light flux at red wavelengths, various filters are used to make the lamp appear blue. The light from the two lamps is combined by a 45° beam splitter in a 1:1 ratio and applied to the spectrograph.[8]

For stability reasons echelle spectrographs are often operated with fiber optics. For flat fielding it is important to note that the light source must be fed into the fiber under the same geometrical conditions as the target light. Otherwise, the sensitivity functions of the individual orders are not adequately corrected (see Fig. 12.1) and the quality of spectral rectification could be potentially degraded especially for wide spectral lines (see Sect. 14.8).

For space applications, no fiber optics are used and usually there is no possibility to accommodate a flat screen. In this case integrating spheres are often used. This makes use of the fact that on an ideal diffuse reflecting inner surface of a hollow sphere, the radiation density is constant when it is illuminated by a light source attached to the inner side. The physical background of this behaviour is the Lambert cosine law the brightness of an illuminated surface depends on the angle of incidence of the incoming irradiance:

$$\frac{dI}{dA} = L \cos\theta.$$

For clarification we first consider the irradiance coming from a unit area dA (Fig. 12.2). For Lambert surfaces the radiance viewed from any angle remains

---

[8]Alternatively sometimes combinations of LEDs of different colors are proposed. However, it should be noted that these sources have typical spectral bands which can be as wide as an entire echelle order. In addition, the combined fluxes of all flat LEDs must cover the entire echelle range, which is hardly feasible in contrast to halogen lamps with an almost black body curve.

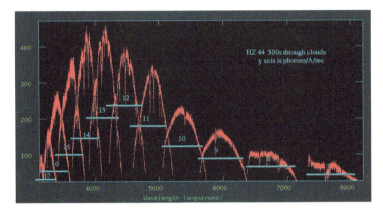

Fig. 12.1 The extracted spectrum of a standard star before rectification obtained with the Blue & Red Channel echelle spectrographs at the MMT telescope. Wavelength ranges of various orders are indicated by blue bars and their respective numbers (image by Craig Foltz, courtesy of the MMT Observatory)

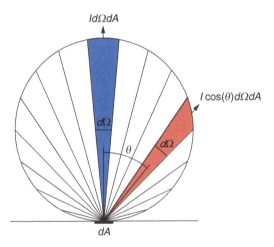

Fig. 12.2 2D-sketch of an integration sphere. The area dA emits light into all directions but different intensities depending on the viewing angle θ (from Wikipedia)

constant. This is because the solid angle (apparent size) depends on the viewing angle of the observed area as well as the radiance (power per unit solid angle and projected source area) emitted. In simple terms, both viewing angles cancel out because they are identical.

First we consider the radiance in the normal direction: I photons/$(s \cdot cm^2 \cdot sr)$. The number of photons per second emitted into the vertical solid angle in Fig. 12.2 is $I d\Omega \cdot dA$. The number of photons per second emitted into the solid angle θ is $I \cdot cos\theta \cdot d\Omega \cdot dA$. Second, in Fig. 12.3 we consider what an observer directly

Fig. 12.3 Observed intensity
from an integration sphere
seen by a normal and
off-normal observer. dA_0 is
the observation aperture and
$d\Omega$ under which dA is
observed

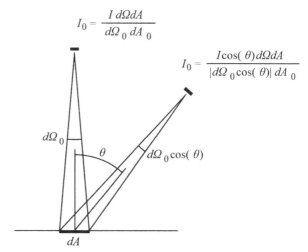

$$I_0 = \frac{I\,d\Omega dA}{d\Omega_0\,dA_0}$$

$$I_0 = \frac{I\cos(\theta)d\Omega dA}{|d\Omega_0\cos(\theta)|\,dA_0}$$

above the area element sees using an aperture dA_0. The observer will detect $I d\Omega dA$ photons per second, and hence, will measure the radiance

$$I_0 = \frac{I d\Omega dA}{d\Omega_0 dA_0}. \tag{12.14}$$

If we now consider an observer viewing dA at an angle θ with the same aperture dA_0 the surface dA will span a solid angle $d\Omega_0 \cos\theta$. The observer will detect $I \cos\theta d\Omega dA$ photons per second. However, dA is viewed at angle θ and the observer will now measure the radiance

$$I_0 = \frac{I \cos\theta d\Omega dA}{d\Omega_0 \cos\theta dA_0}. \tag{12.15}$$

Equations 12.14 and 12.15 are identical. The reflectance of a Lambert surface remains constant independent of the viewing angle:

$$I_0 d\Omega_0 dA_0 = I d\Omega dA.$$

In practice, one can use a hollow sphere with a lamp attached to the inner surface and measure a constant intensity at an aperture in the sphere. The light intensity is then independent of the observation direction and hence constant. The inner sphere (covered with an appropriate reflective material) behaves like a compact flat field of constant intensity.

An example is the calibration assembly on board the James Webb Telescope. The instrument allows the data calibration to be carried out in orbit as well as the control of several important imaging parameters of different optical elements. Different continuous lamps on the inner sphere surface are used for flat fielding as well as

Fig. 12.4 Schematic view of
the Radiometric Calibration
Spectral Source of the
NIRSpec Calibration
Assembly (Taubert et al.
2009)

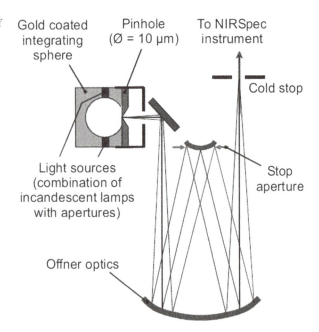

Fig. 12.5 NIRSpec
calibration sphere (Astrium)

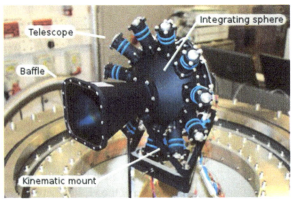

spectral lamps for wavelength calibration (Keyes et al. 2006). Figure 12.4 shows the
schematic structure of the calibration unit and Fig. 12.5 shows the calibration sphere
in the laboratory.

12.3.9 Remarks on the Response Function

With the above considerations on flat fields, an important question arises:

Is it necessary to obtain the sensitivity function (response function) of the telescope–spectrograph–camera system by measuring the spectrum of a standard star and account for offsets with the original spectrum?

The sensitivity function would be important if <u>absolute</u> measurements of stellar light flux are the prime observational goal. However, for high accuracy this is only possible at sites with very stable atmospheric conditions and even there only with standard-star observations or considerable effort via flux calibrations with black body radiators (Tüg 1974, 1980; Tüg et al. 1977a, b). Essentially the flat field already provides the response function; it is simply folded with all other sensitivity functions. However, getting real fluxes, e.g., in erg/s/Å involves calibrating those "other" sensitivity functions, too. This task is highly complex (see again the paper series of Tüg et al.). In addition, a highly transparent and stable atmosphere is required to eliminate parameter fluctuations coming from this source and to obtain reasonable accuracies. So just getting flat fields one is still quite far from the final goal of getting true fluxes. In many applications, one only needs a rectified spectrum, in which case it makes no sense to determine the response function. For high spectral resolution only a few sampling points might be available in the spectrum of a star. This is the case for many Wolf-Rayet stars with extremely broad emission lines and late-type stars full of narrow absorption lines both without any visible photospheric continuum (see Sect. 14.8). For a line analysis, it can then be helpful first to divide the spectrum by a continuum fit of a standard star with narrow absorption lines, then simulate the resulting (e.g. WR) spectrum with a low-order fit and reduce the data accordingly. The continuum fit of the standard star then provides the <u>relative</u> instrumental response function of the measuring system of telescopic optics, spectrograph and detector. However, even for this procedure one should realize that it is very difficult to observe two stars identically; there will always be differences in the continuum shape, due to differences in atmosphere, centering the image, etc. This response function can then be applied to all types of stars. We expressly point out that this method leads to better results only for extremely broad lines. One should also keep in mind that this double fitting procedure, each with fitting functions of lowest possible order also can lead to higher errors in the determination of the continuum, which in turn destroys any accuracy gain intended. However, for echelle spectroscopy a response function might be very useful. Each echelle order exhibits its own sensitivity function (see, e.g., Fig. 12.1) and an adequate continuum rectification might become a delicate if not impossible procedure. An inadequate rectification then exhibits wave-like features in the continuum. In order to avoid this, a response function might deliver an additional rectification support. As we already mentioned at the end of Sect. 5.13, the variable efficiency performance and the non-linear dispersion, are the price to pay for the highly effective capture of the entire spectral range.

In general, the problem drives the decision for a data reduction procedure, not vice versa.

12.4   The Data Reduction Recipe

We now have obtained a master dark and a master flat and can determine the equation for the final CCD reduction. We start with Eq. 12.1 for a data image:

$$s(x, y) = B(x, y) + t_{data} \cdot D(x, y) + t_{data} \cdot \overline{G}(x, y) \cdot I(x, y) + noise.$$

Subtracting the master dark and dividing by the normalized master flat, we get

$$\frac{\overline{F}}{\overline{F}_M(x, y)} \cdot \{s(x, y) - \overline{D}_M(x, y)\} = t_{data} \cdot \overline{G}(x, y) \cdot I(x, y) + noise$$

$$I(x, y) = \frac{1}{\overline{G} \cdot t_{data}} \cdot \frac{s(x, y) - \overline{D}_M(x, y)}{\overline{F}_M(x, y)/\overline{F}}. \qquad (12.16)$$

Hence, our reduction procedure **without** bias is:

We subtract the master dark $\overline{D}_M(x, y)$ from the chip signal $s(x, y)$ and divide by the Master-Flat $\overline{F}_M(x, y)$ which is normalized to the average flat $\overline{F}$. The product $\overline{G} \cdot t_{data}$ represents the average camera sensitivity multiplied by the exposure time. This constant can be finally determined only via an absolute calibrated reference source.

If working with a master bias and a master dark (after bias subtraction) the equation can be modified to

$$I(x, y) = \frac{1}{\overline{G} \cdot t_{data}} \cdot \frac{s(x, y) - t_{data}/t_{dark} \cdot \overline{D}'_M(x, y) - \overline{B}(x, y)}{\overline{F}_M(x, y)/\overline{F}}. \qquad (12.17)$$

Hence, our reduction procedure **with** bias is:

We subtract the master bias and the bias reduced master dark $\overline{D}'_M(x, y)$ (weighted by the ratio of data to dark exposure time) from the chip signal $s(x, y)$. Then we divide by the master flat $\overline{F}_M(x, y)$ which is normalized to the average flat $\overline{F}$. These considerations are valid for all applications, including photometric observations and absolute calibration.

For spectroscopic applications in which the spectral continuum is normalized to unity, Eqs. 12.16 and 12.17 can be significantly simplified. Because of the spectral rectification (see MIDAS data reduction in Appendix A) $\overline{G} \cdot t_{data}$ can be considered as constant. This also applies to the rectification $\overline{F}_M(x, y)/\overline{F}$. $\overline{F}$ can be neglected, as well, as long as the spectra are later rectified.

The CCD reduction with a master dark $\overline{D}_M$ and a master flat $\overline{F}'_M$ but **without bias** then simplifies to

$$I(x, y) = \frac{s(x, y) - \overline{D}'_M(x, y)}{\overline{F}_M(x, y)}. \qquad (12.18)$$

And for a reduction **with bias** we obtain

$$I(x, y) = \frac{s(x, y) - t_{data}/t_{dark} \cdot \overline{D}'_M(x, y) - \overline{B}(x, y)}{\overline{F}_M(x, y)}. \qquad (12.19)$$

We thus have found two different procedures, a slow and a fast one, which can be chosen freely according to the observer's requirements.

1. **The slow method:** The darks for the original data are all exposed with the time t_{data} and the darks for flats are all exposed with t_{flat}. According to Eq. 12.11 we generate an average raw flat and subtract the master dark $\overline{D}^F_M$ from this flat. Thus we obtain the corresponding master flat $\overline{F}_M$. Then we can reduce the data according to Eq. 12.18.
2. **The fast method:** Only one master dark for an arbitrary exposure time is recorded according to Eq. 12.7. This master dark is scaled by the ratio of t_{data}/t_{flat}. Similarly, the master dark for the flat-field is scaled. According to Eq. 12.19 the data can then be correctly reduced. **This method requires knowledge of the temporal scaling behaviour of the dark fields. It should be linear in the best case.**

We stress again at this point that the above standard procedures should be considered only as a working basis. Guidelines for a reasonable data reduction depend on the details of the particular data set, as explained by Massey and Hanson (2010) (see Sect. 12.3), and not on predetermined procedures.

12.5   Noise Contribution of Bias and Dark Fields

In professional astronomy, CCD chips are cooled with nitrogen to $-196\,°C$ and the dark current is correspondingly low. The two *noise* terms in Eqs. 12.2 and 12.5 correspond only to the readout noise of the camera, plus photon noise in the darks. If we now use $\overline{B}$ and $\overline{D}$ with an exposure time t_{data} for data reduction, the subtraction of bias and dark will add a noise component to the raw data. Because the variance within each pixel of different terms can simply be added, that is

$$\sigma^2_{\overline{DB}} = \sigma^2_{\overline{D}} + \sigma^2_{\overline{B}} = \frac{R^2}{n} + \frac{R^2}{m} \cdot \left(\frac{t_{data}}{t_{dark}} \right).$$

The first term simply represents the averaged read-out noise. The second term is the noise contribution from the dark scaled in time without significant dark-noise. For the case of non-negligible dark current the second R should be replaced by the corresponding noise of the time-scaled dark. Hence we obtain

$$\sigma_{\overline{DB}} = R \cdot \sqrt{\frac{1}{n} + \frac{1}{m} \cdot \left(\frac{t_{data}}{t_{dark}}\right)^2}. \tag{12.20}$$

R is the readout noise of the camera, n is the number of averaged bias images, m is the number of averaged dark frames and t_{data} and t_{dark}, the exposure times for the data and dark image, respectively. But one can also work completely without bias fields. Equation 12.8 shows that a master dark can also be achieved without any bias field if the individual darks have the same exposure time as the data images:

$$\overline{D}(x, y, t_{data}) = \frac{1}{m} \cdot \sum_{i=1}^{m} d_i(x, y, t_{data}).$$

Thus, the above noise contribution reduces to

$$\sigma_{\overline{D}} = R \cdot \sqrt{\frac{1}{m}}.$$

However, the price for this noise reduction are long exposure times for the darks compared to bias exposures. For low-noise contributions of the bias (this is the case for cooled detectors in professional astronomy), one can usually accept a slightly stronger overall noise and can save a lot of time with the inclusion of bias images. For commercially available cameras with Peltier cooling and a correspondingly higher noise in the bias images one should test whether the data can be accurately reduced with the bias images. If the amount of noise is too high, one must reduce with darks, accepting longer procedures.

12.6   The Necessary Flat Field Quality

Of course, the flat-field should not reduce the overall quality going from raw to reduced data, so the question arises what necessary quality of the flat field is required.

As we have already pointed out, the flat should take the sensitivity variations of all individual pixels into account so that they do not negatively affect the spectral data. Large sensitivity deviations inevitably lead to a deterioration of the signal-to-noise ratio in the flat and therefore also in the raw spectrum to be processed. The potential effects are of course the more dramatic the higher the desired quality of the stellar spectrum and its S/N. The noise of the flats should influence the target data noise as little as possible. This is not very critical for data with low S/N (~ 10). For high S/N, however, these effects must be taken into consideration and the quality of the flats should be significantly higher than that of the spectral data. To understand the background to this problem, we start with Eq. 12.19:

$$I(x, y) = \frac{s(x, y) - t_{data}/t_{dark} \cdot \overline{D}'_M(x, y) - \overline{B}(x, y)}{\overline{F}_M(x, y)}.$$

Depending on the number of darks and flats, the corresponding master darks and master flats normally will have a lower noise level than the raw signal. Assuming that the target signal counts are much larger than the dark counts we can therefore neglect the dark noise. The signal is photon noise dominated. Since the signal must always be freed of dark and bias, we can use this dark and bias reduced signal $\tilde{s}$ and the flat signal F for overall error estimations via the principle of error propagation:

$$I = \frac{\tilde{s}}{\overline{F}_M}.$$

Performing error propagation we obtain

$$\sigma_I^2 = \left(\frac{1}{\overline{F}_M}\right)^2 \sigma_{\tilde{s}}^2 + \left(\frac{\tilde{s}}{\overline{F}_M^2}\right)^2 \sigma_{\overline{F}_M}^2$$

and division by $I^2 = \tilde{s}/\overline{F}_M$ delivers

$$\frac{\sigma_I^2}{I^2} = \frac{\sigma_{\tilde{s}}^2}{\tilde{s}^2} + \frac{\sigma_{\overline{F}_M}^2}{\overline{F}_M^2}.$$

The relative errors of the data and the flats are added quadratically to the variance of the reduced data. The second equation term represents the error contribution of the master flat. This explains why the master flat must have the highest possible S/N. This is achieved by the combination of as many individual flats as possible to a master flat. For example, if the signal S/N of the raw spectrum and the flat S/N are the same, the S/N of the reduced spectrum deteriorates by a factor of $\sqrt{2}$ = 1.41. If the flat S/N is twice as strong as the S/N of the raw spectrum, then the reduced spectrum S/N deteriorates only by a factor of $\sqrt{1.5} = 1.22$, and for a 10-fold flat S/N this is about $\sqrt{1.1} = 1.05$. One should hence ensure that the master flat noise is significantly smaller than that of the raw signal.

To use the above considerations in a useful way, one should know the corresponding noise behaviour of the CCD chip. According to photon statistics the pixel noise is $\sigma = \sqrt{\tilde{s}}$. Now, however, even the average sensitivity ratio of all pixels p, which introduces a pixel-to-pixel noise (gain) variation between individual

pixels, see Sect. 10.3.4) must be considered. Consequently the total CCD noise is
$\sigma_{total}^2 = I/g + I^2 p^2$ (see Eq. 10.6). Hence

$$\frac{S}{N} = \sqrt{\frac{I/g}{1 + p^2 I}}. \tag{12.21}$$

For no pixel-to-pixel variations ($p = 0$) we immediately get the already known
relation $S/N = \sqrt{I/g}$. Assuming an extreme pixel-to-pixel variations of 10 %
($p = 0.1$) Eq. 12.21 becomes

$$\frac{S}{N} \approx \frac{1}{g \cdot p}.$$

This means that, assuming a gain of $1e^-$ per ADU, a pixel-to-pixel variation of 10 %
limits the achievable signal-to-noise ratio to only 10 (or even to S/N of 5 for a gain
of 2). Because every single flat field has this very same S/N one can increase the
data S/N only by using a large number of flats, e.g., 100 flats for increasing S/N
from 10 to 100. However, the typical value of p for modern CCDs is between 0.01
and 0.001. For the latter, Eq. 12.21 becomes again $S/N = \sqrt{I/g}$.

   This means that the signal-to-noise limitation depends on the individual value
of p for the CCD in use. If the CCD is of high quality, flats are not required for
decreasing the pixel-to-pixel noise but only for local vignetting effects in the optical
telescope—spectrograph system (see Sect. 12.3.5).

12.7   Cosmic Rays

A non-negligible disturbance of the data is afflicted by atomic nuclei (mostly
protons) and electrons which have been accelerated to extremely high energies and
move through the universe. They move with about 90 % of the speed of light and
have energies of up to 10^{21} electron volts (eV). Some of these cosmic particles come
from the Sun, but most have their suspected origin in our Galaxy or other galaxies
while supernovae are probably the origin of the highest energy particles. Cosmic
radiation is attenuated by the atmosphere. However, reactions in the atmosphere
produce secondary particles which consist of muons on the ground for the most
part. When reaching a detector, these particles usually interfere with individual
pixels and at an angle to a row or grazing incidence entire rows of pixels can be
influenced. On the ground, the incidence rate is about 50 per cm^2 and hour. In
order to eliminate the effects of cosmic particles in the images, one divides a single
recording into at least three exposures and compares them to corresponding changes
in individual pixels. The offending image CR hit can then be illuminated by median
filtering.

12.8   A Quick Exposure Time Estimation

For observation programs a quick estimation of the necessary final exposure time per collapsed pixel is often very useful. This is done with a plot of one column at constant wavelength (spatial direction) of a 2D test exposure of time t'. One then estimates the *FWHM* of the seeing profile along with its peak count value c_p. A good approximation of the total number of counts N in the final collapsed spectrum within about 20 % accuracy is then

$$N'_p = FWHM \cdot c_p.$$

If c_p is expressed in number of detected photons, then one can easily find the required exposure time t to reach a given prescribed $S/N \sim N_p/\sqrt{N_p} = \sqrt{N_p}$, with N_p the number of detected photons. This, of course, is normally significantly higher than in the short test exposure. Using Poisson statistics we then obtain for the most optimistic case:

$$t = t' \cdot \frac{(S/N)^2}{N'_p}.$$

12.9   Wavelength Calibration

After the complete CCD reduction, the 2D spectrum is collapsed to a 1D spectrum. That is, all spectral pixels having a sufficient signal are averaged perpendicular to the dispersion direction. The corresponding procedure is part of every reduction software, so we do not discuss this point in detail.

In order to wavelength calibrate the reduced spectrum independent of the stellar spectrum, one uses light sources with appropriate emission lines. These emissions are narrow in wavelength and occur at well-known laboratory wavelengths. They can be taken from corresponding catalogues (see Appendix E). With such catalogues the laboratory wavelengths can be assigned to corresponding pixel positions in the spectrum. The procedures are part of all relevant calibration tools. We show an example for MIDAS in Appendix A.

12.9.1   Standard Light Sources

In practical terms light sources are needed which provide a sufficient number of lines for the wavelength range to be calibrated. For light in the red wavelength range around Hα neon lamps provide a host of strong lines. If the measurements are limited to this wavelength range at moderate resolution, these extremely cheap

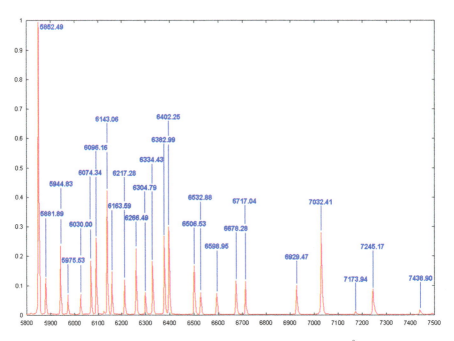

Fig. 12.6 1D calibration spectrum of a neon lamp between 5,800 and 7,500 Å

lamps, found in every baby night lamp, are entirely sufficient. Figure 12.6 shows a
neon spectrum between 5,800 and 7,500 Å.[9]

To provide calibration lines for the entire visual spectrum with a single lamp,
one needs gas fillings of different elements. Typically, commercially available
combinations of thorium and argon (ThAr), neon and argon (NeAr), helium and
argon (HeAr), iron and argon (FeAr) as well as more complex combinations with
suitable mixing ratio are used. The crucial factors are a) a balanced line strength to
avoid individual line intensity overflow in the CCD for a given exposure time and b)
a possible large number of lines to ensure a sufficient determination of the spectral
dispersion behaviour. The larger the wavelength range to be measured and the higher
the dispersion of the spectrograph, the higher the precision that is necessary for line
calibration. So, for echelle spectroscopy, lamps are inevitably needed, which provide
lines over an extremely large wavelength range.

As we already discussed in Sect. 5.13, the angular dispersion changes by a few
percent within one echelle order (see Eqs. 5.3 and 5.4) and one needs sufficiently
many calibration lines for each order. Therefore, for echelle spectroscopy, element
combinations with thorium (e.g., ThAr) are used that deliver uniformly many lines

---

[9]Neon lamps also show a number of lines in the blue wavelength range that are suitable for
calibration. However, they are much weaker than the lines in the red range and the exposure times
must be adjusted accordingly.

Fig. 12.7 2D calibration spectrum of the ThArNe lamp of FEROS (FEROS documentation November 1998)

to all orders. For instance, the FEROS spectrograph utilizes a combination of a ThAr with a Ne lamp and thus can accomplish a peak accuracy of a few meters per second in the present configuration. Figure 12.7 shows a 2D calibration spectrum of FEROS.

For slit spectrographs, it is usually sufficient to position the calibration lamp in front of the long slit (i.e. as close to the telescope focus as possible—see Sect. 4.1.2). The lamp thus illuminates a larger area than the target star, but one chooses an area for calibration which matches the target star spectral area to acquire an averaged 1D calibration spectrum. For fiber-coupled echelle spectrographs it is again important to feed the calibration light under the same optical conditions into the fiber as the starlight because of the non-linear dispersion in the individual orders (see also Sect. 12.3.8),

12.9.2 Laser Frequency Combs

Very sophisticated technology and care are needed especially at high spectral resolution. With ThAr lamps achievable calibration accuracy of a few tens of meters per second is influenced not only by the measurement process but also by contamination of the gas during production as well as by aging effects in the lamp. In addition, line blends and the non-uniform distribution of the calibration lines degrade the measurement accuracy. The accuracy limit for the statistical combination of up to 10,000 lines is approximately 0.3 m/s. For some studies, however, one needs much higher accuracies. These are the determination of fundamental constants (e.g., the fine structure constant) or radial velocities for the discovery of Earth-like extrasolar planets, requiring accuracies of the order of a few centimeters per second.

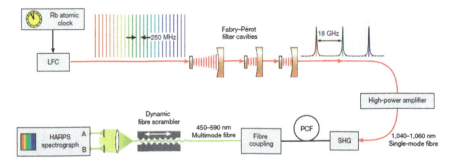

Fig. 12.8 Instrumental setup of the LFC test system at ESO-HARPS spectrograph. The LFC is triggered by a Rubidium clock. A Ytterbium fibre laser serves as the comb generator. Its line spacing is increased with Fabry-Pérot cavities. The amplified and processed comb light is injected into a multimode fibre either through a lens or an integrating sphere. A dynamic scrambler is used for fiber mode handling (Wilken et al. 2012, Reprinted by permission from Macmillan Publishers Ltd: Nature, copyright 2012)

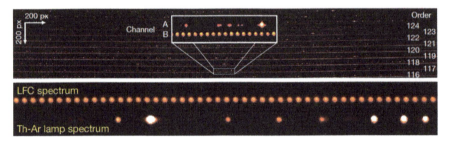

Fig. 12.9 Comparison of LFC and thorium emissions in one echelle order with ESO HARPS at the 3.6 m telescope on La Silla. *Top*: Raw data of one acquisition with HARPS (Wilken et al. 2012). *Bottom*: Enlarged section of an LFC spectrum (Lo Curto 2012, Reprinted by permission from Macmillan Publishers Ltd: Nature, copyright 2012)

Since for very large telescopes the photon noise is correspondingly low and thus the accuracy of high-precision spectroscopy is improved significantly, calibration sources are needed which are approximately ten times better than ThAr lamps. For this purpose, laser combs (laser frequency combs, LFC) can be used. Laser frequency combs translate the accuracy of atomic clocks into optical signals so that they can supply a series of emissions of identical wavelength spacing over the entire optical spectrum. Therefore a calibration spectrum with thousands of lines is provided, whose accuracy is set by fundamental constants (e.g. a rubidium atomic clock). Figure 12.8 shows the working principle of a LFC test setup at the High Accuracy Radial velocity Planet Searcher (HARPS) spectrograph at the 3.6 m telescope on La Silla. And Fig. 12.9 shows a HARPS echelle spectrum (11 orders of short wavelength CCD), obtained with the setup of Fig. 12.8 and compared with a thorium spectrum. The LFC spectrum shows a much denser line distribution than the thorium spectrum. Moreover, it is free of blends and shows a uniform intensity. In comparison, some thorium lines are saturated. The wavelength differences between two adjacent LFC lines correspond to the repetition rate of the exciting femtosecond pulsed laser. The accuracy is consistent with the constancy of the Rb atomic clock. With this approach, calibration accuracies are improved by about an order of magnitude for large telescopes.

Transferring atomic vibrations in atomic clocks to optical spectra, the LFC can achieve calibration accuracies of the order of the natural constants. It might therefore be possible that the optimum in wavelength accuracy is now reached.

Chapter 13
Measurement Errors and Statistics

A Short Story

When Klaus talks about error calculations, Thomas is always cautious. To calculate measurement errors, one has to wrestle with mathematics, particularly statistics, and Thomas does not like that at all. Where do you learn about statistics during university studies—so no idea. Klaus, however, is a large computer and loves to juggle with operators and differential equations. Thomas finds that very suspicious and very often questions him. One day Thomas presented a paper on spectroscopic error considerations at a spectroscopy meeting. He was quite proud that he probably understood the whole content. Klaus watched the paper, full of mathematics, carefully and said: "I do not understand something here". Thomas was a bit upset, but in reality, could not follow the entire idea of the publication (in fact he had no clue). Klaus began to calculate. The whole issue ended with a joint paper published in a refereed journal.

13.1 Basic Remarks

After performing a series of measurements and the presentation of the respective results, one should always ask: What are the measurement errors? For example, if a distance between two points A and B is determined, the measurement error, e.g., 100 ± 1 m, is a critical point to document the accuracy of the method. If this is not done, the measurement has no real value. Even in scientific publications one can sometimes find data series without any indication of the measurement error. In physics, and thus also in spectroscopy, the observer should consider this aspect of his work and always has to conduct his observations and data processing accordingly. We distinguish between systematic errors, drift and statistical errors.

© Springer-Verlag Berlin Heidelberg 2015
T. Eversberg, K. Vollmann, *Spectroscopic Instrumentation*, Springer Praxis Books,
DOI 10.1007/978-3-662-44535-8_13

13.2 Systematic Errors

Each measurement is a comparative process and is performed with a specific instrument. The human eye has intrinsic tendencies—including subjective perceptions influenced by psychological effects. Thus, the moon appears larger on the horizon than at zenith, although this is not true in reality. Human perception, however, needs to perceive danger at the same height quicker than from different heights[1]—a typical systematic error. In the technical context a telescope can be misaligned by a specific value, so that the positioning deviates by a typical value.

An interesting spectroscopical example is the wavelength calibration of spectral absorption lines. In Fig. 13.1 we show the spectrum of the Be star ζ Tauri at Hα. We have highlighted three areas of reduced flux, which can be identified as absorption lines. The blue line on the left is relatively weak and one could now use the centre of Hα and the line on the right side for a wavelength calibration by a linear dispersion fit (see Chap. 12). But in reality, the central "absorption" is merely a reduction in the Hα emission flux (a central reversal) and both line peals "oscillate" in wavelength due to a typical line behavior of Be stars—the disk material above the stellar equator oscillates. The oscillation amplitude is about 1 Å, which causes a systematic calibration error of no less than 1 %, although the actual measurement may be highly accurate.

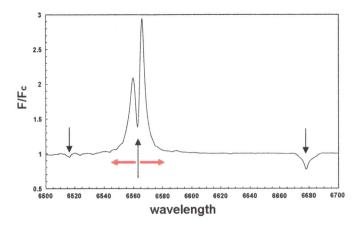

Fig. 13.1 A spectrum of the Be star ζ Tauri around Hα. *Arrows* indicate "identified" absorption lines (see text)

---

[1]In fact, this is not a psychological but a cybernetic problem. For a potential prey, it is vital to detect predator attacks very early. This often happened in ancient times, especially at ground level. A tree, though, usually represented a safe haven. Things in the horizontal direction can, hence, be perceived earlier than that in the vertical direction.

Another example of a systematic error is the determination of the spectral continuum (see Sect. A.4). Experience shows that the choice of the reference continuum points for the continuum varies from person to person, although at the same wavelengths intervals chosen for identical spectra. The continuum noise, which suggests a certain freedom in terms of the intensity, leads to these differences. In reality, noise is a random fluctuation around the true intensity of the stellar flux (see also Sect. 13.4.1). The true intensity can only be determined by averaging mean pixel values in a certain wavelength interval. It is this mathematically correct method that eliminates the subjective error.

In general, systematic errors affect each measurement by external effects. This may be the setting of a mechanical instrument, but also the method of measurement or environmental conditions such as temperature, humidity etc. In addition, systematic errors affect results always in one direction, otherwise it is a stochastic error. Because systematic errors are difficult to identify and can only be eliminated if they are observed, they are often the neglected part of an error analysis. The only way to avoid them is by accurate calibration of the entire measurement system.

13.3 Drift

Sometimes a measurement changes with time in only one direction of its value. For example, the image quality of a telescope can improve from the evening to morning. Or otherwise stable stellar absorption lines move in one wavelength direction during the night. Both effects can be relatively quickly identified as due to temperature: the first is the improvement of seeing conditions inside the observatory and the second is the adaptation of the instrument to ambient temperature. Drift is a systematic error in time and relatively easy to identify. However, measurements often show random fluctuations with a drifting mean. The identification of drift by the comparison of successive data points is not applicable here. A useful alternative is the continuous comparison of the zero point of each single measurement with the zero point at the beginning of all measurements. If the zero point does not always remain constant, there is a systematic error. If one takes, for example, spectra during an observation night, several spectra should be recorded for wavelength calibration (generally at the beginning, the middle and end of each observing night), so that a potential temperature drift can be identified and compensated. This method must be used with great care for photometric observations. Each measurement needs the recording of a standard non-variable star before and after the detection. Especially in amateur astronomy, it is often helpful to compare results of different observers and instruments to compensate systematic errors not noticed by a single observer. Often, measurements show neither a special data behavior nor any drift. By comparison with data which were obtained by other instruments, one can then at least in part compensate for any deviations without an identification of these errors.

13.4   Statistical Errors

If measurements of identical values vary stochastically in time, we talk about statistical errors. A typical example is shown in Fig. 13.2 in the form of a section of a continuous stellar spectrum. Although the stellar continuum is very well defined in nature, the fluxe varies randomly from pixel to pixel and in time around a mean value. The reason for this behavior is the nature of light in the form of discrete photons, the receiving CCD chip and the amplifier of the CCD camera. If a quantity is repeatedly measured under identical conditions, the vast majority of all measured values do not coincide with the "true" value. The values are specifically distributed around the true value. This distribution has a maximum. The individual measured values deviate from this average depending on its distance from the average. The larger the distance the smaller is their number. The resulting distribution curve is valid for a large number of measurements and has been developed by the mathematician Carl Friedrich Gauss, and called a Gaussian normal distribution

$$p(x) = \frac{1}{\sigma\sqrt{2\pi}} e^{-\frac{1}{2}\left(\frac{x-\mu}{\sigma}\right)^2}. \tag{13.1}$$

If the measurement values are distributed around $\mu = 0$ and if $\sigma = 1$, the Gaussian distribution is called a standard normal distribution (Fig. 13.3). The arithmetic mean of many individual measurements are generally calculated by

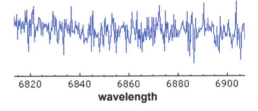

Fig. 13.2 Continuous stellar spectrum (*Courtesy*: Lothar Schanne)

Fig. 13.3
The Gaussian normal distribution

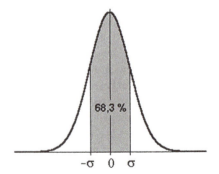

$$\overline{x} = \frac{1}{n} \cdot \sum_{i=1}^{n} x_n.$$

For example, we take the four measurements 2, 3, 5 and 8 and get with $n = 4$

$$\overline{x} = \frac{1}{4} \cdot (2 + 3 + 5 + 8) = 4.5.$$

In this context we ask the following questions:

1. How much do the single measurement values deviate from the average, i.e., how big is the average error of the single measurement?
2. How much does the mean deviate from the true value, i.e., how big is the average error of the mean?
3. How much does a value, estimated from faulty quantities, deviate from the true value, i.e., how big is the average error of the function value?

13.4.1 The Standard Deviation

The Gaussian normal distribution shows the frequency of the measurements as a function of the distance from the arithmetic mean. The average of these deviations is called the mean error of single measurements or the standard deviation σ. Using the measurement values x_i, the arithmetic mean $\overline{x}$ and the number of measurements n we have

$$\sigma = \sqrt{\frac{1}{n-1} \cdot \sum_{i=1}^{n} (x_i - \overline{x})^2}.$$

The inflection points of the distribution are defined by the limits of the interval $\overline{x} \pm \sigma$, with 68.3 % of all measurements lying in this interval. That are 95.4 and 99.7 % of all measured values for the intervals $\overline{x} \pm 2\sigma$ and $\overline{x} \pm 3\sigma$, respectively. However, more measurements lead to a more accurate determination of $\overline{x}$ but not to a reduction of the standard deviation.[2] Let us now again consider our example of the numbers 2, 3, 5 and 8 and their average 4.5. Then we obtain the standard deviation

$$\sigma = \sqrt{\frac{1}{4-1} \cdot \{(2-4.5)^2 + (3-4.5)^2 + (5-4.5)^2 + (8-4.5)^2\}} \approx 2.65.$$

The interval 4.5 ± 2.65 contains 68.3 % of all measurement values.

---

[2]The commonly used terms "variance" or "dispersion" are the square of the standard deviation.

13.4.2 The Standard Deviation of the Average

More important than the standard deviation is the mean error of the mean value for measurement series. It is referred to as the standard deviation of the average. It describes the average deviation of $\overline{x}$ from the average for many measurements,

$$\overline{\sigma} = \sqrt{\frac{1}{n(n-1)} \cdot \sum_{i=1}^{n} (x_i - \overline{x})^2} = \frac{\sigma}{\sqrt{n}}.$$

That is, the more frequently a measurement is repeated, the smaller the deviation. In statistics this rule is known as $\sqrt{n}$ rule. For our example values we obtain

$$\overline{\sigma} = \frac{2.65}{\sqrt{4}} = 1.325.$$

The quantity $\overline{x} \pm \overline{\sigma}$ defines a confidence interval within which the true value lies with a confidence level of 68.3 %. For $\overline{x} \pm 2\overline{\sigma}$ this is 95.4 % and for $\overline{x} \pm 3\overline{\sigma}$ it is 99.7 %. The above equation shows that the accuracy of the average value only increases with the square root of the number of measurements. When we increase the number of measurements by, e.g., a factor of a hundred, then we obtain only a tenfold increase in accuracy. With an increasing number of measurements the value $\overline{\sigma}$ decreases and thus the difference between $\overline{x}$ and the true value. Hence, the true value of our sample measurements lies with a probability of 68.3 % in the interval 4.5 ± 1.325. For completeness, we note again that the above standard deviations are only valid for a large number of measurements ($n > 100$). In practice this is often impossible and one works with $n \approx 3$–10. In this case, one must multiply $\overline{\sigma}$ with a correction term k. Table 13.1 shows this term for various n.

13.4.3 The Average Error of the Function Value

Unknown variables are often not measured directly, but depend on other parameters which in turn are also subject to errors. These errors thus affect the parameter sought initially, according to their functional dependence. We consider the desired function

Table 13.1 Correction factor k for the standard deviation of the average depending on the number n of measurements

	$n = 3$	4	5	6	8	10	20
68.3 %	$k = 1.32$	1.2	1.15	1.11	1.08	1.06	1.03
95.4 %	$k = 19.2$	9.2	6.6	5.5	4.5	4.1	3.4

F, which should be dependent on n parameters $x_1, \ldots, x_n$. We can expand this function $F = F(x_1, \ldots, x_n)$ into a multi-dimensional Taylor series:

$$F(\overline{x}_1 + \Delta x_1, \ldots, \overline{x}_n + \Delta x_n) = F(\overline{x}_1, \ldots, \overline{x}_n) + \sum_{i=1}^{n} \frac{\partial F}{\partial x_i} \Delta x_i + \ldots \quad (13.2)$$

where the points on the right hand side indicate all terms of higher orders. Due to their errors the x_i are random variables and can be expressed by their average $\overline{x}_i$ as follows: $x_i = \overline{x}_i + \Delta x_i$. The derivatives $\frac{\partial F}{\partial x_i}$ are thus calculated at the point $\overline{x}_i$. Under the assumption that the stochastic deviations Δx_i are small, we can neglect the higher powers of the Taylor expansion and can now calculate the variance of the desired F (=var(F)).[3] The variance operator var($*$) is linear, so we first obtain

$$\sigma_F^2 = \text{var}(F) = \text{var}\left(F(\overline{x}_1, \ldots, \overline{x}_n)\right) + \text{var}\left(\sum_{i=1}^{n} \frac{\partial F}{\partial x_i}(x_i - \overline{x}_i)\right).$$

Because $F(\overline{x}_1, \ldots, \overline{x}_n)$ and $\frac{\partial F}{\partial x_i}$ are constant, and using the stochastic rule $\text{var}(k \cdot x_i) = k^2 \text{var}(x_i)$ with k=const. we can write

$$\text{var}(F) = \sum_{i=1}^{n} \left(\frac{\partial F}{\partial x_i}\right)^2 \text{var}(x_i - \overline{x}_i)$$

or with $\text{var}(\overline{x}_i) = 0$

$$\text{var}(F) = \sum_{i=1}^{n} \left(\frac{\partial F}{\partial x_i}\right)^2 \text{var}(x_i) \quad (13.3)$$

and thus for the standard deviation of the multi-dimensional function F

$$\sigma_F = \sqrt{\left(\frac{\partial F}{\partial x_1} \cdot \sigma_{x_1}\right)^2 + \left(\frac{\partial F}{\partial x_2} \cdot \sigma_{x_2}\right)^2 + \ldots + \left(\frac{\partial F}{\partial x_n} \cdot \sigma_{x_n}\right)^2}. \quad (13.4)$$

The above equation describes the famous Gauss error propagation. With the partial derivatives one determines the change of the function value via the change of the respective measured values. Simply expressed: How does F changes if x_1, x_2, ...change, respectively? A basic note: It may happen that some variables do not affect the function value linearly but at higher orders. Such measurements require increased measurement accuracies, since they have a major impact on the final result. For example, the internal stellar pressure changes with the fourth power of

---

[3] Any existing correlations are neglected here!

its radius. Thus, one has to rely on a highly accurate determination of radii to obtain accurate values of internal pressures.

13.5   Statistical Errors of Equivalent Widths

The measurement of the width of a spectral line width in astronomical spectroscopy is the key measurement for determining the amount of light-absorbing material between the light source and the observer. This may be hot plasma of a star, but also material in the interstellar medium. Statistical error bars in the measurement of equivalent width are an important issue for astronomical observations of high temporal and spectral resolution. The uncertainties reduce the limits of detectability in observations of the order of minutes or even seconds. For such a case one can only consider photon noise, i.e., the statistical behavior of countable ensembles. All other errors are of systematic nature and initially not recognisable mathematically. To obtain an equation for the mean error of the equivalent width, we must find an expression for the equivalent width, consisting only of measurable parameters, and then derive the expression using the error propagation law of Gauss (Sect. 13.4.3).

13.5.1   The Equivalent Width of Spectral Lines

In practice, the measurement of noise within a line can be quite difficult because the photon noise is superimposed on stellar line variations. It is therefore necessary to determine an expression for the equivalent width which separates these two parameters and then to determine an expression for the error σ_λ. We start with the definition of the equivalent width (see Fig. 13.4), as

$$W_\lambda = \int_{\lambda_1}^{\lambda_2} \frac{F_c(\lambda) - F(\lambda)}{F_c(\lambda)} d\lambda$$

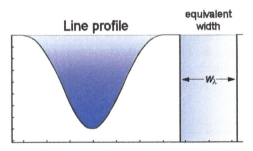

Fig. 13.4 The definition of the equivalent width W_λ. Both areas have the same size

or

$$W_\lambda = \int_{\lambda_1}^{\lambda_2} \left[1 - \frac{F(\lambda)}{F_c(\lambda)} \right] d\lambda, \qquad (13.5)$$

with $F_c(\lambda)$ the radiation flux in the continuum, $F(\lambda)$ the radiation flux in the line at wavelength λ, and $F_\lambda = F_c$ for $\lambda \geq \lambda_2$ and $\lambda \leq \lambda_1$, respectively. According to Eq. 13.5 the equivalent width represents the width of a boxcar of intensity 1 which encloses the same area as the line and the continuum. Moreover, it is obviously not essential to normalize the spectrum corresponding to 1 in order to measure the equivalent width. First, we integrate equation 13.5 separately using the fundamental theorem of integral calculus and we obtain

$$W_\lambda = \Delta\lambda \cdot \left[1 - \overline{\left(\frac{F(\lambda)}{F_c(\lambda)} \right)} \right]. \qquad (13.6)$$

In practice, however, the normalized mean continuum flux is difficult to determine within the line. This is because the photospheric continuum flux is superimposed on the line flux, and thus appropriate reference points for the continuum normalization are not available. Easy to measure, however, are the individual mean values $\overline{F(\lambda)}$ and $\overline{F_c(\lambda)}$. To obtain them, we use the arithmetic mean and substitute the individual fluxes of the line and of the continuum by the respective mean plus corresponding deviations ΔF_i and ΔF_c (see Fig. 13.5). ΔF_i (in contrast to ΔF_c) not only contains noise components but also line information, which can be very intense (the mean $\overline{F}$ corresponds to a rectangle on the line). It can be shown (Vollmann and Eversberg 2006) that

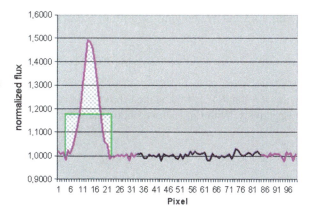

Fig. 13.5 Synthetic emission line with an intensity of 1.5. ΔF_i represents the deviation per pixel from the mean of the whole line. It can be large compared to the line strength. ΔF_c represents the continuum noise which is small compared to the line strength

$$\overline{\left(\frac{F}{F_c}\right)} \approx \frac{\overline{F}}{\overline{F_c}}$$

and Eq. 13.6 then becomes

$$W_\lambda = \Delta\lambda \cdot \left[1 - \frac{\overline{F}}{\overline{F_c}}\right]. \tag{13.7}$$

Thus we can define the equivalent width by the line wavelength interval and the two mean intensities in the line and in the continuum. Since all three parameters are easy to measure, we can now apply the law of error propagation.

13.5.2   The Error of the Equivalent Width

The quantities $F(\lambda)$ and $F_c(\lambda)$ are influenced by statistical errors (photon noise). F_c is generally determined outside the line and interpolated within the line where the line flux F is measured. Now, if we assume that the two corresponding errors are not correlated, we can determine their standard deviations separately. For the expansion of the equivalent width W_λ we obtain

$$W_\lambda = W_\lambda(\overline{F}, \overline{F_c}) + \frac{\partial W_\lambda}{\partial \overline{F}}(F - \overline{F}) + \frac{\partial W_\lambda}{\partial \overline{F_c}}(F_c - \overline{F_c}), \tag{13.8}$$

where $W_\lambda(\overline{F}, \overline{F_c})$ corresponds to Eq. 13.7. The additional terms $\frac{\partial W_\lambda}{\partial \overline{F}}(F - \overline{F})$ and $\frac{\partial W_\lambda}{\partial \overline{F_c}}(F_c - \overline{F_c})$ hence include the present noise by a linear approximation. According to Eq. 13.3 and by applying error propagation we obtain the variance of W_λ as

$$\sigma^2(W_\lambda) = \left[\frac{\partial W_\lambda}{\partial \overline{F}} \cdot \sigma(F)\right]^2 + \left[\frac{\partial W_\lambda}{\partial \overline{F_c}} \cdot \sigma(F_c)\right]^2, \tag{13.9}$$

with $\sigma(F)$ and $\sigma(F_c)$ the standard deviation within the line and in the continuum, respectively. With Eq. 13.7 we obtain the partial derivatives

$$\frac{\partial W_\lambda}{\partial \overline{F}} = -\frac{\Delta\lambda}{\overline{F_c}}$$

and

$$\frac{\partial W_\lambda}{\partial \overline{F_c}} = \frac{1}{\overline{F_c}}(\Delta\lambda - W_\lambda).$$

Fig. 13.6 Synthetic emission line with an intensity of 1:1.05. The uncertainty in the line is determined by the signal-to-noise ratio S/N

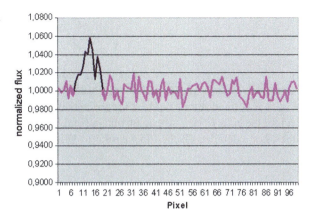

(A) Weak Lines   For weak lines, the line width can be neglected and we have

$$\sigma(F) = \frac{\overline{F}}{S/N} \approx \sigma(F_c) = \frac{\overline{F_c}}{S/N},$$ (13.10)

with S/N the signal-to-noise ratio in the undisturbed continuum (see Fig. 13.6). We obtain

$$\sigma^2(W_\lambda) = \left[\frac{\Delta\lambda}{S/N}\right]^2 \cdot \left[\frac{\overline{F}}{\overline{F_c}}\right]^2 + \left[\frac{\sigma(\overline{F_c})}{\overline{F_c}} \cdot (\Delta\lambda - W_\lambda)\right]^2.$$ (13.11)

The first term corresponds to the photometric uncertainty and the second term to the uncertainty in determining the continuum within the line. According to Eq. 13.7 both terms are identical and we arrive at the intuitively comprehensible result

$$\sigma(W_\lambda) = \sqrt{2} \cdot \frac{(\Delta\lambda - W_\lambda)}{S/N}.$$ (13.12)

(B) Absorption and Emission Lines   For the case of lines with low flux (absorption) and a strong flux (emission; see Fig. 13.5), we use the corresponding Poisson statistics (see Appendix B.2):

$$\sigma(W_\lambda) = \sqrt{\frac{\overline{F_c}}{\overline{F}}} \cdot \sigma(F_c)$$

and obtain

$$\sigma(W_\lambda) = \sqrt{1 + \frac{\overline{F_c}}{\overline{F}}} \cdot \frac{(\Delta\lambda - W_\lambda)}{S/N}. \tag{13.13}$$

In both cases, we can now estimate the standard deviation with the measurable parameters S/N, $\Delta\lambda$, $\overline{F}$ and $\overline{F_c}$ (according to Eq. 13.7 W_λ can be estimated by $\overline{F}$ and $\overline{F_c}$). In addition, Eq. 13.13 represents the general error of equivalent widths and for the case $\overline{F} \approx \overline{f_c}$ we again obtain from Eq. 13.12 the intuitive result for weak lines. For a detailed discussion we refer to Vollmann and Eversberg (2006).

Suggested Readings

- L. Lyons, *A Practical Guide to Data Analysis for Physical Science Students*, Cambridge University Press, 1991
- R. Lupton, *Statistics in Theory and Practice*, Princeton Univ Press, 1993
- G. Cowan, *Statistical Data Analysis*, Oxford University Press, 1997
- S. Brandt, *Data Analysis*, Springer, 1998
- P.R. Bevington, K.D. Robinson, *Data Reduction and Error Analysis for the Physical Sciences*, McGraw-Hill, 2002
- R.J. Barlow, *Statistics: A Guide to the Use of Statistical Methods in the Physical Sciences*, Wiley & Sons, 2008
- J.V. Wall, *Practical Statistics for Astronomers*, Cambridge University Press, 2012

Chapter 14
Massive Stars: Example Targets for Spectroscopy

A Short Story

During his studies in Canada, Thomas was thrilled by stellar winds and now he wanted to explore them with his private telescope. He was dreaming of clumps, outbursts and stellar disks in the wind, the velocity law and new fantastic discoveries with his spectrograph. Klaus was eager to model the non-thermal equilibrium of stellar atmospheres. However, non-spherical inhomogeneous winds were far too complicated for such a task. He first had to perform such calculations for normal O stars. Thomas was strictly against that. He opposed it. But unfortunately Klaus as well was owner of the telescope...

14.1 Some Example Targets

After our purely technical and mathematical considerations about spectroscopy, we finally want to present some worthwhile observing programs. It should be remembered that an outline of all research objectives for astronomical spectroscopy is simply impossible. This would fill several books and is already discussed in the relevant literature. Rather, we want to address the fact that the size and complexity of spectroscopic instrumentation do not change the technical and physical principles. Differences are only in the adaptation to the telescope and the detectors, as we have shown in Chap. 8. This in turn is primarily a question of funding and not of physics. This has pragmatic consequences. Today, professional instrumentation is dominated by heavily oversubscribed telescopes which focus mainly on a limited number of "fashionable" research topics. As a result, time acquisition for massive star research, including extended observation campaigns, becomes more difficult.

Given the recent spectacular instrumental developments for large and small telescopes, astronomical spectroscopy has never been as powerful as today. The general public view is "the bigger the telescope, the better." This is undoubtedly

© Springer-Verlag Berlin Heidelberg 2015

T. Eversberg, K. Vollmann, *Spectroscopic Instrumentation*, Springer Praxis Books,
DOI 10.1007/978-3-662-44535-8__14

correct when we talk about deep-sky observations of low contrast. The spectral signal-to-noise ratio of Quasars and Galaxies observed with small apertures is generally very low and small telescopes hardly play any active role in this research field. This is also valid for professional solar observatories which are well equipped, so that small telescopes can only make small scientific contributions, if at all. Other applications include spectroscopy of planets, nebulae, or studies of the interstellar medium (ISM). But even here generally large instruments are required (the solar-system planets are now examined in situ).

On the other hand, the investigation of stars with small telescopes can fulfil scientific needs. Off-the-shelf, low-cost spectroscopic equipment for such small telescopes can easily be used for scientific investigations of stellar physics, particularly the study of bright emission line stars where line profile analysis of their often fast varying spectra can be performed. For instance, using a standard 10 inch telescope, a signal-to-noise ratio (S/N) of about 100 can be achieved within 30 min for a star of about 8th magnitude in the V band and for a two pixel resolution of about 1 Å. Objects of the order of $V = 10$ mag and fainter require longer exposure times and/or lower S/N. Small telescopes below ~ 1 m aperture can deliver unusual results by taking advantage of their large number and easy access. This is especially valid for spectroscopic long-term campaigns, surveys to support detailed observations by large or space-based telescopes and monitoring of specific spectroscopic parameters over many years. Considering astrophysical studies this is an interesting niche. Basic questions about these objects can easily be illuminated with small instruments. Ongoing detailed questions and measurements can then be carried out with large equipment. The latter could be observations in environments of different metallicity (e.g., Small Magellanic Cloud), particularly high S/N (e.g., high-precision line profile analysis) and remote-distance investigations (e.g. massive stars as standard candles). In the following chapters we will exemplify and examine what may be possible and what physical phenomena play a particular role. For a more detailed overview, we refer to the further readings at the end of this chapter.

14.2 Dots in the Sky

To obtain information about the composition and the three-dimensional structure of stars, spectroscopy is the main tool. Years ago Fraunhofer found absorption lines in the solar spectrum, which are positioned at specific wavelengths and could later be connected to the corresponding element species (Fig. 14.1).

Fig. 14.1 Subsequently colored solar spectrum from Fraunhofer

Fig. 14.2 *Red* or *blue line* shift due to stellar rotation

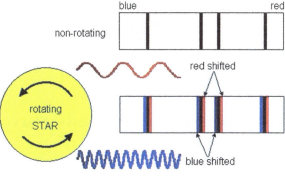

Fig. 14.3 Absorption line broadening by stellar rotation. If the *red absorption line* results from a non-rotating star, the line from a rotating star is broadened as shown in *blue*

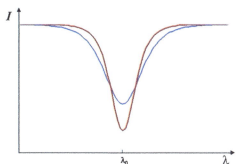

But not only are the spectroscopic findings of various elemental species in the universe directly transferable to all astronomical observations, whether stars or the most distant quasars. In addition, we also get information about the velocities of the absorbing or emitting material relative to the observer via the Doppler effect with $v/c = \Delta\lambda/\lambda_0$. If we choose the rest wavelength λ_0 as a reference point with velocity $v = 0$ in the observation (radial) direction, one can determine the nonrelativistic velocity proportional to the observed wavelength with $\Delta\lambda = \lambda - \lambda_0$. Blue-shifted lines indicate material moving towards the observer and redshifted lines refer to material that is moving away from the observer.

This technique can also be applied to individual stellar absorption lines. Figures 14.2 and 14.3 show the principle for a rotating star. The surface hemisphere which is moving away from the observer emits red-shifted light and the surface hemisphere which is moving towards the observer emits blue-shifted light. The spectral line is broadened by stellar rotation. Since the integrated intensities between the wavelength limits of the line (area between line and continuum) represent the visible light flux and since this amount of light remains constant, the line peak intensity is reduced in the case of line broadening. The blue- and red-shift of the surface thus deliver a parameter for the stellar rotational velocity. Although the star cannot be seen as a two-dimensional image, one can nevertheless obtain additional information to derive the stellar geometry.[1] With spectroscopic investigations one can therefore estimate radial velocities via wavelength shifts as well as identify elements and their abundances. The latter can be performed, e.g., analytically via the curve-of-growth method for mean column densities in a medium.[2]

14.3 The Heavy Weights: Massive Stars

The first things we see when observing distant galaxies, are their brightest stars, mostly blue. Stars with very high luminosities and high temperatures are placed in the upper left part of the Hertzsprung-Russell diagram. From the mass-luminosity relation we also know that $L \sim M^{3.5}$. Apparently, such stars have large masses and they are therefore generally called massive, hot stars. In comparison to the sun the very much higher temperatures of such stars are immediately apparent from a comparison of their spectra (Fig. 14.4). What we first see in galaxies are, hence, massive O and B stars.

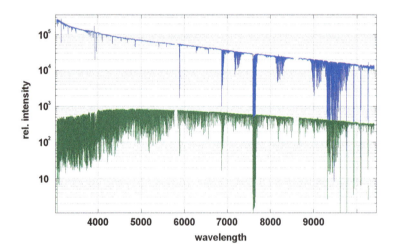

Fig. 14.4 Logarithmic spectral radiation flux of the G5V star HD 59468 ($T_{eff} = 5{,}770$ K, *green*) and the O4V star HD 164794 ($T_{eff} \approx 60{,}000$ K, *blue*) obtained at the VLT. Both spectra are scaled against each other by the Planck intensity at 10,400 Å (UVES Paranal Observatory Project catalogue). In the visual spectral range the O star is about ten magnitudes brighter than the G star

---

[1]Here, we have only considered non-relativistic speeds. However, in spectroscopic studies of distant objects (galaxies, quasars, etc.) we must consider their relativistic velocities. For the observed frequency wavelength λ' this is $\lambda' = \lambda \cdot \sqrt{(1 - \frac{V}{c})/(1 - \frac{V^2}{c^2})}$.

[2]If however layer-dependent abundances are to be estimated, numerical spectra have to be computed and iteratively compared to real spectra.

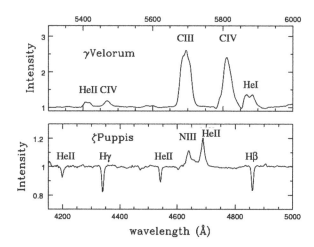

Fig. 14.5 Spectra of the WC8+O binary γ Velorum (*top*) and the O4If-star ζ Puppis (*below*). γ Vel shows various carbon and helium species of different ionization and excitation levels whereas in ζ Pup singly ionized helium (HeII), doubly ionized nitrogen (NIII), and the hydrogen lines Hγ and Hβ can be found (Eversberg et al. 1999)

Normally, stellar spectra show absorption lines, which are produced in a thin layer over the stellar surface, the photosphere. Continuous light from deeper layers is absorbed by elements at specific wavelengths. This is valid for our sun and the vast majority of other stars. In some massive stars, however, emission lines are found in the spectra. An example shows Fig. 14.5 for spectra of the WR star γ Velorum and the O4If star ζ Puppis, both within a distance of about 340 parsecs—virtually in our neighbourhood. In the light of singly ionized helium ζ Pup is about 20 % brighter than in the continuum. A much more extreme example of an emission line can be found in P Cygni (Fig. 14.6). Considering its spectrum one should keep in mind that P-Cygni (with a distance of about 2,000 parsecs, well within our Galactic neighbourhood) is about 13 times brighter in the line peak of hydrogen at 6,562 Å than in continuum light. Elements in its photosphere, though, absorb continuous radiation from the stellar interior only at discrete wavelengths. Hence, emission lines do not originate in the photosphere. There must exist a medium, normally ionized (i.e. a plasma), which generates additional light at these wavelengths. To understand this phenomenon, we need to consider the mechanisms that play a role in the interactions of light and matter in the winds of hot stars in closer detail. These mechanisms are

- Recombination and emission: An ion captures an electron and drops directly into the ground state or it recombines into an excited state. For the latter the transition to the ground state follows in a cascade process. Examples are Hα ($n = 2 \rightarrow n = 3$) and infrared lines in winds of hot stars.
- Pure absorption: An already excited state is excited to an even higher energy state. A transition to a lower energy state is accompanied by spontaneous

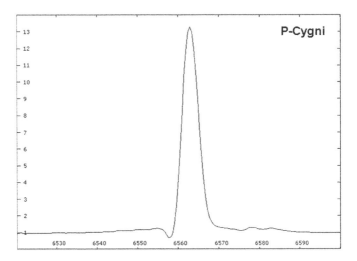

Fig. 14.6 Hydrogen emission of the star P-Cygni (*Courtesy*: C. Buil)

emission. This results in pure absorption of photons of the wavelength of
the excitation photon. In low density plasmas (e.g., stellar winds) collisional
excitation is negligible and radiative transfer processes dominate the energy
transport.

- Line scattering: In this special case of an atomic absorption of a photon
  the atom will be excited and spontaneously falls back to its original state,
  while emitting a photon of identical wavelength. This absorption is called
  (monochromatic) photon scattering because incoming and outgoing photons of
  identical wavelength are indistinguishable. Nevertheless, physically this process
  is pure absorption. For the special case that the transition takes place between the
  ground state and the first excited state, such line scatterings are called resonance
  scattering. The photon is in "resonance" with the first atomic transition. The
  corresponding spectral lines are called resonance lines. Most of the P-Cygni
  profiles are produced by resonance scattering.

- Electron or Thomson scattering: These free-free transitions are due to scattering
  on free electrons and are not tied to a fixed wavelength. They are continuous
  and define the continuum in hot stars in which all elements are almost entirely
  ionized.[3]

These effects give clues for the physics of the observed stars and their winds and
can also supply information on their three-dimensional structure. At higher spectral
resolution we also find absorption lines in the spectrum of the WR+O binary
γ Velorum (Fig. 14.5, see also the WR+O spectrum of WR 140 in Fig. 14.32).
They are, however, all produced in the photosphere of the O companion. Pure WR

---

[3]This is in contrast to H^- transitions which define the continuum in cooler stars.

spectra show mainly emissions of elements such as helium and carbon, which are nuclear-burning products from the stellar core, blown away by the wind. Thus, the hydrostatic stellar "surface" (no longer defined by a photosphere) for most WR stars is apparently hidden behind an optically thick wind (a more detailed consideration is given in Sect. 14.8). In contrast, the winds of O-stars are apparently optically thin because we can see many photospheric absorption lines in their spectra. Only in Of-stars such as ζ Pup (Fig. 14.5) can some emissions in hydrogen, helium and nitrogen be found. They are produced near the stellar photosphere. Only here is the wind optically thick enough to reveal detectable scattering processes (a more detailed consideration is given in Sect. 14.7).

In principle the entire facing hemisphere is observed through a telescope with its spectrograph. Now if the star has a radially symmetric wind, the plasma scatters stellar light in the observer's direction according to the above line-scattering process, which would be lost without scattering. The scattering process increases the solid angle of the star as seen by the observer (figuratively the stellar surface is "unrolled") and an additional amount of light at the scattering wavelength is the result. Different radial velocities relative to the observer average out; we see an emission at rest wavelength of the scattering process.[4] On the other hand, the material in front of the star absorbs stellar light, reduces its intensity and we see an absorption. However, this absorption is often blue-shifted, which requires a component of outward moving stellar wind. With such observations, we have a casual proof for a spherically symmetric wind moving radially symmetrically outwards.

In principle however, one should always keep in mind that spectroscopically we can only measure events in velocity space with respect to the line-of-sight. The wavelength shift of any absorption or emission reflects a velocity in the (radial) direction of observation via the Doppler effect. Information about the real geometry in the spatial domain are obtained with additional parameters (rotation speed, stellar diameter, wind speed etc.). For clarification, the illustrations 14.7 and 14.8 show iso-wavelength contours for a rotating disk of gas whose particles move on Keplerian orbits above the stellar equator and for an equatorial disk with pure radial expansion (Figs. 14.7 and 14.8) Each packet of wind material which is located along the lines of equal radial velocity provides identical wavelength shifts in the recorded spectrum. Light of red- or blueshifted atomic lines or certain line components can thus have different geometric origins.

---

[4]It is generally well known that WR lines are significantly shifted relative to their true motion, mostly red-shifted, but sometimes blue-shifted.

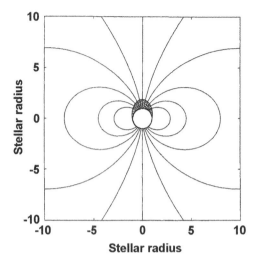

Fig. 14.7 Isowavelength-shift contours, $\frac{\lambda_0-\lambda}{\lambda} \cdot 10^{-3} = $ constant, on a Keplerian rotating disc viewed by an observer inclined at $60°$ to the rotation axis and positioned at infinity in the x–z plane. The hatched region is the area of the disc occulted by the star. Both axes are labeled in units of stellar radii (Wood et al. 1993, reproduced with permission ©ESO). Each packet of wind material along the contour lines delivers identical wavelength shifts. For instance, photons which are emitted at the two different x–y-locations $(-10, -3)$ and $(-2, -4)$ deliver a spectral signal at the same wavelength position

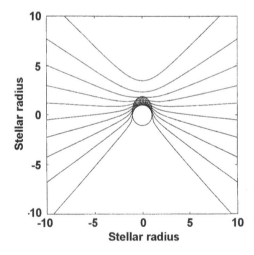

Fig. 14.8 As for Fig. 14.7, but for the disc velocity being purely in expansion (Wood et al. 1993, reproduced with permission ©ESO)

14.4   Winds That Sail on Starlight

The key parameters for all stars are the stellar mass, radius, composition and age. The greater the mass and the smaller the stellar radius the higher is the effective temperature. This is valid for main-sequence stars spending most of their live in the helium-burning stage and representing about 90 % of all stars. If the temperature is sufficiently high and the gas density low enough, radiation pressure dominates gravitation. The corresponding threshold luminosity is:

$$L_{ed} = \frac{4\pi \cdot G \cdot c}{\overline{\kappa}} \cdot M.$$

With $\overline{\kappa}$ the average opacity of the medium.[5] The gravitational constant is G and the stellar mass M. L_{ed} is the so-called Eddington luminosity or the Eddington limit. Up to this luminosity the star normally remains in hydrostatic equilibrium. When the luminosity exceeds the Eddington limit the atmosphere (and possibly even the whole star) becomes unstable and extreme, variable mass loss in the form of a radiatively driven wind flowing outwards is the result. Often the emission lines are broadened, have a flat line peak (plateau) and do not show a Gaussian shape. In addition to that, as in Fig. 14.6, a small blue-shifted absorption is sometimes observed beside the strong (in this case P-Cygni Hα) emission. As already mentioned, this particular line shape can be explained by an unstable atmosphere, moving outwards at high velocity and high mass-loss rate. Figure 14.9 illustrates this explanation. The increasing intensity towards the line center or rest-wavelength is the result of the interaction of light with material in an optically thin wind, whose spectral effects we will examine in more detail in Sect. 14.8.

Generally speaking, emission lines are convincing arguments for the presence of ionized material in the direct vicinity of stars. They can be found in the light of hydrogen and various stages of ionization of many elements, e.g., Helium, silicon, carbon and nitrogen. These elements can have only one ultimate source— the interior of the star. With their relatively high apparent brightness in the sky and an interesting physical wind phenomenon, massive stars ($M > 8M_{\odot}$) are important for studies with smaller telescopes and candidates for repeated measurements. On the other hand, they offer themselves as an indicator for measurements with large telescopes in order to examine the validity of the physical processes and phenomena in other environments such as our own Galaxy.

---

[5]The average opacity $\overline{\kappa}$ of a medium is calculated with the wavelength depended absorption coefficient $a(\lambda)$. It is integrated over the distance d the light travels in the medium. $\overline{\kappa} = \int_{o}^{d} a(r, \lambda)dr$. If photons of wavelength λ travel through a medium of opacity κ and density ρ along the path r the light intensity will be reduced by the amount $I(r) = I_0 e^{-\kappa \rho r}$.

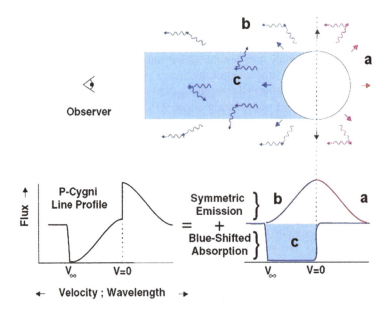

Fig. 14.9 How to produce a P-Cygni profile (Owocki 1998). In region **a** photons are scattered by material which moves away from the observer. Depending on the scattering angle between the line-of-sight and the wind trajectory the scattered photons are more or less redshifted and, accordingly, generate the red emission wings (strong/weak intensities = many/few scatterings). In region **b** the scattering is carried out analogously by material moving towards the observer. It produces blue emission wings. In region **c** photons are absorbed by material moving towards the observer and generate a blue-shifted absorption with respect to the rest wavelength. The absorption increases with the amount of radial absorbing material and ends at terminal wind velocity, V_∞. The two scattering and absorption processes in the stellar wind are observed simultaneously, and therefore the two emission and absorption components in the spectrum add up to a P-Cygni profile

14.5   The Velocity Law

The expansion behaviour for stellar winds $V(r)$ is described by the so-called velocity law. As long as the wind speed is not directly detectable by observational effects (e.g., wind-dependent sub-structures in the line, so-called clumping or blobs, see Sect. 14.9) we depend on theoretical analysis. With various assumptions for a radiation-driven gas, Castor and Lamers (1979) introduced a commonly used Ansatz. This is

$$V(r) \approx V_0 + (V_\infty - V_0) \cdot \left(1 - \frac{R_\star}{r}\right)^\beta .$$

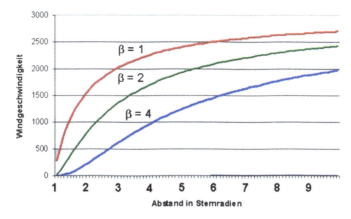

Fig. 14.10 The velocity law for stellar winds. $V_\infty = 3{,}000$ km/s. *Red:* $\beta = 1$. *Green:* $\beta = 2$. *Blue:* $\beta = 4$

At the stellar surface ($r = R_\star$) the velocity V_0 is very small ($V_0 < 1$ km/s) and because of very high wind velocities observed, we can approximate to

$$V(r) \approx V_\infty \cdot \left(1 - \frac{R_\star}{r}\right)^\beta \tag{14.1}$$

with V_∞ the terminal velocity, $R_\star$ the stellar radius, r the distance from the star center and the exponent β the so-called "beta value". Figure 14.10 shows the outwardly directed wind velocity versus distance from the star for $V_\infty = 3{,}000$ km/s. The wind follows an accelerated motion, which asymptotically approaches the terminal velocity V_∞. The velocity law is parameterized by the value of β. The larger β the slower is the wind at a certain distance r. β is usually obtained from theoretical model calculations. Hot, massive O stars dramatically accelerate their winds due to extreme radiation pressure. These winds are well represented by $\beta = 0.8$–1.2. For cooler stars, like the sun, where the wind is driven by a pressure gradient, high acceleration cannot be achieved and we find $\beta > 5$. In Sect. 14.9 we introduce a method to directly determine β.

14.6   Aspheric Geometries: Be Star Disks as Prototypes

In 1866 the Jesuit Angelo Secchi, director of the Collegio Romano at the Pontifical Gregorian University in Rome discovered strong hydrogen emission lines in the spectrum of the B star γ Cassiopaiae (Fig. 14.11). This finding has since been accounted for in the Hertzsprung-Russell classification with an index "e" for emission lines. Since then, γ Cas is considered as a Be star of spectral type B0.5 IVe (Lesh 1968). This star has been widely studied; the primary knowledge of Be stars

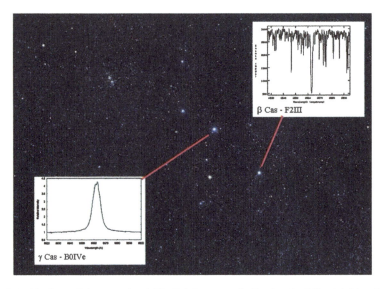

Fig. 14.11 Hα lines of the second and third brightest stars in Cassiopeia β Cas (*right*) and γ Cas (*left*). The latter is the apparently brightest Be star in the sky (*Courtesy*: C. Buil)

comes from it. The origin of the (wavelength) unshifted Hα emission with often undisplaced temporary central reversal is an extended disk out of stellar plasma (solar abundance with metals exhibited by e.g. FeII lines) above the equator with an extension of about 20 stellar radii. This disk is embedded in a spherically-symmetric stellar wind. Figure 14.12 shows an artist's sketch of such a star. Be stars are B-type stars with permanent Hα emission, indicated as Be for luminosity class III to V[6]

The physical explanation for the disks of Be stars is complex and still in debate today. For simplicity one can consider the inclination of the rotation axis of a Be star relative to the observer (pole-on (0°), edge-on (90°) and intermediate angles). This angle also has an impact on the stellar disk Hα emission, because of the deviation from spherical symmetry (Fig. 14.13). In addition, one must take into account whether the disk gas is transparent (optically thin, low opacity) or opaque (optically thick, high opacity). This is important since for the case of optically thin material one can see the <u>whole</u> emitting volume, thus each emitter. On the other hand, in the optically thick case one sees only the outer surface of the emitting volume.

---

[6]B[e] stars show forbidden emission lines in their spectra. The group is very heterogeneous and contains, e.g., massive supergiants, pre-main sequence stars and symbiotic stars. Because of this variety Lamers et al. (1998) propose to use the name "B[e] phenomenon" rather than B[e] stars dividing these objects into the five classes (a) B[e] supergiants or *sgB[e] stars*, (b) pre-main sequence B[e]-type stars or *HAeB[e] stars*, (c) compact planetary nebulae B[e]-type stars or *cPNB[e] stars*, (d) symbiotic B[e]-type stars or *SymB[e] stars* and (e) unclassified B[e]-type stars or *unclB[e] stars*.

Fig. 14.12 Artist's sketch of a Be star (*Courtesy*: Bill Pounds/STScI). An extended plasma disk of solar abundances is located above the equator of a hot B type star. The disc is geometrically very thin and has a relatively small opening angle of a few degrees as seen from the centre of the star. The plasma material of approximately 10^4 K is excited by the strong UV radiation of the central star and recombines by radiating visual light. The disk particles move on Keplerian orbits with a typical orbital period of several years. The radially symmetric outmoving stellar wind with a temperature of about 10^5 K is not shown

The material optical density, the dynamics and the simultaneous consideration of both parameters significantly complicate the interpretation of the observed spectra. And even more: The disk particles rotate in Keplerian orbits with a typical period of a few years. This rotation introduces strongly different radial velocity distributions for each viewing direction (see Fig. 14.7). Considering the additional radially symmetric fast outmoving wind, the spectral interpretation becomes challenging (see Fig. 14.8) because all effects are simultaneously observed.

If we consider the radial velocity field (i.e., with respect to the direction of observation) of the Keplerian disk when looking towards its edge (inclination 90°, edge-on) in more detail, we consider different locations in the disk: For radial velocities $RV \cong 0$ we look right towards the center of the star. The disk edge is projected in front of the star and absorbs its light. The light emitting material column is large (the radial disk extension is up to 20 $R_\star$), but the corresponding solid angle is relatively small (vertical view towards the disk edge). For intermediate radial velocities, we look along the stellar surface. The disk is also visible but does not absorb starlight. The light emitting volume and its surface are large. If we observe high radial velocities we look tangentially along the disk. The observed disk solid angle is located beside the star and does not absorb the light. The light emitting surface and the volume are small.

For all these geometric considerations the opacities of the respective spectral lines have to be considered for an appropriate spectral interpretation.

Optically thin lines:

- The **line center** represents the geometrical position of the Keplerian disk in front of the star. Because of its orbit geometry its radial velocity (RV) = 0. The material column in front of the star contributes relatively little light to the disk emission. The line profile shows a local minimum at this wavelength position. The line is optically thin and, hence, does not absorb stellar light. A real absorption below the stellar continuum is not observed.

- The **line flanks** represent mean orbital velocities and produce maximum emission contributions due to a relatively large solid angle.

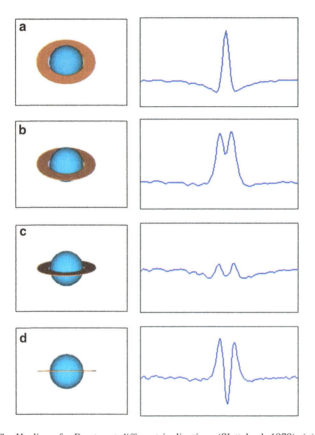

Fig. 14.13 The Hα line of a Be star at different inclinations (Sletteback 1979). (**a**) Observation towards high latitudes. The photospheric absorption is almost entirely superposed by the disk recombination emission. The latter is visible as a broad depression due to rapid stellar rotation and relatively high atmospheric pressure (many free electrons → Stark effect). The outer wavelength regions of the line, hence, come from geometrically inner regions at the photosphere, while the central wavelength ranges are produced in the geometrically distant disk material. (**b**) Observations towards average latitudes. The photospheric absorption is completely superposed by electron scattering in the line wings (see Fig. 14.14). Electron scattering exists in all line regions but is covered by disk emissions around the line center. The central disk emission is reduced by self-absorption of the disk (shell effect). (**c**) Observation towards low latitudes. The disk emission decreases. The self-absorption by the disk grows. (**d**) Observation towards the equator. The central reversal becomes real non-wavelengthshifted absorption (the disk movement has no observable radial component)

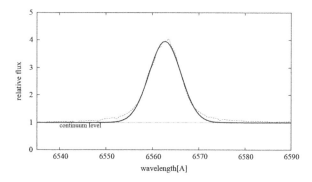

Fig. 14.14 Comparison of the Hα line of the Be star γ Cas (*dashed*) with an appropriate Gaussian plus constant fit to line and continuum (*solid*). The *dotted line* represents the continuum within the line. The line wings show significantly larger intensities due to electron scattering (Nemravová et al. 2012, reproduced with permission ©ESO)

- The **line wings** represent regions of maximum radial velocities. The line intensities decrease towards the continuum because there are only few emitting particles at large radial velocity.

Optically thick lines:

- In the **line center** (RV = 0) the emission is small because of a small emitting surface area. The optically dense material absorbs stellar light and absorption below the continuum is possible.
- The **line flanks** (average RV) produce maximum emission again because of a maximum emitting surface.
- The **line wings** (high RV) fall towards the continuum again because of a small emitting solid angle.

The essential difference between optically thick and optically thin material thus manifests itself in the central region of the line, which, because of the disk temperature of about 10^4 K cannot fall to zero. The extended wings of the line are to some extent caused by the scattering of stellar and disk photons on free electrons of high thermal velocity (Thomson scattering, Fig. 14.14). This occurs in all directions, and the line is thermally broadened by the photons. The underlying photospheric line broadening can be explained by thermal energy (Doppler center) and the Stark effect (line flanks; both define the Voigt profile), rotation (inclination dependent: 0–400 km/s) or electron scattering. At high inclination rotational broadening dominates (it smears-out the Voigt profile). For a geometric interpretation we thus must take various line effects into account.

In addition, we find temporal variations in the Hα line of one and the same star. The so-called V/R variations are cyclical changes of the intensity ratio of violet (blue) and red line peaks in Hα (see Fig. 14.13b). The period is typically a few years (see Fig. 14.15). Already in the early 1990s Okazaki (1991) connected this abnormality to density variations within the disk. These significant density

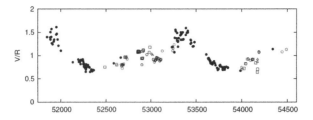

Fig. 14.15 Long-term V/R, Hα emission strength of ζ Tau during almost 10 years. *Filled circles* denote simple double-peaked profiles, *empty circles* the triple-peaked profiles, *squares* the asymmetric double-peaked profiles, *triangles* the quadruple-peaked profiles, and *asterisks* the unclassified profiles (Ruždjak et al. 2009, reproduced with permission ©ESO)

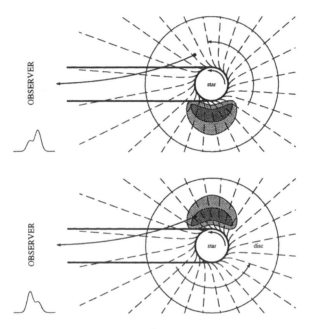

Fig. 14.16 Observations of V/R variations in Hβ and their simplistic interpretation for γ Cas. The polar and equatorial wind components are projected into the equatorial plane. *Gray areas* represent regions of higher density which rotate around the star on time scales of the V/R variations in the Balmer lines. The *dashed lines* correspond to the trajectories of individual wind particles in the polar wind according to Eq. 14.1. The *solid lines* represent the path of the dense material. The structure starts near the equatorial plane, close to the star and migrates outwards, influenced by radial and rotational velocities. The azimuthal density enhancement ends in the line of sight to the observer only for a specific starting point. *Above*: The density enhancement moves away from the observer: we see V < R. *Below*: The density enhancement moves towards observers: we observe V > R (Telting and Kaper 1994, reproduced with permission ©ESO)

differences have since been referred to as a "one-armed disk" (Fig. 14.16). The reason for the one-armed density fluctuations is presumably the kinematic behavior of the wind particles. A considerable proportion of matter moves in elliptical orbits triggered by wind instabilities. These orbits precess. This precession across the entire disk forces the particle paths to converge at one point and thus generate a higher particle density. This region of higher density in turn rotates around the star. Since the recombination emission depends on the density squared ($I \sim \rho^2$) the density fluctuations have significant impact on the line intensities. This means there will be more material on the red line wing during one epoch (dense material moves away from the observer) and more material on the blue line wing during the other epoch (dense material moves towards the observer).

The hydrogen emission is produced in the relatively cool 10,000 K gas disk. Using data from the International Ultraviolet Explorer (IUE) in the late eighties, P-Cygni profiles in UV lines of high excitation energy were discovered in Be stars. Be stars have therefore also radially symmetric winds with temperatures of up to 100,000 K, which flow outwards with nearly 2,000 km/s. The key question is how the stable and relatively cool disk without a radial velocity component can permanently survive within a fast and ten-times hotter wind. Since then various explanations have been suggested. These include a fast-rotating star that throws off material into the equatorial plane, where it forms a gas disk. Alternatively, binary stars have been proposed which form an accretion disk (but not all Be stars have a companion). Today, it is assumed that both the wind and the disk are generated by stellar radiation pressure. The origin of the disk remains unanswered in this model. Be-star rotation periods are about a day and the surface speed is only slightly below the critical speed at which the star is disrupted by centrifugal forces. Be stars have therefore the form of a highly flattened ellipsoid (Fig. 14.17).

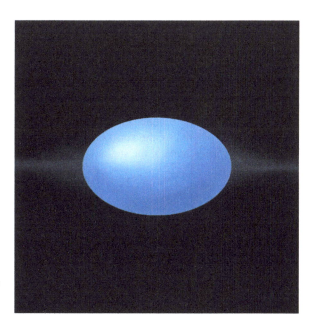

Fig. 14.17 The rotationally induced severely flattened Be star α Eridani (Achernar) (from Wikipedia).

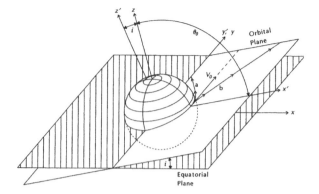

Fig. 14.18 Orbital trajectories of particles in the wind of a fast rotating Be star for the surface origin at different polar angles θ_0. The orbital plane is tilted against the y-axis by the angle i. The *dashed line* shows the original velocity vector tangential to the stellar surface. Trajectory (**a**) represents the case for large rotational velocities in the wind whereas trajectory (**b**) represents the case for low rotational wind speeds. Only trajectory (**a**) for large rotational velocities crosses the stellar equator and trajectories from the other hemisphere. When the southern and norther wind components meet above the equator with supersonic velocities, a shock of sudden material density increase is the result (Bjorkmann and Cassinelli 1993)

A disk scenario has been developed by designing a semi-analytical Be star model for the gas and the stellar wind together. Due to the fast rotation the wind particles follow an inclined orbital trajectory and eventually cross the equatorial plane (Bjorkmann and Cassinelli 1993). There they collide with gas from the other hemisphere and now form a compressed material disk above the equator. Figure 14.18 describes this effect and Fig. 14.19 shows the results of this wind-compressed disk (WCD) model for different rotational velocities of the Be star.

An interesting effect in the model code for fast rotators has initially not been considered—the strong flattening of the star by its fast rotation. Due to the ellipsoidal form of Be stars, the radiation is stronger at the poles than at the equator (Fig. 14.20). This effect is called gravity darkening. The reduced equatorial radiation flux has been explained by the von-Zeipel theorem. It sets the light flux in relation to the effective gravitation at the stellar surface (v. Zeipel 1924). The theorem is

$$F(P) = -\frac{L(P)}{4\pi G M_\star(P)} g_{eff}$$

with the luminosity L, the pressure P, the stellar mass $M_\star$ and the effective gravitational acceleration g_{eff} at the stellar surface. As already mentioned, due to fast stellar rotation, centrifugal forces in Be stars according to $F = m \cdot \Omega^2 \cdot r$ ($\Omega =$ angular speed, $\rho =$ radial distance from the rotation axis) introduce a significant deviation from spherical geometry towards ellipsoid forms (Fig. 14.17). The local

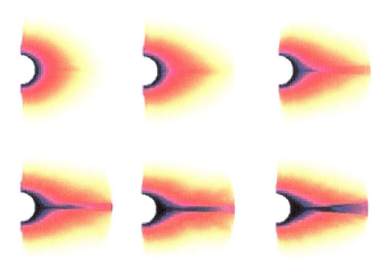

Fig. 14.19 Equatorial disk material density of a rotating Be star for various rotational velocities (200, 250, 300—*top row*, left to right and 350, 400, 450 km/s—*bottom row*, left to right). The critical rotational velocity when the star is disrupted by centrifugal forces is about 500 km/s (Bjorkmann and Cassinelli 1993)

Fig. 14.20 Sketch of gravity darkening for a hot B star and different rotational velocities from 0 to almost 500 km/s. The colours indicate increasing radiation flux from blue to white which is significantly enhanced at the poles for fast rotation

effective gravitation in hydrodynamic equilibrium can then be expressed as

$$g_{eff} = \frac{G \cdot M_{\star}}{R_{\star}^2} - \frac{V_{rot}^2 \cdot \cos^2(\varphi)}{R_{\star}}$$

with the stellar mass $M_{\star}$, the stellar radius $R_{\star}$, the gravitation constant G and the stellar latitude φ. With g_{eff} for a certain stellar latitude one, hence, can estimate the local radiation flux F for fast rotating stars. On the other hand, the radiation flux depends on the fourth (!) power of the effective temperature and we have $T_{eff} \sim g_{eff}^{1/4}$ (Maeder 1999). As a result the equatorial temperatures are lower than at the poles because of reduced gas pressure. Pictorially speaking, the stellar surface experiences reduced gravity at larger distances from the stellar center. This in turn results in a reduced temperature at the equator. For this reason, flattened stars of a given

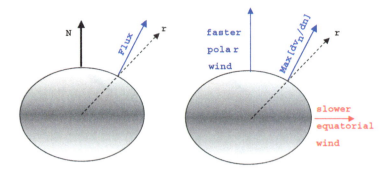

Fig. 14.21 Radiative force from a fast rotating star. *Left*: The stellar oblateness introduces a poleward tilt of the radiative flux vector from the stellar surface. *Right*: The wind speed scales with surface escape speed, which is smaller at the equator due to reduced local gravity (see Fig. 14.20), indicated by the maximum speed gradient in surface normal direction $\text{Max}[dV_n/dn]$. Both effects result in a poleward tilt of the velocity gradient preventing the creation of an equatorial material disk (Owocki 2004, image courtesy Stan Owocki)

luminosity appear as later spectral types than they are in reality when observed at larger inclination (near edge-on). The stellar flattening however means that the vector of the polar wind components prevents the formation of a disk. This is shown in Fig. 14.21. The situation can be highlighted by a simple image: *Imagine a Be star, which is as flat as a postage stamp. Because of the gravity darkening, radiation comes only from the surfaces. Then there is no radiation pressure towards the edge which could form a disk* (Owocki 1996). The question of how a stable and relatively cool equatorial material disk can survive within a fast radial-symmetric outmoving wind that is about 10 times hotter is still unknown.

As already explained, the various atomic ionization stages are also an indicator of the prevailing temperatures. If we now assume a decreasing temperature with distance from the star, and if we put the latter in relation to the excitation energies of various spectral lines, we then at least qualitatively reach conclusions about the radial distances at which the various elemental species arise. This is illustrated in Fig. 14.22, which has been obtained from interferometric measurements of γCas.

About 20 years ago X-rays were detected in γ Cas. According to the required temperatures it was entirely unclear what mechanisms if any can produce such light of extreme energies in a Be star—even in the stellar chromosphere such temperatures should not occur. Simultaneous measurements by the Rossi X-Ray Timing Explorer (RXTE) in X-rays and the Hubble Space Telescope (HST) in the UV have shown that magnetic flares collide with material of the Be star disk. The resulting collision shocks take place at temperatures of up to 100 million K (Fig. 14.23; see also Smith et al. 1998; Smith and Robinson 1999).

Many different physical processes have been observed in Be stars and a consistent overall model has not yet been developed. Due to the extremely complex problems and phenomena at all time scales, additional observations are still highly

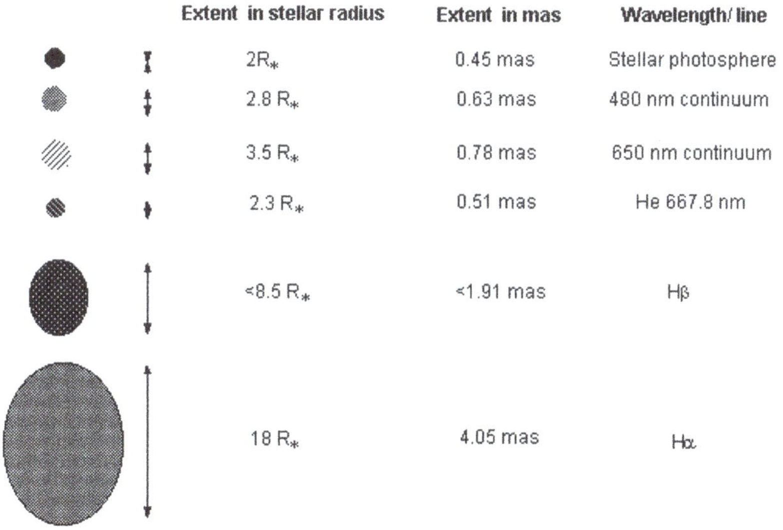

		Extent in stellar radius	Extent in mas	Wavelength/line
●	↕	$2R_*$	0.45 mas	Stellar photosphere
◉	↕	$2.8\,R_*$	0.63 mas	480 nm continuum
▨	↕	$3.5\,R_*$	0.78 mas	650 nm continuum
◕	↓	$2.3\,R_*$	0.51 mas	He 667.8 nm
⬤	↕	$<8.5\,R_*$	<1.91 mas	Hβ
⬭	↕	$18\,R_*$	4.05 mas	Hα

Fig. 14.22 Different regions of the Be star γ Cas at different wavelengths estimated by inter-ferometric observations (Stee et al. 1998, reproduced with permission ©ESO). The reason for the elliptical shape which appears pronounced at larger distances from the star, is the system inclination and the equatorial disk

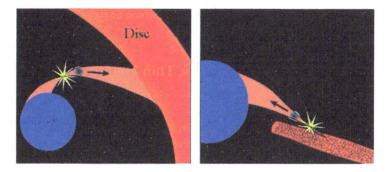

Fig. 14.23 Interaction between a magnetic flare and the disk in γ Cas (Smith et al. 1998)

desirable. We want to emphasize again that Be stars are not uncommon in the sky and relatively bright. Even with small telescopes spectroscopically valuable data can be obtained. This is especially valid for long-term monitoring campaigns. As an example Fig. 14.24 shows the Pleiades and the spectra of their brightest stars. Even in this limited number of stars we find four Be stars with Hα emission.

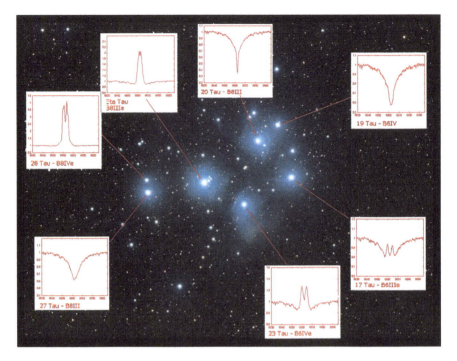

Fig. 14.24 The Pleiades and the Hα lines of the seven brightest Be stars in this star cluster (Buil 2006)

14.7 O Stars: Extreme Radiators, Thin Winds and Rotating Shocks

O stars are the intrinsically brightest known stable stellar objects in the Universe. They show strong and fast winds of relatively low density, which are driven by their intense radiation field. In the hottest, brightest O-stars we find luminosities which are a million times higher than the luminosity of our Sun and mass loss rates reaching $5 \cdot 10^{-6}$ solar masses per year, and more in some cases. That is, a bright O-star can lose one solar mass in less than a million years. Terminal wind velocities of 2,000 km/s and more are found, and surface temperatures of up to 60,000 K are measured. An example is the apparently brightest, hot O supergiant in the sky, ζ Puppis. It is a V = 2.26 magnitude star in the southern constellation Quarterdeck (Puppis), has an approximately 20-fold solar diameter with about 50 solar masses and an effective temperature of about 42,000 K. Its terminal wind speed is about 2,250 km/s and it loses a solar mass about every 300,000 years.

O stars exhibit a variety of spectroscopic line effects, which can only be explained by dramatic changes in their winds. In addition, O stars provide the greatest contribution to the radiation field of their corresponding host galaxies (they are the first stars we see in galaxy photos). As seen in Fig. 14.5 Of stars show relatively

Fig. 14.25 HeII line of the O4If supergiant ζ Puppis (see Fig. 14.5) at a 2-pixel resolution of 0.03 Å. The average spectra for two nights show an obvious change in the central line reversal (Eversberg et al. 1998), probably related to modulations on the 5.1-day rotation period (Moffat and Michaud 1981)

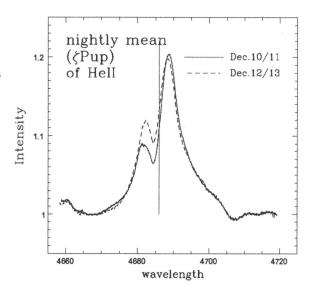

weak emissions with peaks at only ∼20 % of the continuum intensity. Although Of stars have moderately strong winds compared to normal O stars, they are weaker than WR winds (see their relatively weak emission lines). Because one can still see photospheric absorptions in most Of stars we have a clear indicator for a low optical wind density.

14.7.1   Discrete Absorption Components and Co-rotating Interaction Regions

In some O stars (mainly O supergiants) we find HeII emission at 4,686 Å. This line is the result of the atomic transition $n = 4 \rightarrow n = 3$. The upper level has an excitation energy of 54.5 eV and is therefore generated at temperatures of over 600,000 K, just above the photosphere of the stellar wind.[7] This line has frequently been the aim of various studies and it was shown that it is often variable over periods of days (Fig. 14.25). The cause of this periodic variability has not yet been conclusively found. But different observations of OB stars (i.e. all O stars and the brighter B stars) hint at a mechanism that is also known in the solar wind.

With the launch of the IUE satellite, UV lines of silicon and carbon were accessible for extensive studies, not performable from the ground. For example,

---

[7]The temperature concept is inaccurate in this regard and we introduce it only as a reference. Because of the very thin plasma in non-thermal equilibrium the energy transfer is not performed by particle collisions but by radiation transfer from the stellar short-wave UV radiation. The energy transfer is therefore not calculated with the Planck function but with the general source function.

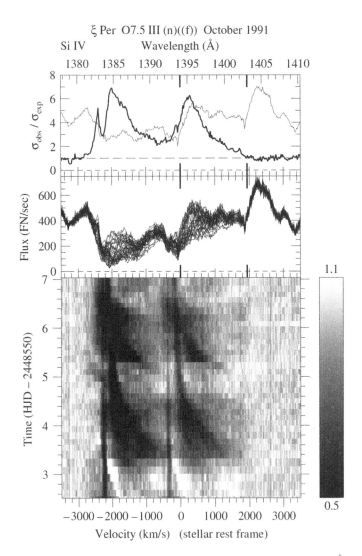

Fig. 14.26 Temporal behavior of the UV doublet of Si IV at 1,394 and 1,402 Å of ξ Persei observed with IUE. *Above*: Average spectrum (*thin line*) to generate the gray scale plot of the residuals and the corresponding variability (*thick line*). *Middle*: Plot of all recorded spectra versus time. *Below*: Excess residuals compared to the average spectrum plotted in time as grayscale intensities. Multiple DACs are visible (Kaper et al. 1997)

temporal changes of the lines were determined by averaging all successively recorded spectra of number N, say. Then this average spectrum was subtracted from all single individual spectra, providing a series of N so-called residual spectra, which only contain the spectral effects deviating from the mean. These residuals can be time-varying excess absorptions or emissions. They can then be stacked in time to estimate the time-dependent behavior of the line changes. In Fig. 14.26

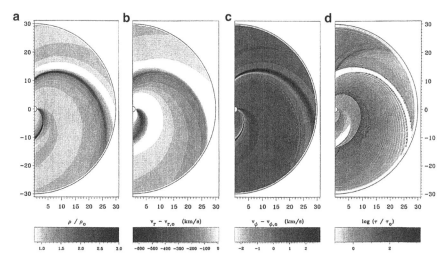

Fig. 14.27 Hydrodynamic simulation of DACs caused by collisions between low and high speed currents, starting at the surface of the star. Slowly outward flowing material is overtaken by faster wind material, is compressed and then forms high density spiral structures due to stellar rotation. (**a**) Density, (**b**) radial velocity, (**c**) azimuthal velocity, (**d**) radial optical density. All normalized to the undisturbed wind (Cranmer and Owocki 1996)

such residuals are plotted as intensity gray-scales for the above mentioned IUE measurements. The time series of the SiIV doublet shows periodic excess absorption in the lines that moves slowly from close to line centre to blue wavelengths, reaching terminal speeds of the P-Cygni profile within some days. The vanishing signature is then succeeded by the following one. These semi-periodic excess absorption features are called *Discrete Absorption Components* (DACs) within the absorption trough of the P Cygni profile from which these structures emerge. These observations point to corotating material which periodically moves away from the star. An explanation has been suggested by means of hydrodynamic models under the assumption that the stellar rotation leads to collisions between low and high speed flows along synchronously rotating interaction regions (the Ulysses spacecraft saw exactly this in the solar wind). The resulting spiral-form shock fronts dramatically compress the material in the wind and increase the optical density, which then appears as a noticeable absorption in the lines (Fig. 14.27). These fronts are called *Corotating Interaction Regions* (CIR). Figure 14.28 shows a corresponding sketch. The similarity to the spiral density enhancements above the solar equator (Fig. 1.2) is striking and suggests that this could be a widespread effect in all stars with winds.

Fig. 14.28 Sketch of four
Corotating Interaction
Regions. *White*: Low density.
Yellow: Medium density. *Red*:
High density. *Blue*: Shock
center. *Courtesy*: Steven
Cranmer

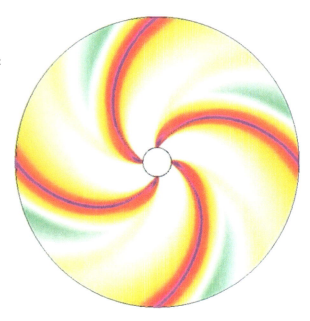

14.7.2 Turbulent Wind Clumps

First indications that the winds of massive stars are further structured were found
by Robert (1992) in optical WR spectra—later elaborated by Lepine and Moffat
(1999)—and by corresponding X-ray observations (Chlebowski et al. 1989; Hillier
et al. 1993). This is important because local densities in the wind (clumps) have
significant impact on the mass loss rates to be determined, which are usually
estimated from presumably unperturbed Hα lines or from radio fluxes.[8] The light
energy emitted from a volume is calculated as

$$dE = j_v dV dt d\nu d\Omega$$

with j_v the line emissivity, dV the emission volume, dt the temporal emission
interval, $d\nu$ the line frequency interval and $d\Omega$ the emission solid angle. The
emissivity generally depends on the product of the interacting ions and electrons:

$$j_v = N_i \cdot N_e \cdot \alpha_{rad} \cdot h\nu.$$

---

[8]This is valid for recombination lines as Hα but not for resonance lines.

with α the recombination coefficient to the upper energy level and N_i and N_e the ion and electron number densities of the emitting volume. The line intensity along the respective path s is calculated by

$$dI_\nu(s) = j_\nu ds.$$

The potential cause for an incorrect mass loss rate determination is therefore the fact that the line intensity is determined by the product of N_i and N_e, i.e., with N_i being proportional to N_e, the density square of the medium ($I \sim \rho^2$). For clarification, let us consider a stellar wind with one single density enhancement. With the same amount of material this clump would emit more recombination or free-free light than a non-structured homogeneous wind because if its higher density. In other words, if the wind is clumped the observed amount of light is then generated from less material. The true mass-loss rate would be lower than without clumps in the wind (Moffat and Robert 1994). Clumps are also a "tracer" for direct information on the behavior of the wind speed. Determining their wavelength positions at different times gives the corresponding velocities and their acceleration and are therefore excellent indicators for the β value in the velocity law and the source distance from the stellar surface (see Sect. 14.9).

Usually mass loss rates are determined from Hα line models. In main-sequence stars these are absorption lines. In Of-star spectra Hα is also in emission that normally occurs relatively close to the star ($\sim < 2R_\star$) at high densities and whose strength depends on these densities. The photospheric and wind temperatures are then about 30,000–50,000 K. Because corresponding line variations were not observed in Hα, one assumed for a long time that O star winds are not disturbed by clumps. Only by extremely high signal-to-noise measurements ($>1,000$ in the continuum) on very short time scales of a few minutes could one show for the Of star ζ Puppis that the HeII line at 4,686 Å exhibits stochastically varying wind features (Fig. 14.29). The line is created just above the photosphere, and the structure is therefore likely to continue into more distant regions in Hα. Since mass loss rates, however, generally are determined by Hα line models, their values must be corrected for $I \sim \rho^2$.

Stochastically clumped winds in visual light were so far only detected in the two Of-stars ζ Puppis (Eversberg et al. 1998) and HD 93129 (Lépine and Moffat 2008). Corresponding variations have previously been discovered by X-ray observations at around 1,000,000 K. In normal O stars, however, no such line structures are found. One now could infer that their winds are unstructured. It is assumed, however, that this is merely a selection effect. Unlike Of stars, O stars (without the "f" designation) usually show no emission lines and therefore the appropriate method of residual analysis is not applicable. However, in evolutionary terms Of and WR stars are closely related and the latter show wind clumps, as well. Therefore, it is generally

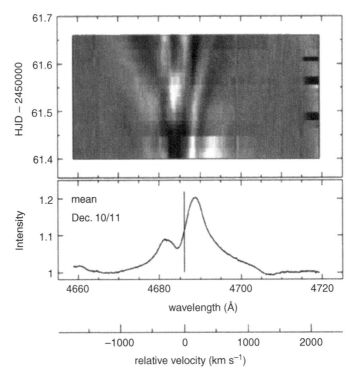

Fig. 14.29 Residuals for ζ Puppis during a single observing night (Eversberg et al. 1998). The HeII line at 4,686 Å shows time dependent signatures. The exposure times are 10 min each. Cyclical variations are not seen on this timescale. Obviously the residuals move from the line center to the red and blue wings during the observation. These clumps represent stochastic density enhancements and turbulence in the wind and provide dramatic local velocity and density differences and thus variations in light intensity. Wavelength shifts indicate Doppler velocities. Different clumps on the blue line wing move at different observation angles to the line-of-sight towards the observer. The same is valid for the red line wing for material moving away from the observer

assumed that normal O stars also have such wind structures. In fact, it is very likely a universal phenomenon among all hot-star winds. In any case, we can presume for good reasons that ζ Pup shows a spectacular combination of spiral structures and clumps in its wind (Fig. 14.30).

Fig. 14.30 The wind geometry of ζ Pup derived from various observations. Corotating Interaction Regions rotate around the star embedded in a wind with stochastic density structures (clumps)

14.8 Wolf–Rayet Stars: Massive, Small Hot Stars Below Thick Winds

Wolf-Rayet (WR) stars have baffled astronomers for a long time. In 1867 the first three were visually discovered using a spectroscope by Charles Wolf (1827–1918) and Georges Rayet (1839–1906) with the 40 cm telescope at the Observatory in Paris. Carlyle Beals then realized that an extended and rapidly expanding wind is the cause of these emissions (Beals 1929). As shown in Fig. 14.31 their spectra do not normally exhibit absorption lines but show only emission lines of different atomic transitions.

Among all stable, hot stars, WR stars show the strongest mass loss via wind mechanisms. Typical mass-loss rates are $(2-5) \times 10^{-5}$ solar masses per year, and the winds can reach terminal velocities of up to 2,500 km/s and more. Thus, these stars can lose a whole solar mass within less than one hundred thousand years only by their winds. Due to a very short lifetime of typically 0.3 million years, there are only about 300 known (but several thousand expected: Shara et al. 2009) Wolf-Rayet stars in our Galaxy. In most WR spectra one can also see that no photospheric absorption lines and thus no photospheric stellar continuum is visible. The star must therefore be hidden behind an optically thick cloud and a respective HR classification of the underlying star is impossible. Therefore, a separate classification was introduced, based on wind emission lines.

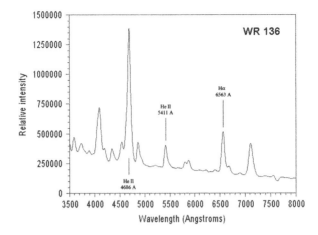

Fig. 14.31 Broadband emission line spectrum of the WN6 star WR 136 = HD 192163 (*Courtesy*: C. Buil). The observed "pseudo continuum" is the last electron scattering at the outer limit of an optically thick wind. The underlying hidden hydrostatic photosphere supplies the photons which are scattered in the optically thick inner wind portion and leave the wind towards the observer at the optical thickness $\tau = 1$. If we instead could see the hydrostatic WR photosphere, we could detect broad and weak absorption lines. In some WR stars with weaker winds than average, such absorptions were detected

- WN stars show strong lines of nitrogen and helium, but weak if any hydrogen.
- WC stars show strong emission lines of carbon and oxygen as well as helium, but no hydrogen.
- The extremely rare WO stars show strong oxygen emissions in addition to the lines of WC stars.

The observation of carbon, nitrogen and oxygen provides an indication that WR stars are massive stars in advanced stages of stellar evolution. Obviously, the elements of the CNO cycle from previous H-burning on the main sequence migrate from the stellar center to its surface and are transported into the wind. In addition, the absence of hydrogen in WC stars suggests that they are in an evolutionarily more advanced stage of He-burning than WN-types, where we often find weak hydrogen lines. Comprehensive descriptions of the WR nature can be found in, e.g., Crowther (2007) and Meynet et al. (2011).

A special spectral feature in some WR emission lines are flat line profiles, which are not found anywhere else in this form. A corresponding example is shown in Fig. 14.32 for the line of doubly ionized carbon CIII 5696 Å. The WR line shape for optically thin lines can be approximated by a series of superposed spherical shells. Consider a spherical, uniform optically thin shell around the star which expands radially and uniformly with velocity v from the star. The emission per unit solid angle of each small sphere volume $I(\Omega)$ of thickness dr is then constant. Although the observer can only observe the expansion Doppler-shifted light flux $J(v)$, radiation flux conservation applies to each infinitesimal sphere volume.

Fig. 14.32 Mean spectrum of the WR+O binary WR 140 = HD 192793, WC7pd + O5.5fc (Fahed et al. 2011) based on constant spectra avoiding periastron passage. *Bars* at the top indicate the ranges of various WR emission lines. O star photospheric absorption lines along with interstellar features are also indicated (Marchenko et al. 2003)

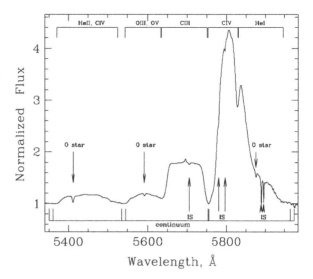

$$I(\Omega)d(\Omega) = J(v)dv.$$

The solid angle Ω changes with the angle θ between emitter trajectory and observation direction as

$$d\Omega = 2\pi \sin\theta d\theta$$

by integrating the azimuth angle $\phi \to 2\pi$. The observed emitter velocity also depends on the trajectory $v(\theta) = v\cos\theta$ and we obtain the derivative $dv = v\sin\theta d\theta$. We obtain

$$d\Omega = -\frac{2\pi}{v}dv$$

and, hence, get

$$J(v) = -\frac{2\pi}{v}I(\Omega). \tag{14.2}$$

$J(v)$ is thus constant, in accordance to the initial conditions (uniform radial expansion of v, constant emission per unit solid angle $I(\omega)$). The result is a flat-topped line profile. If we instead consider a series of such shells at different distances from the star which are all expanding at the same rate, the result will also be a flat line profile within the limits of $-v_\infty$ and $+v_\infty$. Here we should note that WR winds are optically thick only in their inner regions below about two stellar radii, depending on their strength. Only above this distance will photons not be further scattered ("last scattering surface"): the wind becomes transparent and the lines are optically thin. Flat line profiles are therefore generated at large stellar distances according to the above considerations. Therefore, even wind regions behind the star can be observed (with the exception of a small section at very low stellar distances) and thus we can detect $+v_\infty$ and $-v_\infty$. This is exactly what is observed in WR stars. Flat line profiles are rather generated by low excitation stages in the outer wind region near the terminal velocity V_∞. The edges of the line plateaus are thus a direct measure of the terminal velocity in the stellar wind.

Other line shapes can be similarly approximated. For example, if we consider relatively short distances above the photosphere, the wind velocity is not constant but accelerates relatively fast in accordance with the velocity law equation 14.1. And if we consider optical thin winds at short distances from the stellar surface,[9] we can approximate any emission $J(v)$ by a sum of different flat profiles (Fig. 14.33). Again, the strength of recombination lines such as the Hα disk emission in Be stars, depends on the square of the plasma density ($J(v) \sim \rho^2$). Therefore, dense plasma regions close to the star deliver a stronger line profile than the outer regions. Gaussian-like emission profiles are then the result. In this case the wind is thus approximated by a series of flat profiles getting broader with distance from the star,

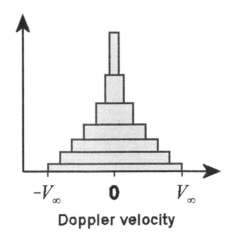

Doppler velocity

Fig. 14.33 Approximation of an emission line profile by a series of flat sections which are formed in different layers of an accelerated stellar wind. In an optically thin wind all velocity ranges of the wind can be detected. By considering each small speed interval dv according to Eq. 14.2 one obtains a flat line within the limits $-v$ and $+v$ for each interval. Since the wind velocity increases according to Eq. 14.1 until terminal velocity v_∞ is reached, the line interval successively broadens accordingly up to $-v_\infty$ and $-v_\infty$. Narrow profile areas occur close to the star at lower speeds. Their intensities are larger (higher plasma density) according to Eq. 14.2. Broad profile areas, however, arise from the outer wind at high speeds where their intensities are low (low plasma density). The line fluxes of all areas are added because of an assumed low optical density wind. Some WR lines miss the narrow but intense contributions from inner wind layers because of their high optical density resulting in flat-top lines

---

[9]This is only valid for optically thin winds down to the photosphere. WR winds are optically thick below $\sim 2R_*$.

whereas their intensities gradually decrease. In the same way we can approximate the absorption part of a P-Cygni profile. The absorption, however, only acts in the observer's direction, i.e. within the velocity speed range from 0 to $-V_\infty$. Close to the star, the light is completely absorbed due to the high wind density. Only in the outer regions does the absorption decrease slowly.

A long unresolved question is the so-called momentum problem in WR winds. When the momentum of the stellar wind is compared with radiation momentum, it turns out in WR stars that the radiation momentum is much smaller than that of the wind. Thus, the observed winds can actually not be accelerated by the stellar radiation. The parameter

$$\eta = \frac{\dot{M} \cdot V_\infty}{L/c}$$

is an efficiency indicator for the momentum transfer in the wind. The product $\dot{M} \cdot V_\infty$ is the momentum flux (rate of momentum transfer at constant velocity) of the wind material with the mass loss rate $\frac{dM}{dt} \equiv \dot{M}$ and the terminal velocity V_∞. The quotient L/c of luminosity and the speed of light, however, is the total stellar photon momentum. One now would expect that the available photon momentum corresponds to the observed wind momentum ($\eta = 1$). However, in optically thin O star winds this number depends on how much of the photon flux is absorbed. Here η is always smaller than 1. For WR stars, η is about 2–10. Typical values for Galactic WN stars are around 3 but for extreme cases one finds 10–20 (Hamann et al. 2006). For WC stars the numbers are more between 10 and 20 (Sander et al. 2012). The explanation for this "momentum problem" is multiple scattering in the wind (Fig. 14.34). η refers to a single scattering process in which a number

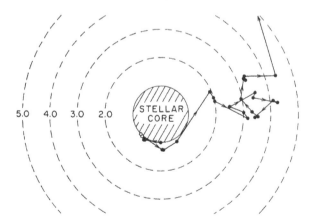

Fig. 14.34 Principle of multiple scattering in an expanding stellar wind of spherical symmetry. Photons from the star reach a velocity region in the wind where they are scattered isotropically by a line. For each scattering process, momentum is transfered depending on the scattering angle. After the process the photon has a longer wavelength and can transfer momentum at lines of this new wavelength (Abbott and Lucy 1985)

of photons transfer their momentum only once to an interacting atom. When the photon transfers its momentum at a certain line wavelength it normally can not transfer its momentum again. After the scattering process the re-emitted photon has a shorter wavelength, it is "reddened", since it has transferred energy and the appropriate atomic energy level with its specific wavelength is no longer available. The physical mechanism that allows additional scatterings is the wind itself. It is accelerating (Fig. 14.10), and the photon can now provide additional momentum at a different wavelength (according to $\Delta E = hc/\Delta\lambda$ the photon then transmits a longer wavelength) to the doppler-shifted line in the wind again. The wind "sucks" the photons outwards. However, instead of "photon reddening" a more accurate expression has been delivered by Owocki et al. (2004) by introducing the expression "photon tiring". *The key issue for the dynamics is the finite amount of energy in the radiation, which limits the amount of work it can do in lifting material out of the gravitational potential, and giving it kinetic energy associated with its terminal speed. It's much like a person who gets "tired" after doing too much work. Of course, such a person might also look pretty "red" to an observer. But that's a consequence of being tired, not a cause of it* Owocki (2013). Since the hot star radiation has its maximum in the UV, where thousands of lines in the so-called "iron forest" are available, multiple scattering is highly efficient for hot stars.[10]

In Sect. 14.7.2 we have already noted that Wolf-Rayet winds show a stochastic structure. Unlike Of stars these structures show much higher S/N and can be detected with small instruments. Figure 14.35 shows the WR nebula M1-67 and its origin,

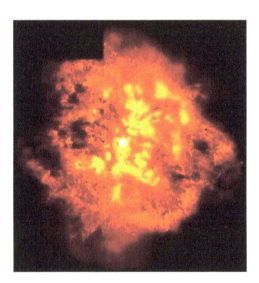

Fig. 14.35 Image of the nebula M1-67 in the light of H-alpha around the WN8 star WR 124. The WR star has been masked (Grosdidier et al. 2001)

---

[10]Gayley and Owocki (1995) pointed out that the "momentum problem" is actually an "opacity problem" of having enough lines spread over the flux spectrum. Further theoretical considerations can be found, e.g., in Gayley et al. (1995), Gräfener et al. (2011), Owocki et al. (2004) and the references therein.

Fig. 14.36 CIII residuals of a single observing night in γ Velorum (Lépine et al. 1999, ©AAS. Reproduced with permission)

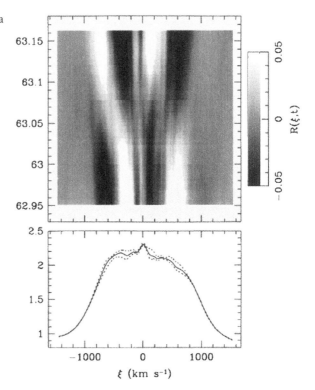

the Wolf-Rayet star WR 124. The high-resolution HST image shows clumping in the wind, whose origin is probably turbulence via smooth wind interaction with a slower wind from preceding evolutionary stages (Luminous Blue Variables or Red Supergiants might play a role). This is confirmed by spectroscopic investigations of various lines, which arise in the direct stellar vicinity. Figure 14.36 shows CIII residuals of the WR+O binary γ Velorum during one night with individual exposure times of about 10 min. As for the residuals of ζ Puppis in Fig. 14.29 we observe density enhancements and turbulence in the wind which are produced right above the surface of the star (HeII requires high excitation temperatures and is thus generated close to the star). From Fig. 14.36 one might expect only a few clumps, since only three residuals moving away from the rest wavelength are identified. A close examination of the wind by means of wavelet analysis (Lépine and Moffat 1999) however showed that at least 10,000 wind clumps create the observed spectral features. It is assumed that the wind of ζ Puppis also has a lot more clumps than the observation of thirteen spectral line residuals implies (Fig. 14.29).[11]

---

[11]Later observations of ζPuppis with the XMM X-ray satellite by Naze et al. (2011) showed no short-term variations above the 1 % detection limit. This implies the presence of at least 10,000 elementary clumps in the wind.

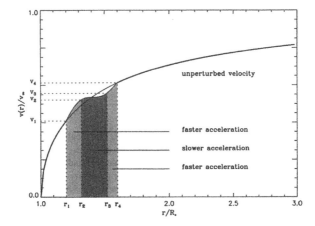

Fig. 14.37 The effect of a sinusoidal velocity perturbation in a line-driven stellar wind (Lamers and Cassinelli 1999). The acceleration between radii r_1 and r_2 and between r_3 and r_4 are greater and between r_2 and r_3 smaller than in the undisturbed wind. This acceleration differences produce shocks with sudden density increases

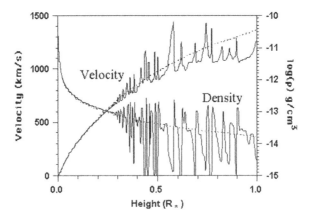

Fig. 14.38 Velocity and density behaviour leading to turbulence in the stellar wind (Owocki 1998)

The source of density fluctuations in the stellar wind comes from supersonic shocks caused by small disturbances in an accelerated medium (Fig. 14.37). Such disturbances may introduce identical wind speeds at different distances from the star. This however, can introduce momentum transfer from photons into a line back towards the star (backscattering, Fig. 14.38). In any case, turbulence in the wind increases the momentum transfer efficiency. The outflowing wind not only "sucks" the light globally outwards but the turbulent structure also multiplies this process accompanied by local deviations from radial direction.

The wind structure continues far into the interstellar medium (ISM). As already seen, Fig. 14.35 shows a high-resolution HST image of WR 124 (WN8), which reveals a dramatic picture of highly structured Hα clouds with shock waves and clumps of different sizes, which move outward. WR stars push away turbulent winds, whose structure extends from the surface far into the ISM and enriches it with heavy elements.

14.9   Clumps as Wind Tracers

In a few Wolf-Rayet and O stars the velocity law can be directly measured. Indicators for O stars were already mentioned in (Sect. 14.7.2) on wind clumps. These are turbulent density enhancements in the stellar wind which manifest themselves as line excess emission structures moving away from the line center in time. If we now assume that these clumps "swim" with the wind (cf. Lepine and Moffat 1999), they are a direct indicator of the wind velocity behavior. To detect such features, relatively high S/N and resolving power are necessary. On the other hand, one requires necessarily short exposure times to track such features during a night. This is especially difficult for Of stars with comparably weak emission lines like HeII 4686 (see Fig. 14.29). But if we can find clumps and determine their wavelengths and thus speed performance over time, we can draw conclusions about the β value of the velocity law. In order to track clumps in time we calculate residuals (see Sect. 14.7.1) and determine their accelerations $a \equiv \dot{V}$ in wavelength or velocity space. Figure 14.39 shows time series of grey-scale residuals for ζ Puppis (Fig. 14.29) now as intensity plots with respective clump tracking in time (dashed lines).

Assuming a constant acceleration[12] we can now compare the observed clump movement with that theoretically expected. To do so, we consider the (constant) acceleration against the respective speed intervals for each clump and compare it with the velocity law for a fixed β and different trajectory angles Θ. Thus, we can examine what values of β and Θ represent the observed clumps best. Figures 14.40 and 14.41 show this comparison for ζ Puppis and WR 140.

O star winds are apparently much more strongly accelerated ($\beta \approx 1$) than WR winds ($\beta > 10$). Given the observed clumps just above the Of photosphere we can assume that the entire wind is structured, as for WR stars. The clump disappearance after short distances might be due to material cooling. Optically thick WR winds make direct measurements just above the photosphere impossible. It seems that their clumps appear only at larger distances and remain visible until the terminal

---

[12]Given the non-linear velocity law (Eq. 14.1), constant acceleration of clumps is only an approximation. This approximation is acceptable with respect to the short time interval and, consequently, short radial distances covered in the wind.

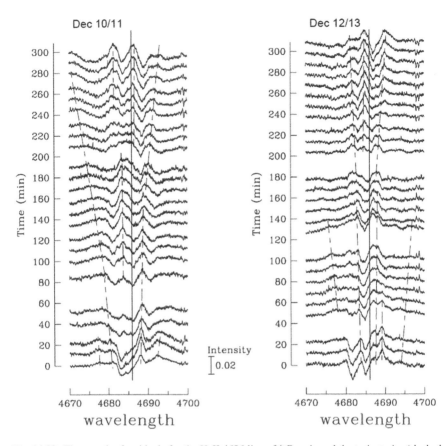

Fig. 14.39 Time stack of residuals for the HeII 4686 line of ζ Puppis and the trajectories (*dashed lines*) of individual clumps during two different observing nights (Eversberg et al. 1998). Tracking the very weak residuals in the wind of this 2nd mag star with short exposure times and sufficient S/N and spectral resolving power required a 4 m-class telescope

velocity is reached. This is depicted in Fig. 14.42 by modelling clumps depending their length scales and their velocity-stretches. However, this conclusion might be biased by the forming region of corresponding lines depending on their excitation energy and stratified wind temperatures. The wind acceleration is much lower than for O stars. However, they reach significantly higher terminal velocities.

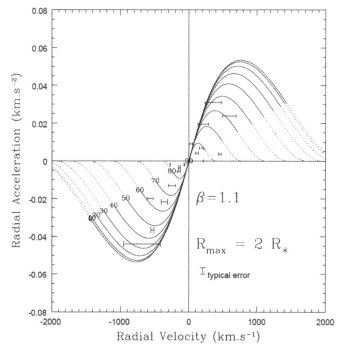

Fig. 14.40 Projected radial acceleration $\dot{V}$ vs. radial velocity V for individual clumps in the HeII4686 line of the Of star ζ Puppis. Each model curve represents a radial velocity vector (clump trajectory) with respect to the line of sight from $\theta = 0°$ (*lower curve*, motion towards the observer) to $\theta = 180°$ (*upper curve*, moving away from the observer). *Model lines* are based on $v(t) = V(t) \cdot \cos\theta$ and $V(r) = V_\infty(1 - R_*/r)^\beta$. *Horizontal bars* indicate the interval where individual clumps are observed and tracked. Because of the projection angle θ the observed clumps show different accelerations and are observed during different time intervals. All clumps together are best matched by a model with $\beta = 1.1$. Lower or higher values of β would result in larger or smaller model amplitudes in $\dot{V}$, respectively (compare with Fig. 14.41). The *solid lines* are continued outward by *dotted lines*, which go beyond the maximum distance for observed clumps from the stellar surface R_{max}. No clumps are found beyond R_{max}. On the other hand, some clumps are detected directly above the photosphere ($V = 0$). The selected terminal velocity is $V_\infty = 2{,}250$ km/s, while the assumed stellar radius R_* is 18 solar radii. The typical 1σ error for the acceleration is indicated (Eversberg et al. 1998)

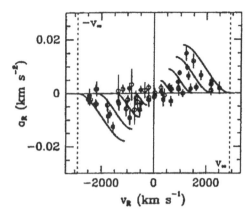

Fig. 14.41 Projected radial acceleration $\dot{V}$ vs. Radial velocity V for individual clumps as in Fig. 14.40 for the CIII line of the Wolf-Rayet star WR 140. The model values are based on $\beta = 12$. The selected terminal velocity is $V_\infty = 2{,}900$ km/s, the assumed stellar radius $R_\star$ is 14 solar radii. 1σ errors for the acceleration are indicated (Robert 1992). In contrast to the Of star ζ Puppis the WC7 star in WR 140 exhibits a significantly smaller wind acceleration but a much higher terminal velocity. This is a direct effect of multiple scattering in the optically dense wind. Because of the high optical density below $2R_\star$, clumps can only be detected after the wind has already been accelerated. They remain visible far out into the wind which gradually approaches V_∞ which in principle requires infinite time

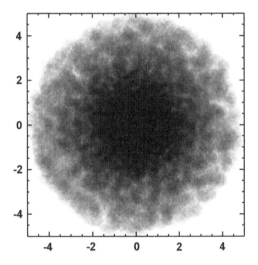

Fig. 14.42 Depiction of randomly generated isotropic clumps in WR winds in front of a uniformly illuminated source from theoretical calculations (Sundqvist et al. 2012). The axes indicate geometric distances in stellar radii. Note that the wind becomes optically thick below $r \sim 2R_\star$

14.10   A Short Remark on Evolution

Classical WR stars have typical masses of 10–20 solar masses whereas O stars have typical masses of 20–60 solar masses, sometimes even reaching more than 100 solar masses.[13] On the other hand, WR stars live just a few 100,000 years because of the very high-energy burning and lose about one solar mass by stellar wind within a hundred thousand years. Before a star now simply disappears because of its mass-loss it needs to change the course of its lifetime. Already Chandrasekhar (1934) discussed the origin of the WR stars and the exact scenario of evolution is still being debated today. However, there is general agreement on the so-called Conti scenario (Conti 1976). Current evolutionary models for massive stars assume Wolf-Rayet stars to be descendants of O stars. However, the WR phenomenon among massive stars has a relatively short lifetime of about $(3 - 5) \times 10^5$ years, making WR stars relatively rare, even though all O stars at solar metallicity are believed to pass through a WR phase. In our Galaxy there are presently only about 300 WR stars known. This is based on observations and assumes that WR stars evolve from O types. A very simplified evolution scenario is

$$O \rightarrow Of \rightarrow WN \rightarrow WC \rightarrow SN.$$

For a detailed schematic scenario for solar metallicities and individual massive stars of different masses from the Zero Age Main Sequence, we refer to Langer et al. (1994), Crowther et al. (1995), Crowther (2007) and Meynet et al. (2011). In addition to various modifications and distinctions for different atomic abundances and phases where Luminous Blue Variables (LBV) play a role, one assumes that massive Of stars with emission lines (i.e. relatively strong winds) develop from normal O stars. The Of star then presumably becomes a nitrogen-rich, then a carbon-rich WR star and ends in a supernova when the final (central) fuel supply is exhausted.

Since WR star photospheres are invisible due to their dense wind, it makes sense (because of the Conti scenario) to obtain parameters such as mass, luminosity, effective temperature, diameter and mass loss rate through the detour of evolutionary considerations by examining Of stars. Vice versa, it might be possible to transfer WR wind phenomena to the optically thin and therefore low contrast winds of O stars. This particularly applies to clumps in O and WR winds (Figs. 14.29 and 14.36) which affect the stellar mass-loss rates.

---

[13]In contrast to the classical helium burning WR stars, some hydrogen burning WR stars on the main sequence have very high masses beyond 60 $M_\odot$.

14.11 Dance of the Giants

A highly interesting phenomenon for the physical diagnosis of massive stars is the collision of massive stellar winds in, e.g., O+O and WR+O binaries. The wind velocities of the individual components of about 2,000 km/s can already exceed the speed of sound in the interstellar medium. If such a wind meets another stellar wind of similar velocity, this results in an abrupt increase of plasma density and temperature in a supersonic shock. In addition, if one of the two components has a significantly larger wind momentum flux, it pushes back the other wind component and forms a wind shock-cone around the star with the weaker wind. At the cone-tip highly excited elements occur due to high temperatures. This can be seen in WR+O binaries. An artist's illustration of a WR+O shock cone is shown in Fig. 14.43.

The hot cone-tip has typically about 3,000 solar luminosities (Hamann 2011). The additional light produced by the shock cone appears as excess emission in the lines of excited element species. This excess emission can move in wavelength and time, depending on the geometry.[14] One example is the highly eccentric WR+O star WR 140. The system has intermediate inclination and longitude of periastron with respect to the observer ($i = 55°, \omega = 44.6°$, Fahed et al. 2011), so that the excess emission migrates from blue to red wavelengths during periastron passage. Before the passage the excited shock material moves towards the observer and after periastron away from him or her. The material then flows along the cone, cools down and elemental species of lower excitation energies become visible. Once the temperature has dropped sufficiently, dust forms and slightly darkens the system by 0.15 mag in V. The material manifests itself thereby as a spiral-form tail due to the rotation of the binary system (Fig. 14.44). Only with the development of modern observational techniques was it possible to examine this

Fig. 14.43 Artist's illustration of the WR+O binary WR 140. A shock cone created by supersonic wind compression wraps around the O component (*Courtesy*: Gemini Telescope Consortium)

---

[14]This is only valid if a radial velocity component in the corresponding observation is present.

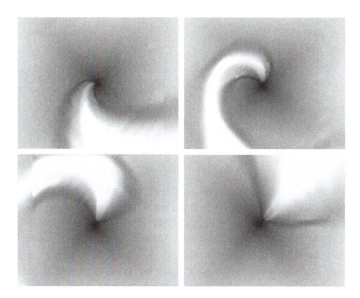

Fig. 14.44 Density simulation of the WR 140 system and its shock-cone on a scale of 333 AU. The stars move clockwise. *Upper left*: Phase 0.0 at periastron. *Upper right*: Phase 0.2. *Lower left*: Phase 0.039. *Lower right*: Phase 0.08 (Walder and Folini 2002)

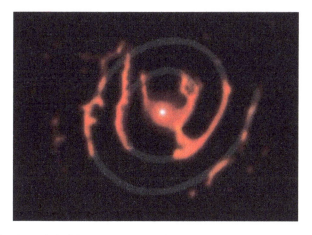

Fig. 14.45 The dust spiral of WR 112, calculated from spectroscopic observations (*gray*) and the detected material from IR observations (Marchenko et al. 2002)

interpretation of spectroscopic imaging techniques directly. Figure 14.45 shows an infrared image of the WR+O binary WR 112 and the calculated track of the dust spiral. The correspondence between spectroscopic and photographic data is striking and confirms the confidence in modern spectroscopic analytical methods for objects that are not geometrically resolvable.

Fig. 14.46 Schematic view
of the geometric model by
Lührs (1997) taken from
Bartzakos et al. (2001) (their
figure 2)

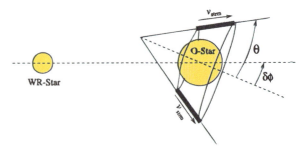

To obtain quantitative information on the shock cone of massive binary stars,
different parameters must be considered via a model. A successful (phenomenological) cone model was developed by Lührs (1997), which provides a relatively simple
approximation of the shock cone emission. This so-called "Lührs model" is shown
in Fig. 14.46 (viewed perpendicular to the orbital plane). This model parameterizes
the inclination i, the outflow velocity of the wind plasma along the conical flanks
v_{strm}, the half opening angle of the shock cone θ and its tilt against the connecting
line between both components $\delta\phi$ because of the coriolis force in the rotating frame
of reference. The Lührs model equations provide expressions for the full-width-at-
half-maximum of the excess emissions $FWHM_{ex}$ and for its mean radial velocity
RV_{ex} for a circular orbit. They are

$$FWHM_{ex} = C_1 + 2v_{str}\sin\theta\sqrt{1 - \sin^2 i\,\cos^2(\phi - \delta\phi)} \qquad (14.3)$$

$$RV_{ex} = C_2 + v_{str}\cos\theta\sin i\,\cos(\phi - \delta\phi) \qquad (14.4)$$

with the constants C_1 and C_2, the orbital inclination i, the orbital phase ϕ and the
already mentioned v_{strm}, θ and $\delta\phi$. Because of the varying line-of-sight orientations,
the excess emission on the line peak (produced in the shock-cone) wanders back and
forth during the orbit and also changes its width, depending on the fixed value for i
and the variable ϕ (Fig. 14.47). One can now compare all values for different orbital
positions ϕ with the model for $FWHM_{ex}$ and RV_{ex} and determine the corresponding
cone parameters by a fit. The examined excess emissions, however, deliver a very
small flux, which makes a possibly high S/N necessary for a physical analysis.
Proper phase-dependent analysis requires prolonged observations for which only
smaller telescopes are normally available. However, this often results in relatively
low S/N and/or reduced spectral resolution. Nevertheless, the excess has to be
sufficiently well separated from the line in order to determine its width and position
as precisely as possible. In contrast to residuals from the mean of all observations
(see Sect. 14.9, Fig. 14.39), residuals from the global minimum of all spectra can
provide a better tool for eccentric orbits where little or no excess emissions are
observable near apastron. For circular orbits (eccentricity $e \approx 0$) a detailed analysis
of all data obtained is needed. Figure 14.48 shows a corresponding procedure for

Fig. 14.47 CIII line at 5,696
Å of the Wolf-Rayet star
WR 79 during an 18 day
observing run. Single
spectra are indicated with
their Julian date. During the
orbital period of about 9 days
the double-peak excess
emission moves twice from
one side of the line plateau to
the other side and back (Hill
et al. 2000)

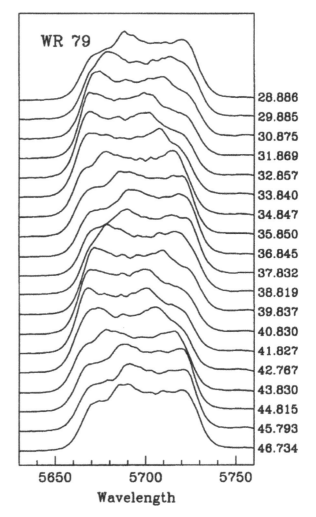

the CIII 5696 Å excess in WR 42 and WR 79. The respective line fits are shown in
Fig. 14.49.

In contrast to the short 8–9 day periods of the WR 42 and WR 79 circular
orbits, the colliding-wind WR+O binary system WR 140 (HD 193793) has a 7.9-
year period where the stellar separation varies between ∼2 AU at periastron to
∼30 AU at apastron. The binary is frequently considered a textbook example of the
colliding-wind phenomenon. It consists of a carbon-sequence Wolf-Rayet (WC7pd)
star orbiting a more massive and luminous O5.5fc giant companion. The system's
high eccentricity and rather favorable inclination help to probe different regions
of the Wolf-Rayet wind and, at the same time, the profound change of conditions
in the wind-wind collision zone at the times of periastron passage. This change is

Fig. 14.48 The CIII 5696
line profiles (*solid*) of WR 42
and WR 79 and their assumed
undisturbed flat-top lines
without excess (*dotted*). Four
different intensity levels for a
best $FWHM_{ex}$ data fit have
been defined (horizontal bars
in the excess mission) and
chosen for maximum data
consistency (Hill et al. 2000)

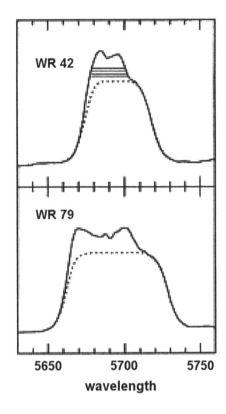

mainly reflected in rapid formation of dust clouds and can be detected as gigantic
IR outbursts occurring on a strictly periodic (once per orbit) timescale. Figure 14.44
shows results of 3D hydrodynamic calculations.

The orbit of WR 140 has an advantageous inclination to the line-of-sight. Shock
cone material changes its direction from blue to red during periastron passage.
Before passage the material moves towards the observer and afterwards it moves
away from the observer. Figure 14.50 shows the CIII 5696 line during the 2009
periastron passage from a contiguous 3.5-month campaign obtained by the 1.93 m
telescope at OHP.

Although the Lührs-Model has been designed for circular orbits one can adapt it
for very eccentric binary orbits. This has been done for WR 140 with its eccentricity
of almost 0.9. Fits for RV_{ex} and $FWHM_{ex}$ are shown in Fig. 14.51. A recurring
complexity for eccentric orbits is the variable tilt angle $\delta\phi$. At periastron passage,
the speed in the Keplerian orbit increases and the Coriolis force grows. Therefore
$\delta\phi$ increases significantly. To determine this tilt one can modify the Lührs equations
by the true anomaly and account for the rapid movement at periastron. Then it is
possible to approximate the eccentric by a circular orbit. For WR 140, it is found
that the shock cone changes its tilt during periastron passage by about 20°.

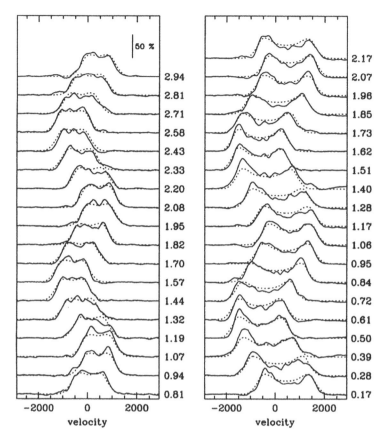

Fig. 14.49 Fit of synthetic Lührs cone profiles (*dotted*) to a time series of WR 42 (*left*) and WR 79 (*right*). The periods are about 8 days (WR 42) and 9 days (WR 79), both orbit eccentricities are assumed to be 0.0 (circular orbits). All Lührs parameters have been fitted simultaneously to all collected line profiles (Hill et al. 2000)

At this point we refer again to the relatively simple Fig. 14.48. It shows the importance of an accurate and dedicated data reduction procedure to obtain maximum information out of the acquired data. Not only has the best definition of $FWHM_{ex}$ been examined but also the underlying wind profile has been consistently defined by the data via different methods. That are, e.g., average of four the excess emission flank intensities (in continuum units) for best definition of $FWHM_{ex}$ (Fig. 14.48) and simultaneously fitting all the profiles in a given data set allowing all the parameters in Eqs. 14.3 and 14.4 to vary. To get an idea of the potential scope of this work, it helps to study the description of the sophisticated data processing by Hill et al. (2000). We refer again to the statement of Massey and Hansen about data reduction in Sect. 12.3. Data reduction and processing should be tailored to the respective data set. There is no "gold standard".

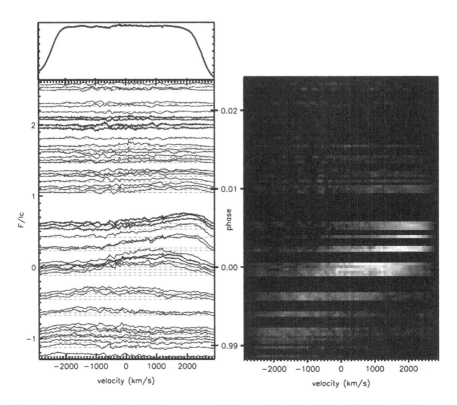

Fig. 14.50 *Left*: The CIII 5696 line excess emission as a function of the orbital phase. This was obtained by subtracting a reference profile (*top panel*; nearly flat-top) from all the spectra. *Right*: Same as left but plotted in greyscale (Fahed et al. 2011)

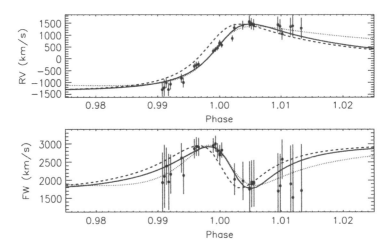

Fig. 14.51 Fit of the radial velocity and width of the CIII 5696 excess emission using the original (*dotted line* for a circular orbit) and a modified Lührs model (*full line* for a realistic eccentric orbit). The *dashed line* shows the solution from Marchenko et al. (2003) (Fahed et al. 2011)

14.12 So What...?

Fast spherical winds with or without turbulence, extensive stable hydrogen disks over the equator and cyclically rotating wind shocks of high temperature are amazing phenomena. But given the fact that the absolute number of massive stars in comparison to all other stars is negligible because of their short lifetime of a few million years[15] and their rarity in a stellar population via the Initial Mass Function (IMF),[16] one might ask why the knowledge of the physics of these big stars is so important. In principle, it is the mass of these objects, which ensures that

- the CNO cycle dominates the stellar burning of hydrogen to helium,
- heavy elements are transported to the surface and thus enrich the interstellar medium,
- structures are then generated in the interstellar medium,
- the enormous luminosities dominate the radiation field of galaxies,
- massive stars thus have a significant share in the evolution of galaxies and
- thus creating the preconditions for life in space.

The enrichment of the ISM with elements from the interior of massive stars is also observable at large scales. Such an accumulation of mass and energy is shown in Fig. 14.52 with NGC 7635 (Bubble Nebula) in Cassiopeia covering an

Fig. 14.52 HST image of the Bubble nebula NGC 7635 (Moore et al. 2002, ©AAS. Reproduced with permission). *Inset*: Amateur image

---

[15]In our Galaxy, only a few hundred WR stars are known until today.

[16]The IMF describes the distribution of stellar masses in a newly developing stellar population. The IMF follows the power law $\frac{dN}{dM} = M^{-\alpha}$. A detailed investigation of the IMF variation for the Galactic field is given by Kroupa (2000).

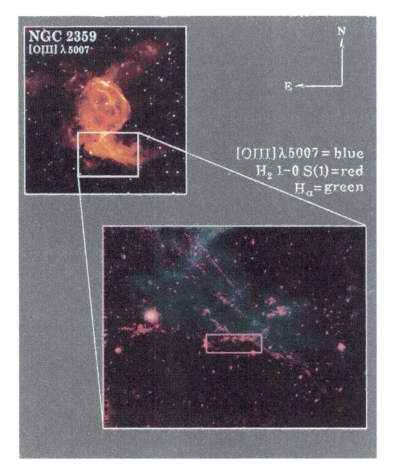

Fig. 14.53 Discovery of molecular Hydrogen in an interstellar wind filament of the WR nebula NGC 2359 (St-Louis et al. 1998, ©AAS. Reproduced with permission)

area of about 6 light years. The nebula has its origin in a WR star and is thus of completely different nature than planetary nebulae with very old central stars. The Bubble nebula is expanding with about 10–100 km/s. An even bigger WR nebula is shown in Fig. 14.53 with NGC 2359 (Thor's Helmet) in Canis Major with an extent of about 50 light years. The origin of the nebula is the WR star visible in the center of the bubble. Within the strongest filament (pictured below the spherical bubble) St-Louis et al. (1998) discovered molecular hydrogen H_2. As generating mechanisms, excitation through shocks or fluorescence by the strong UV radiation is considered. WR stars are thus potential suppliers of a significant amount of H_2 observed in young starburst galaxies.

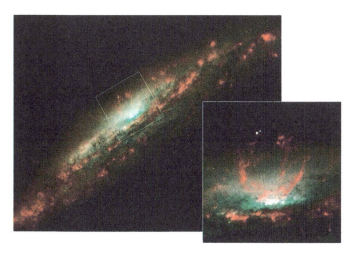

Fig. 14.54 Large-scale winds from star burst regions in NGC 3079 with an extension of about 3,000 light years (Cecil et al. 2001)

The effect of massive stars in galaxies is shown by even more spectacular images from HST. Figure 14.54 shows large-scale stellar winds perpendicular to the galactic plane of the spiral galaxy NGC 3079. Because massive stars trigger the cycle of stellar births and deaths it is not surprising that our solar system is probably born in a Galactic spiral arm and then moved out of it. Only in such an environment can we expect an excess of heavy elements, which can collapse gravitationally to form a star with a planetary system. And only in massive stars, which end in a supernova, can heavy elements be produced.[17] Investigations of massive stars with small telescopes and then deeper measurements in environments of other metalicities and in star formation regions with large telescopes are, hence, important for the understanding of massive stars at all distance scales.

Suggested Readings

- E. Novotny, *Introduction to Stellar Atmospheres and Interiors*, Oxford University Press, 1973
- J.B. Hearnshaw, *The Analysis of Starlight: Two Centuries of Astronomical Spectroscopy*, Cambridge University Press, 2014
- R. Kippenhahn, A. Weigert, A. Weiss, *Stellar Structure and Evolution*, Astronomy and Astrophysics Library, 2nd ed., Springer, 2012

---

[17]In globular clusters with the oldest populations and lowest metallicity no life is therefore expected.

- B.W. Carroll, D.A. Ostlie, *An Introduction to Stellar Astrophysics*, Wiley & Sons, 2010
- A.W. Fullerton, *Observations of Hot-Star Winds* in: Lecture Notes in Physics, Springer 1997, Vol. 497, p.187
- B. Wolf, O. Stahl, A.W. Fullerton (Editors), *Lecture Notes in Physics - Variable and Non-Spherical Stellar Winds in Luminous Hot Stars*, Proceedings of the IAU Colloquium No. 169, Heidelberg, Germany, 1998
- H.J.G.L.M. Lamers, J.P. Cassinelli, *Introduction to Stellar Winds*, Cambridge University Press, 2008
- C. Leitherer, N. Walborn, T. Heckman, C. Norman (Editors), *Introduction to Stellar Winds*, Cambridge University Press, 1999
- J.M. Porter, T. Rivinius, *Classical Be Stars*, Publications of the Astronomical Society of the Pacific, 2003, 115, 1153–1170
- T.W. Hartquist, J.E. Dyson, D.P. Ruffle, *Blowing Bubbles in the Cosmos*, Oxford University Press, 2004
- P.A. Crowther, *Physical Properties of Wolf-Rayet Stars*, Annual Review of Astronomy and Astrophysics, 2007, 45, 177–219
- L. Drissen, C. Robert, N. St-Louis and A.F.J. Moffat (Editors), *Four Decades of Research on Massive Stars*, ASP Conference Series, Vol. 465, 2012

Chapter 15
The Next Step: Polarization

A Short Story

When he worked on his PhD thesis Thomas had to test the William-Wehlau-Spectropolarimeter at the Elginfield Observatory together with his mentor Tony Moffat. During that very time Klaus enthusiastically took part in the observatory test in Ontario along with Thomas and Tony. One night, very repetitive and monotonous tests of the optical components had to be performed and Thomas was already quite bored. When he suggested to shorten the measurements Tony clearly refused to do so. It was early in the morning and Thomas was tired. He was annoyed! But Klaus said dryly. "You should listen to him, he has expertise and knows more than you." Fortunately Thomas followed this advice, and not until much later did this number of measurements turn out to be most important for describing the optical system properties.

15.1 Beyond Spectroscopy

Spectroscopic techniques and their astronomical applications have been highly successful. However, measurements of light intensities (photometrically and spectroscopically for different wavelengths) do not take the complete contents of electromagnetic waves into account. Beside the intensity this is the polarimetric status of the incoming wave, either broadband or wavelength dependent. This status depends on specific physical processes which contribute to the overall image of astronomical targets. One of the most prominent examples are magnetic fields which are considered to be a main driver of various phenomena in, e.g., massive stars (see Chap. 14). Therefore, if we observe stars only with respect to their intensity spectra, we ignore a very large portion of the total information, and it is surprising that it has only been known since 1948 that a majority of stars is partially plane

© Springer-Verlag Berlin Heidelberg 2015
T. Eversberg, K. Vollmann, *Spectroscopic Instrumentation*, Springer Praxis Books,
DOI 10.1007/978-3-662-44535-8_15

polarized. To give a basic idea about how to measure polarized light we discuss some fundamental necessary technical approaches and address some sources for polarized light in massive stars.

15.2   Polarized Light in Astronomy

In nature we find a number of effects which can produce polarized light, some of high interest for applying in optical measurements (Shurcliff 1962):

- *Stark effect*. If a plasma emits light in the region of a strong (uniform) electric field **F**, this light is polarized in the direction perpendicular to the field direction. Each spectral line is split by the field into several lines, and these exhibit polarization. Light emitted perpendicular to **F** consists of lines that are linearly polarized with the electric vibration either parallel to **F** (the p-component) or perpendicular to it (the s-component); this is called the transverse Stark effect. Light emitted parallel to **F** is unpolarized.

- *Zeeman effect*. If a plasma emits light in the region of a strong (uniform) magnetic field **B** a typical spectral line is split into several lines. Light emitted perpendicular to **B** is linear polarized with vibration direction parallel to **B** (p-component) or perpendicular to **B** (s-component); this is the transverse Zeeman effect. Light emitted parallel to **B** is circularly polarized; this is the longitudinal Zeeman effect.

- *Cerenkov effect*. When relativistic electrons travel through a medium and move (temporarily) faster than the light velocity in this medium, the material will emit light. The wavefront of this light is conical and each ray is linearly polarized with the direction of its electric vibration parallel to the pertinent element of the wavefront.

- *Scattering off small particles*. If a light beam scatters off small particles (Rayleigh scattering) this beam is polarized. The effect is seen as the polarization of the sky and the interstellar medium.

- *Cyclotron radiation*. If non-relativistic electrons are captured by a strong magnetic field **B** and spin around the field lines, they emit Bremsstrahlung which is linearly polarized perpendicular to **B**. The beam angle is directly correlated with the kinetic energy of the electrons and the beam is circularly polarized along the field lines. Note that for relativistic synchrotron radiation only linear polarization occurs.

- *Other methods*. There are several other effects which can produce polarized light. Briefly mentioned are the *grating plus electron beam* (Smith and Purcell 1953), the *Undulator* (Motz et al. 1953), *Light from canal rays*, and *K-capture* of high-energy gamma-rays (Hartwig and Schopper 1959).

According to the above listed polarization effects one can expect several sources of polarization in the Universe. In this context we can distinguish between the polarization of extended sources (the Sun, planets, comets, the interstellar medium, nebulae and galaxies) and point sources (distant stars). A brief list of optical phenomena which are responsible for the observed polarization in astronomical objects is given by Serkowski (1974):

- *reflection from solid surfaces*: Moon, Mars, Mercury, minor planets
- *scattering by small grains* zodiacal light, comets, the outer dust rings of Jupiter, reflection nebulae, atmospheres of late-type stars, spiral galaxies, interstellar polarization of starlight
- *scattering by molecules* (Rayleigh scattering): Jupiter and outer planets, Venus
- *scattering by free electrons* (Thomson scattering): solar corona, envelopes of early-type stars
- *Hanle effect* (resonance scattering of bound electrons in magnetic fields): solar chromosphere and corona
- *Zeeman effect* sunspots and magnetic stars, interstellar medium
- *grey-body magneto-emission* white dwarfs
- *gyro-resonance emission* solar chromosphere and corona
- *synchrotron emission* Jupiter, Crab nebula, pulsars, galactic background radio emission, radio galaxies, quasars

For hot stars, mainly those mechanisms play a role that can survive the tremendous radiation field with temperatures of up to several tens of thousands of degrees. These involve mainly Thomson scattering and the Zeeman effect. In late-type stars, dust plays a role, so that scattering on small grains is not negligible in such stars. So if for example linear polarization is detected in a massive star, one can assume that the emission process is dominated by Thomson scattering and that the emitting medium has a preferred orientation (e.g., stellar wind, which deviates from radial symmetry). A typical example are material disks around Be stars. And circularly polarized light observed in a star is an obvious indicator for a magnetic field. With polarimetric measurements one has an additional probe to determine the global and local geometries of stars.

15.3 Description of Polarization with the Stokes Parameters

Most information we can gain about stars is obtained by measuring their radiation. Because of the enormous distances, stellar light reaches us in the form of a parallel bundle of radiation and no two-dimensional information is obtainable at least in most cases, even if astronomers may be on the brink of a resolution break-through. The complete description of a stellar beam is established by the following characteristics:

- the direction of the beam described by its coordinates, e.g., in the equatorial system, Right Ascension α and Declination δ,
- the flux F of the incoming light,[1] and

---

[1]The light flux is measured in energy per second whereas the light intensity has the unit flux per solid angle.

- the status of polarization, described by the Stokes parameters Q, U and V (all in intensity units according to Eqs. 15.1–15.4).

The flux and the polarization are, in general, a function of the wavelength.

In 1852 Sir George Stokes introduced the four Stokes parameters I, Q, U and V to describe polarized light in an easy way. One can choose a coordinate system by considering an electromagnetic wave of a single photon ray[2] where the two mutual perpendicular vectors l and r lie in a fixed plane and $l \times r$ is in the propagation direction of this wave (Fig. 15.1). If furthermore l lies in the plane of the meridian of an equatorial coordinate system, we can describe the components of the electric vector E as a function of time with:

$$E_l = E_{l0} \sin(\omega t - \epsilon_l)$$

$$E_r = E_{r0} \sin(\omega t - \epsilon_r)$$

($\omega =$ angular frequency, $E_{l0}, E_{r0} =$ wave amplitudes, $\epsilon_l, \epsilon_r =$ phases).

From these two equations one can see that the vector E describes an ellipse in the plane of l and r. If Θ is the angle between the long axis of the ellipse and the direction of l, and $\sin \beta$ and $\cos \beta$ are the short and long half-axis of the ellipse,

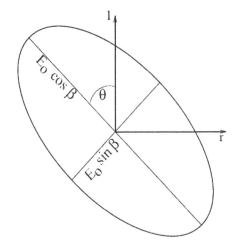

Fig. 15.1 Parameters defining the polarization of a simple wave. The light is coming toward us and l lies in the plane of the meridian of the equatorial coordinate system and directed towards the northern hemisphere (Serkowski 1962)

---

[2]To consider a single photon, we have to use expressions of quantum electro-dynamics: A single photon is only circularly polarized, which follows from its helicity. The polarization vector represents the "spin function" of a photon. Any polarization can be expressed by two orthogonal polarizations in the form of a linear combination of these two directions where the square of the amplitude of the coefficients gives the probabilities for the polarization directions. The complete description of a polarized photon is given by its polarisation matrix (a hermitic tensor of 2nd stage) with Lorentz invariant components.

respectively, one can define the Stokes parameters in the following manner (e.g., Chandrasekhar 1950):

$$I = E_{lo}^2 + E_{ro}^2 = \sqrt{Q^2 + U^2 + V^2} \tag{15.1}$$

$$Q = E_{lo}^2 - E_{ro}^2 = I\cos 2\beta \cos 2\Theta \tag{15.2}$$

$$U = -2E_{l0}E_{r0}\cos(\epsilon_l - \epsilon_r) = I\cos 2\beta \sin 2\Theta \tag{15.3}$$

$$V = 2E_{l0}E_{r0}\sin(\epsilon_l - \epsilon_r) = I\sin 2\beta. \tag{15.4}$$

Note that for another coordinate system, only Θ changes; I and β are *invariant* with respect to a change of the coordinate system and hence also $Q^2 + U^2$ and V are invariants.

15.4 Properties of Stoke Parameters

Because of the *additivity* of the Stokes parameters (Stokes parameters describing light are sums of the corresponding Stokes parameters describing the "simple waves" of which the light is composed) light can always be decomposed into two beams:

1. Unpolarized light with $Q = U = V = 0$ and
2. fully elliptical polarized light of intensity $\sqrt{Q^2 + U^2 + V^2}$.

The degree of polarization is described by

$$P = \frac{\sqrt{Q^2 + U^2}}{I} = \cos 2\beta$$

and the degree of ellipticity by

$$P_v = \frac{|V|}{I} = \sin 2\beta.$$

It is possible to split partially polarized light into two beams of fully plane polarized light. The first E vector then makes an angle Θ to the direction l, described by I_{max}, and the second E vector makes an angle $\Theta + 90°$, described by I_{min}. In the case of electron (Thomson) scattering the scattering process changes the length of the electric vector as described in Fig. 15.2.

The degree of polarization after scattering is

$$P = \frac{I_\perp - I_\parallel}{I_\perp + I_\parallel} = \frac{1 - \cos^2 \chi}{1 + \cos^2 \chi}.$$

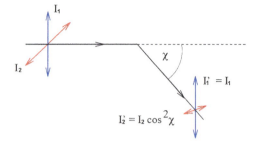

Fig. 15.2 The electron scattering process. The intensity of the electric vector is reduced by $\cos^2 \chi$ for the component parallel to the scattering plane. The perpendicular component is unaltered after scattering. χ is the scattering angle

The polarization position angle, Θ, is the position of the maximized electric vector, which gives the direction of the scattering plane. Thus, one can describe the polarization as a "quasi vector" with magnitude P and direction χ. But these parameters (P, χ) do not follow normal vector addition and one has to use the Stokes parameters which add linearly. The Stokes parameters are sums and differences of the intensity of the radiation field measured along different axes and one can understand them in the following way:

- I is the total intensity of the radiation field.
- Q is the intensity difference between components of the electric vector along two orthogonal directions.
- U is the intensity difference between components of the electric vector along two orthogonal axes rotated through 45° from the Q direction.
- V is the intensity difference between the left and right circularly polarized components.

Since the introduction of the Stokes parameters many polarimetric observations have been performed on planets, comets and galaxies. Sometimes circularly polarised light can be revealed, which can lead to significant success in the estimation of magnetic fields.

15.5 The Mueller Calculus

The Mueller calculus (Mueller 1943) is a matrix-algebraic method of specifying a beam of light and the optical devices encountered by the beam, and computing the outcome. If light passes a number of polarizers and/or retarders, conventional algebraic methods become extremely complicated; the arithmetic required is voluminous and the procedure is different for each different problem. The Mueller calculus on the other hand uses the fact that it is possible to

(a) condense all the necessary parameters for describing a light beam, polarizer, retarder or scatterer *into a single package* simultaneously, and

(b) to provide a set of rules where the result of interposing any number of optical
     elements can be *determined by multiplying the used packages in a standard
     manner using matrices.*

Thus the outcome of any experiment can be determined by a fixed procedure:
selecting the appropriate packages from a table and simply multiplying them
together.

The standard light beam description is simply a four-parameter Stokes vector
(I, Q, U, V) and optical devices (polarizer, retarder, etc.) are described by the so
called *Mueller matrix*, a 4×4 transformation matrix. For ideal devices most of the
elements are zero, which makes the calculation very easy. The individual matrix
deals not only with the composition of the device but also with its *orientation*.
For instance, a linear polarizer with a horizontal transmission axis has a different
Mueller matrix than a linear polarizer which is turned by some degrees.

The Mueller calculus follows the rules for matrix-algebra and the vector representing the incident beam must be written "on the right". For example, the
calculation of the Stokes parameters of a beam which passed a retarder *followed*
by a polarizer is described in matrix-algebra by:

$$
\begin{pmatrix} I' \\ Q' \\ U' \\ V' \end{pmatrix} = \begin{pmatrix} p_{11} & p_{12} & p_{13} & p_{14} \\ p_{21} & p_{22} & p_{23} & p_{24} \\ p_{31} & p_{32} & p_{33} & p_{34} \\ p_{41} & p_{42} & p_{43} & p_{44} \end{pmatrix}_{pol}
$$

$$
\times \begin{pmatrix} r_{11} & r_{12} & r_{13} & r_{14} \\ r_{21} & r_{22} & r_{23} & r_{24} \\ r_{31} & r_{32} & r_{33} & r_{34} \\ r_{41} & r_{42} & r_{43} & r_{44} \end{pmatrix}_{ret} \times \begin{pmatrix} I \\ Q \\ U \\ V \end{pmatrix}
$$

15.6 The Retarder Matrix

A general description of Mueller matrices for retarders and polarizers was given by
Serkowski (1962): By cutting a calcite or quartz crystal parallel to its optical axis
one can produce an optical element which introduces a phase-shift between the two
components of the E vector vibrating in the planes of the l and r direction. The
introduced phase-shifts

$$
\tau = \epsilon_l - \epsilon_r
$$

are 180° and 90° for a $\lambda/2$ and a $\lambda/4$ plate, respectively. If the retardation plate is
rotated counterclockwise by the angle ψ, **the retarder matrix** acquires the form

$$
R = \begin{pmatrix}
1 & 0 & 0 & 0 \\
0 & \cos^2 2\psi + \sin^2 2\psi \cos \tau & (1 - \cos \tau) \cos 2\psi \sin 2\psi & -\sin 2\psi \sin \tau \\
0 & (1 - \cos \tau) \cos 2\psi \sin 2\psi & \sin^2 2\psi + \cos^2 2\psi \cos \tau & \cos 2\psi \sin \tau \\
0 & \sin 2\psi \sin \tau & -\cos 2\psi \sin \tau & \cos \tau
\end{pmatrix}.
$$

In this context one can easily calculate the Mueller matrices for an ideal plate of isotropic, nonabsorbing glass (M_1); for an ideal plate of isotropic, absorbing glass whose transmittance is k (M_2); and for a totally absorbing plate (M_3) :

$$
M_1 = \begin{pmatrix}
1 & 0 & 0 & 0 \\
0 & 1 & 0 & 0 \\
0 & 0 & 1 & 0 \\
0 & 0 & 0 & 1
\end{pmatrix}
$$

$$
M_2 = \begin{pmatrix}
k & k & k & k \\
k & k & k & k \\
k & k & k & k \\
k & k & k & k
\end{pmatrix}
$$

$$
M_3 = \begin{pmatrix}
0 & 0 & 0 & 0 \\
0 & 0 & 0 & 0 \\
0 & 0 & 0 & 0 \\
0 & 0 & 0 & 0
\end{pmatrix}
$$

15.7   The Polarizer Matrix

An ideal polarizer, which transmits only the vibrations in the plane making an angle φ, has the following form:

$$
P = \frac{1}{2} \begin{pmatrix}
1 & \cos 2\varphi & \sin 2\varphi & 0 \\
\cos 2\varphi & \cos^2 2\varphi & \cos 2\varphi \sin 2\varphi & 0 \\
\sin 2\varphi & \cos 2\varphi \sin 2\varphi & \sin^2 2\varphi & 0 \\
0 & 0 & 0 & 1
\end{pmatrix}
$$

In the case of an element which splits the two mutually perpendicular beams that are phase-shifted by the angle τ after passing the retarder (e.g., a Nicol prism), the polarizer matrix acquires a simple form. For the ordinary beam, indicated by $\parallel$, we

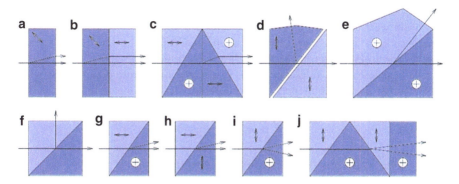

Fig. 15.3 Beam-splitting analyzers. (**a**) Plane-parallel calcite plate; (**b**) double calcite plate; (**c**) double Rochon prism; (**d**) Glan-Foucault prism (modified by Archard and Taylor); (**e**) Glan-Thompson (Foster) prism; (**f**) thin-film polarizing beam-splitter; (**g**) Rochon prism; (**h**) Senarmont prism; (**i**) Wollaston prism; (**j**) three-wedge Wollaston prism. The directions of light beams and the directions of crystallic optical axes are indicated by *arrows*; the circled crosses denote the optical axis perpendicular to the plane of the drawing (Serkowski 1974)

have $\varphi = 0°$ and for the extraordinary beam, indicated by $\perp$, we have $\varphi = 90°$. Then the Mueller matrices become:

$$M_{\parallel} = \frac{1}{2} \begin{pmatrix} 1 & 1 & 0 & 0 \\ 1 & 1 & 0 & 0 \\ 0 & 0 & 0 & 0 \\ 0 & 0 & 0 & 0 \end{pmatrix}$$

$$M_{\perp} = \frac{1}{2} \begin{pmatrix} 1 & -1 & 0 & 0 \\ -1 & 1 & 0 & 0 \\ 0 & 0 & 0 & 0 \\ 0 & 0 & 0 & 0 \end{pmatrix}$$

A number of different beam-splitting analyzers is shown in Fig. 15.3.

15.8   Spectropolarimetry

One might believe that the next step from broadband polarization (photometric polarization measurements) to spectropolarimtetry (spectral polarization measurements), is only small and easy. And indeed, instead of directing both beams through a beam splitter directly onto a detector one can simply position a spectrograph in front of the latter and disperse the light into wavelengths. Meanwhile, spectropolarimetry is one of the astronomical standard applications. The first pioneer instruments for small and medium-sized telescopes, such as the William-Wehlau spectropolarimeter and the Half-Wave Polarimeter (HPOL) of the University of Wisconsin were developed in the 1990s. This also applies to instruments for large telescopes such as the ESO 3.6 m telescope at CASPEC on La Silla. Then the Echelle SpectroPolarimetric Device for the Observation of Stars (ESPaDOnS) at Canada-France-Hawaii Telescope (CFHT) became the spectropolarimetric work horse. Today a number of new spectropolarimetric instruments are planned or already developed. This includes the proposed polarimeter unit for the E-ELT.

15.9   The William–Wehlau Spectropolarimeter

The William-Wehlau-Spektralpolarimeter[3] (Eversberg et al. 1998) is a combination of a retarder consisting of two $\lambda/4$ plates and a Wollaston Prism as polarizer, leading into a CCD spectrograph. The instrument was developed and built at the University of London/Ontario in collaboration with the Universities of Brandon/Manitoba and Montréal/Québec. Figure 15.4 shows the design of the polarimeter unit and Fig. 15.5 the open polarizer in the laboratory.

At the heart of the instrument are the two $\lambda/4$ plates which act as retarders and introduce a 90° shift between the mutually orthogonal components of the partially polarized beam. Figure 15.6 shows one of these plates in the center of a worm while Fig. 15.7 shows both plates installed in the instrument.

The plates are rotatable via stepping motors to different angles ψ. They are controlled by software written in C++ and implemented on a personal computer. After passing the two plates, the retarded beam crosses the polarizer which is here

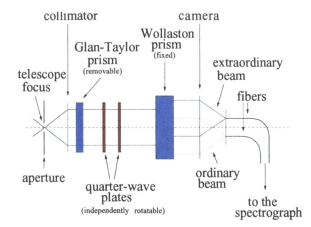

Fig. 15.4 Simple sketch of the William-Wehlau Spectropolarimeter

---

[3]After the principal investigator William Wehlau, Professor at the University of Western Ontario (UWO) (1961–1995), deceased.

Fig. 15.5 View of the
William-Wehlau
Spectropolarimeter in the lab

Fig. 15.6 One of the two $\lambda/4$
retarders

a Wollaston Prism, where the beam is split into an ordinary and an extraordinary
beam. The E-Vector of the ordinary beam is oriented perpendicular to the optical
axis of the crystal and follows Snell's refraction law. Its elemental waves have
spherical shapes. The E-Vector of the extraordinary beam oscillates in the plane
defined by the optical axis and the beam propagation direction. It does not follow
Snell's law and its elemental waves are rotation ellipsoids.

The two beams then reach the two fibers which feed the spectrograph slit. The
light path in the spectrograph is identical with the standard way to obtain a spectrum,
except that we now have two spectra: that of the ordinary beam, $I_o(\lambda)$, and that of the
extraordinary beam $I_e(\lambda)$, aligned parallel on the detector. From these wavelength-
dependent spectra, $I_\parallel$ (parallel) and $I_\perp$ (perpendicular), we can, by combining data

Fig. 15.7 The two λ/4 plates
and their stepping motors in
the polarimeter unit

for various combinations of orientations of the λ/4 retarders, obtain the four Stokes
parameters $I_{total}(\lambda)$, $Q(\lambda)$, $U(\lambda)$ and $V(\lambda)$ via the Mueller calculus.

To test for cross-talk between the beams and for possible non-linearity in the
λ/4 plates, a removable Glan-Taylor prism which produces nearly 100 % linearly
polarized light at all optical wavelengths is installed in front of the retarders. Fast
axes of λ/4 plates must first be aligned with the axis of the Wollaston prism. This
can be automatically done by a computer program.

Following the Mueller calculus and the rules for matrix algebra, we can calculate
the four Stokes parameters for this arrangement with $\tau = 90°$ for both λ/4 plates.
With **A** and **A**$'$ as the Stokes vectors before and after passing the retarder plates **R**$_1$
and **R**$_2$, and the polarizer **P**, we use the *Ansatz*:

$$\mathbf{A}' = P \times R_2 \times R_1 \times A$$

Following this expression for different angular positions of the optical elements,
we get a number of resulting equations for the final, observed Stokes parameter I
that contains all the information we need as a function of I, Q, U and V of the
original beam (A). If we indicate the intensity with the angular values of retarder
one, retarder two and the position of the polarizer (∥ or ⊥), in this order; F as the
time-dependent variation (e.g., seeing, transparency); and G the angular position-
dependent gain factor; we have, for instance:

$$I'_{0,0,\parallel} = 1/2 \cdot (I + Q) \cdot F_{0,0} \cdot G_{\parallel}$$

$$I'_{0,0,\perp} = 1/2 \cdot (I - Q) \cdot F_{0,0} \cdot G_{\perp}$$

$$I'_{45,45,\parallel} = 1/2 \cdot (I - Q) \cdot F_{45,45} \cdot G_{\parallel}$$

$$I'_{45,45,\perp} = 1/2 \cdot (I + Q) \cdot F_{45,45} \cdot G_{\perp}$$

$$I'_{0,45,\parallel} = 1/2 \cdot (I + U) \cdot F_{0,45} \cdot G_{\parallel}$$

$$I'_{0,45,\perp} = 1/2 \cdot (I - U) \cdot F_{0,45} \cdot G_{\perp}$$

$$I'_{0,-45,\parallel} = 1/2 \cdot (I - U) \cdot F_{0,-45} \cdot G_\parallel$$

$$I'_{0,-45,\perp} = 1/2 \cdot (I + U) \cdot F_{0,-45} \cdot G_\perp$$

$$I'_{-45,0,\parallel} = 1/2 \cdot (I + V) \cdot F_{-45,0} \cdot G_\parallel$$

$$I'_{-45,0,\perp} = 1/2 \cdot (I - V) \cdot F_{-45,0} \cdot G_\perp$$

$$I'_{45,0,\parallel} = 1/2 \cdot (I - V) \cdot F_{45,0} \cdot G_\parallel$$

$$I'_{45,0,\perp} = 1/2 \cdot (I + V) \cdot F_{45,0} \cdot G_\perp$$

With these equations we can easily get the intensity-normalized Stokes parameters Q, U and V:

$$\frac{Q}{I} = \frac{R_Q - 1}{R_Q + 1}, \quad \frac{U}{I} = \frac{R_U - 1}{R_U + 1}, \quad \text{und} \quad \frac{V}{I} = \frac{R_V - 1}{R_V + 1} \tag{15.5}$$

with

$$R_Q = \sqrt{\frac{\dfrac{I'_{0,0,\parallel}}{I'_{0,0,\perp}}}{\dfrac{I'_{45,45,\parallel}}{I'_{45,45,\perp}}}} \tag{15.6}$$

$$R_U = \sqrt{\frac{\dfrac{I'_{0,45,\parallel}}{I'_{0,45,\perp}}}{\dfrac{I'_{0,-45,\parallel}}{I'_{0,-45,\perp}}}} \tag{15.7}$$

$$R_V = \sqrt{\frac{\dfrac{I'_{-45,0,\parallel}}{I'_{-45,0,\perp}}}{\dfrac{I'_{45,0,\parallel}}{I'_{45,0,\perp}}}} \tag{15.8}$$

Note that these double ratios are impervious of both time-dependent variations and spatially-dependent gain factors, as long as the two beams are obtained simultaneously on the same part of the detector each time.[4] They should therefore be purely photon-noise limited. The intensity I within a constant is easy to get from a simple addition of corresponding pairs of equations, after appropriate determination of the gain factors by flat-fielding. Also note that any of the angles ψ of the $\lambda/4$ plates can be replaced by $\psi \pm 180°$ with identical results, providing the surfaces of the plates are not inclined to the optical axis (Serkowski 1974). If the $\lambda/4$ plates are not exactly $\tau = 90°$, the above equations must be modified; this can be easily done

---

[4]If there are time variations between observations of different Stokes parameters, then they will not be simultaneous and thus not reflect the true condition of the source at a given instant.

for small deviation from $\tau = 90°$. The design above delivers quasi-simultaneously all four Stokes parameters. Note that one can also use a single $\lambda/2$ retarder instead of two $\lambda/4$ retarders. The former then does not deliver Stokes V for circularly polarised light but only Q and U for the linearly polarized components.

Polarimetric measurements are today a standard tool in professional astronomy. All optical elements for polarimeters can be found on the open market and the construction should be no problem. The procedure is simple. The light from the focus of the telescope is collimated and passes through one retarder or two. The Wollaston prism (or another beam splitter) separates the ordinary from the extraordinary beam and the two beams are then guided to the detector. The design of a spectropolarimeter seems to be straight forward. In reality, however, one needs to take two main issues (physical and technical) into account:

- The low polarization level: In comets polarization degrees of up to 30 % have been observed (30 % of the incident light is polarized). This is extraordinary and detectable even with small telescopes. But the intrinsic light of stellar objects shows a maximum polarization of only about 1 % (up to 10 % linear polarisation including IS grains). Circular polarization from the ISM is extremely small, as is the stellar component normally, which however can reach 50 % or more in some extreme cases (e.g. magnetic CVs). The polarized light flux is thus significantly reduced by large factors compared to the total flux. So one has to use either large telescopes or reduce the spectral resolving power and sufficiently increase the exposure time to measure enough polarized light. Small telescopes of up to about 1 m aperture can, hence, measure only the brightest stars in the sky, some of which have wind geometries that can polarize the light (or stellar winds with preferred geometries for linear polarization, material disks or magnetic fields for circular polarization). For this reason one needs telescopes of at least 5 m aperture in order to achieve very high spectral resolving power and sufficient S/N. However, due to the small number of available big instruments (about a dozen) there is a particular need for low-resolution spectropolarimetric measurements. Therefore, this particular window for smaller apertures is still of great interest.
- Instrumental effects: Great care should be invested in the design of the spectrograph light-feeding unit. The intensity-normalized Stokes parameters Q, U and V are derived by a set of Eqs. 15.5 including expressions with double-ratios for the two beams (parallel and perpendicular) at different retarder positions (Eqs. 15.6–15.8). The denominator of these double-ratios can become very small and slightest intensity deviations for specific retarder positions can introduce significant deviations for the final result. Such small intensity deviations can be introduced by a fiber connection to the spectrograph due to (a) focal ratio degradation (FRD, see Sect. 11.5) and (b) a non-uniform transmissivity at the fiber surfaces. Hence, a direct telescope-instrument link should be preferred. If fiber optics are used one should invest great care in the reduction of FRD (e.g. fiber scrambling, Sect. 11.7) and should possibly image the whole telescope aperture onto the fiber aperture to reduce stellar scintillation (seeing) at the fiber aperture.

The latter again can introduce time-dependent as well as double-beam effects at the CCD detector. These are (a) Doppler shifts between two observations due to radial velocity variations of stellar or terrestrial origin, (b) stellar rotation or variability which may cause some spectral changes, (c) drift in the spectrograph, (d) unknown detector pixel sensitivity above a level of 10^{-3}, (e) different spectra imaging due to optical aberrations and (f) the introduction of small wavelength differences between corresponding pixels of both spectra due to detector misalignment.

15.10 Polarimetric Investigations of Massive Stars

15.10.1 Interstellar Polarization

After the description of possible intrinsic polarization in hot-star binaries by Chandrasekhar (1946, 1947), Hiltner (1949) and Hall (1949) in search of it, actually discovered interstellar linear polarization. The interstellar gas component consists mainly of atoms, ions and molecules of hydrogen. Before the discovery of interstellar polarization it was believed that the geometrical shape of interstellar dust particles was spherically symmetric and, hence, they can not polarize stellar light. But it became clear that the degree of polarization is correlated with the IS extinction due mainly to dust. In addition, the polarization level can be up to 6 %. The highest degree of IS polarization lies in the galactic plane and we find regions where the polarization direction is strongly correlated from star to star (at similar distances). The variation of the degree of polarization between $0.33\,\mu$ ($3,300\,\text{Å}$) and $1\,\mu$ is relatively small. In general, $P(\lambda)$ reveals a fairly flat curve with its maximum between 5,000 and 7,000 Å. For this dependence, Serkowski (1975) found an empirical (but quite accurate) relation between $P(\lambda)/P_{max}$ and λ_{max}/λ which is shown as a plot in Fig. 15.8.

$$\frac{P(\lambda)}{P_{max}} = e^{-1.15\ln^2(\lambda_{max}/\lambda)}.$$

In addition, observations showed that the distribution of the interstellar polarization is not regular. The polarization vectors in many regions of the galactic plane are in parts strongly aligned and not stochastic. This can be understood by the idea that the polarization is introduced by light which is scattered off *non-spherical grains* which are aligned by the magnetic field of our Galaxy with a strength of some 10^{-6} G. This means that measurements of intrinsic stellar polarization can never neglect the influence of the polarization of material between the star and the observer. Such measurements must be corrected for interstellar polarization.

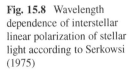

Fig. 15.8 Wavelength
dependence of interstellar
linear polarization of stellar
light according to Serkowsi
(1975)

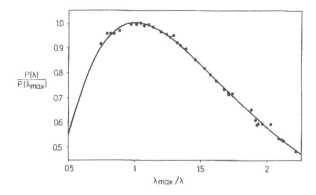

$$\frac{P(\lambda)}{P(\lambda_{max})}$$

15.10.2 Intrinsic Linear Polarization

Intrinsic polarization of early-type stars arises mainly from Thomson (electron) scattering (see Fig. 15.2) of stellar light. The fundamental condition to show any intrinsic polarization and/or polarimetric variability is a non-spherical stellar atmosphere. If uniform spherical geometry prevails, all polarization due to Thomson scattering cancels out. But if any deviation from a radial sphere exists (e.g., an equatorial disk or localized regions of higher ionic or atomic density, so called *blobs*, or only density enhancement above the equator) an observer should be able to detect intrinsic polarization. This is also valid if the source is asymmetric with respect to an outer shell.

Chandrasekhar (1946) predicted intrinsic polarization values for a stellar disk to be zero at the center and as high as $\sim 11\%$ at the limb. Although only a prediction, this effect is thought to be detectable during eclipses. The electric vector is then vibrating tangentially to the stellar surface. To estimate the asymmetric geometry of the star-envelope system via spectropolarimetry one has to consider the expected degree of polarization.

Polarization in early-type stars often yields intrinsic variability in time and sometimes with wavelength.

(A) Wavelength Dependence

There are four different explanations how the observed polarization becomes wavelength dependent. If the star and its scattered light are both in the observer's beam, and the polarized scattered light L_{scat} is small compared to the direct unpolarized starlight $L_\star$, the observed degree of polarization p_{obs} is $p_{obs} = p(\Theta)L_{scat}/(L_\star + L_{scat})$, with the polarization degree $p(\Theta)$. Spectral lines formed at larger distances from the star show a decreasing polarization degree but their polarization vectors lie parallel to the polarization vector of the continuum. All polarization values can,

Fig. 15.9 Sources of spectropolarimetric observations (Nordsieck et al. 1992)

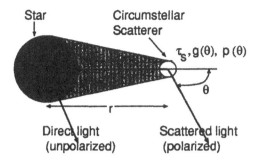

hence, be found on a straight line in the Q-U diagram. Figure 15.9 illustrates the four basic ways of how to make p_{obs} wavelength dependent.

(a) Unpolarized light L_{dil} not originating in the star (e.g., nebular emission from circumstellar material) dilutes the polarized light:

$$p_{obs} = p(\Theta)L_{scat}/(L_\star + L_{scat} + L_{dil}),$$

where $L_{dil}(\lambda)$ is wavelength dependent.

(b) Absorptive opacity may reduce scattered light more than direct starlight:

$$p_{obs} \sim 1 - \tau_a(\lambda)$$

and we observe polarization with features that are weak or do not exist in the stellar spectrum.

(c) Different illumination geometry at different wavelengths (e.g., due to limb darkening):

$$p_{obs} \sim D(\lambda)\sin^2\Theta(\lambda).$$

(d) The scattering process may itself be wavelength dependent, e.g., due to dust or atomic scattering:

$$p_{obs} \sim p_{max}(\lambda)g(\Theta,\lambda)\tau_s(\lambda).$$

However, this is ruled out for hot stars, due to the "grey" Thomson-scattering process.

An axisymmetric scattering envelope leads to

$$p_{obs} \sim \tau_s(1 - 3\gamma)\sin^2 i,$$

where γ is an envelope shape factor (for a spherical envelope $\gamma = 1/3$; $\gamma = 0$ (1) for a flat (oblate) disk), i is the inclination, and $\tau_s = \sigma_T \int_V N_e(R)dV$, with the electron density distribution $N_e(R)$. The Thomson-scattering cross section is

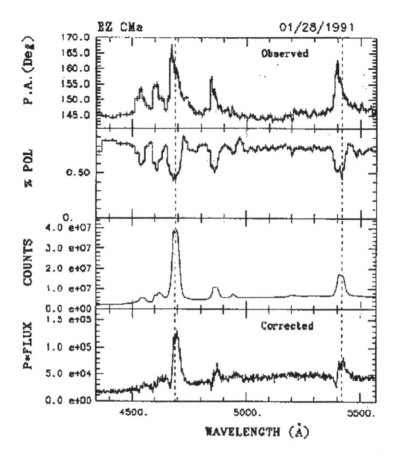

Fig. 15.10 Spectropolarimetric measurements of EZ CMa at the Anglo-Australian Telescope. Total counts, polarization in percent, and position angle in degrees, versus wavelength for EZ CMa, measured. *Error bars* are $\pm 1\sigma$. Changes across line profiles, which are different in flux and polarization, can be seen. P∗FLUX is the spectrum of the polarized counts after correction for interstellar polarization. *Dashed lines* mark the positions of the observed line centers for HeII 4,686 and 5,412 Å in the count spectrum (Schulte-Ladbeck et al. 1992)

wavelength independent and the scattering process does not alter the scattered spectrum. Chandrasekhar already suggested that the best condition for observable polarization from the limbs of early-type stars should be the symmetry breaking effect of an eclipse by a companion. This phenomenon has indeed been found in the 1990s. An example for spectropolarimetric measurements is given in Fig. 15.10 for the WR star EZ CMa. The star is a single rotating WR star probably with CIRs (see Sect. 14.7.1) which produce the (mainly) continuum polarization.

To consider the different aspects of wavelength dependence one can split the WR star into three different zones: a *continuum-forming region*, a *line-emitting region*, and an *electron-scattering region*. If we suppose that the scattering events happen close to the star, whereas the emission-line forming region is at large distances, then

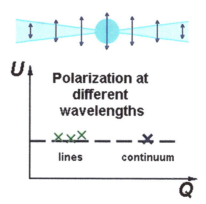

Fig. 15.11 Cartoon depicting a steady-state "disk". The continuum polarization has a direction perpendicular to the disk. Lines formed at increasing radii exhibit decreasing amounts of polarization, but the direction of the line polarization is the same as the continuum polarization. Thus, at a given time, polarizations measured in the continuum and different lines lie along a straight line in the Q-U plane (Schulte-Ladbeck et al. 1992)

the line polarization is almost zero; recombination emission is unpolarized and the radiation is emitted far from the scattering electrons. The result is a reduced level of polarization in the emission lines. Because of this *dilution effect* it is necessary to treat the continuum and line forming regions *together* with the scattering regions.

Figure 15.11 presents a sketch of a radiatively driven WR wind with an equatorial material enhancement. If different lines are formed in different radii (this is a reasonable assumption, as we will see later), then they will also show different amounts of polarization. The geometry of the continuum and line scattering regions is the same and their co-linear polarization vectors lie perpendicular to the equatorial plane. In the Q-U diagram of Fig. 15.11, the polarization at different wavelengths will follow a straight line. This is valid for the continuum as well as for the lines if there is no interstellar polarization component. The correction for interstellar polarization is possible with the "Serkowski law" (Serkowski et al. 1975), which gives the shape of the interstellar polarization as a function of wavelength.

A common problem in observations of intrinsic polarization is that of small contrast. Assuming an interstellar polarization of 1 % and an intrinsic stellar polarization of about 0.1 % with a line-flux to continuum ratio of 10, then the continuum polarization is 1.1 % and the line polarization is 1.01 %. For a 3σ detection of this feature we need an internal accuracy of at least 0.03 % per pixel. This means that a signal-to-noise ratio of 3,000(!) is necessary. So, to observe not only bright but also faint stars and to understand the nature of variability we need a large telescope for a reasonably long run.

Fig. 15.12 Polarization from axisymmetric "clumps". When the polarization is measured at different times in a continuum filter, individual data points are found to fall along a line in the Q-U plane, because the spatial distribution of the clumps has a preferred plane, and thus there exists a preferred direction of the polarization (Schulte-Ladbeck et al. 1992)

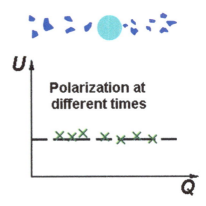

(B) Time-Dependence

As an example Fig. 15.12 shows the sketch of a star with a radiatively driven wind containing clumps at random time intervals in an equatorial plane with an axisymmetric distribution. Integrated over the ensemble of clumps at any time, the direction of the polarization vector is more or less constant with time. If plotted in a Stokes-Q-U diagram the data points for different clumps lie along a straight line. Each data point represents the vectorial sum of a time independent interstellar polarization vector and the intrinsic polarization vector with temporal variations. The polarization position angle Θ will remain constant, due to the geometry. If the interstellar polarization is known, we could estimate the true zero point for the intrinsic coordinate system and hence the direction on the plane of the sky in which the material is concentrated. To obtain the shape of the data points in the Q-U plane and thus to answer the question if there is a special wind geometry *and* variability, many observations are necessary.

1. In the case of *random* variability one *always* observes the preferred axis, but not vice versa. The preferred axis is always visible in the Q-U plane.
2. There will be a certain number of disk-like systems which exhibit random variability because they are seen nearly pole-on. For such systems we can detect random but not a preferred direction of variability.
3. If the atmosphere shows no temporal variability, this only means that the wind is homogeneous; its geometry would remain unknown. A steady-state disk would not show polarization variations.

15.10.3   Intrinsic Circular Polarization

The creation of linearly polarized light is also possible via cyclotron emission. A free electron captured by a magnetic field spins around the field lines and emits Bremsstrahlung which is polarized. The emitted wave oscillates perpendicularly to the field direction. In the direction of the field the electrons emit circularly polarized light. One refers to a transverse and a longitudinal field for the field components perpendicular and parallel to the observer's direction, respectively. This means that if we see circularly (linearly) polarized light, we look "onto" the electrons with their spin parallel (perpendicular) to the line-of-sight. As a result, circularly polarized light is a relatively unambiguous indicator for intrinsic magnetic fields. Linearly polarized light is an indicator of scattering processes as well as magnetic fields, although not as sensitive as circularly polarized light.

The detection of magnetic fields is done through the observation of the Zeeman effect in spectral lines. The energy levels of an atom in an external magnetic field are split into 2J + 1 sublevels. The energy difference of these sublevels is $\Delta E = gehB/4\pi mc$, were g is the Landé factor. As a result, the atomic lines are also split and we call the components for which $\Delta M = \pm 1$ (quantum number M for angular momentum) the σ component and for $\Delta M = 0$ we call them the π components. These components are polarized: For a transverse field, the π components are linearly polarized parallel to the field and σ components are polarized perpendicular to it. For a longitudinal field the σ components have opposite circular polarizations, whereas the π components are not visible. This situation is represented in Fig. 15.13 from Landstreet (1979).

The differences between the analyzed line profiles in panel (c) of Fig. 15.13 are the basis for the detection of fields which are too weak to distort the profiles in (b). Figure 15.13 represents an ideal case of clearly separated longitudinal and transverse fields. The real case is, in general, more complex. As an example, Fig. 15.14 shows the effect of changing the field orientation with time. The profiles show additional complexity due to the rotational broadening of the line.

To perform polarimetric observations for the detection of magnetic fields we need to take fundamental principles into account. If we assume a stellar disk, affected by two magnetc spots of opposite polarities, with no relative velocity to the observer, the sum of their circular polarizations would be reduced or cancelled out (see Fig. 15.15). In the case of non-zero stellar rotation, the relative velocities of the spots are separated due to the effect of circular polarization. Then,

- the measured global magnetic flux is reduced and differential measurements may lead to the determination of the line-of-sight component of the magnetic field, and
- to increase the S/N of Stokes V, one may add the signals from several lines, to increase the quality of magnetic field measurements.

From the technical point of view, the application of Zeeman Doppler imaging has some important requirements. (A) Because of polarization rates of the order of 0.1 percent, the S/N ratio must be high. (B) To see small scale structures, the spectral resolution should be reasonably high. (C) Spurious polarization levels should be reduced to the level of the photon noise.

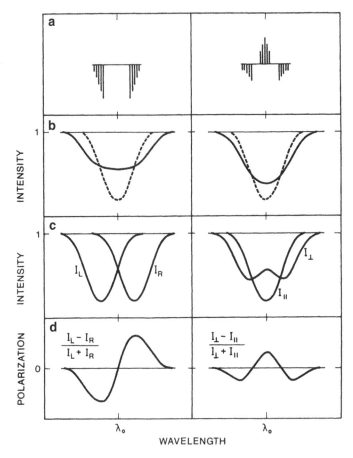

Fig. 15.13 Wavelength dependence of emergent intensity and polarization across a spectral line split by a longitudinal (*left*) and a transverse (*right*) magnetic field. (**a**) The splitting of a spectral line for a $^3P_2 - ^3D_3$ transition, π components above the line and σ components below, with the lengths of the bars indicating the strengths; (**b**) The appearance of the stellar absorption line with a field present (*solid*) and absent (*dashed*); (**c**) *on the left side*: the line profile as seen in right (I_R) and left (I_L) circularly polarized light; (**c**) *on the right side*: the line profile seen in linearly polarized light parallel ($I_\parallel$) and perpendicular ($I_\perp$) to the field. (**d**) The net circular polarization (*left*) ($V = (I_L - I_R)/(I_L + I_R)$) and linear polarization (*right*) ($Q = (I_\perp - I_\parallel)/(I_\perp + I_\parallel)$) across the line (Landstreet 1979, ©AAS. Reproduced with permission)

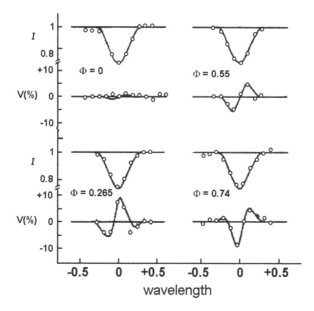

Fig. 15.14 Flux and circular polarization profiles of Fe II 4,520.2 Å in 78 Vir at several phases. The points are observed data, while the smooth curves are the predictions of the model, given in Landstreet (1979), ©AAS. Reproduced with permission. The profiles clearly show the added complexity which arises from the rotational broadening of the line

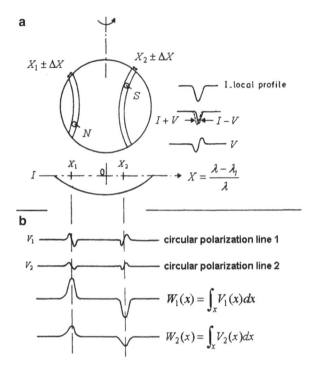

Fig. 15.15 Magnetic spots and circular polarization. The star can be divided into zones of equal velocity. The light emerging from each zone has a particular Doppler shift. In the spectrum, the contributions from different zones to a given spectral line are separated by their Doppler shift. Circular polarization due to the first magnetic spot appears at X_1, and the second at X_2 in the V profiles. The same is true for all spectral lines. (**a**) On the right, the two profiles $I + V$ and $I - V$ correspond to the two states of circular polarization. Their difference gives Stokes V. It has a particular shape easy to recognize. Each stellar spot contributes such a signal to the spectrum at the appropriate wavelength. (**b**) The observed circular polarization in the stellar spectrum. V_1 and V_2 correspond to the lines 1 and 2, respectively. W_1 and W_2 are the integrals of V_1 and V_2, respectively (Semel 1989, reproduced with permission ©ESO)

Suggested Readings

- J.C. del Toro Iniesta, *Introduction to Spectropolarimetry*, Cambridge University Press, 2008
- J. Trujillo-Bueno, F. Moreno-Insertis, F. Sanchez (Eds.), *Astrophysical Spectropolarimetry*, Cambridge University Press, 2001
- R.T. Chornock, *Astrophysical Applications of Spectropolarimetry*, Proquest, Uni Dissertation Publishing, 2011

Chapter 16
Epilogue: Small Telescopes Everywhere

A Short Story

In January 2009, the relatively brief periastron passage of the long-period Wolf-Rayet + O binary star WR 140 took place, long awaited by stellar-wind astronomers. Thomas discussed this event with some colleagues in Canada and made this system popular in the amateur community to obtain data together. This was in his mind when he visited his long-time friend Johan Knapen at the Astronomical Institute of the Canaries on Tenerife. When approaching the island, he saw the local observatory from the plane and thought that one should leave bad weather in Central Europe behind with mobile equipment. This he told Johan during a beer and he in turn calmy replied: "You can also do that with amateur astronomers at Teide observatory. We have a 50 cm-telescope for that!"

16.1 Small versus Big

As we have shown in Chap. 14 many scientific observation programs still require small telescopes. Long-term observations cannot normally be carried out with increasingly large telescopes due to limited observation time. In contrast, the number of smaller professional telescopes does not increase significantly. Professional astronomy moves towards "Big Science". Before a complex of smaller units will be built, often money is available for a more suitable large telescope. This gap is now filled more and more by very small robotic telescopes of less than 1m aperture. In addition, it is important for the scientific community to recognize that amateur spectroscopists have advanced into scientific areas for about 10 years (see, e.g., Fahed et al. 2011; Morel et al. submitted; Morel et al. 2011; Miroshnichenko et al. 2013). Figure 16.1 shows an example for the data quality that can be delivered by experienced amateur spectroscopists.

© Springer-Verlag Berlin Heidelberg 2015
T. Eversberg, K. Vollmann, *Spectroscopic Instrumentation*, Springer Praxis Books,
DOI 10.1007/978-3-662-44535-8_16

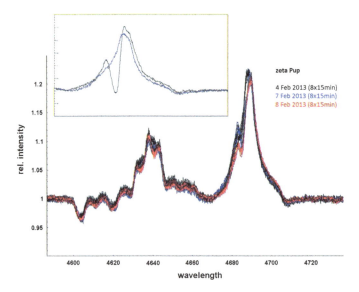

Fig. 16.1 Variability of the He II 4686 and N III complex lines of ζPuppis over 4 days. *Inset*: Extreme values for the HeII central reversal depth. Celestron C11 telescope with LHIRES III off-the-shelf spectrograph, exposure time 15 min, $R \approx 13,000$, S/N ≥ 450 recorded 4 km from the centre of Melbourne, Australia (Heathcote 2013, for comparison with a 4 m class telescope see also Fig. 14.25)

The extraordinary market performance of the past 30 years has led to telescopes becoming cheaper and cheaper in all sizes. In the 1970s, a standard 14-inch telescope cost approximately $US 30,000. Today these systems are about five times cheaper! And who does not want to build a photometer or spectrograph? Such systems can be found off-the-shelf for relatively little money (see Chap. 8). For university departments with a limited budget, it might be attractive to set up an entire cluster of 50 cm telescopes plus auxiliary instrumentation to keep whole student groups busy with data acquisition and reduction instead of a large telescope for a seven-digit sum of money. Apart from measurements of fast phenomena at very high spectral resolution ($R > 30,000$) and high signal-to-noise ratio ($S/N > 1,000$) it should be clear from our discussions that for many spectroscopy fields, the aperture does not need to be the driver for good science. Even astronomers with a private observatory can, of course depending on local weather conditions, provide data at almost any time and without special preparation. For performing scientifically useful work with small telescopes, two strategies offer themselves which should match the corresponding research objective—monitoring and campaigns.

- **(A) Monitoring**—The investigation of many astronomical phenomena requires measurements over months to years. This especially requires continuous observations as often as possible, which should be carried out with stable and fully tested systems. For professional astronomers, however, this is a problem because even at small observatories one can get regular observing runs over a maximum of a few weeks. Under real conditions, it is impossible for a researcher to, e.g., investigate the period of the oscillating Kepler disk of a Be star, as well as its detailed behavior. This requires accurate observations over at least two periods, i.e., up to about 20 years. A serious amateur observer or professional groups can perform this with small telescopes, as long as a corresponding continuity is ensured both in technology and all corresponding procedures. A prototype example is the eclipsing binary ϵ Aurigae, a bright star eclipsing once every 27 years. An F-type primary star is then occulted by an extended cool disk out of dust of about 1,000 K with a B5V in its center. During a purely amateur campaign some hundred spectra were obtained during the two year eclipse event covering various spectral lines, e.g., K 7699 Å, Na D 5890/5896 Å, Mg II 4481 Å (Leadbeater 2011; Leadbeater et al. 2012; Mauclaire et al. 2013). Figure 16.2 (left) shows the K 7699 line behaviour for the three year campaign. An interesting result is an obvious structure within the disk. Figure 16.2 (right) shows the behaviour of the K 7699 excess absorption equivalent width during eclipse. The steplike increase and decrease as well as a lack of symmetry hints towards a possible elliptical system or spiral density waves produced by the tidal influence of the F-star structures. Similar features are also seen in the Na D line behaviour. Note that almost all measurements have been performed with off-the-shelf spectrographs at telescopes of 20–40 cm aperture.

- **(B) Campaigns**—Normally amateur astronomers do not fight with bad equipment but with bad weather. For telescopes smaller than 1 m aperture, this is often reversed in professional astronomy. Sometimes one can find telescopes at world-class sites that are no longer state-of-the-art. It is therefore reasonable, to bring both sides together and perform professional campaigns with amateurs and their tested technology. This has been done in an exemplary fashion for the periastron passage of the WR+O binary star WR 140 in 2009 (see Chap. 14) as well as

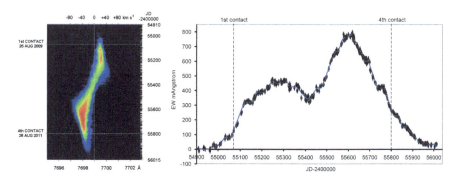

Fig. 16.2 The K 7699 line behaviour of epsilon Aur observed over about 3 years during eclipse. The interstellar line component has been removed (*colours* indicate the absorption line strength increasing from *blue* to *red*). Tick marks on the right-hand y-axis indicate actual measurements, intermediate rows are interpolated (Leadbeater et al. 2012)

for some Oe and B stars from December 2008 to March 2009. Professionals and amateurs came together for a global observation campaign. These included twenty colleagues from Portugal, England, Holland, Belgium, Canada, France and Germany, who worked at the 50 cm telescope at Teide Observatory in Tenerife, at Observatoire de Haute Prevence as well as at backyard observatories in Portugal, England and Germany. The backyard observatories had telescopes of 30 cm aperture using mostly commercial spectrographs (Fahed et al. 2011; Morel et al. 2011, in Sect. 14.11 as well as in Figs. 14.50 and 14.51 we have already presented some of the results). In 2011 professionals and amateurs studied the Be star δ Scorpii at the Teide IAC80. They investigated the influence of the compact companion on the Be star disk during periastron passage. The one-week measurements were performed with a commercially available LHIRES III Littrow spectrograph (Miroshnichenko et al. 2013). in 2013 again at IAC80 a 4-month ProAm campaign with nine teams examined the Wolf-Rayet stars WR 134, WR 135 and WR 137. The goal was the measurement of periodic and stochastic changes in the emission lines of the stellar winds (Aldoretta et al. 2014). In addition, further analysis on the stars HD 160529 (A hypergiant) and HD 316285 (LBV) were conducted (Richardson et al. in preparation). The instrument used for this purpose was a commercial echelle spectrograph. These core observations were complemented by a large number of support observations, including at Keck observatory.

Both monitoring and campaigns need support by professional astronomers. Especially amateurs without a scientific education need appropriate information on the physics of the target objects and about corresponding spectroscopic phenomena. For the continuous determination of spectral parameters during monitoring (e.g., equivalent widths), the necessity of uniform procedures (e.g., continuum definition, measurement intervals, error-bar determination) cannot be understated. In particular, the determination of measurement errors is often a neglected aspect in the amateur domain. Each campaign must be sufficiently prepared and appropriate procedures must be understandable for every campaigner. Today, at least for amateurs, a ProAm community has been established at least in Europe, providing data for research from both private backyard telescopes as well as from professional observatories (Tenerife, Pic du Midi).[1] The two major amateur groups, internationally working, maintaining a regular online exchange, organizing conferences and adviced by professionals, are the Spectroscopy Section of the GERMAN AMATEUR ASTRON-OMY ASSOCIATION (VdS) and the ASTRONOMICAL RING FOR ACCESS TO SPECTROSCOPY (ARAS) in France.

With Fig. 1.4 in the prologue of this book we have highlighted the amazing progress made in recent decades in terms of the quality of modern CCD detectors. Telescopes of a few inches in diameter provide images, which were unattainable

---

[1] When finalizing this book, an amateur spectroscopy community is just starting in the southern hemisphere (New Zealand, Australia, Brazil).

even with the largest equipment 50 years ago. Now we realize that this high level of performance has reached optical spectroscopy. In particular, because of the now commercially available and inexpensive off-the-shelf spectrographs, new possibilities for professional institutes in practical spectroscopy have opened up. Regardless of the contents of this book, this refers also to spectroscopy of other celestial bodies (planets, comets, nebulae, galaxies). The first of these are bright enough for fast high quality recordings and the last are stable over time, so that long exposures are likely to be applicable. This means that spectroscopy now reaches a broad astronomical basis and the corresponding impact should reasonably expand astronomical research despite the presence of modern big-size and robotic telescopes. From our point of view, this depends solely on whether the modern research community recognizes this opportunity and actively uses the enthusiasm of spectroscopic amateur aficionados.

This book aims to address this very point.

Chapter 17
Acknowledgements

A Short Story

"Thar she blows!" was the email from Tony Moffat when returning the English fine tuning for one of the big chapters within only some days. Tony is our scientific hero and an important native speaker proof-reading the entire book. When checking the corrections we discovered his comment "This paragraph above has zero English errors, which is quite unusual for you guys. You always have a few in a text like this! I hope you didn't lift this directly from some other source! If so, then you should put it in quotations and give a reference." Having been one of his students Thomas was well aware of Tony's affinity for classical literature and his lasting concern for good science. To calm him down we answered: "Hi Moby Dick! Don't worry, we plagiarize the whole book".

When we started writing this book, many spectroscopic facts were still unclear to us and we had to rely on help from others. In particular, that included discussions where we could argue with friends and then relax together. First, we need to mention all colleagues in the Section Spectroscopy of the German Association of Astronomers (VdS) who willingly discuss their knowledge and answer questions in the group's discussion forum and during the annual section conferences. At the regular Round Table Astro Meeting in Bochum we could enthusiastically discuss (not only) spectroscopic issues and cool-down again with pizza and beer. There our knowledge curve quietly increased, although some members still believe that the universe is beer-bottle-shaped. We are grateful for the benevolence of Hans-Werner Eurskens and for his practical support with his skills as a mechanic. We thank Karin and Wolfgang Holota for their friendly discussions about optics and various instrumental planning issues. Karin's cake will be unforgotten. Many thanks go to the true pragmatist Lothar Schanne, who helped us finding technical solutions for many problems. We also thank Otmar Stahl who always answered our questions with rocket speed emails. We are grateful to Daniel Sablowski for many discussions

© Springer-Verlag Berlin Heidelberg 2015
T. Eversberg, K. Vollmann, *Spectroscopic Instrumentation*, Springer Praxis Books,
DOI 10.1007/978-3-662-44535-8__17

and suggestions. Our thanks go to Guenter Gebhard for his penetrating critical remarks that are often made us sweating. He provided his MIDAS tutorial, which we recorded almost unchanged in the appendix. We also thank Richard Walker. He made his calibration catalogue available for the appendix. Our thanks also go to the workshop of the physics department of Bochum University. They are, inter alia, Wolfgang Conrad, Georg Schäfer, Christian Vilter and Clementz Wirtz. For many mechanical problems we knew us there in good hands and have had a lot of fun during our internship at the lathe and the broom. Johan Knapen has corrected parts of the book text. For this and for his friendship and time consuming massive support of our observing campaigns we thank him heartily. We are very thankful to Wolfhard Schlosser for the many years of lasting goodwill and confidence in our projects. He was often our teacher, had always opened the laboratory door and thereby shared his critical thoughts with us. A special role has our friend and colleague, Norbert Reinecke. His helpful comments and good suggestions had a significant impact on our work. With his enthusiasm he constantly motivated us and we are very pleased to have his support.

This book is dedicated to two persons close to us:

First we thank Hans-Siegfried Nimmert for many years of support of our projects, for his ever-pragmatic answers to various technical questions and for his motivation to go deep into the mathematical details.

And we are grateful to our "nitpicker" Tony Moffat. As one of the world-leading massive star researchers he has carefully read the entire script, corrected and eliminated text errors and provided helpful comments and suggestions ("Do it! Now!"). He is our guide for how to make good science. He delivered the decisive contribution to the quality of our book. In addition, he is the main driver of our professional-amateur-campaigns on massive stars and thus shows a possible path to the future of small telescopes.

Finally, we want to thank our wives Britta and Susanne for their endless patience. We greatly appreciate their lasting willingness to endure cryptic dialogues on anamorphic magnification factors and dispersion elements.

Appendix A
The MIDAS Data Reduction

A Short Story

We build our measuring instruments by ourselves, but the costs for various components are always a problem. Who has a thousand Euros ready to go for a highly efficient grating? You ask colleagues if they have a mirror mount for you or can provide a cheap slit. An impossible situation! Eventually we needed a new part for our spectrograph. We called a manufacturer and were connected to a low-rank technician who wasn't too helpful. Mutual understanding among professionals is high, and we now share our various experiences. We "whine" about our financial suffering and describe our technical needs. On the other hand, we hear things like "I have here a large grating in storage that we don't need anymore. I can send it to you..." Hence, our scrounging was highly successful and we refer to such requests as "High level begging".

A.1   The MIDAS Environment

The astronomical data reduction packages MIDAS and IRAF both supply all necessary steps for high quality data reduction. They are complex and require extensive training. However, the respective user manuals are not easy to understand so that the average novice in a research institute can not easily get started. Most upcoming researchers benefit from the "generation contract" within the community. The more experienced help the beginners and, hence, within an institute someone is always approachable for questions. It is not our intention to discuss all detailed reduction tasks. At this point we only want to deliver the main and most important parts for the very first steps in MIDAS (the IRAF structure is somewhat different but equivalent) and want to give a very first impression of its command structure, some commands and steps. For program details we refer to the MIDAS user manual.

© Springer-Verlag Berlin Heidelberg 2015
T. Eversberg, K. Vollmann, *Spectroscopic Instrumentation*, Springer Praxis Books,
DOI 10.1007/978-3-662-44535-8

Working with MIDAS requires some adjustment by the majority of (Windows) users, and they need to get used to the different working steps in the UNIX environment. After learning the first steps, MIDAS is very comfortable to work with and one quickly gets reliable results. First, we give a quick overview about what MIDAS is and how it works, and then we describe some details of different reduction steps.

ESO-MIDAS has already been developed in the 1980s and quickly became one of the main tools for astronomical image and data processing in UNIX operating system environments. It is fully interactive and can perform user-defined procedures and algorithms. It has a modular design so that it can be adapted to different environments.[1] It can be ported to different machines, works with standard languages such as FORTRAN and C, and can handle the XWindow system. It can be programmed via simple interface routines and its code is open source for easy implementation of new software. The basic structure consists of several parts. For instance, it permits a backup of all historical operation commands during one session. Additional applications can be made via the corresponding interfaces for FORTRAN77, C procedures and for the MIDAS Control Language (MCL). These are ordered in different parts according to their importance. The main areas are core applications like "Image and Graphic Display", general image processing, "Table File System", a "Fitting Package" to process non-linear functions and Data I/O (Input/Output) for data transfer to other storage media. MIDAS procedures can be written in MCL and be grouped in so-called "jobs", without having to recompile the program. The "MCL Debugger" enables error fixing in the source code. In this chapter we demonstrate a simplified MIDAS reduction procedure for spectroscopic data, based on a tutorial by Günter Gebhard. We want to emphasize that the steps presented here only reflect a small selection of all MIDAS procedures and we refer to the corresponding MIDAS User Manual.

A.1.1   Nomenclatura

MIDAS expects all image recordings in fits format, the standard file format used in astronomy and of CCD software on the market. Images and graphics are imaged in separate displays. Internally MIDAS treats images and tables in the specific formats *.bdf and *.tbl, respectively, which can be converted into appropriate fits formats at any time for use outside MIDAS. MIDAS has specific packages, the so-called "contexts" for various applications and ESO instruments. The contexts can be easily adapted to other instrumentation, off-the-shelf or not.

To describe the MIDAS procedures, we write all MIDAS commands in uppercase and all parameters as well as image and file names in italics. MIDAS commands

---

[1]This includes UNIX emulation in Windows using virtual machines, e.g., VMware.

always consist of two parts, the COMMAND/QUALIFIER. For instance, the command
OVERPLOT/SYMBOL *15 4108,0.5 2.0*
draws an upward pointing arrow in the graphics window at the coordinates x = 4108, y = 0.5 in double size.

One can either write the full COMMAND/QUALIFIER command or use the first three letters. OVE/SYM is the same as OVERPLOT/SYMBOL. MIDAS distinguishes uppercase and lowercase letters for files but not for commands. For clarity, here we write MIDAS commands always in uppercase letters and the monitor output in bold face.

A.1.2   Start, Help and End

We start MIDAS in an xterm window with the command >Inmidas. With "help" we can call descriptive help texts. We end MIDAS again with "bye" and go back to the UNIX/Linux shell. When starting MIDAS, a directory "midwork" is created in which the login.prg file is executed, if it is there. login.prg contains optional MIDAS commands that are executed immediately when MIDAS starts.

A.1.3   Image Import

We start MIDAS in the User directory with
    user@linux:~ > inmidas
and see the welcome window (Fig. A.1). From the user directory we first change to the directory in which the raw data are stored.
    Midas 001> CHA/DIRE *data*

```
            ESO-MIDAS version 01FEBpl1.1 on linux
 *************************************************************************
 **                                                                   **
 **          Copyright (C) 1996, European Southern Observatory        **
 **                                                                   **
 **   ESO-MIDAS comes with ABSOLUTELY NO WARRANTY; for details type   **
 **   '@ license w'. This is free software, and you are welcome to    **
 **   redistribute it under certain conditions; type '@ license c'    **
 **   for details.                                                    **
 **                                                                   **
 *************************************************************************
```

Fig. A.1 MIDAS starting window

We are now in the data directory, but MIDAS does not reveal this by itself. We need to request this information with the UNIX command pwd (*present working directory*). MIDAS only understands UNIX commands with a dollar prefix.

Midas 002> $ pwd

/home/user/Daten

The directory content can also be called with the corresponding LINUX command.

Midas 003> $ ls

cflat.fts dark0020.fts dark0014.fts neon1.fts rigel.fts sir1.fts re0317.fts bias.fts

The listed images are

cflat.fts	CCD flat field
dark0020.fts	Dark with 20s exposure time
dark0014.fts	Dark with 14s exposure time
neon1.fts	Neon lamp
rigel.fts	Spectrum of Rigel
sir1.fts	Spectrum of Sirius
star1.fts	Spectrum of another star
bias.fts	Bias field

Note that these data are only example images we use for explanation. To actively follow our steps it is necessary to have such calibration and stellar data (not necessarily Rigel and Sirius) available. We first work with the two images star1.fts and the corresponding dark field d0020.fts. Both are stored in a self-created directory, e.g., /home/user/data. First, we want to look at some pictures. We first open a display

Midas 004> CREATE/DISPLAY

and then load the images with

Midas 005> LOAD/IMAGE *star1.fts*

We now follow a simplified data reduction algorithm as introduced in Chap. 12[2] by first taking our example spectrum star.fts (Fig. A.2) and subtract a suitable dark field (Fig. A.3).[3] The stellar spectrum to be reduced should have the same exposure time as the dark.[4] To find the exposure time, we open the image descriptor (i.e. the image content) of the spectrum star.fts (Fig. A.4).

Midas 006> READ/DESCRIPTOR *star1.fts o_time*

The exposure time indicated in the descriptor is 20 seconds and we need to choose the d0020.fts image for reduction.

Midas 007> COMPUTE/IMAGE *as1 = star1.fts − d0020.fts*

---

[2]In this chapter we focus on the fundamental MIDAS commands and neglect the exact data reduction steps as described in Chap. 12.

[3]The conversion to bdf is internally done by MIDAS.

[4]Note that the dark can also be scaled by a scaling-factor, as described in detail in Sect. 12.3.4.

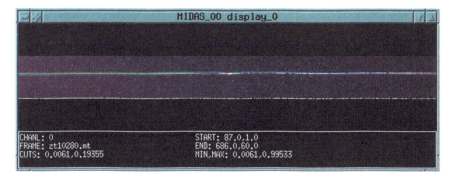

Fig. A.2 Spectrum of a star

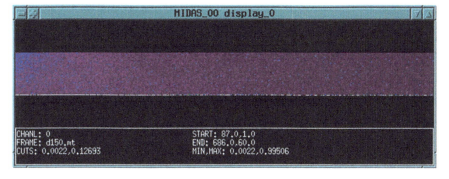

Fig. A.3 Dark field

```
frame: stern1.fts  (data = R4, format = FITS)
O_TIME:
     1901.1534246575      0.0000000000000      0.0000000000000
     0.0000000000000      0.0000000000000      0.0000000000000
     20.000000000000
```

Fig. A.4 Example of an image descriptor for the exposure time

The image as1.bdf is now the spectrum without dark background. Often it is necessary to adjust the image size by cutting entire rows and/or columns. We then first have to estimate the image size (see A.1.5). The image size is 772×60 pixel. Let us now assume we have to deal with overexposed rows. Then we extract an image which is correspondingly cut in the y-direction.

Midas 008> EXTRACT/IMAGE *bs1 = as1 [<,2:>,59]*

The notations "<" and ">" mean $-\infty$ and $+\infty$, respectively, i.e. from the left to the right image edge. We apply the same cutting procedure for our flat.

Midas 009> EXTRACT/IMAGE *flate = cflat [<,2:>,59]*

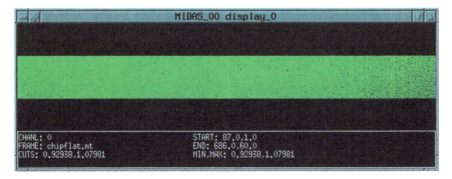

Fig. A.5 Our flat field

We now normalize *flate* by its mean value. The mean can be found again with the image statistics command (see A.1.6 and mean value in Fig. A.8).

Midas 010> COMPUTE/IMAGE *nflat* = *flate* / 9.987945e−01

We divide *bs1* by the flat.

Midas 011> COMPUTE/IMAGE *cs1* = *bs1* / *nflat*

Note that images can be overwritten without warning. For instance, we can perform the following reduction procedure:

Midas 012> COMPUTE/IMAGE *as2* = *as1* − *bias.fts*

Now we divide the result as2 by the flat field flat.fts (Fig. A.5).

Midas 013> COMPUTE/IMAGE *as3* = *as2* / *nflat.fts*

All three arithmetic operations created a new image (as1, as2 and as3). One can also overwrite the original image by the example commands *as1* = *as1* − *bias.fts*. This should be avoided, though, for safety reasons and for documentation of the steps. In principle, the raw image of the stellar spectrum is now prepared for the production of a one-dimensional spectrum, and below we will illustrate different versions. First, however, we present some important and often performed operations.

A.1.4   The Display

With the command CREATE/DISPLAY we can open a standard image window for 512×512 pixels. In order to adapt the window, we need to determine the format of our images. We get this again from the image descriptor (Fig. A.6).

Midas 014> READ/DESCR *cs1*

The image has 772×58 pixels. The parameter LHCUTS represents the image intensities. The lowest value is −0.003 and the highest value is 0.92. We now open a display with size 800×80 pixels and can then load the image.

Midas 015> CREATE/DISPLAY *0 800,80*

```
frame: stern1.fts  (data = R4, format = FITS)
NAXIS:                    2
NPIX:                   772          60
START:           1.0000000000000         1.0000000000000
STEP:            1.0000000000000         1.0000000000000
IDENT:
CUNIT:
LHCUTS:          0.000000      0.000000      0.000000      0.000000
```

Fig. A.6 Example image descriptor

```
frame: cs1.bdf  (data = R4)
NAXIS:                    2
NPIX:                   772          58
START:           1.0000000000000         2.0000000000000
STEP:            1.0000000000000         1.0000000000000
IDENT:
CUNIT:
LHCUTS:          0.000000      0.000000     -0.2998207E-02  0.9234945
```

Fig. A.7 Example image descriptor for the image size

```
frame: flate.bdf  (data = R4)
complete area of frame
minimum, maximum:                9.293849e-01    1.079806e+00
at pixel (754,8),(1,40)
mean, standard_deviation:        9.987945e-01    1.720531e-02
3rd + 4th moment:                0.0126176       2.98679
total intensity:                 43951
median, 1. mode, mode:           9.982401e-01    9.296798e-01    1.001056e+00
total no. of bins, binsize:      256             5.898866e-04
# of pixels used = 44776 from 1,1 to 772,58 (in pixels)
```

Fig. A.8 Image statistics

A.1.5   Image Size Estimation

To determine the image size, we again use the image descriptor (Fig. A.7).
   Midas 016> READ/DESCRIPTOR *star1.fts*

A.1.6   Image Statistics

Midas 017> STATISICS/IMAGE *flate.bdf*
   Using this command we get the minimum and maximum pixel intensity values
and the corresponding pixel numbers, the mean pixel value and the standard
deviation over the entire image, 3rd + 4th moment, total intensity, etc. (Fig. A.8).

```
frame: cs1.bdf  (data = R4)
       plane_no 1 loaded
cursor #0
     frame pixels           world coords        intensity
         9        46      9.00000    47.0000      0.455804
       738        47      738.000    48.0000      0.877758
```

Fig. A.9 Cursor output

A.1.7   Copies of the Original Image

As an example, we copy the image flat.fts including cutting the first and last two lines.[5]

Midas 018> EXTRACT/IMAGE *flate = flat.fts [<,2:>,59]*

A.1.8   Image Rotation

To further extract the data, the image should be exactly horizontal. We measure the positions of the two ends of the spectrum with the cursor. One can consecutively determine different image positions by pressing the left mouse button (Fig. A.9).

Midas 019> GET/CURSOR

Pressing the right mouse button finishes the procedure. With the output positions, we can now rotate the image accordingly, but need the corresponding angle of rotation for the MIDAS command. This rotation angle is calculated by

$$\varphi = \arctan\left(\frac{\Delta y}{\Delta x}\right)$$

Midas 020> COMPUTE *atan((47-46) / (738-9))*
0.7859498E-01
We now perform the rotation.
Midas 021> REBIN/ROTATE *cs1 es1 -0.7859498E-01*

A.1.9   The MIDAS Descriptor

The graphics window shows a legend on the right side (Fig. A.10). This can be suppressed by
Midas 022> SET/GRAPHICS *pm=0*

---

[5]If the file names already vary after the first letter, one can use the tab key after the first letter.

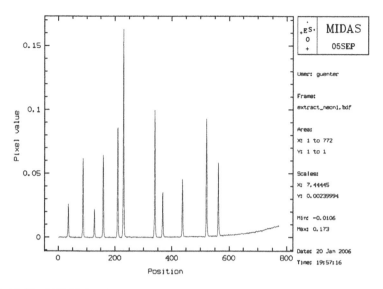

Fig. A.10 The MIDAS descriptor

A.1.10 Look-Up Tables (LUT)

By default the settings of the system shows intensities as grayscale, but we can use different look-up tables (LUT) for color images. A LUT change is carried out through the command LOAD/LUT.

Midas 023> LOAD/LUT *rainbow*

Now we load the image again. It is however unnecessary to re-enter the complete command, but we simply type the appropriate number.

A.1.11 Positioning the Graphic Window

By default the graphics window is placed in the upper left corner of the monitor. One can also decide for a different position, e.g., the point (5,180).[6]

Midas 024> CREATE/GRAPHICS 0 600,400,5,180

---

[6]By writing the row CREATE/DEFAULT CREATE/GRAPHICS ? *800,400,5,5* into the file login.prg each graphics window will have the size of 800 × 400 pixel and will be opened at the coordinates (5,5).

A.2   Spectrum Extraction

A.2.1   AVERAGE/ROW

Before we reduce the two-dimensional image to a one-dimensional spectrum we
need to be sure that the image contains only the stellar information. Since our
algorithm only eliminated instrument interference, introduced by the instrument
(flat field) and the CCD camera (dark and bias field), the sky background or scattered
light are still present in the image data. Both, however, must be eliminated. We select
two image regions next to the spectrum. We determine the intensities of different
pixels or even better the average value of all pixels and then subtract it from the
entire image. Our spectrum is positioned approximately on the line 47 and the region
between the corners (2,2) and (770,40) (Fig. A.11) as well as (2,54) and (770,59)
(Fig. A.12) should contain only noise.

Midas 025> STATISTICS/IMAGE *es1 [2,2:770,40]*

Midas 026> STATISTICS/IMAGE *es1 [2,54:770,59]*

It is now important not to calculate the average of all pixel values, but the median,
because extreme deviations such as cosmic rays and hot pixels are then sorted out.
Such pixels would otherwise seriously affect the mean and distort the results. In our
example, the median is only 0.0006 and is almost negligible.

Midas 027> COMPUTE/IMAGE *gs1 = es1 − 0.0006*

The spectral signal on the chip must now be collapsed perpendicular to the
dispersion direction to a 1D spectrum. To visualize the pixel boundaries for

```
frame: es1.bdf   (data = R4)
area [2,2:770,40] of frame
minimum, maximum:                 -1.823950e-03     3.115081e-03
at pixel (338,23),(605,40)
mean, standard_deviation:          5.277409e-04     5.917163e-04
3rd + 4th moment:                  0.0285777        3.01925
total intensity:                   15.8275
median, 1. mode, mode:             5.039032e-04  -1.814266e-03    6.396685e-06
total no. of bins, binsize:        256           1.936875e-05
# of pixels used = 29991 from 3,2 to 771,40 (in pixels)
```

Fig. A.11 Image statistics for the region [2,2:770,40]

```
frame: es1.bdf   (data = R4)
area [2,54:770,59] of frame
minimum, maximum:                 -1.518485e-03     3.253763e-03
at pixel (377,58),(392,55)
mean, standard_deviation:          9.446378e-04     6.942128e-04
3rd + 4th moment:                  0.0845274        2.61251
total intensity:                   4.35856
median, 1. mode, mode:             9.296935e-04  -1.509127e-03    6.763244e-06
total no. of bins, binsize:        256           1.871470e-05
# of pixels used = 4614 from 3,54 to 771,59 (in pixels)
```

Fig. A.12 Image statistics for the region [2,2:770,40]

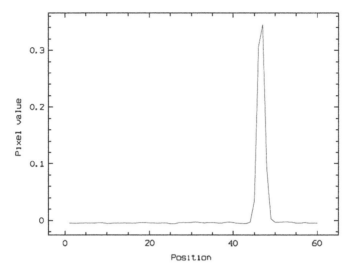

Fig. A.13 Column plot perpendicular to the dispersion direction

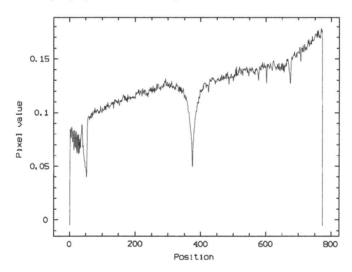

Fig. A.14 Collapsed 1D spectrum

this procedure, we plot a vertical section of the spectrum (perpendicular to the dispersion) at a given × position (Fig. A.13).

Midas 028> PLOT/COLUMN *gs1 5*

We now see that the spectrum is located between the lines 42 and 50. Within these boundaries we can now collapse the 2D spectrum in order to obtain a 1-dimensional spectrum (Fig. A.14).

Midas 029> AVERAGE/ROW *hs1 = gs1 [42,50]*

Midas 030> PLOT/ROW *hs1*

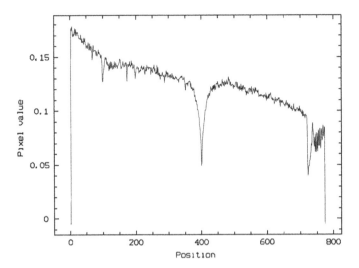

Fig. A.15 Flipped image of Fig. A.14

We have now created an uncalibrated one-dimensional spectrum. However, we can see that the oxygen molecular complex is located to the left of the hydrogen line, although the latter has a larger laboratory wavelength. We need to reflect the spectrum to put increasing wavelengths in the increasing x-direction.

Midas 031> FLIP/IMAGE *hs1*

The flip command does not changed the spectral information, but the x-axis now starts at 800. This must also be changed by manipulating the image descriptor. To start again at 0 with an increment of 1 both the "start value" and the "steps" need to be changed. We then plot the image again (Fig. A.15).

Midas 032> WRITE/DESCRIPTOR *hs1 start/d/1/1 0.0*

Midas 033> WRITE/DESCRIPTOR *hs1 step/d/1/1 1.0*

A.2.2   EXTRACT/AVERAGE

We have subtracted the light of the sky as a lump sum average. MIDAS provides a more accurate method, which is especially suitable for long-slit spectrographs. We demonstrate this on a spectrum of Regulus. For doing so we plot column no. 5, change to the plotting colour red and overplot column number 500 (Fig. A.16).

Midas 034> PLOT/COLU *re30170.fts 5*

Midas 035> SET/GRAP *color=2*

Midas 036> OVERPL/COLU *re30170.fts 500*

We already mentioned above that MIDAS offers different contexts for different instruments. We are now working with the Context Long, offered for the processing

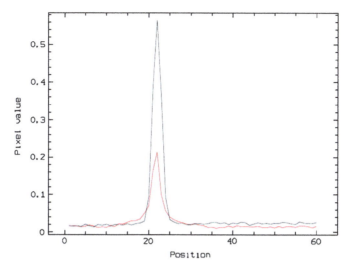

Fig. A.16 Plot of the two image columns no. 5 and 500

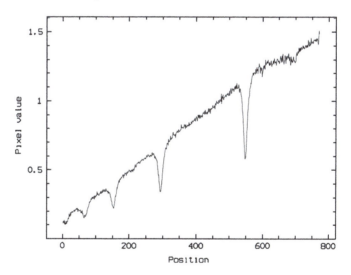

Fig. A.17 Stellar spectrum, background reduced with EXTRACT/AVERAGE

of long-slit spectra. However, in this context slitless and fiber spectra can be reduced as well.

Midas 037> SET/CONTEXT *long*

As regions for the reduction of the sky background, we choose lines 2–15 and 30–45. Lines 17–27 are covered by the spectrum. After extracting the background we plot the respective spectrum again (Fig. A.17).

Midas 038> EXTRACT/AVERAGE *re30170.fts reg 17,27 2,15,30,45*

Midas 039> plo reg

The command EXTRACT/AVERAGE subtracted a defined background average between defined lines. In fact, the command contains three commands of the above AVERAGE/ROW procedure (the two windows next to the spectrum and the spectrum itself).

A.2.3   EXTRACT/LONG

The command SKYFIT/LONG provides a more elegant method to subtract the sky background. It statistically analyzes the regions adjacent to the spectrum and calculates a global fit to these two regions. This is especially important if the image has a strong background intensity gradient.

Midas 040> SKYFIT/LONG *re30170.fts sky 2,15,30,45*

The new image sky.bdf now contains the fitted background and we can use it to edit the spectrum. The fit eliminated cosmic rays, hot pixels and any intensity gradient. For that the read-out noise in analog-to-digital-units (ADU) needs to be considered as well as the CCD gain in electrons per ADU.

Midas 041> SET/LONG *gain=0.0024 ron=1.5*

Midas 042> EXTRACT/LONG *re30170.fts reg sky 17,27*

The result of EXTRACT/LONG is shown in black and that of EXTRACT/AVERAGE in red, shifted by 0.1 relative pixel values (Fig. A.18). We point out that a spectrum of only a few pixels width should be reduced via EXTRACT/LONG, since the information for statistical image analysis is too small.

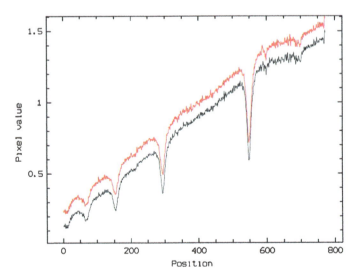

Fig. A.18 Sky reduced images. *Black*: Reduced with EXTRACT/LONG. *Red*: Reduced with EXTRACT/AVERAGE

A.3 Wavelength Calibration

Basically, two different methods are used for wavelength calibration. The first and simplest one is the identification of two or more stellar photospheric or atmospheric (telluric) absorption lines in the spectrum and the second, and more reliable, is the allocation of emission lines of a calibration lamp to the corresponding pixel position of the target spectrum.

A.3.1 Calibration with Two Absorption Lines

If a spectrum shows only two usable absorption lines for calibration, we have no choice other than to calibrate the spectrum with a linear dispersion function. We demonstrate this with a spectrum of Sirius `hs1` (Fig. A.19). The strong absorption at pixel 400 is the Hα line. Its laboratory wavelength is 6,562.73 Å. At pixel 720 we find the first line of molecular oxygen at 6,867.2 Å. To find the exact pixel position, we now use a Gaussian fit of these lines. We do this again with the left cursor button and finish with the right one (Fig. A.20).

Midas 043> CENTER/GAUSS *gcurs ? a*

The exact pixel positions are thus 399.563 and 726.285. The dispersion relation is $\lambda = m \cdot x + \lambda_0$, with the wavelength λ of pixel x, the linear dispersion m and the wavelength of the first pixel λ_0. λ_0 and m can be found in the header of the fits-image as descriptors *start* and *step*. Our input values are $x_1 = 399.563$

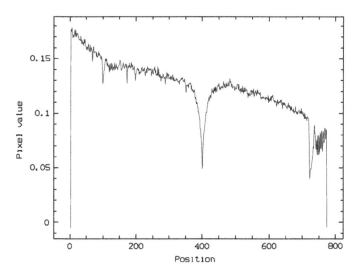

Fig. A.19 Raw spectrum of Sirius

start	end	center	pixel_value	FWHM
383.858	418.480	399.563	0.49291E-01	11.700
717.328	735.551	726.285	0.39818E-01	23.165

Fig. A.20 Estimated pixel positions with CENTER/GAUSS

$x_2 = 726.285$
$y_1 = 656.7$
$y_2 = 6867.2$
and we now obtain $m = \Delta\lambda/\Delta x$ and $\lambda_0 = \lambda_1 - mx_1$

A short MIDAS script can calculate the above equation. Using the editor we create a file called *twolines.prg*.

```
! twolines.prg
! This program will generate a wavelength scale out of two pixel numbers of
Halpha (6563) and O2 (6867).
!
! Call: @@ ZweiLinien P1 P2 P3
!
! P1 Image name
! P2 Pixel number of Halpha
! P3 Pixel number of O2
!
DEFINE/PARA P1 ? I "image name : "
DEFINE/PARA P2 ? N "Halpha : "
DEFINE/PARA P3 ? N "O2 : "
!
DEFINE/LOCAL m/r/1/1 0 ! linear dispersion
DEFINE/LOCAL l0/r/1/1 0 ! start wavelength
DEFINE/LOCAL dx/r/1/1 0 ! pixel difference
!
dx = P2 - P3
IF dx .equation 0 THEN
WRITE/OUT "I cannot divide by zero"
RETURN
ENDIF
m = (6562.7 - 6867.2)/dx
WRITE/OUT "Linear dispersion : m Angstrom/pixel"
l0 = 6562.7 - m * P2
COPY/II P1 lP1
WRITE/DESCRIPTOR lP1 start/d/1/1 l0
WRITE/DESCRIPTOR lP1 step/d/1/1 m
PLOT/ROW lP1
RETURN
```

Such MIDAS scripts should be stored in the directory /home/user/midwork.

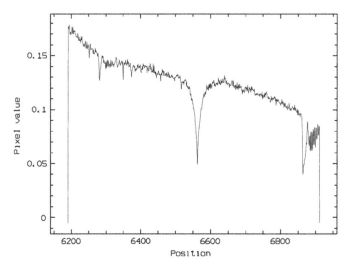

Fig. A.21 Calibrated spectrum `lhs1`

The command
Midas 044> @@ twolines *hs1 399.559 726.285*
now produces the calibrated spectrum `lhs1` and prints it (Fig. A.21).

A.3.2   Many Spectral Absorption Lines

If more usable absorption lines for wavelength calibration are present in the
spectrum, MIDAS may fit polynomials of higher order to the dispersion relation.
However, laboratory wavelengths of all lines must be available in a corresponding
table. An interactive procedure is impossible. As an example, we use again the
spectrum of Regulus. First, we create a one-column table wl.dat with an editor
including all laboratory wavelengths of the hydrogen lines.
3770.6
3797.9
3835.4
3889.1
3970.1
4101.7
4340.5
4861.3
6562.8

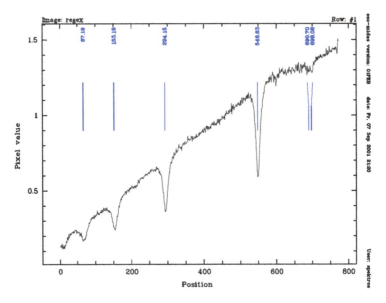

Fig. A.22 Line identification with a line table

Midas works with its own table format in which we transform wl.dat. For a single-column table, this is simple and we give the column of wl.dat the name WAVE.

Midas 045> CREATE/TABLE *hydro 1 ? wl*

Midas 046> NAME/COLUMN *hydro #1 :WAVE*

We are still within the CONTEXT LONG. We determine the line positions and then plot the result (Fig. A.22).

Midas 047> SEARCH/LONG *reg .03 30 ? ? ? abs*

Midas 048> PLOT/SEARCH

For accurate line identification we can adjust the parameters "Threshold" (= 0.03) and "Width" (= 30) to the spectrum. This adjustment is quickly done by using the "up arrow" button and calling the commands 45 and 46 again. From left to right, we now find the absorption lines H8 , Hϵ, Hδ and Hγ. For identification, MIDAS now needs the corresponding wavelengths from table hydro.tbl. We choose hydro.tbl and identify the lines with the mouse.

Midas 049> SET/LONG lincat = hydro

Midas 050> IDENTIFY/LONG reg

With the left mouse button we identify at least two lines and MIDAS identifies all remaining catalogue wavelengths by itself (Fig. A.23). Missing wavelengths will automatically be skipped.

Now we command the respective calibration fit (Fig. A.24).

Midas 051> CALIBRATE/LONG *0.5 2*

```
No. of selections:      5
*** INFO: Position cursor and press left mouse button or any key (not RETURN)
          Use second left mouse button or space bar to exit
    X COORD.    Y COORD.    Sequence     IDENT         X
 0.547348E+03 0.117843E+01      4          *        547.60
IDENT          ?: 4340
 0.658729E+02 0.386769E+00      1          *         64.58
IDENT          ?: 3889
```

Fig. A.23 Line identification with the curser

```
Number of lines in catalog (total, selected)   : 9 , 9
Number of lines in table line (total, selected) : 5 , 5
Y =   1 --  4 lines out of  5 -- RMS = 0.087006 wav. units

Initial Wav. : 3829.289
Final Wav.   : 4552.18
Step         : 0.935
```

Fig. A.24 Line calibration with CALIBRATE/LONG

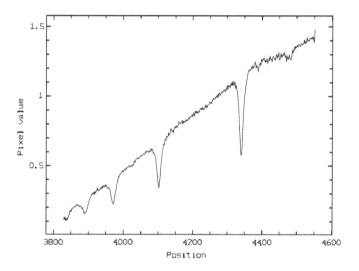

Fig. A.25 Line calibration with REBIN

Finally, we transfer this calibration to our spectrum and force a constant step size (Fig. A.25).

Midas 052> REBIN/LONG *reg rreg*
Midas 053> REBIN/LINEAR *rreg lreg .5*

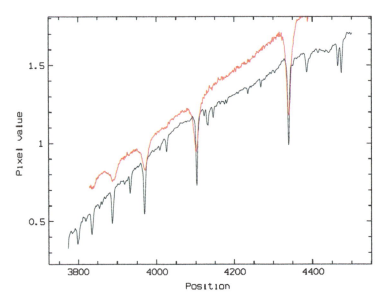

Fig. A.26 Spectrum of Rigel in *black*. For comparison, a spectrum of Regulus is shown in *red* (*Courtesy*: D. Goretzki)

A.3.3 Prism Spectra

Unlike grating spectra, prism spectra have a non-negligible non-linear dispersion in the form of a polynomial. No context for prism spectroscopy has been developed in MIDAS. However, it is quite possible to find a dispersion relation, if sufficiently many lines as reference points are available. A spectrum of Rigel (Fig. A.26) shows the corresponding procedure as an example. The spectrum has some hydrogen lines. One can calibrate the spectrum with CALIBRATE/LONG with the corresponding parameter settings (tolerance = 5 and polynomial degree = 2). The result is shown in black and a spectrum of Regulus for comparison is shown in red.

One can see that both calibrations differ at the spectral interval limits. The table compares the literature values with those from the spectrum of Rigel.

Ion	Measurement	Literature
CaII	3932.9	3933
HeI	4026.2	4026
HeI	4465.2	4471
MgII	4474.0	4481

```
search lines
------------

input image  : extract_neon1.bdf
output table : line.tbl
input parameters
        search window :   8 pixels
    detection threshold :        0.01 DN
            average on :   0 scan-lines
                step of :   1 scan-lines
     centering method : gaussian fit
   search for emission lines
     no. of detections:        11
```

Fig. A.27 Detection of catalogue calibration lines

```
No. of selections:    11
*** INFO: Position cursor and press left mouse button or any key (not RETURN)
          Use second left mouse button or space bar to exit
   X COORD.   Y COORD.    Sequence      IDENT         X
  0.877044E+02 0.841405E-01      2              *     86.75
IDENT          ?:
6266
   0.521828E+03 0.957014E-01     10              *    521.01
IDENT          ?:
6678
```

Fig. A.28 Line positions found with the cursor

A.3.4   The Use of a Comparison Spectrum

In order to obtain a constant and defined spectral resolution, a slit spectrograph is essential (see Chap. 2). A spectrum is recorded from a corresponding calibration lamp at the same position of the star and reduced to a one-dimensional spectrum and then calibrated as shown in steps $047 - 051$. The corresponding line catalog must be available. Here we call it neon.tbl. The spectrum is first collapsed to 1D and then defines the emission lines with the line catalogue (Figs. A.27 and A.28).

Midas 054> AVERAG/ROW extract neon1 = neon1.fts 42,50
Midas 055> FLIP/IMAG extract neon1.bdf
Midas 056> WRITE/DESC extract neon1 start 1.0
Midas 057> WRITE/DESC extract neon1 step 1.0
Midas 058> SET/LONG lincat = neon
Midas 059> SEARCH/LONG extract neon1 0.01 search lines
From 11 detected lines 2 are identified (Figs. A.28 and A.29).
Midas 060> IDENTIFY/LONG *extract neon1*
and then the calibration relation can be calculated (Fig. A.30).

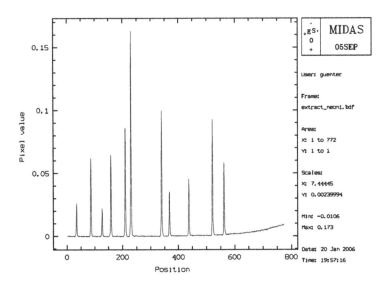

Fig. A.29   Collapsed calibration spectrum

```
Updated keyword LINCAT = MID_ARC:neon.tbl
**** Column already exists
Number of lines in catalog (total, selected)   : 26 , 26
Number of lines in table line (total, selected) : 11 , 11
Y =   1 -- 11 lines out of 11 -- RMS = 0.020225 wav. units

Initial Wav. : 6185.093
Final Wav.   : 6915.678
Step         : 0.948
```

Fig. A.30   Calibration relation calculated with CALIBRATE/LONG

Midas 061> CALIBRATE/LONG
Finally, we transfer it to our spectrum of Sirius and enforce a constant step size.
Midas 062> REBIN/LONG *hs1 rebhs1*
Midas 063> REBIN/LINE *rebhs1 linhs1*

A.3.5   Spectral Resolving Power

One can also estimate the spectral resolving power with a calibration lamp. This is determined from the ratio of the wavelength to its line full width at half maximum (FWHM). Figure A.31 shows a portion from the spectrum of a neon lamp.

With CENTER/GAUSS gcurs we can now produce a Gaussian fit and MIDAS immediately provides the respective line parameters.

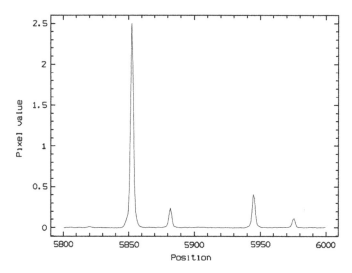

Fig. A.31   200 Å interval of a neon calibration lamp

start	end	center	pixel value	FWHM
5858.894	5844.824	5852.547	2.5016	2.4587
5877.082	5887.034	5881.883	0.23992	2.4784
5938.853	5951.207	5944.823	0.40690	2.4750
5981.062	5971.110	5975.418	0.10934	2.4594

The spectral resolving power is, hence, about $6{,}000/2.5 = 2{,}400$.

A.4   Rectification

So far, all sample spectra have an intensity distribution, which essentially depends on the temperature of the star (Planck function), the earth's atmosphere, the vignetting in the telescope and spectrograph, the sensitivity function of the CCD chip and (for grating spectrographs) the blaze function of the grating. To eliminate these effects and for the sake of a uniform spectral analysis, spectra are rectified/normalized, i.e. the entire continuum is defined by the intensity value of 1. For doing so, individual points are determined in the continuum and MIDAS fits a function as a synthetic continuum (Fig. A.32).

Midas 064> NORMALIZE/SPEC *lreg nor*

Then the whole spectrum is divided by this fit (a spline of third order is usually the best choice (see Appendix B.4) and normalized to 1 (Fig. A.33).

Midas 065> COMPU/IMA *nreg = lreg / nor*

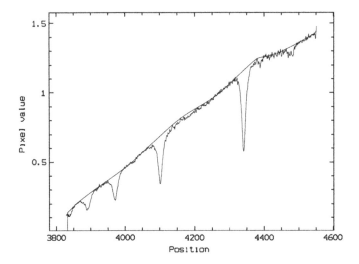

Fig. A.32 Continuum fit for consecutive rectification

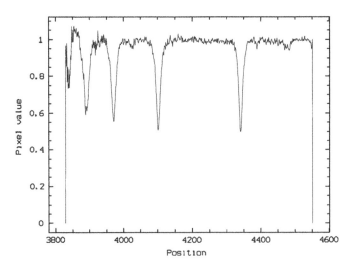

Fig. A.33 Rectified spectrum

A.5   Spectral Analysis

A.5.1   The Equivalent Width

For didactic reasons, we describe the measurement of the equivalent width in
MIDAS. For a description of the mathematical background we refer to Sect. 13.5.1.
The equivalent width (EW) of a line is defined by the area between the line itself and

X_start (pix/world)		X_end (pix/world)		Pixel sep.
Line+Cont.	Continuum	Line	Line/Cont	Equiv. w.
4312.69	968.229	4370.90	1084.65	0.500000
50.1579	58.3015	-8.14369	-0.139682	8.13123

Fig. A.34 Estimation of the line equivalent width

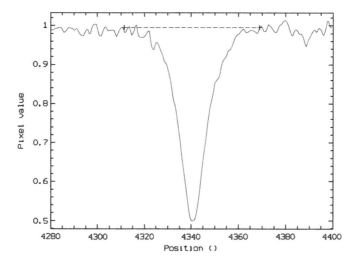

Fig. A.35 Estimation of the line equivalent width

the continuum. It is a measure of the amount of the absorbing or emitting material at the target star, and for circumstellar or interstellar lines, the amount of material between the target star and the observer. We show the determination of the line equivalent width for the Hα line of Regulus. First we plot only the short wavelength interval around Hα and then integrate the line with the cursor (Figs. A.34 and A.35).

Midas 066> SET/GRA *xaxis = 4280, 4400*

Midas 067> INTEGRATE/LINE *nreg*

We determine an EW value of about 8.1 Å. One should realize that the spectrum is noisy and therefore one decimal point for EW is sufficient in this case.

A.5.2 Measuring the Signal-to-Noise Ratio

We now describe the measurement of the signal-to-noise ratio. For a description of the mathematical background we refer to Sect. 10.7. The signal-to-noise ratio (S/N) is normally determined and specified for the continuum and can be estimated in line-free regions of the spectrum. In our example, the wavelength interval 4,180–4,250 Å of the spectrum nreg.bdf is suitable (Fig. A.36 and A.37).

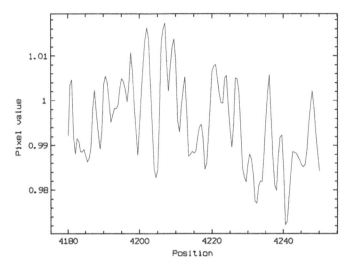

Fig. A.36 Estimation of the line equivalent width

```
frame: nreg.bdf  (data = R4) (desc = ZFormat)
area [4180:4250] of frame
minimum, maximum:                   9.722524e-01    1.017298e+00
at pixel (824),(757)
mean, standard_deviation:           9.949313e-01    9.612208e-03
3rd + 4th moment:                   0.171098        2.39671
total intensity:                    140.285
median, 1. mode, mode:              9.934505e-01    9.723407e-01    9.880627e-01
total no. of bins, binsize:         256             1.766513e-04
# of pixels used = 141 from 703 to 843 (in pixels)
```

Fig. A.37 Fitting parameters

Midas 068> STATIS/IMAG *nreg.bdf [4180:4250]*

In a normalized spectrum the standard deviation σ of the continuum signal corresponds to its so-called noise level. Thus, the S/N can be identified with the ratio of signal to continuum σ. In our example, we measure a mean continuum value of 9.949313e−01 but for a perfect normalization, this value should be exactly 1. With that we can now calculate the S/N.

Midas 069> *comp 9.949313e-01 / 9.612208e-03*

103.5070

Our S/N is about 100.

A.5.3   Spectral Co-adding

MIDAS enables the co-adding of several normalized spectra. However, for doing so, all spectra must have an identical increment. We show this with two spectra of Sirius, which cover different but overlapping wavelength intervals (Fig. A.38).

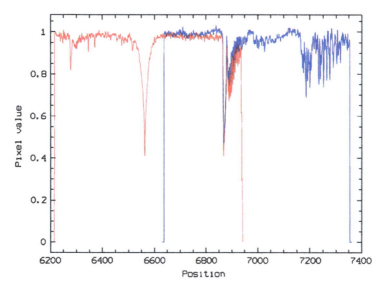

Fig. A.38 Wavelength overlap of two spectra

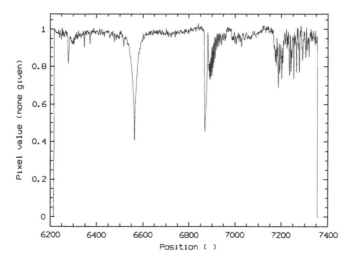

Fig. A.39 After merging the two spectra

For this example, the co-adding wavelength 6,750 Å is a good choice. We cut both spectra so that the overlap is relatively small. MIDAS then calculates a corresponding transition region (Fig. A.39).

Midas 070> EXTRAC/IMA *left = sir2.fts [<:6760]*
Midas 071> EXTRAC/IMAG *right = sir1.fts [6740:>]*
Midas 072> MERGE/SPEC *left right sirius*
Midas 073> plo Sirius

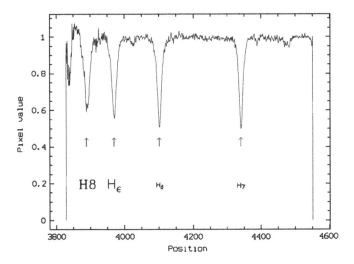

Fig. A.40 Labeling within the MIDAS graphic window. Co-adding the two spectra

Finally, we convert the coupled spectrum into a 1-dimensional fits format.
Midas 074> OUTDISK/FITS *sirius.bdf sirius.fts*

A.5.4 *Window Texts*

One can insert text at freely elected positions with the cursor. To do this, one uses
the left mouse button within the graphics window and the previously written text
appears at that position. This text can also be repositioned again by clicking again
and terminating with the right mouse button at the new position. The font is selected
by using the command SET/GRA font = ?. Greek letters are defined as in LATEX
(Fig. A.40).
  Midas 075> LABEL/GRA Hγ
  Midas 076> LABEL/GRA Hδ
  Midas 077> LABEL/GRA Hϵ ? ? 2
  Midas 078> SET/GRA *font = 1*
  Midas 079> LABEL/GRA *H-8*
Special characters, for example arrows, are positioned via a coordinate input.
With that we can see in our example that our wavelength calibration was relatively
good. For instance, symbol 15 is an upward pointing arrow.
  Midas 080> SET/GRA *color = 4*
  Midas 081> OVERPL/SYM *15 4340,.4 2*
  Midas 082> OVERPL/SYM *15 4101,.4 2*
  Midas 083> OVERPL/SYM *15 3970,.4 2*
  Midas 084> OVERPL/SYM *15 3889,.4 2*

Once texts and labels are set, one cannot move or change them again. It makes sense to keep the corresponding commands in a script.

A.5.5 Exporting Reduced Spectra: Fits/ASCII/Postscript

The export to a fits format has already been briefly described above.

Midas 085> OUTDISK/FITS *sirius.bdf sirius.fts*

However, the notes in the graphics window are not exported. The export as ASCII file can be performed in any operating system. The 1-dimensional spectrum is then first converted into a table format and then printed as ASCII. But first one needs to tell MIDAS the file name. As an example, we export the normalized spectrum of Rigel.

Midas 086> COPY/IT *nrig nrig wavelength*

Midas 087> ASSIGN/PRINT *file nrig.dat*

Midas 088> PRINT/TABL *nrig N*

Midas 089> $ls nrig.dat

1 3.79489e+03 9.47901e-01
2 3.79539e+03 9.33670e-01
3 3.79589e+03 9.18116e-01
4 3.79639e+03 8.77109e-01
5 3.79689e+03 8.34155e-01
6 3.79739e+03 7.89637e-01
7 3.79789e+03 7.54711e-01
8 3.79839e+03 7.42803e-01

. . .

Notes and other texts are not exported.

A.5.6 Postscript

Images and graphics in PostScript format are very popular in science and are especially used in the LATEX environment. Export into this format is very simple in MIDAS .

Midas 090> COPY/GRA *postscript*

Note, the command must be carried out literally. The resulting file can then be renamed and printed.

```
postscript.3l:pscript.3l # postscript device A3 format landscape
postscript.3p:pscript.3p # postscript device A3 format portrait
#
# *** da kommt mein eigener Drugger ***
lp:pscript                 # der Standarddrucker unter Linux
Epos360:Skript             # mein Epson im Economymode
#
# *** here for the null device
null:null                  # null device
#
# *** The following lines MUST remain ones of the file. No reading will
#     be done beyound that point.
#**************************************************************************
unknown:null               # end of device file  ! this must remain the last line
#**************************************************************************
```

Fig. A.41 Output for postscript format

A.5.7   Printing

MIDAS must know the printer for printing graphic content.

The file /midas/01FEBpl1.1/systab/ascii/plot/agldevs.dat contains an instrument list of all the peripherals that can be addressed by MIDAS. At the end of this list the ROOT can add an appropriate printer. The printer name must also be listed in the /etc/printcap directory. Now we can send the image to the printer (Fig. A.41).

Midas 091> COPY/GRA *eps360*

The picture fills exactly one A4 page, even if the graphics window has a different format.

Midas 092> BYE

user@linux:~ >

Appendix B
Important Functions and Equations

B.1 The Bessel Function

The German astronomer and mathematician Friedrich Wilhelm Bessel examined a homogeneous 2nd order differential equation, which has many applications in physics and engineering including the diffraction pattern of optical elements. This differential equation is

$$x^2 y'' + xy' + (x^2 - n^2)y = 0.$$

The solution of this equation is obtained via a power series and subsequent comparison of the coefficients:

$$J_n(x) = \sum_{k=0}^{\infty} \frac{(-1)^n}{k!\,\Gamma(n + k + 1)} \left(\frac{x}{2}\right)^{n+2k}$$

with the so-called Gamma function

$$\Gamma(x) \stackrel{def}{=} \lim_{x \to \infty} \frac{n!\,n^{(x-1)}}{x(x + 1)(x + 2)\ldots(x + n - 1)}.$$

Thus, the intensity function $I(r)$ of a diffraction pattern is described by a mathematically complete Bessel function of first kind $J_0(x)$ and 0th order:

$$I(r) = \left(\frac{J_1(r)}{r}\right)^2.$$

The distance between the first two minima of the intensity function $I(r)$ defines the diameter of the diffraction image, which determines the spectral resolution in

© Springer-Verlag Berlin Heidelberg 2015
T. Eversberg, K. Vollmann, *Spectroscopic Instrumentation*, Springer Praxis Books,
DOI 10.1007/978-3-662-44535-8

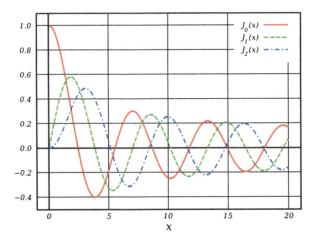

Fig. B.1 Several Bessel functions $J_n(x)$ of various orders n. The function $J_0(x)$ delivers the description of the diffraction function (from Wikipedia)

spectroscopy. This diameter is measured in radians (Fig. B.1). Together with the wavelength λ and the diameter of the optics D it is

$$\alpha = 2.44 \cdot \frac{\lambda}{D}.$$

The linear diameter for optics with focal length f is then

$$d = 2.44 \cdot \frac{\lambda \cdot f}{D}.$$

A helpful rule of thumb for 5,500 Å is

$$\alpha["] = \frac{13.8}{D[cm]}$$

or

$$d[\mu m] = 0.67 \cdot \frac{f}{D}.$$

B.2   The Poisson Distribution

If one repeatedly performs a random experiment, which only delivers two possible results (e.g., "Yes" or "No") and in which the probability of occurrence is low compared to the number of the events, then the Poisson distribution is a good approximation for the corresponding probability distribution. For example, if photons reach a detector, this is a purely random event. To make quantitative statements about this process, one needs to count these events within a given time under constant conditions. The variations show a specific distribution function (in this case, the Poisson distribution). This indicates how likely it is to count n photons in time t, if the expectance value is k. This is

$$P(n,t,k) = \frac{k^n}{n!} e^{-k}.$$

By adding up all n particles, the corresponding probability must be 1, of course. With the definition of the exponential function $e^x = \sum_n \frac{x^n}{n!}$ we obtain

$$\sum_n P(n,t,k) = e^{-k} \sum_n \frac{k^n}{n!} = e^{-k} \cdot e^k = 1$$

and with the definition of the mean value $< n >$ we obtain

$$< n > = \sum_n n P(n,t,k) = e^{-k} \sum_n \frac{n \cdot k^n}{n!} = k \cdot e^{-k} \sum_n \cdot \frac{k^{n-1}}{(n-1)!} = k.$$

The expectance value k is therefore equal to the average of n. For each distribution, including the Poisson distribution, the variance σ^2 is given by $\sigma^2 = \langle n^2 \rangle - \langle n \rangle^2$. For the Poisson distribution we have

$$\langle n^2 \rangle = \sum_n n^2 P(n) = e^{-k} \sum_n n \cdot \frac{(n-1)+1}{n!} \cdot k^n$$

$$= e^{-k} \left[\sum_n k^2 \frac{k^{n-2}}{(n-2)!} + \sum_n k \frac{k^{n-1}}{(n-1)!} \right] = k^2 + k.$$

For $\langle n \rangle^2$, the result is $\sigma^2 = k \Rightarrow \sigma = \sqrt{k}$. The Poisson distribution thus depends only on the expectance value. The Poisson distribution is generally valid even for non-integer n. For this case, we define $n \equiv x$ and we obtain

$$P(x,k) = \frac{k^x}{\Gamma(x+1)e^{-k}}$$

by using the Gamma function from Appendix B.1 which is $\Gamma(x+1) = n!$ for integer numbers $x = n$. It should be noted that the sum changes to an integral if we change from discrete (integer) n to continuous (any) x:

$$\langle n \rangle^2 = \sum_{n=0}^{\infty} n P(n) \to \langle x \rangle^2 = \int_{n=0}^{\infty} x P(x) dx = k.$$

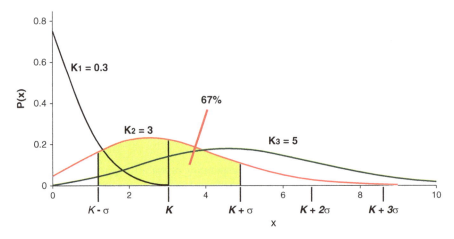

Fig. B.2 Several Poisson distributions $P(x)$ with different values of k (0.3 , 3 and 5), and depiction of the probability for the interval $[k_2 - \sigma, k_2 + \sigma]$

In Fig. B.2 three typical Poisson distributions are shown. For $k \leq 1$ the Poisson distribution is strongly asymmetric and the argument x for the maximum value P_{max} of the distribution is always smaller than k. For curve 1 in Fig. B.2 we have $x_{max} = 0$ and $k = 0.3$. For large k, the Poisson distribution becomes a Gaussian distribution with $\sigma = \sqrt{k}$ (curve 3):

$$\lim P_{Poiss} = \frac{1}{\sqrt{2\pi k}} e^{-(n-k)^2/2k}.$$

The probability of finding a value x within the standard deviation $\sigma = \sqrt{k}$ of the average value k can be calculated via $W(k,\sigma) = \int_{k-\sigma}^{k+\sigma} P(x,k)dx$. We obtain (see Fig. B.2):

$$k \pm \sigma \Rightarrow W(k,\sigma) = 0.67$$

$$k \pm 2\sigma \Rightarrow W(k,\sigma) = 0.95$$

$$k \pm 3\sigma \Rightarrow W(k,\sigma) = 0.997$$

However, in spectroscopy we measure photons, discrete particles, and instead of continuous values for x we have to change to discrete values of n. The Poisson distribution for discrete n is called the binomial distribution.

B.3   The Fresnel Equations

When light passes a boundary layer between two media, one part of the light beam is reflected and the other transmitted. A quantitative calculation succeeded that of Jean Fresnel for isotropic and non-ferromagnetic materials. Considering an electromagnetic wave, the ratio between the electric and magnetic field strength is

$$H = n \cdot \sqrt{\frac{\epsilon_0}{\mu_0}} \cdot E.$$

The constants ϵ_0 or μ_0 are the fundamental dielectric and magnetic constants and n is the refractive index of the irradiated material. At normal incidence[1] we have

$$E' = \frac{n - n''}{n + n''} \cdot E \text{ and } E'' = \frac{2n}{n + n''} \cdot E.$$

The reflected light is indicated by a straight line whereas the diffracted light is indicated by two lines. For reflection, the wave intensity is the square of the amplitude:

$$\left(\frac{E'}{E}\right)^2 = \left(\frac{n - n''}{n + n''}\right)^2 = R.$$

The reflectivity R is thus the ratio of reflected to incident wave intensity and the transmissivity T is the ratio of transmitted to incident wave intensity. Because of the conservation of energy we have $R + T = 1$ and thus

$$T = 1 - R = \frac{4nn''}{(n + n'')^2}.$$

The oblique wave incidence can be represented by the individual components, which oscillate perpendicularly and parallel to the plane of incidence (1 for vertical and 2 for parallel).

$$R_1 = \frac{\sin^2(\beta'' - \beta)}{\sin^2(\beta'' + \beta)}$$

$$R_2 = \frac{\tan^2(\beta'' - \beta)}{\tan^2(\beta'' + \beta)}$$

$$T_1 = \frac{\sin 2\beta \cdot \sin 2\beta''}{\sin^2(\beta'' + \beta)}$$

$$T_2 = \frac{\sin 2\beta \cdot \sin 2\beta''}{\sin^2(\beta'' + \beta) \cdot \cos^2(\beta'' - \beta)}$$

with the angles of incident, reflection and dispersion β, β' and β''.

---

[1] The Fresnel equations also include dielectric materials, magnetic permeability and polarisation. For a detailed general discussion we refer to the corresponding literature.

B.4   The Spline Function

To normalize the spectrum of a continuum to 1, a continuous function has to be
fitted to the continuum of the original spectrum and then divided into the original
spectrum. For the fit we use reference points between the spectral lines which, apart
from the noise, are well defined in intensity. Here we encounter a difficulty. In
the wavelength intervals of emission or absorption lines there are no continuum
reference points available. Nevertheless, to find an appropriate continuum function
within the lines, we need a continuous function that has sufficient stability within the
spectral lines and which reliably approximates the real continuum. Such a function
is a *spline*[2]

To find a function that passes exactly through a predetermined number of points
one has to interpolate. For interpolation of n intervals with $n+1$ nodes one needs to
find a function of order n, which passes exactly through all the specified points.
However, classical polynomials tend to oscillate between the supporting points
(e.g., atomic lines). To avoid oscillations one uses continuous splines of 3rd order
(cubic splines), which are partially composed of individual polynomials and have
the lowest possible curvature. It is also required that the curve be continuous up
to the second derivative at all junctions. For each interpolation interval between
two nodes, four coefficients of the cubic function must be found. For $n+1$ nodes
there are $4n$ necessary coefficients. To determine the function coefficients, we use
the requirement of identical function values at the n junctions and thus obtain
$2n$ interpolation conditions. In addition at all connection points, continuity of the
function as well as its first and second derivative must be ensured. Hence, we get
another $2(n-1)$ interpolation conditions. Thus, we have $2n + 2n - 2 = 4n - 2$
interpolation conditions. To determine the $4n$ coefficients two conditions are still
missing. We get them from the requirement that the function derivatives (slopes)
should not change at the two function ends. Thus, the second derivative of the
function is zero there.

We start with $n+1$ different x nodes (x_i, y_i) sorted in ascending order, with
$i = 1,\ldots,n$, for which we seek the spline $f(x)$ (respectively $I(\lambda)$) in the definition
of the spectral continuum). For each section i we define the third-order polynomial
$s_i(x)$:

$$s_i(x) = a_i(x - x_i)^3 + b_i(x - x_i)^2 + c_i(x - x_i) + d_i.$$

With the different interval lengths $h_i = x_{i+1} - x_i$ we get for the interval i the
following conditions for s_i and its first and second derivatives of s_i' and s_i'':

$$s_i(x_i) = d_i = y_i \tag{B.1}$$

---

[2]The term comes from the constriction of ships, where one determined the shape of the planking
using a flexible ruler, called spline, for minimum mechanical stress.

$$s_i(x_{i+1}) = a_i h_i^3 + b_i h_i^2 + c_i h_i + d_i = y_{i+1} \tag{B.2}$$

$$s'(x_i) = c_i \tag{B.3}$$

$$s_i'(x_{i+1}) = 3a_i h_i^2 + 2b_i h_i + c_i \tag{B.4}$$

$$s''(x_i) = 2b_i = \kappa_i \tag{B.5}$$

$$s_i''(x_{i+1}) = 6a_i h_i + 2b_i = \kappa_{i+1} \tag{B.6}$$

The coefficients can now be expressed by the known function values y_i and y_{i+1} as well as the unknown function curvatures κ_i and κ_{i+1}. From Eq. B.5 we get with the requirement for constant function gradients $s''(x_i) = 0$

$$b_1 = \kappa_i/2.$$

Hence, we obtain from Eq. B.6

$$a_i = \frac{\kappa_{i+1} - \kappa_i}{6h_i}$$

and then with Eq. B.2

$$c_i = \frac{y_{i+1} - y_i}{h_i} - \frac{h_i}{6}(2\kappa_i + \kappa_{i+1}).$$

From Eq. B.1 we directly obtain d_i. We repeat this procedure for the adjacent interval $i - 1$ and we obtain

$$s_{i-1}'(x_i) = 3\frac{\kappa_i - \kappa_{i-1}}{6h_{i-1}}h_{i-1} + 2\frac{\kappa_{i-1}}{2}h_{i-1} + \frac{y_i - y_{i-1}}{h_{i-1}} - \frac{h_{i-1}}{6}(2\kappa_{i-1} + \kappa_i)$$

or

$$s_{i-1}'(x_i) = \frac{2\kappa_i + \kappa_{i-1}}{6}h_{i-1} + \frac{y_i - y_{i-1}}{h_{i-1}}.$$

We require that the 1st derivative of s should also be continuous. Thus we obtain

$$\frac{2\kappa_i + \kappa_{i-1}}{6}h_{i-1} + \frac{y_i - y_{i-1}}{h_{i-1}} = -\frac{2\kappa_i + \kappa_{i+1}}{6}h_i + \frac{y_{i+1} - y_i}{h_i}$$

and after re-arranging

$$\kappa_{i-1}h_{i-1} + \kappa_i 2(h_i + h_{i-1}) + h_i\kappa_{i+1} = \frac{6}{h_i}(y_{i+1} - y_i) - \frac{6}{h_{i-1}}(y_i - y_{i-1}). \tag{B.7}$$

If we now substitute Eq. B.7 for each interval $i = 1$ to $i = n - 1$, we obtain a system of equations, which according to Eq. B.5, depend only on the second derivatives of the desired interpolation at the points $x_1, \ldots, x_n$. This system of $n - 1$ equations with $n - 1$ unknown curvatures κ_i is solvable in principle.

In the beginning $\kappa_o = \kappa_n = 0$ was required. This requirement includes the problem that we have used two adjacent intervals for the first derivative of s_i. For the edge points this is impossible, though. At these points we have to determine reasonable assumptions for s, e.g., a linear extrapolation for s_0 or s_n from s_1 and s_2 or s_{n-1} and s_{n-2}, respectively. With the other κ_i we obtain a system of equations, which we can write as an equation matrix equation:

$$
\begin{pmatrix}
2(h_0 + h_1) & h_1 & 0 & 0 & 0 \\
h_1 & 2(h_1 + h_2) & h_2 & \vdots & \vdots \\
0 & h_2 & 2(h_2 + h_3) & \vdots & \vdots \\
\vdots & \ddots & \ddots & \ddots & 0 \\
0 & \cdots & 0 & h_{n-2} & 2(h_{n-2} + h_{n-1})
\end{pmatrix}
\begin{pmatrix}
\kappa_1 \\
\kappa_2 \\
\kappa_3 \\
\vdots \\
\kappa_{n-1}
\end{pmatrix}
$$

$$
= 6
\begin{pmatrix}
\frac{y_2 - y_1}{h_1} - \frac{y_1 - y_0}{h_0} \\
\frac{y_3 - y_2}{h_2} - \frac{y_2 - y_1}{h_1} \\
\frac{y_4 - y_3}{h_3} - \frac{y_3 - y_2}{h_2} \\
\vdots \\
\frac{y_n - y_{n-1}}{h_{n-1}} - \frac{y_{n-1} - y_{n-2}}{h_{n-2}}
\end{pmatrix}
$$

In order to solve the equation system, κ_1 (the curvature in the first node) is expressed with the first equation by κ_2. This expression is again implemented in the second equation so that κ_2 is expressed by κ_3. This forward substitution is carried along until the last equation and we can thus compute κ_{n-1} directly. By backward substitution one then works back to determine all other κ. The calculation procedure is thus as follows:

1. Define n, x_i and y_i for all $i = 0$ to $i = n$.
2. Define for $i = 0$ to $i = n - 1$: $h_i = x_{i+1} - x_i$ and $\epsilon_i = \frac{6}{h_i}(y_{i+1} - y_i)$
3. Define by forward substitution: $\mu_i = 2(h_0 + h_1)$ and $\rho_i = \epsilon_1 - \epsilon_0$
4. Consecutive repetition for $i = 2$ to $i = n - 1$: $\mu_i = 2(h_i + h_{i-1}) - \frac{h_{i-1}^2}{\mu_{i-1}}$ and
   $\rho_i = \epsilon_i - \epsilon_{i-1} - \frac{\rho_{i-1} h_{i-1}}{\mu_{i-1}}$
5. Now backward substitution: $\kappa_n = 0$
6. Consecutive repetition for $i = n - 1$ to 1: $\kappa_i = \frac{\rho_i - h_i \kappa_{i+1}}{\mu_i}$ and $\kappa_0 = 0$

In contrast to other polynomials, the effort to solve these equations as well as the required computer memory increase only linearly with the number of nodes. This and the very good interpolation properties of splines make it the first choice for the production of smooth interpolation. Figures B.3 and B.4 illustrate the difference between an interpolation with a classical equation of third degree and one with a cubic spline.

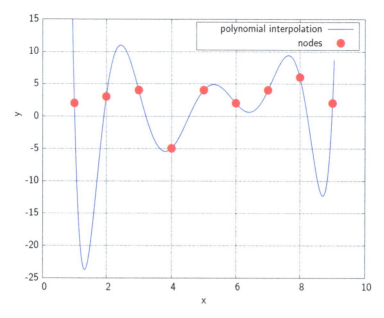

Fig. B.3 Classical interpolation with a 3rd degree polynomial

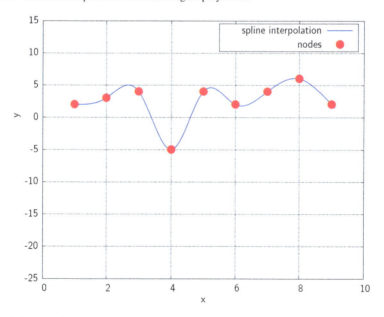

Fig. B.4 Interpolation with a cubic spline

We now recognize the impact of our requirement $\kappa_o = \kappa_n = 0$. This requirement has a significant influence on the oscillation behaviour of the entire function and makes the spline so much better than other methods.

B.5   The Continuous Fourier Transform

B.5.1   Rules for the One-Dimensional Fourier Transform

$$X(\omega) = \int_{-\infty}^{\infty} x(t)e^{-j\omega t}\,dt$$

$$x(t) = \frac{1}{2\pi}\int_{-\infty}^{\infty} X(\omega)e^{j\omega t}\,d\omega$$

Linearity	$\alpha x(t) + \beta y(t) \leftrightarrow \alpha X(\omega) + \beta Y(\omega)$
Symmetry	If $x(t) \leftrightarrow X(\omega)$, then $X(t) \leftrightarrow 2\pi x(-\omega)$
Scaling	$x(\alpha t) \leftrightarrow \frac{1}{|\alpha|}X(\frac{\omega}{\alpha})$; $x(-t) \leftrightarrow X^*(\omega)$ if $x(t)$ is real
Shift	$x(t - t_0) \leftrightarrow e^{-j\omega t_0}X(\omega)$; $e^{j\omega_0 t}x(t) \leftrightarrow X(\omega - \omega_0)$
Modulation	$\cos(\omega_0 t)x(t) \leftrightarrow 1/2 X(\omega - \omega_0) + 1/2 X(\omega + \omega_0)$
Conjugation	$x^*(t) \leftrightarrow X^*(-\omega)$. If $x(t) = x^*(t)$, then $X^*(\omega) = X(-\omega)$
Derivation	$\frac{d^n}{dt^n}x(t) \leftrightarrow (j\omega)^n X(\omega)$; $(-jt)^n x(t) \leftrightarrow \frac{d^n}{d\omega^n}X(\omega)$
Folding in the time domain	$x(t) * y(t) = \int_{-\infty}^{\infty} x(t - \tau)y(\tau)d\tau \leftrightarrow X(\omega)Y(\omega)$
Multiplication in the time domain	$x(t)y(t) \leftrightarrow \frac{1}{2\pi}X(\omega) * Y(\omega)$
Correlation	$r_{xy}^E(t) = \int_{-\infty}^{\infty} y(t + \alpha)x^*(\alpha)d\alpha \leftrightarrow Y(\omega)X^*(\omega)$
Parseval's theorem	$\int_{-\infty}^{\infty} |x(t)|^2\,dt = \frac{1}{2\pi}\int_{-\infty}^{\infty} |X(\omega)|^2\,d\omega$

B.5.2   Correspondences of the One Dimensional Fourier Transform

$x(t)$	$X(\omega)$
$\delta(t)$	1
1	$2\pi\delta(\omega)$
$sgn(t)$	$\frac{2}{j\omega}$
$rect(t/T)$	$T\,si(\omega T/2)$
$tri(t/T)$	$T\,si^2(\omega T/2)$
$si(\omega_0 t)$	$\frac{\pi}{\omega_0}rect(\frac{\omega}{2\omega_0})$
$\cos(\omega_0 t)$	$\pi[\delta(\omega - \omega_0) + \delta(\omega + \omega_0)]$
$\sin(\omega_0 t)$	$-j\pi[\delta(\omega - \omega_0) - \delta(\omega + \omega_0)]$
$e^{-a|t|}$	$\frac{2a}{\omega^2 + a^2}$
$\varepsilon(t)e^{-at}$	$\frac{1}{j\omega + a}$
$e^{j\omega_0 t}$	$2\pi\delta(\omega - \omega_0)$
$\sum_{n=-\infty}^{\infty}\delta(t - nT)$	$\omega_0\sum_{k=-\infty}^{\infty}\delta(\omega - k\omega_0)$ mit $\omega_0 = 2\pi/T$

Appendix C
Diffraction Indices of Various Glasses

Prismatic spectra are caused by the wavelength-dependent refractive index $n(\lambda)$. This behavior dependents on the glass type of the corresponding prism. Figure C.1 illustrates the indices of refraction for all Schott glass types for the wavelength interval 4,000–8,500 Å. Their values for specific wavelengths are listed in Tables C.1, C.2, C.3, C.4, C.5 and C.6. The corresponding values for the entire interval 2,483–23,254 Å are available at Schott or on their website.

© Springer-Verlag Berlin Heidelberg 2015

601

T. Eversberg, K. Vollmann, *Spectroscopic Instrumentation*, Springer Praxis Books,
DOI 10.1007/978-3-662-44535-8

Fig. C.1 Graph of the
refraction indices of all
available Schott glass types
for the wavelength interval
4,000–8,500 Å (Schott). For
visual wavelengths the
indices of refraction are
between 1.4 and 2. They
increase towards shorter
wavelengths for all glasses

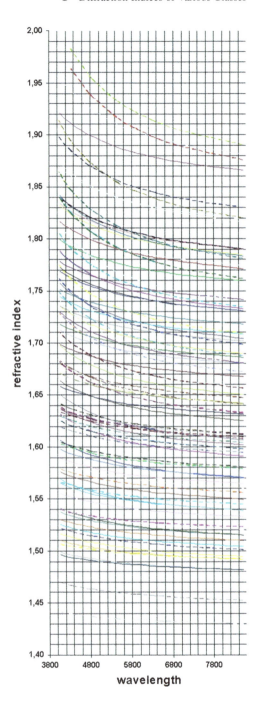

Table C.1 Refraction indices n at different wavelengths (8,521–5,893 Å) for all Schott glasses (Schott)

Glass	8,521 Å	7,065 Å	6,563 Å	6,438 Å	6,328 Å	5,893 Å
F2	1.60671	1.61227	1.61503	1.61582	1.61656	1.61989
F5	1.59093	1.59616	1.59875	1.59948	1.60017	1.60328
K10	1.49389	1.49713	1.49867	1.49910	1.49950	1.50129
K7	1.50394	1.50707	1.50854	1.50895	1.50934	1.51105
KZFS12	1.68071	1.68717	1.69033	1.69122	1.69206	1.69583
KZFSN5	1.64075	1.64644	1.64920	1.64998	1.65070	1.65397
LAFN7	1.73264	1.73970	1.74319	1.74418	1.74511	1.74931
LF5	1.57014	1.57489	1.57723	1.57789	1.57851	1.58132
LLF1	1.53845	1.54256	1.54457	1.54513	1.54566	1.54803
N-BAF10	1.65849	1.66339	1.66578	1.66645	1.66708	1.66990
N-BAF4	1.59452	1.59926	1.60157	1.60222	1.60282	1.60556
N-BAF51	1.64059	1.64551	1.64792	1.64860	1.64924	1.65211
N-BAF52	1.59801	1.60254	1.60473	1.60535	1.60593	1.60852
N-BAK1	1.56421	1.56778	1.56949	1.56997	1.57041	1.57241
N-BAK2	1.53234	1.53564	1.53721	1.53765	1.53806	1.53988
N-BAK4	1.56034	1.56400	1.56575	1.56624	1.56670	1.56874
N-BALF4	1.57065	1.57447	1.57631	1.57683	1.57731	1.57946
N-BALF5	1.53885	1.54255	1.54430	1.54479	1.54525	1.54730
N-BASF2	1.65007	1.65607	1.65905	1.65990	1.66070	1.66430
N-BASF64	1.68982	1.69578	1.69872	1.69955	1.70033	1.70384
N-BK10	1.49127	1.49419	1.49552	1.49589	1.49623	1.49775
N-BK7	1.50980	1.51289	1.51432	1.51472	1.51509	1.51673
N-F2	1.60667	1.61229	1.61506	1.61584	1.61658	1.61990
N-FK5	1.48137	1.48410	1.48535	1.48569	1.48601	1.48743
N-FK51A	1.48165	1.48379	1.48480	1.48508	1.48534	1.48651
N-K5	1.51507	1.51829	1.51982	1.52024	1.52064	1.52241
N-KF9	1.51507	1.51867	1.52040	1.52089	1.52134	1.52337
N-KZFS11	1.62540	1.63069	1.63324	1.63395	1.63462	1.63762
N-KZFS2	1.54944	1.55337	1.55519	1.55570	1.55617	1.55827
N-KZFS4	1.60199	1.60688	1.60922	1.60987	1.61049	1.61324
N-KZFS8	1.70416	1.71099	1.71437	1.71532	1.71622	1.72029
N-LAF2	1.73064	1.73627	1.73903	1.73981	1.74054	1.74383
N-LAF21	1.77434	1.78019	1.78301	1.78380	1.78454	1.78785
N-LAF33	1.77138	1.77751	1.78049	1.78134	1.78213	1.78567
N-LAF34	1.75962	1.76515	1.76780	1.76855	1.76924	1.77236
N-LAF35	1.73086	1.73620	1.73876	1.73948	1.74015	1.74317
N-LAF36	1.78435	1.79076	1.79390	1.79478	1.79561	1.79935
N-LAF7	1.73272	1.73972	1.74320	1.74419	1.74511	1.74931
N-LAK10	1.70815	1.71328	1.71572	1.71641	1.71705	1.71990
N-LAK12	1.66772	1.67209	1.67419	1.67478	1.67533	1.67779

Table C.2 Refraction indices n at different wavelengths (8,521–5,893 Å) for all Schott glasses (Schott)

Glass	8,521 Å	7,065 Å	6,563 Å	6,438 Å	6,328 Å	5,893 Å
N-LAK14	1.68612	1.69077	1.69297	1.69358	1.69415	1.69669
N-LAK21	1.63143	1.63538	1.63724	1.63776	1.63825	1.64040
N-LAK22	1.64141	1.64560	1.64760	1.64816	1.64868	1.65103
N-LAK33A	1.74186	1.74707	1.74956	1.75025	1.75090	1.75380
N-LAK34	1.71787	1.72277	1.72509	1.72574	1.72634	1.72904
N-LAK7	1.64220	1.64628	1.64821	1.64875	1.64925	1.65150
N-LAK8	1.70181	1.70668	1.70897	1.70962	1.71022	1.71289
N-LAK9	1.68033	1.68497	1.68716	1.68777	1.68834	1.69089
N-LASF31A	1.92093	1.91050	1.89950	1.89822	1.88815	1.88300
N-LASF40	1.87393	1.86275	1.85114	1.84981	1.83935	1.83404
N-LASF41	1.86872	1.85949	1.84972	1.84859	1.83961	1.83501
N-LASF43	1.84106	1.83137	1.82122	1.82005	1.81081	1.80610
N-LASF44	1.83405	1.82594	1.81731	1.81630	1.80832	1.80420
N-LASF45	1.84237	1.83068	1.81864	1.81726	1.80650	1.80107
N-LASF46A	1.95645	1.94129	1.92586	1.92411	1.91048	1.90366
N-LASF9	1.89845	1.88467	1.87058	1.86898	1.85650	1.85025
N-PK51	1.54010	1.53704	1.53372	1.53333	1.53019	1.52855
N-PK52A	1.50720	1.50450	1.50157	1.50123	1.49845	1.49700
N-PSK3	1.56688	1.56302	1.55885	1.55835	1.55440	1.55232
N-PSK53A	1.63445	1.63007	1.62534	1.62478	1.62033	1.61800
N-SF1	1.76224	1.74919	1.73605	1.73457	1.72308	1.71736
N-SF10	1.77578	1.76191	1.74800	1.74643	1.73430	1.72828
N-SF11	1.84235	1.82533	1.80841	1.80651	1.79192	1.78472
N-SF14	1.81570	1.79986	1.78405	1.78228	1.76859	1.76182
N-SF15	1.74182	1.72933	1.71677	1.71536	1.70438	1.69892
N-SF4	1.80668	1.79158	1.77647	1.77477	1.76164	1.75513
N-SF5	1.71106	1.69998	1.68876	1.68750	1.67763	1.67271
N-SF56	1.84126	1.82460	1.80800	1.80614	1.79179	1.78470
N-SF57	1.91440	1.89423	1.87432	1.87210	1.85504	1.84666
N-SF57HT	1.91440	1.89423	1.87432	1.87210	1.85504	1.84666
N-SF6	1.86506	1.84738	1.82980	1.82783	1.81266	1.80518
N-SF64	1.74912	1.73657	1.72392	1.72249	1.71142	1.70591
N-SF66		1.98285	1.95739	1.95459	1.93322	1.92286
N-SF6HT	1.86506	1.84738	1.82980	1.82783	1.81266	1.80518
N-SF8	1.72948	1.71775	1.70589	1.70455	1.69413	1.68894
N-SK11	1.57946	1.57530	1.57081	1.57028	1.56605	1.56384
N-SK14	1.61988	1.61542	1.61059	1.61003	1.60548	1.60311
N-SK16	1.63773	1.63312	1.62814	1.62756	1.62286	1.62041
N-SK2	1.62562	1.62073	1.61547	1.61486	1.60994	1.60738
N-SK4	1.63042	1.62568	1.62059	1.61999	1.61521	1.61272

Table C.3 Refraction indices n at different wavelengths (8,521–5,893 Å) for all Schott glasses (Schott)

N-SK5	1.60530	1.60100	1.59635	1.59581	1.59142	1.58913
N-SSK2	1.64232	1.63691	1.63112	1.63045	1.62508	1.62229
N-SSK5	1.68079	1.67471	1.66824	1.66749	1.66152	1.65844
N-SSK8	1.63923	1.63335	1.62713	1.62641	1.62068	1.61773
N-ZK7	1.52238	1.51869	1.51470	1.51423	1.51045	1.50847
P-LASF47	1.84064	1.83112	1.82110	1.81994	1.81078	1.80610
P-PK53	1.54029	1.53673	1.53288	1.53243	1.52880	1.52690
P-SF67		1.96401	1.93985	1.93717	1.91675	1.90680
P-SK57	1.60359	1.59917	1.59440	1.59384	1.58935	1.58700
SF1	1.76201	1.74916	1.73610	1.73462	1.72310	1.71736
SF10	1.77579	1.76198	1.74805	1.74648	1.73430	1.72825
SF2	1.68233	1.67249	1.66238	1.66123	1.65222	1.64769
SF4	1.80589	1.79121	1.77636	1.77468	1.76167	1.75520
SF5	1.71069	1.69986	1.68876	1.68750	1.67764	1.67270
SF56A	1.84092	1.82449	1.80800	1.80615	1.79180	1.78470
SF57	1.91366	1.89393	1.87425	1.87204	1.85504	1.84666
SF57HHT	1.91366	1.89393	1.87425	1.87204	1.85504	1.84666
SF6	1.86436	1.84707	1.82970	1.82775	1.81265	1.80518
SF6HT	1.86436	1.84707	1.82970	1.82775	1.81265	1.80518
LITHOTEC-CAF2	1.44149	1.43947	1.43727	1.43702	1.43494	1.43385
LITHOSIL-Q	1.46959	1.46667	1.46348	1.46310	1.46005	1.45844
P-SF8	1.72950	1.71778	1.70591	1.70457	1.69414	1.68893
N-SF2	1.68273	1.67265	1.66241	1.66125	1.65222	1.64769
N-KZFS5	1.68318	1.67511	1.66667	1.66570	1.65803	1.65412

Table C.4 Refraction indices n at different wavelengths (5,876–4,047 Å) for all Schott glasses (Schott)

Glass	5,876 Å	5,461 Å	4,861 Å	4,800 Å	4,358 Å	4,047 Å
F2	1.62004	1.62408	1.63208	1.63310	1.64202	1.65064
F5	1.60342	1.60718	1.61461	1.61556	1.62381	1.63176
K10	1.50137	1.50349	1.50756	1.50807	1.51243	1.51649
K7	1.51112	1.51314	1.51700	1.51748	1.52159	1.52540
KZFS12	1.69600	1.70055	1.70951	1.71065	1.72059	1.73017
KZFSN5	1.65412	1.65803	1.66571	1.66668	1.67512	1.68319
LAFN7	1.74950	1.75458	1.76464	1.76592	1.77713	1.78798
LF5	1.58144	1.58482	1.59146	1.59231	1.59964	1.60668
LLF1	1.54814	1.55099	1.55655	1.55725	1.56333	1.56911
N-BAF10	1.67003	1.67341	1.68000	1.68083	1.68801	1.69480
N-BAF4	1.60568	1.60897	1.61542	1.61624	1.62336	1.63022
N-BAF51	1.65224	1.65569	1.66243	1.66328	1.67065	1.67766
N-BAF52	1.60863	1.61173	1.61779	1.61856	1.62521	1.63157
N-BAK1	1.57250	1.57487	1.57943	1.58000	1.58488	1.58941
N-BAK2	1.53996	1.54212	1.54625	1.54677	1.55117	1.55525
N-BAK4	1.56883	1.57125	1.57591	1.57649	1.58149	1.58614
N-BALF4	1.54739	1.54982	1.55451	1.55510	1.56016	1.56491
N-BASF2	1.66446	1.66883	1.67751	1.67862	1.68838	1.69792
N-BASF644	1.70400	1.70824	1.71659	1.71765	1.72690	1.73581
N-BK10	1.49782	1.49960	1.50296	1.50337	1.50690	1.51014
N-BK7	1.51680	1.51872	1.52238	1.52283	1.52668	1.53024
N-F2	1.62005	1.62408	1.63208	1.63310	1.64209	1.65087
N-FK5	1.48749	1.48914	1.49227	1.49266	1.49593	1.49894
N-FK51A	1.48656	1.48794	1.49056	1.49088	1.49364	1.49618
N-K5	1.52249	1.52458	1.52860	1.52910	1.53338	1.53734
N-KF9	1.52346	1.52588	1.53056	1.53114	1.53620	1.54096
N-KZFS11	1.63775	1.64132	1.64828	1.64915	1.65670	1.66385
N-KZFS2	1.55836	1.56082	1.56553	1.56612	1.57114	1.57580
N-KZFS4	1.61336	1.61664	1.62300	1.62380	1.63071	1.63723
N-KZFS8	1.72047	1.72539	1.73513	1.73637	1.74724	1.75777
N-LAF2	1.74397	1.74791	1.75562	1.75659	1.76500	1.77298
N-LAF21	1.78800	1.79195	1.79960	1.80056	1.80882	1.81657
N-LAF33	1.78582	1.79007	1.79833	1.79937	1.80837	1.81687
N-LAF34	1.77250	1.77621	1.78337	1.78427	1.79196	1.79915
N-LAF35	1.74330	1.74688	1.75381	1.75467	1.76212	1.76908
N-LAF36	1.79952	1.80400	1.81277	1.81387	1.82345	1.83252
N-LAF7	1.74950	1.75459	1.76472	1.76602	1.77741	1.78854
N-LAK10	1.72003	1.72341	1.72995	1.73077	1.73779	1.74438
N-LAK12	1.67790	1.68083	1.68647	1.68717	1.69320	1.69882
N-LAK14	1.69680	1.69980	1.70554	1.70626	1.71237	1.71804

Table C.5 Refraction indices n at different wavelengths (5,876–4,047 Å) for all Schott glasses (Schott)

Glass	5,876 Å	5,461 Å	4,861 Å	4,800 Å	4,358 Å	4,047 Å
N-LAK21	1.64049	1.64304	1.64790	1.64850	1.65366	1.65844
N-LAK22	1.65113	1.65391	1.65925	1.65992	1.66562	1.67092
N-LAK33A	1.75393	1.75737	1.76398	1.76481	1.77187	1.77845
N-LAK34	1.72916	1.73235	1.73847	1.73923	1.74575	1.75180
N-LAK7	1.65160	1.65425	1.65934	1.65998	1.66539	1.67042
N-LAK8	1.71300	1.71616	1.72222	1.72297	1.72944	1.73545
N-LAK9	1.69100	1.69401	1.69979	1.70051	1.70667	1.71239
N-LASF31A	1.88281	1.87853	1.87757	1.87656	1.87298	1.86572
N-LASF40	1.83385	1.82946	1.82849	1.82745	1.82380	1.81643
N-LASF41	1.83484	1.83100	1.83014	1.82923	1.82599	1.81936
N-LASF43	1.80593	1.80200	1.80113	1.80020	1.79691	1.79018
N-LASF44	1.80405	1.80060	1.79983	1.79901	1.79609	1.79006
N-LASF45	1.80087	1.79640	1.79541	1.79436	1.79066	1.78325
N-LASF46A	1.90341	1.89781	1.89657	1.89526	1.89064	1.88143
N-LASF9	1.85002	1.84489	1.84376	1.84255	1.83834	1.82997
N-PK51	1.52849	1.52711	1.52680	1.52646	1.52527	1.52278
N-PK52A	1.49695	1.49571	1.49544	1.49514	1.49408	1.49184
N-PSK3	1.55224	1.55048	1.55008	1.54965	1.54811	1.54482
N-PSK53A	1.61791	1.61595	1.61550	1.61503	1.61334	1.60979
N-SF1	1.71715	1.71247	1.71144	1.71035	1.70651	1.69889
N-SF10	1.72806	1.72314	1.72206	1.72091	1.71688	1.70891
N-SF11	1.78446	1.77860	1.77732	1.77596	1.77119	1.76182
N-SF14	1.76157	1.75606	1.75485	1.75356	1.74907	1.74022
N-SF15	1.69872	1.69425	1.69326	1.69222	1.68854	1.68122
N-SF4	1.75489	1.74959	1.74842	1.74719	1.74286	1.73432
N-SF5	1.67253	1.66848	1.66759	1.66664	1.66330	1.65661
N-SF56	1.78444	1.77868	1.77741	1.77607	1.77137	1.76213
N-SF57	1.84635	1.83956	1.83807	1.83650	1.83099	1.82023
N-SF57HT	1.84635	1.83956	1.83807	1.83650	1.83099	1.82023
N-SF6	1.80491	1.79883	1.79749	1.79608	1.79114	1.78144
N-SF64	1.70571	1.70119	1.70020	1.69914	1.69544	1.68806
N-SF66	1.92248	1.91414	1.91232	1.91039	1.90368	1.89064
N-SF6HT	1.80491	1.79883	1.79749	1.79608	1.79114	1.78144
N-SF8	1.68874	1.68448	1.68354	1.68254	1.67904	1.67203
N-SK11	1.56376	1.56188	1.56146	1.56101	1.55939	1.55597
N-SK14	1.60302	1.60101	1.60056	1.60008	1.59834	1.59467
N-SK16	1.62032	1.61824	1.61777	1.61727	1.61548	1.61167
N-SK2	1.60729	1.60513	1.60465	1.60414	1.60230	1.59847
N-SK4	1.61262	1.61052	1.61005	1.60954	1.60774	1.60393
N-SK5	1.58904	1.58710	1.58666	1.58619	1.58451	1.58094

Table C.6 Refraction indices n at different wavelengths (5,876–4,047 Å) for all Schott glasses (Schott)

Glass	5,876 Å	5,461 Å	4,861 Å	4,800 Å	4,358 Å	4,047 Å
N-SSK2	1.62219	1.61985	1.61933	1.61877	1.61678	1.61264
N-SSK5	1.65833	1.65574	1.65517	1.65455	1.65237	1.64785
N-SSK8	1.61762	1.61515	1.61460	1.61401	1.61192	1.60759
N-ZK7	1.50840	1.50671	1.50633	1.50592	1.50445	1.50129
P-LASF47	1.80593	1.80203	1.80116	1.80023	1.79696	1.79028
P-PK53	1.52683	1.52522	1.52486	1.52447	1.52309	1.52017
P-SF67	1.90644	1.89841	1.89666	1.89480	1.88833	1.87574
P-SK57	1.58691	1.58492	1.58447	1.58399	1.58227	1.57862
SF1	1.71715	1.71245	1.71141	1.71031	1.70647	1.69888
SF10	1.72803	1.72309	1.72200	1.72085	1.71681	1.70887
SF2	1.64752	1.64379	1.64297	1.64210	1.63902	1.63289
SF4	1.75496	1.74969	1.74853	1.74730	1.74300	1.73456
SF5	1.67252	1.66846	1.66756	1.66661	1.66327	1.65664
SF56A	1.78444	1.77866	1.77740	1.77605	1.77136	1.76220
SF57	1.84636	1.83957	1.83808	1.83650	1.83102	1.82038
SF57HHT	1.84636	1.83957	1.83808	1.83650	1.83102	1.82038
SF6	1.80491	1.79884	1.79750	1.79609	1.79117	1.78157
SF6HT	1.80491	1.79884	1.79750	1.79609	1.79117	1.78157
LITHOTEC-CAF2	1.43381	1.43289	1.43268	1.43246	1.43167	1.43003
LITHOSIL-Q	1.45838	1.45699	1.45668	1.45634	1.45512	1.45244
P-SF8	1.68874	1.68447	1.68353	1.68252	1.67901	1.67200
N-SF2	1.64752	1.64380	1.64298	1.64210	1.63902	1.63282
N-KZFS5	1.65398	1.65072	1.65000	1.64922	1.64649	1.64087

Appendix D
Transmissivity of Various Glasses

In a telescope-spectrograph system, light passes through a series of optical elements. As a result, the overall efficiency of standard spectrographs is of the order of 50 %, for echelle spectrographs even only 10 %. It is therefore essential to know the efficiency of all individual optical elements. Figure D.1 illustrates the refraction indices for all Schott glasses in the wavelength interval 4,000–8,500 Å (25 mm glass thickness). The corresponding values for specific wavelengths are shown in Tables D.1, D.2, D.3, D.4, D.5, and D.6. The corresponding values for the entire interval 2,500–25,000 Å are available from Schott or directly on their website.

© Springer-Verlag Berlin Heidelberg 2015   609
T. Eversberg, K. Vollmann, *Spectroscopic Instrumentation*, Springer Praxis Books,
DOI 10.1007/978-3-662-44535-8

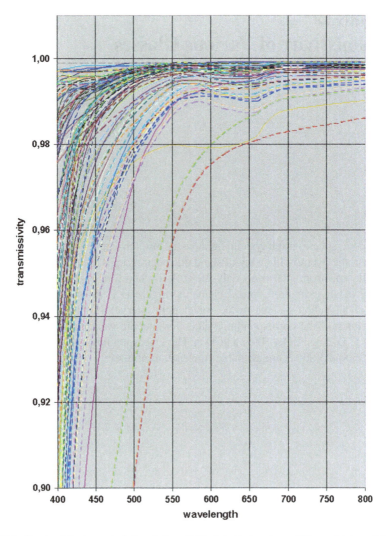

Fig. D.1 Graph of the transmissivity of all available Schott glass types (25 mm thickness) for the wavelength interval 4,000–8,500 Å (Schott). The majority of all glasses become opaque below about 500 nm. Only some glasses are transparent in UV wavelengths

Table D.1 Transmissivity n at different wavelengths (10,600–5,460 Å) for all Schott glasses of 25 mm thickness (Schott)

Glas	10,600 Å	7,000 Å	6,600 Å	6,200 Å	5,800 Å	5,460 Å
F2	0.999	0.999	0.999	0.999	0.999	0.999
F5	0.999	0.999	0.998	0.998	0.998	0.998
K10	0.998	0.999	0.998	0.997	0.997	0.997
K7	0.998	0.998	0.998	0.998	0.998	0.998
KZFS12	0.998	0.997	0.997	0.997	0.996	0.996
KZFSN5	0.999	0.998	0.998	0.998	0.998	0.998
LAFN7	0.998	0.998	0.998	0.998	0.998	0.998
LF5	0.999	0.999	0.999	0.999	0.999	0.999
LLF1	0.998	0.999	0.998	0.998	0.999	0.999
N-BAF10	0.998	0.998	0.996	0.996	0.996	0.996
N-BAF4	0.998	0.998	0.996	0.996	0.997	0.997
N-BAF51	0.997	0.997	0.996	0.996	0.997	0.996
N-BAF52	0.998	0.997	0.996	0.996	0.996	0.996
N-BAK1	0.998	0.999	0.998	0.998	0.998	0.998
N-BAK2	0.999	0.998	0.998	0.998	0.998	0.998
N-BAK4	0.998	0.999	0.998	0.998	0.998	0.998
N-BALF4	0.997	0.999	0.998	0.998	0.998	0.998
N-BALF5	0.996	0.998	0.997	0.997	0.998	0.998
N-BASF2	0.999	0.996	0.994	0.994	0.995	0.994
N-BASF64	0.994	0.988	0.982	0.979	0.979	0.980
N-BK10	0.998	0.998	0.997	0.997	0.997	0.997
N-BK7	0.999	0.998	0.998	0.998	0.998	0.998
N-F2	0.998	0.997	0.996	0.996	0.997	0.997
N-FK5	0.999	0.998	0.998	0.997	0.998	0.998
N-FK51A	0.998	0.998	0.998	0.998	0.999	0.999
N-K5	0.998	0.998	0.997	0.997	0.998	0.998
N-KF9	0.998	0.999	0.998	0.998	0.998	0.998
N-KZFS11	0.999	0.998	0.997	0.997	0.997	0.997
N-KZFS2	0.996	0.998	0.998	0.998	0.998	0.998
N-KZFS4	0.998	0.998	0.997	0.997	0.997	0.997
N-KZFS8	0.999	0.998	0.998	0.998	0.998	0.997
N-LAF2	0.999	0.998	0.997	0.997	0.997	0.998
N-LAF21	0.998	0.998	0.998	0.998	0.998	0.998
N-LAF33	0.998	0.998	0.997	0.997	0.997	0.997
N-LAF34	0.998	0.998	0.998	0.998	0.998	0.998
N-LAF35	0.998	0.998	0.998	0.998	0.998	0.998
N-LAF36	0.998	0.998	0.998	0.997	0.997	0.996
N-LAF7	0.998	0.997	0.995	0.994	0.992	0.988
N-LAK10	0.998	0.999	0.998	0.998	0.997	0.998
N-LAK12	0.997	0.997	0.996	0.995	0.996	0.996

Table D.2 Transmissivity n at different wavelengths (10,600–5,460 Å) for all Schott glasses of 25 mm thickness (Schott)

Glas	10,600 Å	7,000 Å	6,600 Å	6,200 Å	5,800 Å	5,460 Å
N-LAK14	0.998	0.998	0.998	0.997	0.997	0.998
N-LAK21	0.998	0.998	0.996	0.996	0.997	0.997
N-LAK22	0.998	0.998	0.997	0.996	0.997	0.997
N-LAK33A	0.998	0.998	0.998	0.998	0.998	0.998
N-LAK34	0.998	0.999	0.999	0.998	0.998	0.999
N-LAK7	0.998	0.998	0.998	0.998	0.998	0.998
N-LAK8	0.998	0.998	0.998	0.998	0.998	0.998
N-LAK9	0.998	0.998	0.998	0.998	0.998	0.998
N-LASF31A	0.996	0.996	0.995	0.994	0.995	0.994
N-LASF40	0.998	0.998	0.998	0.997	0.997	0.995
N-LASF41	0.998	0.998	0.998	0.997	0.998	0.997
N-LASF43	0.998	0.998	0.998	0.997	0.996	0.995
N-LASF44	0.998	0.998	0.998	0.998	0.998	0.998
N-LASF45	0.997	0.997	0.995	0.994	0.994	0.993
N-LASF46A	0.999	0.996	0.994	0.993	0.993	0.991
N-LASF9	0.998	0.995	0.994	0.993	0.992	0.990
N-PK51	0.998	0.997	0.996	0.997	0.998	0.998
N-PK52A	0.998	0.997	0.997	0.998	0.999	0.999
N-PSK3	0.999	0.998	0.997	0.997	0.997	0.997
N-PSK53	0.998	0.998	0.997	0.997	0.998	0.998
N-PSK53A	0.998	0.998	0.997	0.997	0.998	0.998
N-SF1	0.998	0.996	0.994	0.995	0.996	0.994
N-SF10	0.996	0.993	0.990	0.991	0.991	0.989
N-SF11	0.999	0.994	0.992	0.992	0.994	0.991
N-SF14	0.999	0.994	0.991	0.992	0.994	0.992
N-SF15	0.998	0.995	0.993	0.994	0.994	0.994
N-SF4	0.999	0.995	0.993	0.993	0.993	0.990
N-SF5	0.998	0.996	0.995	0.995	0.996	0.995
N-SF56	0.998	0.994	0.992	0.992	0.993	0.990
N-SF57	0.999	0.991	0.987	0.988	0.990	0.986
N-SF57HT	0.999	0.992	0.988	0.989	0.991	0.987
N-SF6	0.998	0.993	0.991	0.991	0.992	0.989
N-SF64	0.998	0.994	0.992	0.992	0.994	0.993
N-SF66	0.996	0.991	0.987	0.983	0.976	0.963
N-SF6HT	0.999	0.994	0.991	0.992	0.992	0.990
N-SF8	0.997	0.995	0.993	0.993	0.994	0.993
N-SK11	0.998	0.998	0.998	0.998	0.998	0.999
N-SK14	0.998	0.998	0.998	0.998	0.998	0.998
N-SK16	0.998	0.998	0.998	0.997	0.998	0.998
N-SK2	0.998	0.998	0.998	0.998	0.998	0.998

Table D.3 Transmissivity n at different wavelengths (10,600–5,460 Å) for all Schott glasses of 25 mm thickness (Schott)

Glas	10,600 Å	7,000 Å	6,600 Å	6,200 Å	5,800 Å	5,460 Å
N-SK4	0.997	0.998	0.998	0.998	0.998	0.998
N-SK5	0.999	0.998	0.998	0.997	0.998	0.998
N-SSK2	0.997	0.998	0.998	0.997	0.998	0.998
N-SSK5	0.996	0.997	0.997	0.997	0.997	0.996
N-SSK8	0.997	0.998	0.996	0.996	0.997	0.997
N-ZK7	0.998	0.998	0.998	0.998	0.998	0.998
P-LASF47	0.999	0.998	0.998	0.998	0.998	0.998
P-PK53	0.998	0.997	0.997	0.998	0.998	0.999
P-SF67	0.994	0.983	0.981	0.978	0.971	0.954
P-SK57	0.999	0.999	0.999	0.999	0.999	0.999
SF1	0.998	0.998	0.998	0.998	0.998	0.998
SF10	0.999	0.998	0.997	0.997	0.998	0.998
SF2	0.998	0.998	0.998	0.998	0.998	0.998
SF4	0.998	0.998	0.998	0.998	0.998	0.998
SF5	0.998	0.998	0.998	0.998	0.998	0.998
SF56A	0.999	0.998	0.997	0.998	0.998	0.998
SF57	0.999	0.998	0.998	0.998	0.998	0.998
SF57HHT	0.999	0.999	0.999	0.999	0.999	0.999
SF6	0.999	0.999	0.998	0.998	0.999	0.998
SF6HT	0.999	0.999	0.998	0.998	0.999	0.998
LITHOTEC-CAF2	0.999	0.999	0.999	0.999	0.999	0.999
LITHOSIL-Q	0.999	0.999	0.999	0.999	0.999	0.999
P-SF8	0.999	0.995	0.994	0.994	0.995	0.994
N-SF2	0.999	0.995	0.994	0.994	0.995	0.994
N-KZFS5	0.999	0.998	0.997	0.997	0.997	0.997

Table D.4 Transmissivity n at different wavelengths (5,000–4,000 Å) for all Schott glasses of 25 mm thickness (Schott)

Glas	5,000 Å	4,600 Å	4,360 Å	4,200 Å	4,050 Å	4,000 Å
F2	0.999	0.998	0.997	0.996	0.995	0.994
F5	0.998	0.996	0.996	0.995	0.994	0.993
K10	0.996	0.996	0.995	0.995	0.995	0.994
K7	0.997	0.996	0.996	0.996	0.996	0.996
KZFS12	0.994	0.988	0.977	0.963	0.933	0.919
KZFSN5	0.997	0.994	0.991	0.987	0.980	0.976
LAFN7	0.998	0.993	0.986	0.976	0.950	0.937
LF5	0.998	0.998	0.998	0.997	0.997	0.997
LLF1	0.998	0.998	0.998	0.998	0.998	0.997
N-BAF10	0.992	0.987	0.981	0.976	0.959	0.950
N-BAF4	0.994	0.988	0.983	0.976	0.959	0.946
N-BAF51	0.994	0.988	0.982	0.976	0.963	0.954
N-BAF52	0.992	0.987	0.981	0.975	0.959	0.950
N-BAK1	0.997	0.996	0.996	0.996	0.996	0.996
N-BAK2	0.998	0.997	0.997	0.997	0.997	0.997
N-BAK4	0.998	0.996	0.995	0.995	0.993	0.992
N-BALF4	0.997	0.994	0.993	0.992	0.988	0.985
N-BALF5	0.997	0.995	0.994	0.991	0.986	0.983
N-BASF2	0.988	0.980	0.971	0.954	0.915	0.891
N-BASF64	0.976	0.967	0.959	0.950	0.933	0.924
N-BK10	0.996	0.996	0.996	0.996	0.996	0.996
N-BK7	0.998	0.997	0.997	0.997	0.997	0.997
N-F2	0.994	0.989	0.985	0.980	0.959	0.946
N-FK5	0.997	0.997	0.997	0.997	0.998	0.998
N-FK51A	0.998	0.997	0.997	0.997	0.997	0.997
N-K5	0.997	0.996	0.996	0.996	0.996	0.995
N-KF9	0.998	0.996	0.995	0.994	0.990	0.986
N-KZFS11	0.996	0.993	0.991	0.990	0.988	0.987
N-KZFS2	0.997	0.995	0.992	0.990	0.987	0.985
N-KZFS4	0.995	0.990	0.987	0.984	0.981	0.979
N-KZFS8	0.994	0.988	0.982	0.976	0.967	0.963
N-LAF2	0.993	0.985	0.976	0.965	0.944	0.933
N-LAF21	0.995	0.989	0.983	0.976	0.959	0.950
N-LAF33	0.994	0.987	0.980	0.973	0.962	0.957
N-LAF34	0.996	0.992	0.987	0.981	0.971	0.967
N-LAF35	0.997	0.994	0.990	0.987	0.980	0.976
N-LAF36	0.992	0.985	0.976	0.967	0.954	0.946
N-LAF7	0.971	0.937	0.901	0.857	0.782	0.752
N-LAK10	0.995	0.991	0.985	0.976	0.963	0.959
N-LAK12	0.994	0.987	0.983	0.981	0.977	0.976

Table D.5 Transmissivity n at different wavelengths (5,000–4,000 Å) for all Schott glasses of 25 mm thickness (Schott)

Glas	5,000 Å	4,600 Å	4,360 Å	4,200 Å	4,050 Å	4,000 Å
N-LAK14	0.997	0.994	0.991	0.988	0.984	0.981
N-LAK21	0.995	0.990	0.987	0.985	0.982	0.979
N-LAK22	0.995	0.992	0.990	0.989	0.987	0.985
N-LAK33A	0.998	0.994	0.991	0.988	0.981	0.976
N-LAK34	0.998	0.995	0.992	0.989	0.983	0.981
N-LAK7	0.997	0.994	0.991	0.988	0.981	0.977
N-LAK8	0.998	0.995	0.992	0.988	0.981	0.977
N-LAK9	0.997	0.994	0.991	0.988	0.983	0.980
N-LASF31A	0.988	0.974	0.963	0.950	0.933	0.924
N-LASF40	0.987	0.973	0.954	0.937	0.905	0.891
N-LASF41	0.994	0.985	0.976	0.967	0.954	0.948
N-LASF43	0.990	0.980	0.967	0.954	0.933	0.919
N-LASF44	0.996	0.991	0.986	0.980	0.967	0.963
N-LASF45	0.983	0.965	0.946	0.924	0.877	0.857
N-LASF46A	0.980	0.959	0.937	0.905	0.847	0.815
N-LASF9	0.980	0.959	0.933	0.901	0.831	0.799
N-PK51	0.997	0.995	0.994	0.994	0.994	0.994
N-PK52A	0.998	0.997	0.996	0.996	0.997	0.997
N-PSK3	0.996	0.995	0.994	0.994	0.995	0.994
N-PSK53	0.997	0.994	0.993	0.992	0.988	0.985
N-PSK53A	0.997	0.994	0.993	0.992	0.988	0.985
N-SF1	0.987	0.976	0.963	0.946	0.896	0.867
N-SF10	0.978	0.963	0.946	0.924	0.867	0.837
N-SF11	0.981	0.967	0.946	0.919	0.852	0.815
N-SF14	0.984	0.971	0.963	0.946	0.910	0.891
N-SF15	0.988	0.977	0.964	0.941	0.887	0.857
N-SF4	0.978	0.959	0.933	0.896	0.821	0.787
N-SF5	0.990	0.982	0.973	0.963	0.928	0.905
N-SF56	0.980	0.963	0.941	0.905	0.837	0.799
N-SF57	0.971	0.949	0.919	0.872	0.782	0.733
N-SF57HT	0.972	0.951	0.928	0.896	0.831	0.793
N-SF6	0.977	0.961	0.946	0.919	0.857	0.821
N-SF64	0.984	0.971	0.957	0.934	0.882	0.852
N-SF66	0.928	0.887	0.831	0.758	0.592	0.504
N-SF6HT	0.980	0.966	0.954	0.937	0.901	0.877
N-SF8	0.985	0.976	0.965	0.950	0.919	0.901
N-SK11	0.998	0.996	0.995	0.994	0.992	0.990
N-SK14	0.997	0.995	0.994	0.993	0.991	0.990
N-SK16	0.996	0.994	0.992	0.992	0.990	0.988
N-SK2	0.996	0.993	0.993	0.994	0.994	0.994

Table D.6 Transmissivity n at different wavelengths (5,000–4,000 Å) for all Schott glasses of 25 mm thickness (Schott)

Glas	5,000 Å	4,600 Å	4,360 Å	4,200 Å	4,050 Å	4,000 Å
N-SK4	0.997	0.994	0.993	0.993	0.992	0.990
N-SK5	0.998	0.996	0.995	0.994	0.993	0.992
N-SSK2	0.997	0.994	0.992	0.990	0.985	0.981
N-SSK5	0.993	0.987	0.982	0.976	0.963	0.959
N-SSK8	0.994	0.987	0.982	0.975	0.959	0.950
N-ZK7	0.997	0.995	0.994	0.992	0.991	0.990
P-LASF47	0.995	0.990	0.985	0.980	0.971	0.967
P-PK53	0.998	0.996	0.995	0.994	0.994	0.994
P-SF67	0.901	0.810	0.707	0.574	0.364	0.276
P-SK57	0.998	0.996	0.996	0.995	0.994	0.994
SF1	0.997	0.994	0.990	0.984	0.971	0.967
SF10	0.996	0.991	0.984	0.967	0.910	0.862
SF2	0.997	0.995	0.993	0.990	0.985	0.981
SF4	0.996	0.992	0.987	0.980	0.963	0.954
SF5	0.997	0.995	0.993	0.989	0.983	0.980
SF56A	0.996	0.990	0.980	0.959	0.896	0.857
SF57	0.994	0.987	0.971	0.941	0.882	0.847
SF57HHT	0.996	0.991	0.985	0.971	0.941	0.924
SF6	0.996	0.991	0.982	0.967	0.933	0.915
SF6HT	0.996	0.992	0.987	0.977	0.954	0.941
LITHOTEC-CAF2	0.999	0.999	0.999	0.999	0.999	0.999
LITHOSIL-Q	0.999	0.999	0.999	0.999	0.999	0.999
P-SF8	0.989	0.980	0.971	0.959	0.937	0.924
N-SF2	0.990	0.984	0.979	0.970	0.944	0.928
N-KZFS5	0.994	0.990	0.986	0.983	0.978	0.976

Appendix E
Line Catalogues for Calibration Lamps

E.1 Line Catalogue Sources

- THORIUM-ARGON Line Catalogue
  FOCES High Resolution Echelle Spectra at 2.2 m Calar Alto
  http://www.caha.es/pedraz/Foces/thar_lamp.html
- NOAO SPECTRAL ATLAS
  ThAr Photron Lamp at 2.1 m KPNO
  http://www.noao.edu/kpno/specatlas/thar_photron/thar_photron.html
- NOAO SPECTRAL ATLAS
  ThAr Westinghouse Lamp at 2.1 m KPNO
  http://www.noao.edu/kpno/specatlas/thar/thar.html
- NOAO SPECTRAL ATLAS
  FeAr Lamp at 2.1 m KPNO
  http://www.noao.edu/kpno/specatlas/fear/fear.html
- NOAO SPECTRAL ATLAS
  HeNe Lamp at 2.1 m KPNO
  http://www.noao.edu/kpno/specatlas/henear/henear.html
- NOAO SPECTRAL ATLAS
  CuAr Lamp at 2.1 m KPNO
  http://www.noao.edu/kpno/specatlas/cuar/cuar.html
- High Resolution Line Atlas
  He-Ar-Fe-Ne Lamp at ESO 1.5 m
  http://www.ls.eso.org/lasilla/Telescopes/2p2T/E1p5M/memos_notes/Line-atlas-feros.html
- CFHT Coudé Comparison Arc Spectral Atlases
  ThAr, FeAr, CdNe and ThNe Lamps at 3.6 m CFHT
  http://www.cfht.hawaii.edu/Instruments/Spectroscopy/Gecko/CoudeAtlas/

© Springer-Verlag Berlin Heidelberg 2015
T. Eversberg, K. Vollmann, *Spectroscopic Instrumentation*, Springer Praxis Books,
DOI 10.1007/978-3-662-44535-8

- HET HRS Th-Ar and Echelle Tilt Atlas
  Th-Ar Lamp at 9 m Hobby-Eberly Telescope
  http://het.as.utexas.edu/HET/hetweb/Instruments/HRS/hrs_thar.html

E.2   Line Catalogue for the Glow Starter RELCO SC480

A visible spectrum of the RELCO SC480 low-cost glow starter[1] has been obtained
with an off-the-shelf SQUES echelle spectrograph (Huwiler 2014) at $R \sim 20,000$.
The wavelength range of 3,888–8,136 Å is distributed over 30 echelle orders. 2D-
spectra are shown in Figs. E.1, E.2, and E.3. All usable emission lines are indicated
in Figs. E.4, E.5, E.6, E.7, E.8, and E.9 (some lines occur twice due to overlapping
orders. Their wavelengths are labeled in different colours). The corresponding
catalogue is shown in Tables E.1, E.2, E.3, E.4 and E.5. In order 29 (7,600–7,900
Å), only two usable emission lines are generated by noble gases. However, they are
supplemented by a strong oxygen triplet. Line pairs which appear scarcely resolved
are referred to as a "blend".

---

[1] The off-the-shelf RELCO starter costs only some cents.

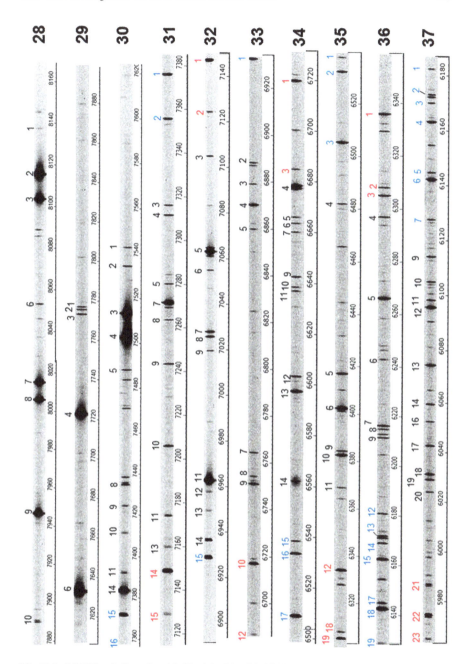

Fig. E.1 SQUES echelle orders 28–37 of the RELCO SC480 glowstarter

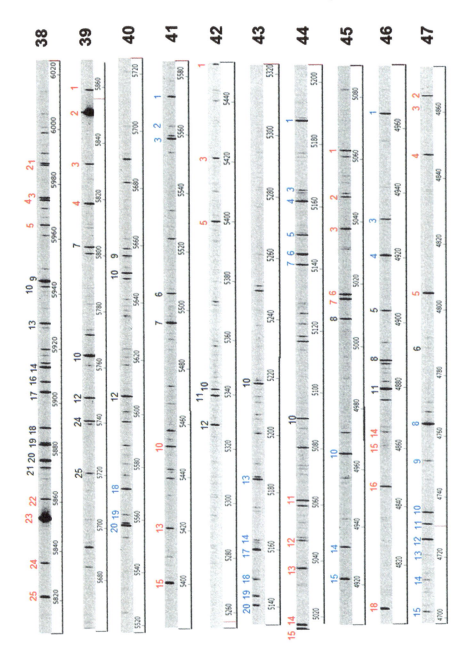

Fig. E.2 SQUES echelle orders 38–47 of the RELCO SC480 glowstarter

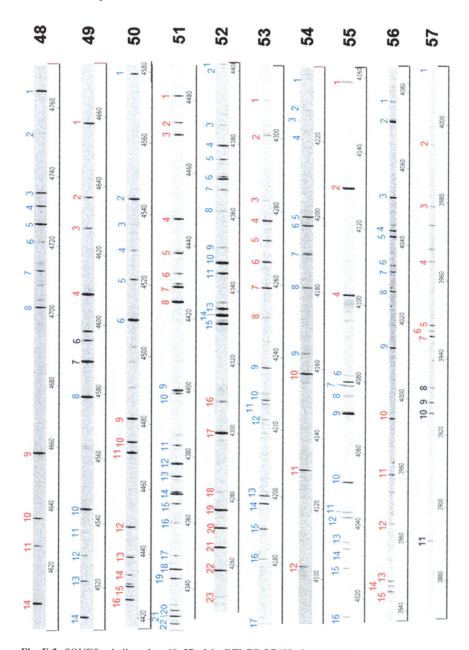

Fig. E.3 SQUES echelle orders 48–57 of the RELCO SC480 glowstarter

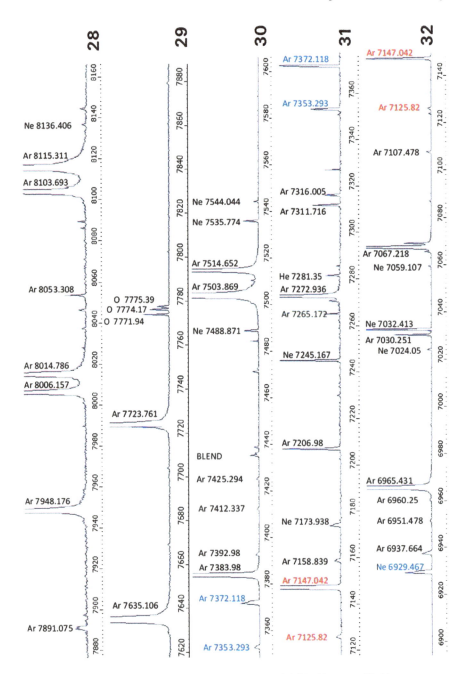

Fig. E.4 SQUES echelle spectrum of the glowstarter RELCO SC480 (orders 28–32)

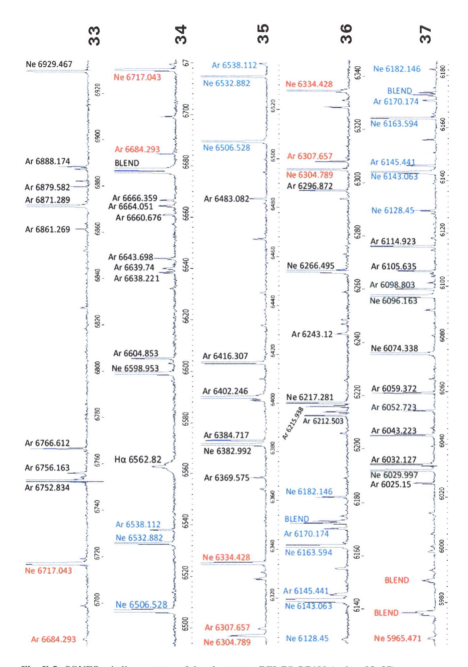

Fig. E.5 SQUES echelle spectrum of the glowstarter RELCO SC480 (orders 33–37)

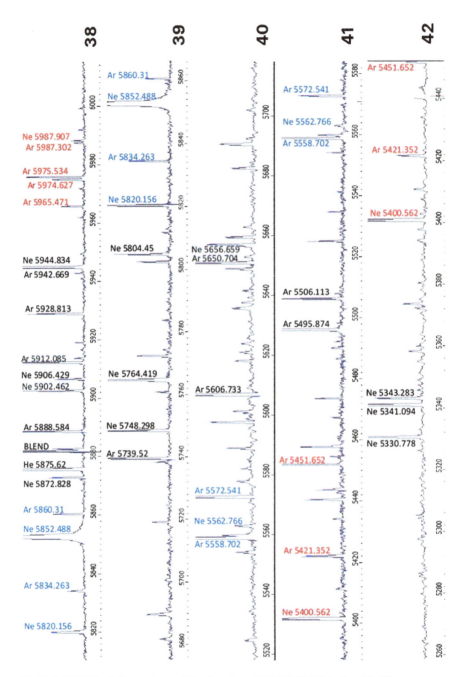

Fig. E.6 SQUES echelle spectrum of the glowstarter RELCO SC480 (orders 38–42)

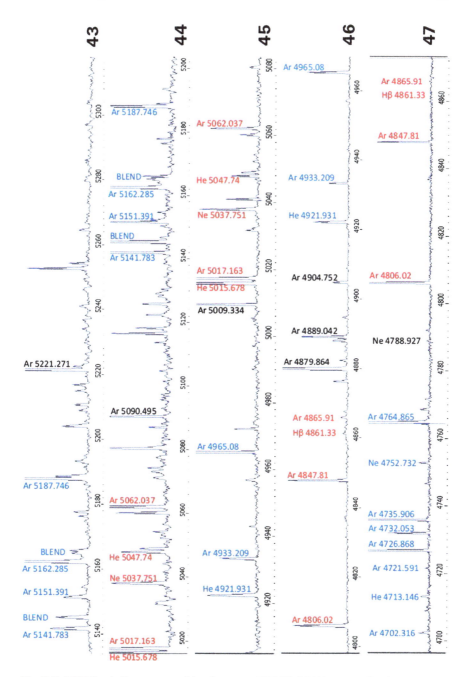

Fig. E.7 SQUES echelle spectrum of the glowstarter RELCO SC480 (orders 43–47)

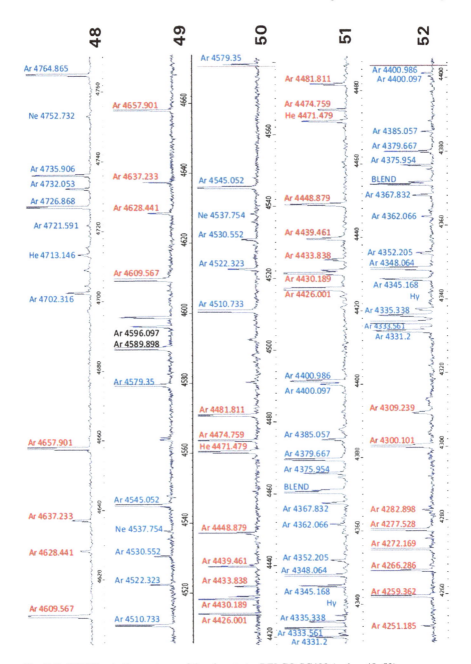

Fig. E.8 SQUES echelle spectrum of the glowstarter RELCO SC480 (orders 48–52)

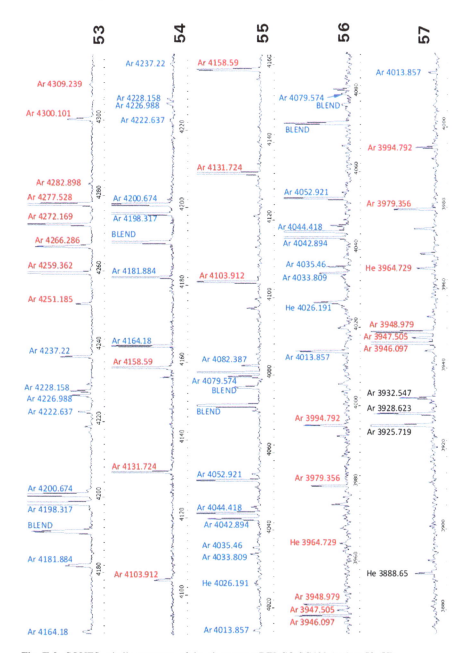

Fig. E.9 SQUES echelle spectrum of the glowstarter RELCO SC480 (orders 53–57)

Table E.1 RELCO SC480 glowstarter line catalogue 8,136.406–6,483.082 Å (SQUES echelle spectrograph)

Order	Line	Wavelength (Å)	Element	Order	Line	Wavelength (Å)	Element
28	1	8136.406	Ne	32	3	7107.478	Ar
28	2	8115.311	Ar	32	5	7067.218	Ar
28	3	8103.693	Ar	32	6	7059.107	Ne
28	6	8053.308	Ar	32	7	7032.413	Ne
28	7	8014.786	Ar	32	8	7030.251	Ar
28	8	8006.157	Ar	32	9	7024.05	Ne
28	9	7948.176	Ar	32	11	6965.431	Ar
28	10	7891.075	Ar	32	12	6960.25	Ar
29	1	7775.39	O	32	13	6951.478	Ar
29	2	7774.17	O	32	14	6937.664	Ar
29	3	7771.94	O	32	15	6929.467	Ne
29	4	7723.761	Ar	33	1	6929.467	Ne
29	6	7635.106	Ar	33	2	6888.174	Ar
30	1	7544.044	Ne	33	3	6879.582	Ar
30	2	7535.774	Ne	33	4	6871.289	Ar
30	3	7514.652	Ar	33	5	6861.269	Ar
30	4	7503.869	Ar	33	7	6766.612	Ar
30	5	7488.871	Ne	33	8	6756.163	Ar
30	8	Blend	Ar	33	9	6752.834	Ar
30	9	7425.294	Ar	33	10	6717.043	Ne
30	10	7412.337	Ar	33	12	6684.293	Ar
30	11	7392.98	Ar	34	1	6717.043	Ne
30	14	7383.98	Ar	34	3	6684.293	Ar
30	15	7372.118	Ar	34	4	Blend	Ar/Ne
30	16	7353.293	Ar	34	5	6666.359	Ar
31	1	7372.118	Ar	34	6	6664.051	Ar
31	2	7353.293	Ar	34	7	6660.676	Ar
31	3	7316.005	Ar	34	9	6643.698	Ar
31	4	7311.716	Ar	34	10	6639.74	Ar
31	5	7281.35	He	34	11	6638.221	Ar
31	7	7272.936	Ar	34	12	6604.853	Ar
31	8	7265.172	Ar	34	13	6598.953	Ne
31	9	7245.167	Ne	34	14	6562.82	H?
31	10	7206.98	Ar	34	15	6538.112	Ar
31	11	7173.938	Ne	34	16	6532.882	Ne
31	13	7158.839	Ar	34	17	6506.528	Ne
31	14	7147.042	Ar	35	1	6538.112	Ar
31	15	7125.82	Ar	35	2	6532.882	Ne
32	1	7147.042	Ar	35	3	6506.528	Ne
32	2	7125.82	Ar	35	4	6483.082	Ar

Table E.2 RELCO SC480 glowstarter line catalogue 6,416.307–5,562.766 Å (SQUES echelle spectrograph)

Order	Line	Wavelength (Å)	Element	Order	Line	Wavelength (Å)	Element
35	5	6416.307	Ar	37	19	6029.997	Ne
35	6	6402.246	Ar	37	20	6025.15	Ar
35	9	6384.717	Ar	37	21	Blend	Ne/Ar
35	10	6382.992	Ne	37	22	Blend	Ne
35	11	6369.575	Ar	37	23	5965.471	Ne
35	12	6334.428	Ne	38	1	5987.907	Ne
35	18	6307.657	Ar	38	2	5987.302	Ar
35	19	6304.789	Ne	38	3	5975.534	Ne
36	1	6334.428	Ne	38	4	5974.627	Ne
36	2	6307.657	Ar	38	5	5965.471	Ne
36	3	6304.789	Ne	38	9	5944.834	Ne
36	4	6296.872	Ar	38	10	5942.669	Ar
36	5	6266.495	Ne	38	13	5928.813	Ar
36	6	6243.12	Ar	38	14	5912.085	Ar
36	7	6217.281	Ne	38	16	5906.429	Ne
36	8	6215.938	Ar	38	17	5902.462	Ne
36	9	6212.503	Ar	38	18	5888.584	Ar
36	12	6182.146	Ne	38	19	Blend	Ne/Ar
36	13	Blend	Ar	38	20	5875.62	He
36	14	6170.174	Ar	38	21	5872.828	He
36	15	6163.594	Ne	38	22	5860.31	Ar
36	17	6145.441	Ar	38	23	5852.488	Ne
36	18	6143.063	Ne	38	24	5834.263	Ar
36	19	6128.45	Ne	38	25	5820.156	Ne
37	1	6182.146	Ne	39	1	5860.31	Ar
37	2	Blend	Ar	39	2	5852.488	Ne
37	3	6170.174	Ar	39	3	5834.263	Ar
37	4	6163.594	Ne	39	4	5820.156	Ne
37	5	6145.441	Ar	39	7	5804.45	Ne
37	6	6143.063	Ne	39	10	5764.419	Ne
37	7	6128.45	Ne	39	12	5748.298	Ne
37	9	6114.923	Ar	39	24	5739.52	Ar
37	10	6105.635	Ar	40	9	5656.659	Ne
37	11	6098.803	Ar	40	10	5650.704	Ar
37	12	6096.163	Ne	40	12	5606.733	Ar
37	13	6074.338	Ne	40	18	5572.541	Ar
37	14	6059.372	Ar	40	19	5562.766	Ne
37	16	6052.723	Ar	40	20	5558.702	Ar
37	17	6043.223	Ar	41	1	5572.541	Ar
37	18	6032.127	Ar	41	2	5562.766	Ne

Table E.3 RELCO SC480 glow-starter line catalogue 5,558.702–4,596.097 Å (SQUES echelle spectrograph)

Order	Line	Wavelength (Å)	Element	Order	Line	Wavelength (Å)	Element
41	3	5558.702	Ar	46	1	4965.08	Ar
41	6	5506.113	Ar	46	3	4933.209	Ar
41	7	5495.874	Ar	46	4	4921.931	He
41	10	5451.652	Ar	46	5	4904.752	Ar
41	13	5421.352	Ar	46	8	4889.042	Ar
41	15	5400.562	Ne	46	11	4879.864	Ar
42	1	5451.652	Ar	46	14	4865.91	Ar
42	3	5421.352	Ar	46	15	4861.33	H?
42	5	5400.562	Ne	46	16	4847.81	Ar
42	10	5343.283	Ne	46	18	4806.02	Ar
42	11	5341.094	Ne	47	2	4865.91	Ar
42	12	5330.778	Ne	47	3	4861.33	H?
43	10	5221.271	Ar	47	4	4847.81	Ar
43	13	5187.746	Ar	47	5	4806.02	Ar
43	14	Blend	Ar	47	6	4788.927	Ne
43	17	5162.285	Ar	47	8	4764.865	Ar
43	18	5151.391	Ar	47	9	4752.732	Ne
43	19	Blend	Ne/Ar	47	10	4735.906	Ar
43	20	5141.783	Ar	47	11	4732.053	Ar
44	1	5187.746	Ar	47	12	4726.868	Ar
44	3	5165.773	Ar	47	13	4721.591	Ar
44	4	5162.285	Ar	47	14	4713.146	He
44	5	5151.391	Ar	47	15	4702.316	Ar
44	6	Blend	Ne/Ar	48	1	4764.865	Ar
44	7	5141.783	Ar	48	2	4752.732	Ne
44	10	5090.495	Ar	48	3	4735.906	Ar
44	11	5062.037	Ar	48	4	4732.053	Ar
44	12	5047.74	He	48	5	4726.868	Ar
44	13	5037.751	Ne	48	6	4721.591	Ar
44	14	5017.163	Ar	48	7	4713.146	He
44	15	5015.678	He	48	8	4702.316	Ar
45	1	5062.037	Ar	48	9	4657.901	Ar
45	2	5047.74	He	48	10	4637.233	Ar
45	3	5037.751	Ne	48	11	4628.441	Ar
45	6	5017.163	Ar	48	14	4609.567	Ar
45	7	5015.678	He	49	1	4657.901	Ar
45	8	5009.334	Ar	49	2	4637.233	Ar
45	10	4965.08	Ar	49	3	4628.441	Ar
45	14	4933.209	Ar	49	4	4609.567	Ar
45	15	4921.931	He	49	6	4596.097	Ar

Table E.4 RELCO SC480 glowstarter line catalogue 4,589.898–4,198.317 Å (SQUES echelle spectrograph)

Order	Line	Wavelength (Å)	Element	Order	Line	Wavelength (Å)	Element
49	7	4589.898	Ar	51	20	4335.338	Ar
49	8	4579.35	Ar	51	21	4333.561	Ar
49	10	4545.052	Ar	51	22	4331.2	Ar
49	11	4537.754	Ne	52	1	4400.986	Ar
49	12	4530.552	Ar	52	2	4400.097	Ar
49	13	4522.323	Ar	52	3	4385.057	Ar
49	14	4510.733	Ar	52	4	4379.667	Ar
50	1	4579.35	Ar	52	5	4375.954	Ar
50	2	4545.052	Ar	52	6	Blend	Ar
50	3	4537.754	Ne	52	7	4367.832	Ar
50	4	4530.552	Ar	52	8	4362.066	Ar
50	5	4522.323	Ar	52	9	4352.205	Ar
50	6	4510.733	Ar	52	10	4348.064	Ar
50	9	4481.811	Ar	52	11	4345.168	Ar
50	10	4474.759	Ar	52	13	4335.338	Ar
50	11	4471.479	He	52	14	4333.561	Ar
50	12	4448.879	Ar	52	15	4331.2	Ar
50	13	4439.461	Ar	52	16	4309.239	Ar
50	14	4433.838	Ar	52	17	4300.101	Ar
50	15	4430.189	Ar	52	18	4282.898	Ar
50	16	4426.001	Ar	52	19	4277.528	Ar
51	1	4481.811	Ar	52	20	4272.169	Ar
51	2	4474.759	Ar	52	21	4266.286	Ar
51	3	4471.68	He	52	22	4259.362	Ar
51	4	4448.879	Ar	52	23	4251.185	Ar
51	5	4439.461	Ar	53	1	4309.239	Ar
51	6	4433.838	Ar	53	2	4300.101	Ar
51	7	4430.189	Ar	53	3	4282.898	Ar
51	8	4426.001	Ar	53	4	4277.528	Ar
51	9	4400.986	Ar	53	5	4272.169	Ar
51	10	4400.097	Ar	53	6	4266.286	Ar
51	11	4385.057	Ar	53	7	4259.362	Ar
51	12	4379.667	Ar	53	8	4251.185	Ar
51	13	4375.954	Ar	53	9	4237.22	Ar
51	14	Blend	Ar	53	10	4228.158	Ar
51	15	4367.832	Ar	53	11	4226.988	Ar
51	16	4362.066	Ar	53	12	4222.637	Ar
51	17	4352.205	Ar	53	13	4200.674	Ar
51	18	4348.064	Ar	53	14	4198.317	Ar
51	19	4345.168	Ar	53	15	Blend	Ar/Hf

Table E.5 RELCO SC480 glowstarter line catalogue 4,181.884–3,888.65 Å (SQUES echelle spectrograph)

Order	Line	Wavelength (Å)	Element	Order	Line	Wavelength (Å)	Element
53	16	4181.884	Ar	55	16	4013.857	Ar
53	17	4164.18	Ar	56	1	Blend	Ar/?
54	1	4237.22	Ar	56	2	Blend	Ar/Ar
54	2	4228.158	Ar	56	3	4052.921	Ar
54	3	4226.988	Ar	56	4	4044.418	Ar
54	4	4222.637	Ar	56	5	4042.894	Ar
54	5	4200.674	Ar	56	6	4035.46	Ar
54	6	4198.317	Ar	56	7	4033.809	Ar
54	7	Blend	Ar/Hf	56	8	4026.36	He
54	8	4181.884	Ar	56	9	4013.857	Ar
54	9	4164.18	Ar	56	10	3994.792	Ar
54	10	4158.59	Ar	56	11	3979.356	Ar
54	11	4131.724	Ar	56	12	3964.729	He
54	12	4103.912	Ar	56	13	3948.979	Ar
55	1	4158.59	Ar	56	14	3947.505	Ar
55	2	4131.724	Ar	56	15	3946.097	Ar
55	4	4103.912	Ar	57	1	4013.857	Ar
55	6	4082.387	Ar	57	2	3994.792	Ar
55	7	4079.574	Ar	57	3	3979.356	Ar
55	8	Blend	Ar/?	57	4	3964.729	He
55	9	Blend	Ar/Ar	57	5	3948.979	Ar
55	10	4052.921	Ar	57	6	3947.505	Ar
55	11	4044.418	Ar	57	7	3946.097	Ar
55	12	4042.894	Ar	57	8	3932.547	Ar
55	13	4035.46	Ar	57	9	3928.623	Ar
55	14	4033.809	Ar	57	10	3925.719	Ar
55	15	4026.36	He	57	11	3888.65	He

Appendix F
Manufacturers and Distributors

F.1 Spectrographs

- ASTRO SPECTROSCOPY INSTRUMENTS
  Erich-Weinert-Str. 19, 14478 Potsdam, Germany
  http://en.astro-spec.com
- BAADER PLANETARIUM
  Zur Sternwarte, 82291 Mammendorf, Germany
  http://www.baader-planetarium.de
- OPTOMECHANICS RESEARCH, INC.
  P.O. Box 87, Vail, AZ 85641, USA
  http://www.echellespectrographs.com
- PATON HAWKSLEY EDUCATION LTD
  59, Wellsway, Keynsham, Bristol BS311PG, GB
  http://www.patonhawksley.co.uk
- PRINCETON INSTRUMENTS
  3660 Quakerbridge Road, Trenton, NJ 08619 USA
  http://www.princetoninstruments.com
- SANTA BARBARA INSTRUMENT GROUP (SBIG)
  5880 West Las Positas Blvd, Pleasanton, CA 94588, USA
  http://www.sbig.com
- SHELYAK INSTRUMENTS
  73, rue de Chartreuse, 38420 Le Versoud - France
  http://www.shelyak.com

© Springer-Verlag Berlin Heidelberg 2015
T. Eversberg, K. Vollmann, *Spectroscopic Instrumentation*, Springer Praxis Books,
DOI 10.1007/978-3-662-44535-8

F.2　Fiber Optics

- ASTRO SPECTROSCOPY INSTRUMENTS
  Erich-Weinert-Str. 19, 14478 Potsdam, Germany
  http://en.astro-spec.com
- CERAMOPTEC Industries Inc.
  515A Shaker Rd. East Longmeadow, MA 01028, USA
  http://www.ceramoptec.de
- FOP GmbH
  74564 Crailsheim, Roßfelderstrasse 36, Germany
  http://www.fop.de
- INSOFT FIBER SHOP
  Schwarzenberger Str. 41a 51647 Gummersbach, Germany
  http://www.fiber-shop.de
- LWL SACHSENKABEL GMBH
  Auerbacher Str. 24, 09390 Gornsdorf, Germany
  http://www.sachsenkabel.de
- POLYMICRO TECHNOLOGIES
  18019 N. 25th Avenue, Phoenix, AZ 85023-1200, USA
  http://www.polymicro.com

F.3　Optical Elements: Laboratory Material

- ASTRO SPECTROSCOPY INSTRUMENTS
  Erich-Weinert-Str. 19, 14478 Potsdam, Germany
  http://en.astro-spec.com/
- BERNHARD HALLE Nachfl. GmbH
  Hubertusstraße 10, 12163 Berlin, Germany
  http://www.b-halle.de
- CVI MELLES GRIOT
  200 Dorado Place SE, Albuquerque, NM 87123, USA
  http://www.cvimellesgriot.com
- EDMUND OPTICS Inc.
  101 East Gloucester Pike, Barrington, NJ 08007-1380 USA
  http://www.edmundoptics.com
- GRATINGWORKS Co.
  42 Quarry Rd, Acton, MA 01720 USA
  http://www.gratingworks.com
- HORIBA SCIENTIFIC
  2, Miyanohigashi-cho, Kisshoin, Minami-ku, Kyoto 601-8510, Japan
  http://www.horiba.com

- LASER COMPONENTS GmbH
  Werner-von-Siemens-Str. 15, 82140 Olching, Germany
  http://www.lasercomponents.com
- LOT ORIEL GmbH & Co. KG
  Im Tiefen See 58, 64293 Darmstadt, Germany
  http://www.lot-oriel.com
- MACONY DESIGN
  Gladiolenstr. 16 16348 Wandlitz, Germany
  http://www.macony-design.de
- NEWPORT CORPORATION
  1791 Deere Avenue, Irvine, CA 92606, USA
  http://www.newport.com
- OPTOMETRICS
  8 Nemco Way, Ayer, MA, 01432, USA
  http://www.optometrics.com
- QIOPTIQ Photonics GmbH & Co. KG
  Königsallee 23, 37081 Göttingen, Germany
  http://www.qioptiq.com
- SCHOTT AG
  Hattenbergstrasse 10, 55122 Mainz, Germany
  http://www.schott.com
- SHIMADZU, http://www.shimadzu.com/
- SILL OPTICS GmbH & Co. KG
  Johann-Höllfritsch-Str. 13, 90530 Wendelstein, Germany
  http://www.silloptics.de
- SPECTRUM SCIENTIFIC, Inc.
  16692 Hale Avenue, STE A Irvine, CA 92606 USA
  http://ssioptics.com/
- THORLABS
  56 Sparta Avenue Newton, NJ 07860, USA
  http://www.thorlabs.com

Suggested Reading

- *Chrisphin Karthick, M., Astronomer's Data Reduction Guide: Image processing through IRAF commands, LAP LAMBERT Academic Publishing, 2012*

A Short Story

Klaus and Thomas regularly discuss questions about spectroscopy and observations of massive stars with other colleagues. After some time, several questions about fundamental optics repeated themselves, especially when novice spectroscopists wanted to learn something. The same questions popped up again and again and fingers were chopped sore for answers via email and in discussion forums. This could not go on! One day Thomas said "Actually, we should write some explanatory texts or even a book on the subject, then we can refer to it". Klaus was not thrilled, thinking of all the work behind it—'No way!"—and he refused. Thomas then said: "Ok, then I write it alone . . . "

Bibliography

Abbott, D. C., & Lucy, L. B. (1985). *The Astrophysical Journal, 288*, 679.

Aldoretta, E., et al. (in preparation).

Avila, G., Burwitz, V., Guirao, C., Rodriguez, J., Shida, S., & Baade, D. (2007). ESO-Messenger, No. 129 (p. 62).

Avila, G., Burwitz, V., Guirao, C., & Rodriguez, J. (2007). http://www.eso.org/projects/caos/.

Avila, G., & Guirao, C. (2009). *Tapered optical fibres*. http://spectroscopy.wordpress.com.

Avila, G., Singh, P., & Chazelas, B. (2010). In *SPIE conference proceedings. Ground based and airborne instrumentation for astronomy III* (Vol. 7735, p. 1).

Barker, P. K., Landstreet, J. D., Marlborough, J. M., Thompson, I., & Maza, J. (1981). *The Astrophysical Journal, 250*, 300.

Barker, P. K., Landstreet, J. D., Marlborough, J. M., & Thompson, I. (1985). *The Astrophysical Journal, 288*, 741.

Bartzakos, P., Moffat, A. F. J., & Niemela, V. S. (2001). *Monthly Notices of the Royal Astronomical Society, 324*, 33.

Beals, C. S. (1929). *Monthly Notices of the Royal Astronomical Society, 90*, 202.

Beckert, E., et al. (2008). In *SPIE conference proceedings. Advanced optical and mechanical technologies in telescopes and instrumentation* (Vol. 7018, p. 82).

Behr, A. (1959). *Nachr. Akad. Wissenschaften Göttingen 2, Mathem. Physik, K1*, 7185.

Berek, M. (1970). *Grundlagen der praktischen Optik*. Walter de Gruyter & Co.

Bjorkman, K. S. (1992). In L. Drissen, C. Leitherer, & A. Nota (Eds.), *Nonisotropic and variable outflow from stars. ASP Conference Series* (Vol. 22, p. 71).

Bjorkmann, J. E., & Cassinelli, J. P. (1993). *The Astrophysical Journal, 409*, 429.

Born, M., & Wolf, E. (1959). *Principles of optics*. Pergamon Press

Bouchy, F., Díaz, R. F., Hébrard, G., Arnold, L., Boisse, I., Delfosse, X., et al. (submitted). *Astronomy & Astrophysics*

Bowen, I. S. (1938) *The Astrophysical Journal, 88*, 113.

Buil, C. (2003). http://www.astrosurf.com/buil/us/compute/simspec.xls.

Buil, C. (2006). http://www.astrosurf.com/buil/bestars/m45/img.htm.

Buil, C. (2010). http://www.astrosurf.com/buil/star/epsaur/epsaur.htm.

Bures, J. (2008). *Guided optics: Optical fibers and all-fiber components*. Wiley-VCH. ISBN-10: 3527407960.

Byard, P. L., & O'Brien, T. P. (2000). In M. Iye & A. F. Moorwood (Eds.), *SPIE conference proceedings. Optical and IR telescope instrumentation and detectors* (Vol. 4008, p. 934).

Cassinelli, C. P., & Haisch, B. M. (1974). *The Astrophysical Journal, 188*, 101.

© Springer-Verlag Berlin Heidelberg 2015
T. Eversberg, K. Vollmann, *Spectroscopic Instrumentation*, Springer Praxis Books,
DOI 10.1007/978-3-662-44535-8

Castor, J. I., & Lamers, H. J. G. L. M. (1979). *The Astrophysical Journal Supplement Series, 39*, 481.

Cecil, G., Bland-Hawthorn, J., Veilleux, S., & Filippenko, A. V. (2001). *The Astrophysical Journal, 555*, 338.

Chandrasekhar, S. (1934). *Monthly Notices of the Royal Astronomical Society, 94*, 522.

Chandrasekhar, S. (1946). *The Astrophysical Journal, 103*, 351.

Chandrasekhar, S. (1947). *The Astrophysical Journal, 105*, 424.

Chandrasekhar, S. (1950). *The Astronomical Journal, 55*, 209.

Chini, R., Hoffmeister, V., Kimeswenger, S., Nielbock, M., Nürnberger, D., Schmidtobreick, L., et al. (2004). *Nature, 429*(6988), 155.

Chlebowski, T., Harnder, F. R., & Sciortino, S. (1989). *The Astrophysical Journal, 341*, 427.

Clarke, D., & McLean, I. S. (1974). *Monthly Notices of the Royal Astronomical Society, 167*, 27P.

Conti, P. S. (1976). *Societé Royale des Sciences de Liége*. Mémoires, 9 (p. 193); Discussion (p. 213).

Corcoran, M. F., Hamaguchi, K., Pollock, A. M. T., Russell, C. M. P., Moffat, A. F. J., Owocki, S., et al. (2012). *American Astronomical Society Meeting, 219*, 249.

Cranmer, S. R., & Owocki, S. P. (1996). *The Astrophysical Journal, 462*, 469.

Crowther, P. A., Smith, L. J., Hillier, D. J., & Schmutz, W. (1995). *Astronomy and Astrophysics, 293*, 427.

Crowther, P. A. (2007). *Annual Review of Astronomy & Astrophysics, 45*(1), 177.

Dougherty, S. M., Beasley, A. J., Claussen, M. J., Zauderer, B. A., & Bolingbroke, N. J. (2005). *The Astrophysical Journal, 623*, 447.

Drissen, L., St.-Louis, N., Moffat, A.F.J., & Bastien, P. (1987). *The Astrophysical Journal, 322*, 888.

Drissen, L., Robert, C., & Moffat, A. F. J. (1992). *The Astrophysical Journal, 386*, 288.

Ebert, H. (1889). *Wied. Ann., 38*, 489.

Erickson, F. E., & Rabanus, D. (2000). *Applied Optics, 39*, 4486.

Eversberg, T., Lépine, S., & Moffat, A. F. J. (1998). *The Astrophysical Journal, 494*, 799.

Eversberg, T., Moffat, A. F. J., Debruyne, M., Rice, J. B., Piskunov, N., Bastien, P., et al. (1998). *Publications of the Astronomical Society of the Pacific, 110*, 1356.

Eversberg, T., Moffat, A. F. J., & Marchenko, S. V. (1999). *Publications of the Astronomical Society of the Pacific, 111*, 861.

Eversberg, T. (2011). *Bulletin de la Société Royale des Sciences de Liège, 80*, 469.

Fabricant, D. G., Hertz, E. H., & Szentgyorgyi, A. H. (1994). In D. L. Crawford & E. R. Craine (Eds.), *SPIE conference proceedings. Instrumentation in astronomy VIII* (Vol. 2198, p. 251).

Fahed, R., Moffat, A. F. J., Zorec, J., Eversberg, T., Chené, A. N., Alves, F., et al. (2011). *Monthly Notices of the Royal Astronomical Society, 418*, 2.

Fastie, W. G. (1952). *Journal of the Optical Society of America, 42*, 647.

Federspiel, M. (2002). *VdS-Journal, 8*, 71.

Feger, T. (2012). http://astrospectroscopy.wordpress.com.

Feynman, R. P. (1985). *QED - The strange theory of light and matter*. Princeton, NJ: Princeton University Press.

Gauss, C. F. (1841). *Dioptrische Untersuchungen*. Göttingen: Verlag der Dieterichschen Buchhandlung.

Gayley, K. G., & Owocki, S. P. (1995) *The Astrohysical Journal, 446*, 801.

Gayley, K. G., Owocki, S. P., & Cranmer, S. R. (1995). *The Astrohysical Journal, 442*, 296.

Gillet, D., et al. (1994). *Astronomy & Astrophysics Supplement Series, 108*, 181.

Gräfener, G., Vink, J. S., de Koter, A., & Langer, N. (2011) *Astronomy & Astrophysics, 535*, 56.

Gray, F. D. (1992). *The observation and analysis of stellar photospheres*. Cambridge Astrophysics Series 20.

Green, J. C., Froning, C. S., Osterman, S., Ebbets, D., Heap, S. H., Leitherer, C., et al. (2012). *The Astrohysical Journal, 744*, 60.

Grosdidier, Y., Moffat, A. F. J., Blais-Ouellette, S., Joncas, G., & Acker, A. (2001). *The Astrophysical Journal, 562*, 753.

Gross, H., Zügge, H., Peschka, M., & Blechinger, F. (2007). *Handbook of Optical Systems, Aberration Theory and Correction of Optical Systems*, Wiley-VCH, (Vol. 3).

Grupp, F. (2003). *Astronomy & Astrophysics, 412*, 897.

Guinouard, I., et al. (2006) In E. Atad-Ettedgui, J. Antebi, & D. Lemke (Eds.), *SPIE conference proceedings. Optomechanical technologies for astronomy* (Vol. 6273, p. 3).

Hall, J. S. (1949). *Science, 109*, 165.

Hamann, W.-R., Gräfener, G., & Liermann, A. (2006). *Astronomy & Astrophysics, 457*, 1015.

Hamann, W. R. (2011). Private communication.

Harries, T. J. (1995). *Spectropolarimetry as a probe of stellar winds*. Ph.D. thesis. University College London.

Harries, T. J., & Howarth, I. D. (1996). *Astronomy & Astrophysics, 310*, 553.

Hartwig, G., & Schopper, H. (1959). *Bulletin of the American Physical Society, 4*, 77.

Hayes, D. P. (1984). *Astronomical Journal, 89*, 1219.

Haynes, D. M, Withford, M. J., Dawes, J. M., Haynes, R., & Bland-Hawthorn, J. (2008). In E. Atad-Ettedgui & D. Lemke (Eds.), *SPIE conference proceedings. Advanced optical and mechanical technologies in telescopes and instrumentation* (Vol. 7018, p. 8).

Heathcote, B. (2013). Private communication.

Hecht, E., & Zajac, A. (2003). Optics (4th ed.).

Hill, G. M., Moffat, A. F. J., St-Louis, N., & Bartzakos, P. (2000) *Monthly Notices of the Royal Astronomical Society, 318*, 402.

Hillier, D. J., Kudritzki, R.-P., Pauldrach, A. W. A., Baade, D., Cassinelli, J. P., Puls, J., et al. (1993). *Astronomy & Astrophysics, 276*, 128.

Hiltner, W. A. (1949). *Science, 109*, 165.

Hutley, M. C. (1982). *Diffraction Gratings, Techniques of Physics* (Vol. 6). Academic Press.

Huwiler, M. www.eagleowloptics.com.

James, J. F. (2007). *Spectrograph design fundamentals*. New York: Cambridge University Press.

Jiang, Y. M., Walker, G. A. H., Dinshaw, N., & Matthews, J. M. (1993). In L. A. Balona, H. F. Henrichs, & J. M. Le Contel (Eds.), *Pulsation, rotation and mass loss in early-type stars*. Proc. IAU Symp. No. 163 (p. 232). Dordrecht: Kluwer.

Kaper, L., Henrichs, H. F., Fullerton, A. W., Ando, H., Bjorkman, K. S., Gies, D. R., et al. (1997). *The Astrohysical Journal, 327*, 281.

Kataza, H., Okamoto, Y., Takubo, S., Onaka, T., Sako, S., Nakamura, K., et al. (2000).In M. Iye & A. F. Moorwood (Eds.), *SPIE conference proceedings. Optical and IR telescope instrumentation and detectors* (Vol. 4008, p. 1144).

Kaufer, A., & Pasquini, L. (1998). In S. D'Odorico (Ed.), *SPIE conference proceedings. Optical astronomical instrumentation* (Vol. 3355, p. 844).

Kaufer, A. (1998). *ASP Conference Series* (Vol. 152, p, 337).

Kay, L., & Shepherd, R. (1982). *Journal of Physics E: Scientific Instruments, 16*.

Kayser, H. (1900). *Handbuch der Spectroscope, 1*.

Kenworthy, M. A., Parry, I. R., & Taylor, K. (2001). *Publications of the Astronomical Society of the Pacific, 113*, 215.

Keyes, C. D., Long, K., & Hunter, D. (2006). STScI JWST Configuration Management - JWST-STScI-000851, SM-12.

Kodak Application Note MTD/PS-0233. (2003). *CCD image sensor noise sources*. Eastman Kodak Company

Kotelnikov, V. A. (1933, January 14). *O propusknoj sposobnosti 'efira' i provoloki v elektrosvjazi, (On the transmission capacity of 'ether' and wire in electro-communication). Proceedings of the first all-union conference on the technological reconstruction of the communications sector and the development of low-current engineering*. Moscow.

Krane, K. S. (1988). *Introductory nuclear physics* (p. 537). New York: John Wiley & Sons.

Kroupa, P. (2000, March 20–24). *Astronomische Gesellschaft meeting abstracts*. Heidelberg. Talk no. 11.

Lamers, H. J. G. L. M., Zickgraf, F.-J., de Winter, D., Houziaux, L., & Zorec, J. (1998). *Astronomy & Astrophysics, 340*, 117.

Lamers, H. J. G. L. M., & Cassinelli, J. P. (1999). *Introduction to stellar winds*. Cambridge University Press
Landstreet, J. D. (1979). *The Astronomical Journal, 85*, 611.
Langer, N., Hamann, W.-R., Lennon, M., Najarro, F. Pauldrach, A. W. A., & Puls, J. (1994). *Astronomy and Astrophysics, 290*, 819.
Leadbeater, R. (2010). http://www.threehillsobservatory.co.uk.
Leadbeater, R. (2011). In T. Eversberg & J. H. Knapen (Eds.), *Stellar winds in interaction*, arXiv:1101.1435.
Leadbeater, R., Buil, C., Garrel, T., Gorodenski, S., Hansen, T., Schanne, L. et al. (2012). arXiv:1206.6754.
Lépine, S., Eversberg, T., & Moffat, A. F. J. (1999). *The Astronomical Journal, 117*, 1441.
Lépine, S., & Moffat, A. F. J. (1999). *The Astronomical Journal, 514*, 909.
Lépine, S., & Moffat, A. F. J. (2008). *The Astronomical Journal, 136*, 548.
Lesh, J. R. (1968). *The Astrophysical Journal Supplement Series, 17*, 371.
Lo Curto, G. (2012). ESO Messenger No. 149 (p. 2).
Lührs, S. (1997). *Publications of the Astronomical Society of the Pacific, 109*, 504.
Lupie, O. L., & Nordsieck, K. H. (1987). *Astronomical Journal, 92*, 214.
Maeder, A. (1999). *Astronomy & Astrophysics, 347*, 185.
Marchenko, S. V., et al. (2001). *The Astrophysical Journal, 596*, 1295.
Marchenko, S. V., Moffat, A. F. J., Vacca, W. D., Côté, S., & Doyon, R. (2002). *The Astrophysical Journal, 565*, 59.
Marchenko, S. V., Moffat, A. F. J., Ballereau, D., Chauville, J., Zorec, J., Hill, G. M., et al. (2003). *The Astrophysical Journal, 596*, 1295.
Massey, P., & Hanson, M. M. (2010). arXiv:1010.5270.
Mauclaire, B., Buil, C., Garrel, T., & Lopez, A. (2013). arXiv:1207.0795.
McLean, I. S. (1978). *Monthly Notices of the Royal Astronomical Society*, 186.
McLean, I. S., Coyne, G.V., Frecker, J.E., Serkowski, K., 1979, *The Astrophysical Journal, 228*, 802.
McLean, I. S., Coyne, G. V., Frecker, J. E., Serkowski, K. (1979). *The Astrophysical Journal, 231*, L141.
McComas, D. J., Bame, S. J., Barraclough, B. L., Feldman, W. C., Funsten, H. O., Gosling, J. T., et al. (1998). *Geophysical Research Letters, 25*(1), 1.
Menn, N. (2004). *Practical optics*. Academic Press.
Meynet, G., Georgy, C., Hirschi, R., Maeder, A., Massey, P., Przybilla, N., et al. (2011). In G. Rauw, M. De Becker, Y. Nazé, J.-M. Vreux, & P. Williams (Eds.), *Proceedings of the 39th Liège Astrophysical Colloquium* (Vol. 80, p. 266).
Mihalas, D. (1970). *Stellar atmospheres*. San Francisco: W. H. Freeman and Co.
Miroshnichenko, A.S., Pasechnik, A.V., Manset, N., Carciofi, A.C., Rivinius, Th., Stefl, S., et al. (2013). *The Astrophysical Journal, 766*, 119.
Moffat, A. F. J. (1969). *Astronomy & Astrophysics, 3*, 455.
Moffat, A. F. J., & Robert, C. (1994). *The Astrophysical Journal, 421*, 310.
Morel, T., Alves, F., Bergmann, T., Carrera, L. F. G., Dias, F. M., Eversberg, T., et al. (submitted). *Astronomy and Astrophysics*.
Morel, T., Rauw, G., Eversberg, T., Alves, F., Arnold, W., Bergmann, T., et al. (2011). G. Rauw, M. De Becker, Y. Nazé, J.-M. Vreux, & P. Williams (Eds.), *Proceedings of the 39th Liège Astrophysical Colloquium* (Vol. 80. p. 170).
Morel, T., St.-Louis, N., & Marchenko, S. (sumbitted). *The Astrophysical Journal*.
Moore, B. D., Hester, J. J., & Scowen, P. A. (2000). *The Astronomical Journal, 119*, 2991.
Moore, B. D., Walter, D. K., Hester, J. J., Scowen, P. A., Dufour, R. J., & Buckalew, B. A. (2002). *The Astronomical Journal, 124*, 3313.
Motz, H., Thon, W., & Whitehurst, R. N. (1953). *Journal for Applied Physics, 24*, 826.
Naze, Y., Rauw, G., Herve, A., & Oskinova, L. (2011, June 27–30). In *The X-ray universe 2011*, Conference held in Berlin, Germany. Article id.111.

Newberry, M. (1998–2000). *Measuring the gain of a CCD camera* (Axiom Tech Note 1). Axiom Research Inc.

Nemravová, J., Harmanec, P., Koubský, P., Miroshnichenko, A., Yang, S., Šlechta, M., et al. (2012). *Astronomy & Astrophysics, 537*, 11.

Noguchi, K., Aoki, W., Kawanomoto, S., Ando, H., Honda, S., Izumiura, H., et al. (2002). *Publications of the Astronomical Society of Japan, 54*, 855.

Nordsieck, K. H., Babler, B., Bjorkman, K. S., Meade, M. R., Schulte-Ladbeck, R. E., Taylor, M. J. (1992). In L. Drissen, C. Leitherer, & A. Nota (Eds.), *Nonisotropic and variable outfow from stars*. ASP Conference Series (Vol. 22, p. 114).

Nyquist, H. (1928). *Transactions of the American Institute of Electrical Engineers, 47*, 617.

Okazaki, A. T. (1991). *Publications of the Astronomical Society of Japan, 43*, 75.

O'Shea, D. C. (1985). *Elements of modern optical design*. Wiley Series in Pure and Applied Optics.

Osmer, P. S., et al. (2000). In M. Iye & A. F. Moorwood (Eds.), *SPIE conference proceedings. Optical and IR telescope instrumentation and detectors* (Vol. 4008, p. 40).

Owocki, S. P. (1996). *Quebec – Delaware annual meeting on massive stars*. Private communication

Owocki, S. P. (1998). *Turbulence in line-driven stellar winds*. Workshop on 'Interstellar Turbulence' held January 1998 in Puebla, Mexico.

Owocki, S. P. (2004). In M. Heydari-Malayeri, Ph. Stee, & J.-P. Zahn (Eds.), *Evolution of Massive Stars*. EAS Publications Series (Vol. 13, p. 163).

Owocki, S. P., Gayley, K. G., & Shavic, N. J. (2004). *The Astrophysical Journal, 616*, 525.

Owocki, S. P. (2013). Private communication.

Picht, J. (1955). *Grundlagen der geometrisch-optischen Abbildung*. Berlin: VEB Deutscher Verlag der Wissenschaften.

Pierce, A. K. (1965). *Publications of the Astronomical Society of the Pacific, 77*, 216.

Poeckert, R., Marlborough, J. M. (1977). *The Astrophysical Journal, 218*, 220.

Poeckert, R., & Marlborough, J. M. (1978). *The Astrophysical Journal, 220*, 940.

Pogge, R. W., et al. (2006). In I. S. McLean, & Iye, M. (Eds.), *SPIE conference proceedings. Ground-based and airborne instrumentation for astronomy* (Vol. 6269, p. 62690I).

Pogge, R. W., et al. (2010). In I. S. McLean, S. K. Ramsay, & Takami, H. (Eds.), *SPIE conference proceedings. Ground-based and airborne instrumentation for astronomy III* (Vol. 7735, p. 77350A).

Puls, J., Kudritzki1, R.-P., Herrero, A., Pauldrach, A. W. A., Haser, S.M., Lennon, D. J., et al. (1996). *Astronomy & Astrophysics, 305*, 171.

Pyo, T.-S. (2003). *IRCS echelle spectrograph and data handling*. Subaru Telescope National Astronomical Observatory.

Prinja, R. K., & Howarth, I. D. (1986). *The Astrophysical Journal Supplement Series, 61*, 357.

Ramsey, L. W. (1988) In S. C. Barden (Ed.), *Fibers optics in astronomy*. ASP Conf. Ser. 3, Tucson (p. 26).

Rauw, G., Alves, F., Bergmann, T., Carrera, L. F. G., Dias, F. M., Eversberg, T., et al. (submitted). *Astronomy and Astrophysics*.

Reiner, J. (2002). *Grundlagen der Ophthalmologischen Optik*, Books on Demand.

Reipurth, B., Yu, K. C., Rodríguez, L. F., Heathcote, S., & Bally, J. (1999). *Astronomy and Astrophysics, 352*, L83.

Richardson, N., et al. (in preparation).

Robb, P. N. (1985). *Applied Optics, 24*, 1864.

Robert, C., & Moffat, A. F. J. (1989). *The Astrophysical Journal, 343*, 902.

Robert, C., Moffat, A. F. J., Bastien, P., Drissen, L., & St-Louis, N. (1989). *The Astrophysical Journal, 347*, 1034.

Robert, C., Moffat, A. F. J., Bastien, P., St-Louis, N., & Drissen, L. (1990). *The Astrophysical Journal, 359*, 211.

Robert, C., Moffat, A. F. J., Drissen, L., Lamontagne, R., Seggewiss, W., Niemela, V. S., et al. (1992). *The Astrophysical Journal, 397*, 277.

Robert, C. (1992). *Dissertation Abstracts International* (Vol. 55-03, Section B, p. 0953). PhD thesis, Université de Montréal (Canada).

Robinson, L. B. (1988). *Instrumentation for ground-based optical astronomy*. Springer.

Rosenhauer, K., & Rosenbruch, K. J. (1960). *Optik, 17*, 249–277.

Ruždjak, et al. (2009). *Astronomy & Astrophysics, 506*, 1319.

Sablowski, D. (2012). In *Spektrum* - Mitteilungsblatt der Fachgruppe Spektroskopie in der Vereinigung der Sternfreunde e.V., No. 43. http://spektroskopie.fg-vds.de/pdf/Spektrum43.pdf.

Sahnow, D. J., Friedman, S. D., Oegerle, W. R., Moos, H. W., Green, J. C., & Siegmund, O. H. (1996). In *SPIE conference proceedings. Space telescopes and instruments IV* (Vol. 2807, p. 2).

Sander, A., Hamann, W.-R., & Todt, H. (2012). *Astronomy & Astrophysics, 540*, A144.

Schaerer, D., Schmutz, W., & Grenon, M. (1997). *The Astrophysical Journal, 484*, L153.

Schroeder, D. J. (1967). *Publications of the Astronomical Society of the Pacific, 82*, 1253.

Schroeder, D. J. (1970). *Applied Optics, 6*(11).

Schroeder, D. J. (1987). *Astronomical Optics*, Academic Press.

Schulte-Ladbeck, R. E., Nordsieck, K. H., Nook, M. A., Magalhães, A. M., Taylor, M., Bjorkman, K. S., et al. (1990). *The Astrophysical Journal, 365*, L19.

Schulte-Ladbeck, R. E., Nordsieck, K. H., Taylor, M., Nook, M. A., Bjorkman, K. S., Magalhães, A. M., et al. (1991). *The Astrophysical Journal, 382*, 301.

Schulte-Ladbeck, R. E., Meade, M., & Hillier, D. J. (1992). In L. Drissen, C. Leitherer, & A. Nota (Eds.), *Nonisotropic and variable outflow from stars*. ASP Conference Series (Vol. 22, p. 118).

Schulte-Ladbeck, R. E., Nordsieck, K. H., Taylor, M., Bjorkman, K. S., Magalhães, M. A., & Wolff, M. J. (1992a). *The Astrophysical Journal, 387*, 347.

Schulte-Ladbeck, R. E., Nordsieck, K. H., Code, A. D., Anderson, C. M., Babler, B.L., Bjorkman, K. S., et al. (1992b). *The Astrophysical Journal, 391*, L37.

Schulte-Ladbeck, R. E., Eenens, P. R., & Davis, K. (1995). *The Astrophysical Journal, 454*, 917.

Schwarzschild, M. (1950). *The Astrophysical Journal, 112*, 222.

Seidel, L. (1856). *Astronomische Nachrichten, 43*, 289.

Semel, M. (1989). *Astronomy & Astrophysics, 225*, 456.

Serkowski, K. (1962). *Advances in Astronomy and Astrophysics, 1*, 247.

Serkowski, K. (1970). *The Astrophysical Journal, 160*, 1083.

Serkowski, K. (1974). *Methods of Experimental Physics, 12* (Chap. 8).

Serkowski, K., Mathewson, D. S., & Ford, V. L. (1975). *The Astrophysical Journal, 196*, 261.

Serkowski, K., Mathewson, D. S., & Ford, V. L. (1975). *The Astrophysical Journal, 196*, 261.

Shafer, A. B., et.al. (1964). *Journal of the Optical Society of America, 54*, 7.

Shannon, C. E. (1948). *Bell System Technical Journal, 27*, 379.

Shannon, C. E. (1948). *Bell System Technical Journal, 27*, 623.

Sharma, A. B., Halme, S. J., & Butusow, M. M. (1981). *Optical fibre systems and their components*. Springer.

Shurcliff, W. (1962). *Polarized light*. Cambridge University Press.

Silva, D. R., & Cornell, M. E. (1992). *The Astrophysical Journal Supplement Series, 81*, 865.

Sletteback, A. (1979). *Space Science Review, 23*, 541.

Slijkhuis, S. (2012). In *Spektrum* - Mitteilungsblatt der Fachgruppe Spektroskopie in der Vereinigung der Sternfreunde e.V., No. 43. http://spektroskopie.fg-vds.de/pdf/Spektrum43.pdf.

Smith, S. J., & Purcell, E. M. (1953). *Physical Review, 92*, 1069.

Smith, M. A., Robinson, R. D., & Corbet, R. H. D. (1998). *The Astrophysical Journal, 503*, 877.

Smith, M. A., Robinson, R. D., & Hatzes, A. P. (1998). *The Astrophysical Journal, 507*, 945.

Smith, M. A., & Robinson, R. D. (1999). *The Astrophysical Journal, 517*, 866.

St-Louis, N., Drissen, L., Moffat, A. F. J., Bastien, P., & Tapia, S. (1987). *The Astrophysical Journal, 322*, 870.

St-Louis, N., Moffat, A. F. J., Drissen, L., Bastien, P., & Robert, C. (1988). *The Astrophysical Journal, 330*, 286.

St-Louis, N., Moffat, A. F. J., Lapointe, L., Efimov, Yu. S., Shakhovskoy, N.M., Fox, G. K., et al. (1993). *The Astrophysical Journal, 410*, 342.

St-Louis, N., Dalton, M. J., Marchenko, S. V., Moffat, A. F. J., & Willis, A. J. (1995). *The Astrophysical Journal, 452*, 57.

St-Louis, N., Doyon, R., Chagnon, F., & Nadeau, D. (1998). *The Astronomical Journal, 115*, 2475.

Stee, Ph., Vakili, F., Bonneau, D., & Mourard, D. (1998). *Astronomy & Astrophysics, 332*, 268.

Sundqvist, J. O., Owocki, S. P., & Puls, J. (2012). In *ASP Conference Series* (Vol. 465, p. 119).

Suto, H., & Takami, H. (1997). *Applied Optics, 38*(19).

Taubert, R. D., Monte, C., Baltruschat, C., Schirmacher, A., Gutschwager, B., Hartmann, J., et al. (2009). *Metrologia, 46*(4), 207.

Taylor, M. (1992). In L. Drissen, C. Leitherer, & A. Nota (Eds.), *Nonisotropic and variable outflow from stars. ASP Conference Series* (Vol. 22, p. 57).

Telting, J, H., & Kaper, L. (1994). *Astronomy and Astrophysics, 284*, 515.

Thizy, O. (2007). In *Proceedings of the 26th annual symposium on telescope science, Society for Astronomical Sciences* (p. 31).

Tüg, H. (1974). *Astronomy and Astrophysics, 37*, 249.

Tüg, H. (1980). *Astronomy and Astrophysics, 82*, 195.

Tüg, H., White, N. M., & Lockwood, G. W. (1977a). *Astronomy and Astrophysics, 61*, 679.

Tüg, H., White, N. M., & Lockwood, G. W. (1977b). *Astronomy and Astrophysics, 66*, 469.

Vernet, J., et al. (2011). *Astronomy and Astrophysics, 536*, 105.

Vollmann, K., & Eversberg, T. (2006). *Astronomical Notes, 327*, 862.

Vollmann, K. (2008). http://www.stsci.de/simspec$_$slit$_$e.xls.

Walder, R., & Folini, D. (2002). In *Proc. IAU Symposium No. 212.*

Waldschläger, U. (2014). Private communication.

Walraven, T., & Walraven, J. H. (1972). In S. Lausten & A. Reiz (Eds.), *Proceedings of the Conference on Auxiliary Instrumentation for Large Telescopes* (p. 175).

Whittaker, E. T. (1915). *Procceedings of the Royal Society Edinburgh, 35*, 181.

Whittaker, J. M. (1935). *Interpolatory function theory.* Cambridge University Press.

Widenhorn, R., Blouke, M. M., Weber, A., Rest, A., & Bodegom, E. (2002). In *SPIE conference proceedings. Sensors and camera systems for scientific, industrial, and digital photography* (Vol. 4669).

Wilken, T., Lo Curto, G., Probst, R. A., Steinmetz, T., Manescau, A., Pasquini, L., et al. (2012). *Nature, 485*, 611.

Witt, V. (2005). *Sterne und Weltraum, 10*, 73.

Wood, K., Brown, J. C., & Fox, G. K. (1993). *Astronomy and Astrophysics, 271*, 492.

v. Zeipel, H. (1924). *Monthly Notices of the Royal Astronomical Society, 84*, 702.

Index

© Springer-Verlag Berlin Heidelberg 2015   645
T. Eversberg, K. Vollmann, *Spectroscopic Instrumentation*, Springer Praxis Books,
DOI 10.1007/978-3-662-44535-8

Lightning Source UK Ltd.
Milton Keynes UK
UKOW06n1244040516

273539UK00003B/16/P